AF757493

Advances in Pectin and Pectinase Research

Advances in Pectin and Pectinase Research

edited by

Fons Voragen

Professor in Food Sciences,
Laboratory of Food Chemistry,
Wageningen University, The Netherlands

Henk Schols

Associate professor in Food Sciences,
Laboratory of Food Chemistry,
Wageningen University, The Netherlands

and

Richard Visser

Professor in Plant Sciences,
Laboratory of Plant Breeding,
Wageningen University, The Netherlands

KLUWER ACADEMIC PUBLISHERS

DORDRECHT / BOSTON / LONDON

A C.I.P. Catalogue record for this book is available from the Library of Congress.

ISBN 1-4020-1144-X

Published by Kluwer Academic Publishers,
P.O. Box 17, 3300 AA Dordrecht, The Netherlands.

Sold and distributed in North, Central and South America
by Kluwer Academic Publishers,
101 Philip Drive, Norwell, MA 02061, U.S.A.

In all other countries, sold and distributed
by Kluwer Academic Publishers,
P.O. Box 322, 3300 AH Dordrecht, The Netherlands.

Printed on acid-free paper

All Rights Reserved
© 2003 Kluwer Academic Publishers
No part of this work may be reproduced, stored in a retrieval system, or transmitted in any form or by any means, electronic, mechanical, photocopying, microfilming, recording or otherwise, without written permission from the Publisher, with the exception of any material supplied specifically for the purpose of being entered and executed on a computer system, for exclusive use by the purchaser of the work.

Printed in the Netherlands.

Contents

Preface

The second international symposium on Pectins and Pectinases was organized by Wageningen University and Research Centre and held in Rotterdam, May 6-10, 2001. This successful meeting was attended by around 130 participants from more than 20 countries representing almost all of the groups and industries working worldwide on pectins and pectinases. Following the first meeting on this subject held in December 1995, the symposium definitely forms a platform for researchers and industries working in the field, all within their own discipline and expertise. The symposium demanded a written account and this book is the result of that. It contains all keynote lectures and other oral presentations and provides an update about the current research. Significant progress has been made in the last 5 years.
This book provides an up-to-date insight into the research on pectin and pectic enzymes involved in their biosynthesis, degradation, modification or utilization. The progress in the elucidation of the chemical structure of pectin and mode of action and 3-D structure of the pectin degrading enzymes allows us to identify and influence the functionality of pectins and pectic enzymes, both *in vitro* after isolation as well in the plants themselves (*in planta*). Other contributions deal with new applications of both pectin and pectin-degrading enzymes, while more and more attention is paid to health and nutritional aspects of pectins.
The book provides a 'state of the art' account for both beginners and experienced researchers of almost all disciplines of pectin research.
We hope that it will satisfy your interests in this important and fast developing research field.

Fons Voragen, Henk Schols & Richard Visser, editors

INTERACTION OF A STYLAR PECTIC POLYSACCHARIDE AND A BASIC PROTEIN (SCA) MEDIATES LILY POLLEN TUBE ADHESION

J-C. MOLLET, S-Y. PARK, E.M. LORD
Department of Botany and Plant Sciences, University of California Riverside, CA, 92521 - USA. E-mail, jmollet@citrus.ucr.edu

Abstract

Though pectins are implicated in cell adhesion in plants, this has never been tested directly. We developed an in vitro assay to study pollen tube adhesion to the stylar extracellular matrix (ECM) in lily. The adhesion of pollen tubes to the ECM of the stylar transmitting tract epidermis *in vivo* is proposed to be essential for a proper delivery of the sperm cells to the ovary. Using the assay, we identified two stylar molecules responsible for adhesion, a small protein and a pectic polysaccharide. The combination of at least these two molecules is required for this adhesion event. The 9-kD protein is cysteine-rich with some sequence similarity to lipid transfer protein. We named it stigma/style cysteine-rich adhesin (SCA). The second molecule has been isolated from the style using an imidazole extraction method and is mostly composed of galacturonic acid (70-75 mole%) with arabinosyl, galactosyl, rhamnosyl and glucuronosyl residues. This fraction reacts strongly with JIM5 (monoclonal antibody [MAb] to low esterifed homogalacturonans) and has some reaction with JIM7, LM5 and PAM1 (MAbs to esterified homogalacturonans, β-[1-4]-D-galactans, and blocks of 30 GalA repeat units). Pollen tube adhesion can be significantly reduced with a pre-treatment of this pectic fraction with endopolygalacturonase. All these data implicate a stylar pectic polysaccharide in lily pollen tube adhesion. *In vivo*, immuno-localization data show that SCA and low esterified homogalacturonan are co-localized at the transmitting tract epidermal surface where the pollen tubes adhere. Binding assays reveal that pectin and SCA bind each other in a pH dependent manner and that binding is necessary to produce pollen tube adhesion in the assay. Involvement of pectic polysaccharide and proteins in cell adhesion will be discussed.

1. Introduction

The new view of the plant cell wall as a cellular compartment rather than a rigid and inert network involved in protection and structural support is more and more accepted. This makes the plant cell wall, to some extent, comparable to the extracellular matrix (ECM) of animal cells in spite of their tremendous differences in structure and composition. Throughout plant development, the cell wall is subjected to many chemical and physical changes such as loosening during cell expansion and enzymatic degradation during fruit ripening. These defined wall modifications are spatially and temporally regulated [1], probably by their own cells and/or their neighboring cells.

F. Voragen et al. (eds.), Advances in Pectin and Pectinase Research, 1-13.
© 2003 *Kluwer Academic Publishers. Printed in the Netherlands.*

Thus, cell contact or cell adhesion sites might be involved in the transfer of information, communication and/or signaling between cells. The most studied case of signaling via cell contact is probably the recognition and rejection of self-incompatible pollen that involves interacting molecules from the male and female parts [2, 3]. To determine the importance of adhesion in cell development, we have developed an *in vitro* adhesion bioassay, using lily pollen and stylar extracts, that mimics the adhesion of pollen tubes observed within the style. Using this assay we have isolated the molecules required for this adhesion event. Based on the results obtained with our adhesion system, cell adhesion in general will be discussed with regard to the molecules involved.

2. Pollen tube adhesion in lily

During pollination in lily, pollen grains land on the surface of the stigma, hydrate and produce pollen tubes that travel along the hollow style to deliver the two sperm cells to the ovule (Figure 1). Pollen tubes grow by tip growth ie. secretion of wall material via vesicles targeted to the tip. After passing through the stigma, the growing pollen tubes come in contact with the transmitting tract epidermis (TTE) that lines a canal filled up with an exudate secreted by the TTE cells and composed of lipids, proteins and carbohydrates [4].

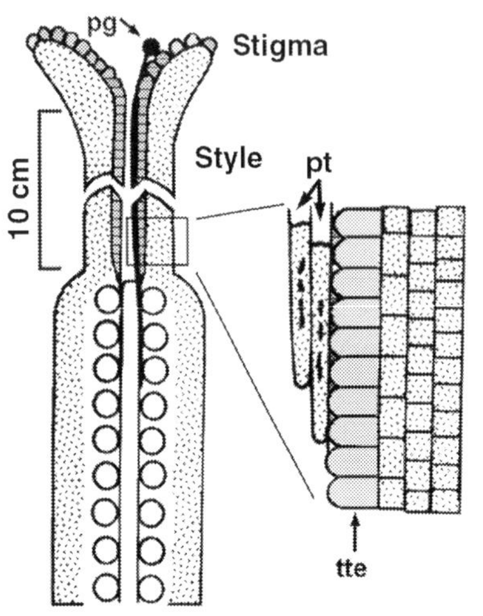

Figure 1. Pollination and adhesion of lily pollen tubes on the transmitting tract epidermis of the stigma and style. pg. pollen grain, pt. pollen tube, tte. transmitting tract epidermis. Reprinted with permission from Elsevier Science, *Trends in Plant Science* [5].

Adhesion between the pollen tube walls and the TTE surface can be observed, after cryo-preservation of pollinated styles preserving the extracellular matrix or ECM (Figure 2A-B). Adhesion occurs also between pollen tubes themselves (Figure 2B). This adhesion between the pollen tubes and the TTE cells has been proposed to be, at least in part, responsible for the fast growth rates of *in vivo* grown pollen tubes compared to the pollen tubes grown in an artificial liquid medium [6].

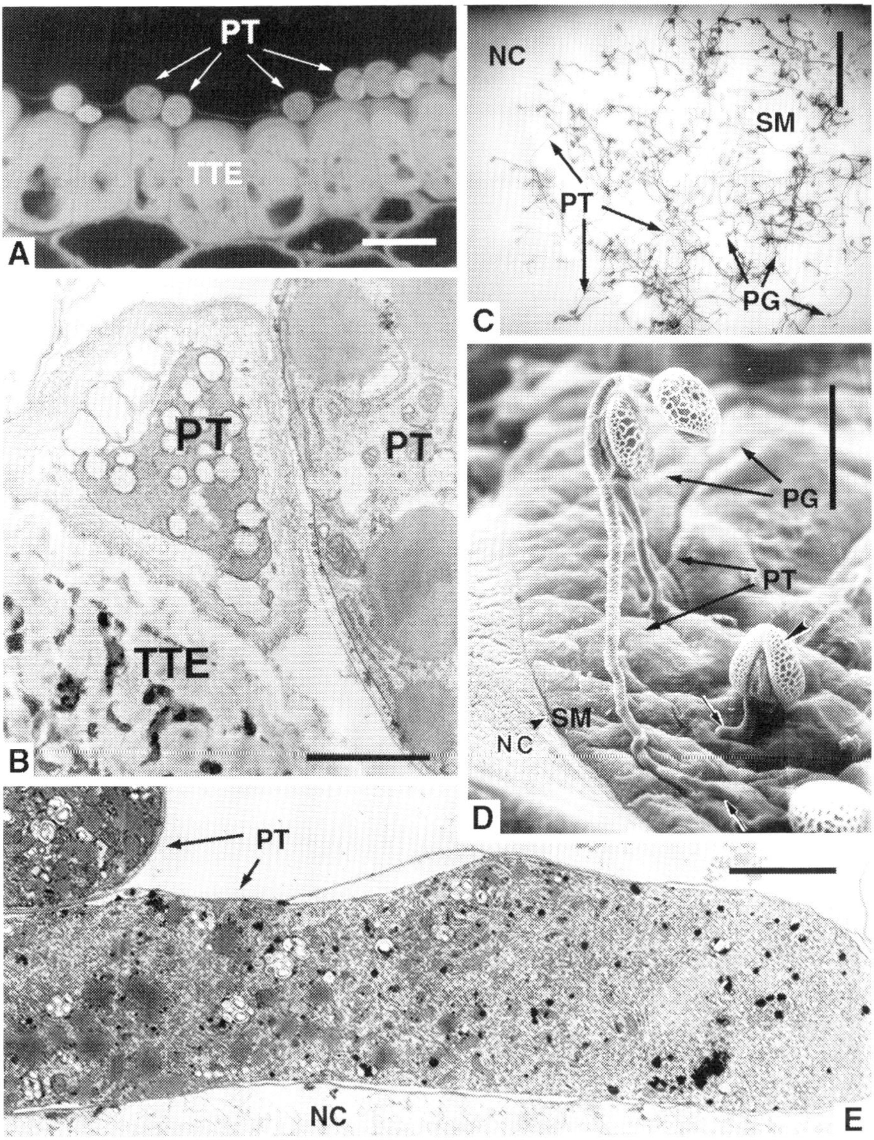

Figure 2. In vivo (A-B) and *in vitro* (C-E) adhesion of lily pollen tubes to the TTE and the *in vitro* artificial stylar matrix, respectively. A. Cross section of a lily pollinated style showing the pollen tubes adhering to the TTE surface. Scale bar = 25 μm. B. TEM view of pollen tubes in cross section adhering to the TTE surface and to each other. Scale bar = 5 μm. Reprinted with permission from Springer, *Sexual Plant Reproduction* [7]. C. Stereomicroscope view of the *in vitro* adhesion assay after staining with coomassie blue. Pollen tubes adhere to the stylar matrix but not to the surrounding nitrocellulose membrane. Scale bar = 1 mm. D. SEM view of pollen tubes adhering to the stylar matrix. Note the pollen tubes adhere at the tip. Scale bar = 50 μm. Reprinted with permission from Springer, *Sexual Plant Reproduction* [8]. E. TEM view of pollen tube tip adhering to the *in vitro* stylar matrix. Scale bar = 5 μm. Reprinted with permission from ASPB, *Plant Cell* [9]. NC. nitrocellulose membrane, PG. pollen grain, PT. pollen tube, SM. stylar matrix composed of SCA and stylar pectin, TTE. transmitting tract epidermis.

To determine the nature of the stylar molecules involved in this adhesion, an *in vitro* adhesion bioassay was developed [8]. Pre-germinated pollen tubes are incubated with nitrocellulose membranes impregnated with the molecules isolated from styles and/or stigmas as shown in Figure 3. If adhesion occurs, the number and length of pollen tubes can be measured under a stereomicroscope after staining with coomassie blue (Figure 2C). Typically, pollen tubes adhere via their tips to this artificial matrix within 2h of incubation, while the pollen grains and the pollen tubes back from the tips float in the liquid medium (Figure 2D). In this assay, pollen tubes adhering to the stylar matrix are also able to adhere to each other and grow (Figure 2E), as observed *in vivo*.

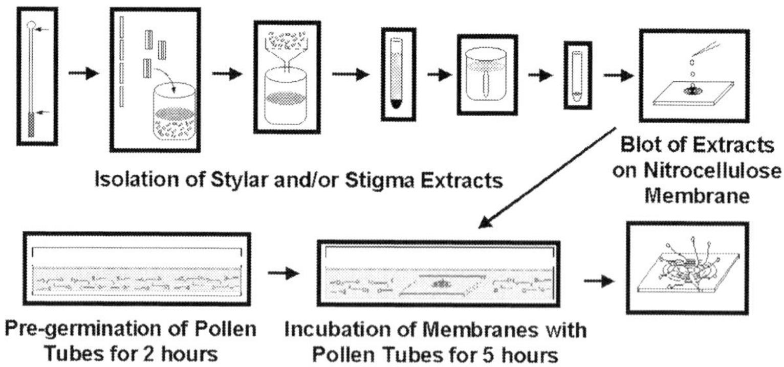

Figure 3. Diagram of the *in vitro* pollen tube adhesion assay method

3. Stylar molecules required for lily pollen tube adhesion

3.1. A 9-KD POLYPEPTIDE (SCA)

Preliminary data suggested that at least two molecules of different molecular weights were involved in pollen tube adhesion based on a size fractionation of a stylar extract. The low molecular weight molecule was purified from stigmas and styles following the method shown in Figure 4. It corresponds to a 9 kD, basic (pI 8.6), cysteine-rich protein with some sequence similarity with plant lipid transfer protein including eight conserved cysteine residues (LTP, Figure 5) [9]. We named this protein SCA for style/stigma cysteine-rich adhesin. *In vitro*, LTPs are able to transfer lipids between organellar membranes [10] but there is no evidence of this function *in vivo*. Since LTPs are secreted into the plant ECM [11], it is unlikely that they play a role in cytoplasmic lipid transfer [12]. *In vivo* and *in vitro*, LTPs have been shown to have anti-microbial activity [13]. We are currently testing if SCA has antimicrobial activity. SCA is not produced by the male gametophyte, ie pollen tube and pollen grain [14]. However, immunolocalization of SCA on pollinated styles reveals that SCA is detected on the wall of the TTE but also on the wall of the pollen tubes [9]. SCA is secreted by the TTE cells and probably binds to the pollen tube walls during their travel along the style. The exact role of SCA in adhesion is still unclear. SCA is not adhesive alone but needs another large molecule to allow pollen tube adhesion.

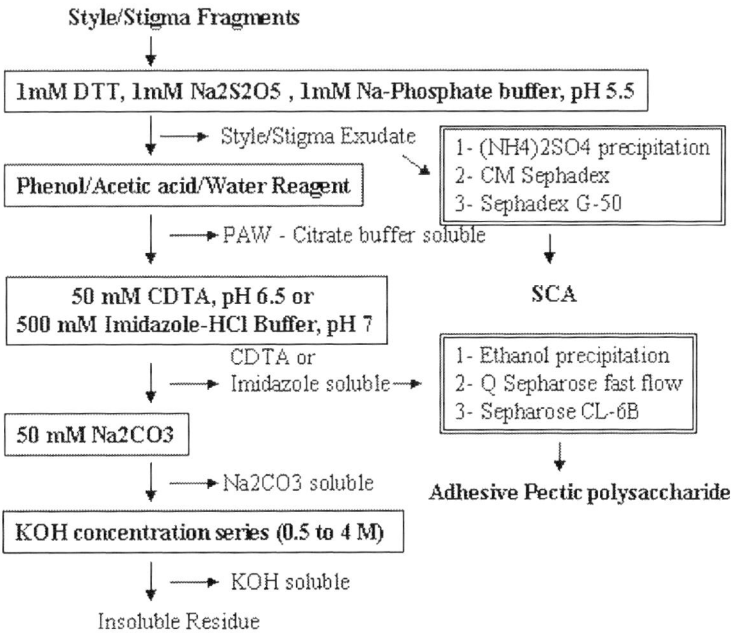

Figure 4. Flow chart of the method used for the fractionation and purification of SCA and pectin from lily styles and/or stigma

```
Lily 9 kD                MAR-SS-AVCF--LLLLAF-LIGTA-S-A
Rice                     :::AQLVL:AVVAA:::A--PH:A-V-:
Maize                    :::TQQL::VATAVV--:LV:LAA:T:E:
Arabidopsis              ::GVMKLA-:L--::ACMIVA-:PIT:N:

               ▼
Lily 9 kD      -ITCGQVDSDLTSCLGYARKGGVIPP-GCCAGVRTLNNLAKTTPDRQ
Rice           -:::::N:AVGP::T:::G:AGPSAA-::S:::S:KAA:SN:A::R
Maize          A:S::::A:AIAP:IS:::GQ:AG:SA:::S:::S:::A:R::A::R
Arabidopsis    ALS::S:N:N:AA:I::VLQ:::::A-::S::KN::SI:::::::

Lily 9 kD      TACNCLKSLVNP-SLGLNAAIVAGIPGKCGVNIPYPIRMQTIDCNKVR
Rice           ::::::NAARG-IK::::GNA:S::S::::SV::T:SASIH:SR:S
Maize          A:::::NAAAG-VS::::GNA:S::S::::S:::T:STS:::SR:S
Arabidopsis    Q::::IQGAARALGS::::GRA::::KA:::::::K:STS:N:KT::
```

Figure 5. Amino acid sequence alignment of SCA deduced from the cDNA compared to several plant LTPs. The arrow indicates the N terminus of the mature protein. Identical amino acids are indicated by colons and dashes represent spaces introduced to maximize alignment. Cysteine residues are in bold, lysines are double underlined and arginines are single underlined - Reprinted with permission from ASPB, *Plant Cell* and modified from [9].

3.2. A STYLAR PECTIN

The second molecule involved in adhesion has been chemically extracted from style fragments using the method described in Figure 4. None of the stylar extracts allow pollen tube adhesion on their own. In combination with SCA, the most adhesive stylar extract was detected in the imidazole-HCl fraction (Figure 6).

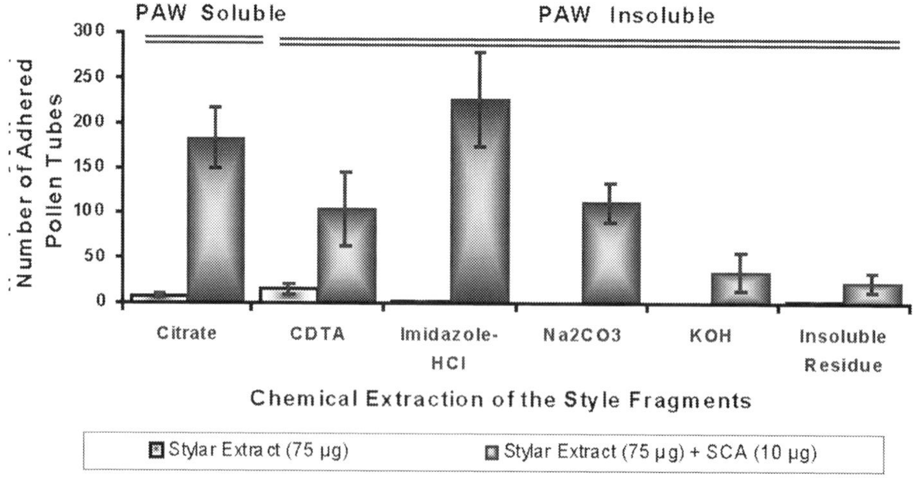

Figure 6. Fractionation of the stylar adhesive molecule using an in vitro adhesion assay. The different fractions were obtained by sequential chemical extraction as shown in Figure 4. Reprinted with permission from ASPB, *Plant Cell* and modified from [15].

The imidazole extract was further fractionated by ethanol precipitation, anion exchange (Q sepharose fast flow) and size exclusion (Sepharose CL-6B) chromatography using 100 mM imidazole-HCl, pH 7 as eluent. Adhesion assays, in combination with SCA, demonstrated that the most adhesive fractions were detected between 40-60% ethanol precipitation and eluted between 300-400 mM NaCl from the Q Sepharose [15]. After size exclusion chromatography, 50 µg of the most adhesive fraction in combination with 5 µg of SCA are sufficient to allow 401 ± 53 pollen tubes to adhere to the artificial matrix (Table 1). The estimated size of this fraction on a sepharose CL-6B was 1,500,000 MW compared to dextran standards. This fraction is a galacturonic acid enriched polysaccharide (Table 2) with a molar ratio of GalA to Rha of 11. It reacts strongly with JIM5 (MAb to low esterified homogalacturonan) and has some reaction with JIM7, LM5 and PAM1 (MAbs to esterified homogalacturonan, β-[1-4]-D-galactan and a stretch of 30 GalA units, respectively) (Figure 7) indicating the presence of homogalacturonan and rhamnogalacturonan 1 regions. The pectic fraction reacts also with JIM13 and JIM8 (Mabs to carbohydrate epitopes of arabinogalactan-proteins) but does not react with the Yariv phenylglycoside, a diagnostic tool to detect arabinogalactan-proteins. This implies that arabinogalactan side chains may also be part of the RG1 region. Proteinase K pre-treatment on the pectic fraction does not affect the number of pollen tubes adhering to the matrix [15]. Endopolygalacturonase treatments on the pectic fraction [15] or on the pectic fraction combined with SCA result in a significant decrease in the number of pollen tubes adhering to the nitrocellulose membrane (Table 1). All these data suggest that the second molecule required for lily pollen tube adhesion is a pectic polysaccharide.

TABLE 1. Adhesion of lily pollen tubes on a matrix coated with stylar pectin and SCA after their treatments with endopolygalacturonase.

Fraction and Enzymatic Treatment	Number of Adhered Pollen Tubes	
Stylar Pectin [a] (50 μg) + SCA (5 μg)	401 ± 53	(n = 5)
PGase treatment on the stylar pectin [b]		
Stylar Pectin 50 μg (PGase treated) + SCA (5μg)	51 ± 16	(n = 3)
Stylar Pectin 50 μg (boiled PGase) + SCA (5 μg)	311 ± 43	(n = 2)
PGase treatment on the stylar pectin combined with SCA [c]		
[Stylar Pectin (50 μg) + SCA (5 μg)] PGase treated	18 ± 6	(n = 2)
[Stylar Pectin (50 μg) + SCA (5 μg)] boiled PGase	356 ± 62	(n = 2)

[a] Stylar pectin was obtained after ethanol precipitation, Q sepharose and Sepharose CL-6B chromatography of the imidazole extract.
[b] Stylar pectin was incubated with endopolygalacturonase 2 (PGase treated) or boiled endopolygalacturonase 2 (boiled PGase) and combined with SCA. Reprinted with permission from ASPB, *Plant Cell* and modified from [15].
[c] Stylar pectin and SCA were combined and incubated with endopolygalacturonase 2 (PGase treated) or boiled endopolygalacturonase 2 (boiled PGase).
(n) number of replicates.

TABLE 2. Chemical composition of the stylar pectic polysaccharide from the imidazole extract fractionated by ethanol precipitation, Q Sepharose fast flow and Sepharose CL-6B chromatography.

Carbohydrate	Protein	Glycosyl composition (mol%)								
% (w/w)	% (w/w)	Ara	Rha	Fuc	Xyl	Man	Gal	Glc	GlcA	GalA
98.6	0.5	4.6	6.6	0.8	1.5	0.8	4	0.8	7.6	73.3

Reprinted with permission from ASPB, *Plant Cell* and modified from [15]

4. Pectin in adhesion

Pectins and especially low esterified homogalacturonan have been implicated in intercellular adhesion at the middle lamella (see review [24]) based mainly on immuno-localization data. An antibody specific for blocks of 30 galacturonic acids (PAM1) bound in the contact zone between two cells in *Arabidopsis* suspension cultures [17]. Mutants deficient in cuticle showing abnormal fusion of organs revealed low esterified pectic polysaccharide in the junction zone [25, 26], as well. However, methyl esterified homogalacturonan [27] and neutral side chains of RG1 [28] have also been implicated in cell attachment. Decrease in arabinose and galactose content has also been reported during the dis-adherence of walls during the fruit ripening [29]. All these genetic, immuno-localization and biochemical data demonstrate that pectins are good candidates as adhesive molecules but the region of the pectic polysaccharide involved is still a question of debate. In our assay system, the adhesive pectic fraction contains low esterified homogalacturonan and RG1 regions. The presence of RG2 and xylogalacturonan is not known yet. The determination of the fine structure of our adhesive pectic polysaccharide with specific enzymes will help to define the region that is clearly involved in cell adhesion.

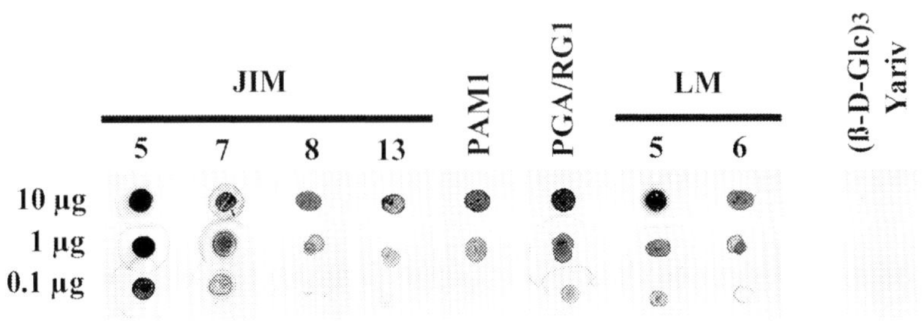

Figure 7. Dot-Blot immuno-assays of the stylar pectin fraction isolated from lily styles with the imidazole method after ethanol precipitation, Q sepharose fast flow and sepharose CL-6B. JIM5, MAb to low esterified homogalacturonan [16]; JIM7, MAb to high esterified homogalacturonan [16]; JIM8 and JIM13, MAbs to carbohydrate epitopes of arabinogalactan-proteins [17, 18]; PAM1, phage against blocks of 30 GalA [19]; PGA/RG1, polyclonal Ab to RG1 regions [20]; LM5, MAb to (1-4)-β-D-galactan [21]; LM6, MAb to (1-5)-α-L-arabinan [22] and $(\beta\text{-D-Glc})_3$ Yariv phenylglycoside, a red synthetic probe that binds arabinogalactan-proteins [23].

The cohesion of homogalacturonan in cell corners is thought to be maintained by calcium bridges [30]. It appears that pectin behavior in solution with calcium may be more complex than the "egg box" model [31]. In our assay, the role of calcium is difficult to evaluate as it is a requirement in the germination medium to promote pollen tube growth, a prerequisite for adhesion. In addition, an important component in our adhesion assay is a small basic polypeptide (SCA).

5. Protein - pectin interaction

5.1 SITE OF PROTEIN - PECTIN BINDING

In addition to the structural polysaccharides, a wide range of proteins, enzymes and glycoproteins exist (see review [32]) that can be bound to the wall by covalent linkages or anionic interactions. For example, extensins can be cross-linked in the wall to pectins via rhamnogalacturonan [33]. In our system, a 9 kD basic protein and a pectic polysaccharide act together to promote lily pollen tube adhesion. SCA and the stylar pectin appear to bind to each other by ionic interaction rather than covalent cross-linking as the binding can be disrupted by a slight modification of pH over the isoelectric point of SCA [15]. Recently, two cationic amino acids (lysine and arginine) from two different wall enzymes (pectate lyase and peroxidase) have been shown to be responsible for the binding of these proteins to the homogalacturonan region of pectins [31, 34]. SCA possesses five lysines and five arginines (Figure 5) that may also be involved in the binding of SCA to the stylar pectin. In solution, basic proteins can also modify the physico-chemical properties of pectins as does calcium with homogalacturonan [35]. SCA may have a similar action on the adhesive stylar pectin but probably in a more specific manner, as substitution of SCA with other small basic peptides like cytochrome c or polylysine did not allow for pollen tube adhesion [15]. Other reports mention that cysteine-rich domains of proteins can mediate specific carbohydrate binding [36] as do lectins through "deep or shallow pockets" in their three dimensional structure [37]. We are using a variety of methods, ie synthesis of commercial peptides for competition studies, to determine the binding site of SCA and pectin.

5.2 SIGNALING BY PROTEIN- PECTIN INTERACTION

Other proteins or proteoglycans located at the cell wall/plasma membrane interface have also been reported to bind to pectins (arabinogalactan-proteins [AGPs] and wall associated kinases [WAKs]). These two molecules might be involved in signaling at the plasma membrane. AGPs have been reported to bind and/or to be co-purified with pectin [38] and they can be anchored to the plasma membrane through a glycosylphosphatidyl inositol anchor (GPI anchor) [39, 40] that may transfer signals to the cell [41]. WAKs have been also reported to be covalently linked to pectin [42]. These data suggest that pectins and other wall polysaccharides might be somehow linked together to form a network able to perceive or transmit signals.

6. Pollen tube wall and enzymes

The pollen tube cell is highly polarized. Even the wall at the tip, the site of secretion and growth, is different from the wall back from the tip (Figure 8). Based on immuno-localization using monoclonal antibodies, low and esterified pectins were found to be the main components at the tip in lily [43] and in other species [44]. In *Arabidopsis*, only esterified pectins were detected at the tip of *in vivo* grown pollen tubes [45]. In lily, arabinogalactan-proteins have been also detected in the tip region [43]. Back from the tip, the wall is mainly composed of low esterified pectins, callose and low amounts of cellulose as well as arabinogalactan-proteins [43]. In lily, rhamnogalacturonan 2 has also been detected [46] as well as β-(1-4)-D-galactans (Mollet, unpublished data). Pectins are secreted at the tip in the esterified form as detected by immuno-localization of the Golgi vesicles using JIM7 and de-esterified or at least partially de-esterified in the wall (Figure 8). The binding site of the SCA/stylar pectin complex on the pollen tube is not known. The pectic polysaccharide of the tube wall may interact with the pectic polysaccharide of the TTE via SCA bridging (Figure 8) or SCA may bind a receptor in the pollen tube plasma membrane at the tip.

Pollen tubes might also be involved in the modification of the matrix. Indeed, during tube growth, many genes coding for enzymes able to modify pectic polysaccharides are highly expressed including pectate lyases, polygalacturonases and pectin-esterases [44]. The role of these enzymes in the development and physiology of the pollen tube remains unclear. These enzymes may modify the pollen tube wall during growth or may be secreted to modify the wall of the TTE [47]. As the pollen tubes dis-adhere back from the tip, pectolytic enzymes might be involved in degrading the connection of the tube wall to the TTE matrix. Recently, it has been shown that tobacco pollen tubes are able to modify and incorporate molecules from the ECM of the transmitting tract in the style on their travel to the ovary [48]. The degradation of the stylar pectin with these enzymes in lily might also produce an elicitation response by oligogalacturonides as would an oligosaccharin [49] and/or release SCA from the stylar pectin that could then bind to a receptor in the pollen tube.

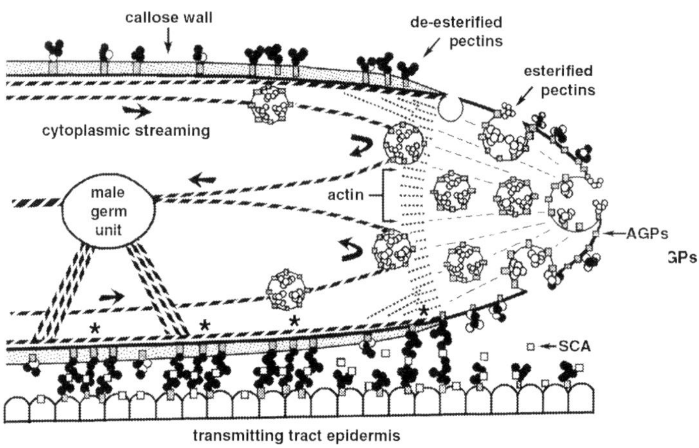

Figure 8. Model for lily pollen tube adhesion to the stylar transmitting tract. Vesicles, produced from the Golgi, are secreted at the tip where pollen tube growth occurs furnishing wall and membrane materials. Vesicles containing esterified pectin and AGPs fuse at the plasma membrane tip. Pectins are then de-esterified, in the wall, especially back from the tip. AGPs are at the plasma membrane. The TTE wall of the style is also composed of low esterified and esterified pectins and AGPs. SCA is secreted by the TTE cells and binds to the stylar pectins. Reprinted with permission from Elsevier Science, *Trends in Plant Science* [5].

7. Conclusion

Two molecules, a small basic protein SCA (an LTP-like protein) and a large pectic polysaccharide cause *in vitro* lily pollen tube adhesion. Pollen tube adhesion can be significantly reduced with a pre-treatment of the pectic fraction with an endopolygalacturonase or with a proteinase K pre-treatment of SCA. Immuno-localization data show that SCA and low esterified homogalacturonan are co-localized at the transmitting tract epidermal surface where the pollen tubes adhere [9, 15]. In addition, binding assays reveal that pectin and SCA bind each other in a pH dependent manner and that binding is necessary to produce pollen tube adhesion in the assay [15]. These two classes of molecules have been known for many years but their functions and/or modes of action are far from being understood. In our system, many questions remain unanswered: what is the binding site for SCA on the stylar pectic polysaccharide ? Is there a membrane receptor in the pollen tube or is SCA directly binding the pectic wall of the pollen tube or its plasma membrane ? Does the adhesion in the middle lamella in all plant tissues involve this same pectic polysaccharide and SCA ? Functional adhesion assays can now be developed for a variety of plant cell types to explore the mechanisms of adhesion in somatic tissues.

8. Acknowledgements

The pectin fractionation and chemical composition were performed in collaboration with Eugene A. Nothnagel and the isolation and purification of SCA was done in collaboration with Linda L. Walling, both at UC Riverside. The authors are grateful to Dean DellaPenna (Michigan State University, USA) for the endopolygalacturonase, to Paul Knox (University of Leeds, UK) for the LM5, LM6 and PAM1 antibodies, to Keith Roberts (John Innes Centre, Norwich, UK) for the JIM antibodies and to Andrew Staehelin (University of Colorado, USA) for the PGA/RG1 antibody. Support for this work was provided by NSF grants IBN 9603826 and 0077886 to E.M.L.

9. References

1. Roberts, K. (1994) The plant extracellular matrix: in a new expansive mood, *Curr. Opin. Cell Biol.* **6**, 1020-1027.
2. Schopfer, C.R., Nasrallah, M.E., and Nasrallah, J.B. (1999) The male determinant of self-incompatibility in *Brassica*, *Science* **286**, 1697-1700.
3. Cock, J.M., Cabrillac, D., Giranton, J-L., Pastuglia, M., Ruffio-Chable, V., Miege, C., Dumas, C., and Gaude, T. (2000) Investigating the molecular mechanism of the incompatibility response in *Brassica*, *Ann. Bot.* **85**, 147-153.
4. Miki-Hirosige, H., Hoek, I.H., and Nakamura, S. (1987) Secretions from the pistil of *Lilium longiflorum*, *Am. J. Bot.* **74**, 1709-1715.
5. Lord, E.M. (2000) Adhesion and cell movement during pollination: cherchez la femme, *Trends Plant Sci.* **5**, 368-373.
6. Lord, E.M., Mollet, J-C., and Park, S-Y. (2001) *In vivo* pollen tube growth: Tube cell adhesion and movement in lily, in A. Geitmann, M. Cresti and I.B. Heath (eds.), *Cell Biology of Plant and Fungal Tip Growth*, NATO Science Series, IOS Press, pp. 187-201.
7. Jauh, G-Y., and Lord, E.M. (1995) Movement of the tube cell in the lily style in the presence of the pollen grain and the spent pollen tube, *Sex. Plant Reprod.* **8**, 168-172.

8. Jauh, G.Y., Eckard, K.J., Nothnagel, E.A., and Lord, E.M. (1997) Adhesion of lily pollen tubes on an artificial matrix, *Sex. Plant Reprod.* **10**, 173-180.
9. Park, S-Y., Jauh, G-Y., Mollet, J-C., Eckard, K.J., Nothnagel, E.A., Walling, L.L., and Lord, E.M. (2000) A lipid transfer like protein is necessary for lily pollen tube adhesion to an *in vitro* stylar matrix, *Plant Cell* **12**, 151-163.
10. Kader, J-C. (1997) Lipid-transfer proteins: a puzzling family of plant proteins, *Trends Plant Sci.* **2**, 66-70.
11. Thoma, S., Hecht, U., Kippers, A., Botella, J., De Vries, S., and Somerville, C.R. (1994) Tissue-specific expression of a gene encoding a cell wall-localized lipid transfer protein from *Arabidopsis, Plant Physiol.* **105**, 35-45.
12. Arondel, V., Vergnolle, C., Cantrel, C., and Kader, J-C. (2000) Lipid transfer proteins are encoded by a small multigene family in *Arabidopsis thaliana*, *Plant Sci.* **157**, 1-12.
13. Segura, A., Moreno, M., and Garcia-Olmedo, F. (1993) Purification and antipathogenic activity of lipid transfer proteins (LTPs) from the leaves of *Arabidopsis* and spinach, *FEBS Lett.* **332**, 243-246.
14. Park, S.Y., and Lord, E.M. Expression studies of SCA in lily and confirmation of its role in pollen tube adhesion (submitted).
15. Mollet, J-C., Park, S-Y., Nothnagel, E.A., and Lord, E.M. (2000) A lily stylar pectin is necessary for pollen tube adhesion to an *in vitro* stylar matrix, *Plant Cell* **12**, 1737-1749.
16. Knox, P.J., Linstead, P.J., King, J., Cooper, C., and Roberts, K. (1990) Pectin esterification is spatially regulated both within cell walls and between developing tissues of root apices, *Planta* **181**, 512-521.
17. Pennell, R.I, Janniche, L., Kjellbom, P., Scofield, G.N., Peart, J.M., and Roberts, K. (1991) Developmental regulation of a plasma membrane arabinogalactan protein epitope in oilseed rape flowers, *Plant Cell* **3**, 1317-1326.
18. Yates, E.A., Valdor, J-F., Haslam, S.M., Morris, H.R., Dell, A., Mackie, W., and Knox, J.P. (1996) Characterization of carbohydrate structural features recognized by anti-arabinogalactan-protein monoclonal antibodies, *Glycobiology* **6**, 131-139.
19. Willats, W.G.T., Gilmartin, P.M., Mikkelsen, J.D., and Knox, J.P. (1999) Cell wall antibodies without immunization: Generation and use of de-esterified homogalacturonan block-specific antibodies from a naive phage display library, *Plant J.* **18**, 57-65.
20. Moore, P.J., Darvill, A.G., Albersheim, P., and Staehelin, A.L. (1986) Immunogold localization of xyloglucan and rhamnogalacturonan I in the cell walls of suspension-cultured sycamore cells, *Plant Physiol.* **82**, 787-794.
21. Jones, L., Seymour, G.B., and Knox, J.P. (1997) Localization of pectic galactan in tomato cell walls using a monoclonal antibody specific to (1→4)-β-D-galactan, *Plant Physiol.* **113**, 1405-1412.
22. Willats, W.G.T., Marcus, S.E., and Knox, J.P. (1998) Generation of a monoclonal antibody specific to (1→5)-α-L-arabinan, *Carbohydr. Res.* **308**, 149-152.
23. Nothnagel, E.A. (1997) Proteoglycans and related components in plant cells, *Int. Rev. Cytol.* **174**, 195-291.
24. Knox, P.J. (1997) The use of antibodies to study the architecture and developmental regulation of plant cell walls, *Int. Rev. Cytol.* **171**, 79-120.
25. Sinha, N., and Lynch, M. (1998) Fused organs in the *adherent1* mutation in maize show altered epidermal walls with no perturbations in tissue identities, *Planta* **206**, 184-195.
26. Sieber, P., Schorderet, M., Ryser, U., Buchala, A., Kolattukudy, P., Metraux, J.-P., and Nawrath, C. (2000) Transgenic *Arabidopsis* plants expressing a fungal cutinase show alterations in the structure and properties of the cuticle and postgenital organ fusions, *Plant Cell* **12**, 721-737.
27. Liners, F., Gaspar, T., and Van Cutsem, P. (1994) Acetyl- and methyl-esterification of pectins of friable and compact sugar-beet calli: Consequences for intercellular adhesion, *Planta* **192**, 545-556.
28. Kikuchi, A., Edashige, Y., Ishii, T., Fujii, T., and Satoh, S. (1996) Variations in the structure of neutral sugar chains in the pectic polysaccharides of morphologically different carrot calli and correlations with the size of cell clusters, *Planta* **198**, 634-639.
29. Gross, K.C., and Sams, C.E. (1984) Changes in cell wall neutral sugar composition during fruit ripening: a species survey, *Phytochemistry* **23**, 2457-2461.
30. Jarvis, M.C. (1984) Structure and properties of pectin gels in plant cell walls, *Plant Cell Environ.* **7**, 153-164.
31. Scavetta, R.D., Herron, S.R., Hotchkiss, A.T., Kita, N., Keen, N.T., Benen, J.A., Kester, H.C., Visser, J., and Jurnak, F. (1999) Structure of plant cell wall fragment complexed to pectate lyase C, *Plant Cell* **11**, 1081-1092.
32. Cassab, G.I. (1998) Plant cell wall proteins, *Annu. Rev. Plant Physiol. Plant Mol. Biol.* **49**, 281-309.

33. Qi, X.Y., Beherens, B.X., West, P.R., and Mort, A.J. (1995) Solubilization and partial characterization of extensin fragments from cell walls of cotton suspension cultures: Evidence for a covalent cross-link between extensin and pectin, *Plant Physiol.* **108**, 1691-1701.
34. Carpin, S., Crevecoeur, M., de Meyer, M., Simon, P., Greppin, H., and Penel, C. (2001) Identification of a Ca^{2+}-pectate binding site on an apoplastic peroxidase, *Plant Cell* **13**, 511-520.
35. MacDougall, A.J., Brett, G.M., Morris, V.J., Rigby, N.M., Ridout, M.J., and Ring, S.G (2001) The effect of peptide-pectin interactions on the gelation behaviour of plant cell wall pectin, *Carbohydr. Res.* **335**, 115-126.
36. Fiete, D.J., Beranek, M.C., and Baenziger, J.U. (1998) A cysteine-rich domain of the "mannose" receptor mediates GalNAc-4-SO_4 binding, *Proc. Natl. Acad. Sci. USA* **95**, 2089-2093.
37. Rini, J.M. (1995) Lectin structure, *Annu. Rev. Biophys. Biomol. Struct.* **24**, 551-577.
38. Baldwin, T.C., McCann, M.C., and Roberts, K. (1993) A novel hydroxyproline-deficient arabinogalactan protein secreted by suspension-cultured cells of *Daucus carota*. Purification and partial characterization, *Plant Physiol.* **103**, 115-123.
39. Youl, J., Bacic, A., and Oxley, D. (1998) Arabinogalactan-proteins from *Nicotiana alata* and *Pyrus communis* contain glycosylphosphatidylinositol membrane anchors, *Proc. Natl. Acad. Sci. USA*. **95**, 7921-7926.
40. Svetek, J., Yadav, M.P., and Nothnagel, E.A. (1999) Presence of a glycosylphosphatidylinositol lipid anchor on rose arabinogalactan proteins, *J. Biol. Chem.* **27**, 14724-14733.
41. Schultz, C.J., Johnson, K.L., Currie, G., and Bacic, A. (2000) The classical arabinogalactan protein gene family of *Arabidopsis*, *Plant Cell* **12**, 1751-1767.
42. Wagner, T., and Kohorn, B.D. (2001) Wall-associated kinases are expressed throughout plant development and are required for cell expansion, *Plant Cell* **13**, 303-318.
43. Jauh, G.Y., and Lord, E.M. (1996) Localization of pectins and arabinogalactan-proteins in lily (*Lilium longiflorum* L.) pollen tube and style, and their possible roles in pollination, *Planta* **199**, 251-261.
44. Taylor, L.P., and Hepler, P.K. (1997) Pollen germination and tube growth, *Annu. Rev. Plant. Physiol. Plant Mol. Biol.* **48**, 461-491.
45. Lennon, K.A., and Lord, E.M. (2000) The *in vivo* pollen tube cell of *Arabidopsis thaliana*, I: tube cell cytoplasm and wall, *Protoplasma* **214**, 45-56.
46. Matoh, T., Takasaki, M., Takabe, K., and Kobayashi, M. (1998) Immunocytochemistry of rhamnogalacturonan II in cell walls of higher plants, *Plant Cell Physiol.* **39**, 483-491.
47. Hadfield, K.A., and Bennett, A.B. (1998) Polygalacturonases: Many genes in search of a function, *Plant Physiol.* **117**, 337-343.
48. Wu, H-M., Wang, H., Cheung, A.Y. (1995) A pollen tube growth stimulatory glycoprotein is deglycosylated by pollen tubes and displays a glycosylation gradient in the flower, *Cell* **82**, 395-403.
49. Fry, S.C., Aldington, S., Hetherington, P.R., and Aitken, J. (1993) Oligosaccharides as signals and substrates in the plant cell wall, *Plant Physiol.* **103**, 1-5.

TOWARDS UNRAVELLING THE BIOLOGICAL SIGNIFICANCE OF THE INDIVIDUAL COMPONENTS OF PECTIC HAIRY REGIONS IN PLANTS

RONALD J.F.J. OOMEN[1], JEAN-PAUL VINCKEN[1-2], MAXWELL S. BUSH[3], MICHAEL SKJØT[4], CHANTAL H.L. DOESWIJK-VORAGEN[2], PETER ULVSKOV[4], ALPHONS G.J. VORAGEN[2], MAUREEN C. MCCANN[3] AND RICHARD G.F. VISSER[1].

[1]*Wageningen University, Laboratory of Plant Breeding, Binnenhaven 5, 6709 PD Wageningen, The Netherlands, e-mail: ronald.oomen@pv.dpw.wau.nl,* [2]*Wageningen University, Laboratory of Food Chemistry, Bomenweg 2, 6703 HD Wageningen, The Netherlands,* [3]*John Innes Centre, Department of Cell Biology, Colney Lane, Norwich NR4 7UH, UK,* [4]*Biotechnology Group, DIAS, 40 Thorvaldsensvej, DK-1871 Frederiksberg C, Denmark*

Abstract

Highly branched pectins, which are comprised of a rhamnogalacturonan (RG I) backbone carrying galactan and arabinan side-chains, are generally referred to as hairy regions. Even though composition of the hairy regions has been well established in many plants, their biological function is still unknown. Developmental studies have already shown distinct antibody labelling patterns for the different epitopes present on the hairy region, suggesting that they may have different functions. This review compares the results from the developmental studies together with those from mutagenized and genetically modified plants with compositional alterations to the hairy region. In particular, the specific degradation of hairy regions, by the introduction of fungal enzymes in potato, enables the assignment of a putative biological function to the constituent polymers of the hairy region. We hypothesize that the most important function of the galactan hairs is to regulate the pore size of the cell wall. The deposition of galactan may restrict the access of modifying enzymes to the wall. *In planta* fragmentation of the RG I backbone shows severe histological modifications in potato tuber tissue. This suggests that the RG I backbone has an important function for normal potato tuber cell division and tissue development.

1. Introduction

In potato (*Solanum tuberosum* L.), pectin is one of the major components of the cell wall, comprising about 56% (Figure 1) of the total cell wall material (Jarvis *et al.*, 1981; Vincken *et al.*, 2000). The pectic polysaccharides are suggested to form a matrix in the primary cell wall in which the load-bearing network of cellulose and cross-linking glycans is embedded (Carpita and Gibeaut, 1993). This pectic network is independent, but coextensive with the cellulose/xyloglucan network, and is suggested to structurally regulate the pore-size of the wall (McCann and Roberts, 1996).

F. Voragen et al. (eds.), Advances in Pectin and Pectinase Research, 15-34.
© 2003 *Kluwer Academic Publishers. Printed in the Netherlands.*

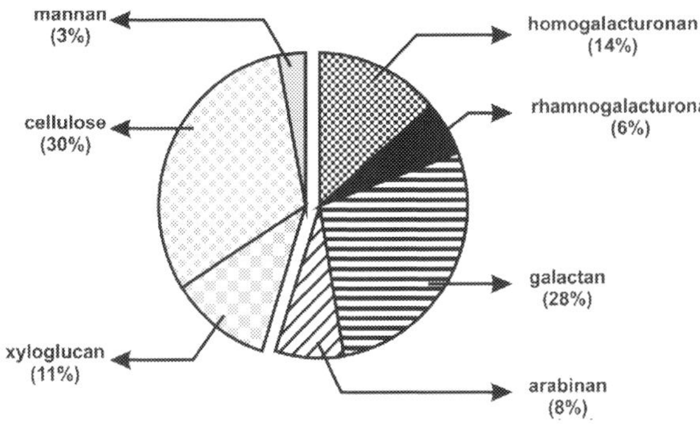

Figure 1.
Polysaccharide composition (% w/w) of the potato cell wall, representing all pectic polysaccharides, (except RG II) in black (56%), and other cell wall polysaccharides in gray {Vincken, Borkhardt, et al. 2000 271 /id}.

The primary structure of the individual pectic polysaccharides is well established. They comprise mainly homogalacturonan (HGA) and rhamnogalacturonan I (RG I). HGA (the smooth region) is composed of unbranched α-1→4-linked galacturonic acid (GalA) residues, which may be methyl-esterified and/or *O*-acetylated (Carpita and Gibeaut, 1993). RG I has a backbone composed of repeating α-(1→2)-L-rhamnose(Rha)-α-(1→4)-D-GalA disaccharide units. Also in the RG I backbone the GalA residues may be *O*-acetylated (Carpita and Gibeaut, 1993). Side-chains, mainly consisting of arabinan and/or galactan, may be attached to the RG I backbone at the *C*-4 position of the Rha residues (Carpita and Gibeaut, 1993; O'Neill *et al.*, 1990; Schols and Voragen, 1994). To date, there is little evidence that arabinan and galactan occur as independent polysaccharides in the cell wall. The macromolecular structure composed of RG I, galactan and arabinan is often referred to as hairy regions (Vries de *et al.*, 1981), with arabinan and galactan comprising the hairs.

However, much less information is available on the assembly of these polysaccharides into higher order structures and on their specific functions in the cell wall. The different pectic polysaccharides have been implicated in regulating cell expansion, organogenesis, textural changes during fruit ripening, and in serving as a source of oligosaccharins which can act as signalling molecules (Aldington and Fry, 1993; Darvill *et al.*, 1992; McCann and Roberts, 1994).

The physical properties of pectin make it suitable as a gelling agent, stabiliser or emulsifier in several food applications (Voragen *et al.*, 1995). Not all plant species are suitable sources of commercial pectin; currently mainly apple pomace and citrus peels, two by-products of food industrial processes, are used for pectin extraction. Commercially useful pectin should preferably have a high degree of methylation and a low proportion of hairy regions. The relatively high content of branched RG I, and low degree of methylation, make potato pectin a less suitable gelling agent for food applications (Ryden and Selvendran, 1990). The ability to modify the structure of potato tuber pectin *in planta* has the potential to increase the industrial applicability of potato fibre, which is a voluminous by-product of the potato starch industry (Chapple and Carpita, 1998; Thakur *et al.*, 1997; Vincken *et al.*, 2000).

Before it is possible to bio-engineer viable plants with pectin structures, which are improved for particular applications, it is necessary to understand the biosynthetic pathways for building and modifying the pectin. Further, it is important to know which pectin structures are present in various plants, plant tissues and cell types.

In this overview, we compare developmental studies, showing the naturally occuring variation in pectin structure, with studies on mutagenized and genetically modified plants. These approaches have helped to clarify the significance of the different hairy region structures for cell wall architecture/properties and developmental processes. Finally, this knowledge may enable the rational modification and future production of any desirable pectin structure *in planta*.

2. Pectin structure may determine tissue characteristics

Plants exhibit species-specific differences in their cell wall structure, together with compositional variations in the pectic polysaccharides, and their relative abundance. Determination of the monosaccharide compositions of different plant polymers provides useful structural information about pectic polysaccharides, but this approach fails to consider the importance of tissue- and cell-specific localisations of particular pectins.

The production of a series of antibodies recognizing different epitopes, each representing a part or particular structure of pectin, has clearly facilitated studies of developmental, tissue, and cell specific localization of the corresponding pectic structures (Knox, 1997; Willats *et al.*, 2000; Willats *et al.*, 2001a).

The monoclonal antibodies LM5 (recognizing (1→4)-β-D-galactan; (Jones *et al.*, 1997), and LM6 (recognizing (1→5)-α-L-arabinan; (Willats *et al.*, 1998), have been used extensively to study the presence and location of their respective epitopes in different plants and tissues at different developmental stages. These experiments have clearly indicated that hairy regions appear in a tightly regulated fashion with respect to cell wall localization, and also in relation to cell differentiation and cell proliferation (Bush *et al.*, 2000; Bush and McCann, 1999; Jones *et al.*, 1997; McCartney *et al.*, 2000; Orfila and Knox, 2000; Willats *et al.*, 1998; Willats *et al.*, 1999; Willats *et al.*, 2000; Willats *et al.*, 2001b).

In potato tubers, pectic epitopes are developmentally regulated during tuberisation and show also tissue-specific localisations in the mature tuber (Bush and McCann, 1999; (Bush *et al.*, 2001). An example of this are the labelling patterns of the LM5 and 2F4 (recognizing a calcium induced conformation of HGA; Liners *et al.*, 1994) antibodies as shown in Figure 2. The distributions of epitopes of galactan and HGA are complementary.

The LM5 and LM6 antibodies which both recognize RG I side-chains nevertheless show differences in labelling patterns. In stolons, the arabinan is enriched in the younger cells at the stolon tip (distal to the hook), while the galactan becomes more abundant in walls of older, elongated cells proximal to the stolon hook. In mature tubers, the arabinan epitopes are located throughout primary walls and middle lamellae of both cortical and perimedullary cell walls except at the expanded middle lamella at cell corners ((Bush *et al.*, 2001). Galactan is more abundantly localized in the primary walls of cortical and perimedullary tissues and in the primary cell wall it is mostly localized to the region close to the plasma membrane.

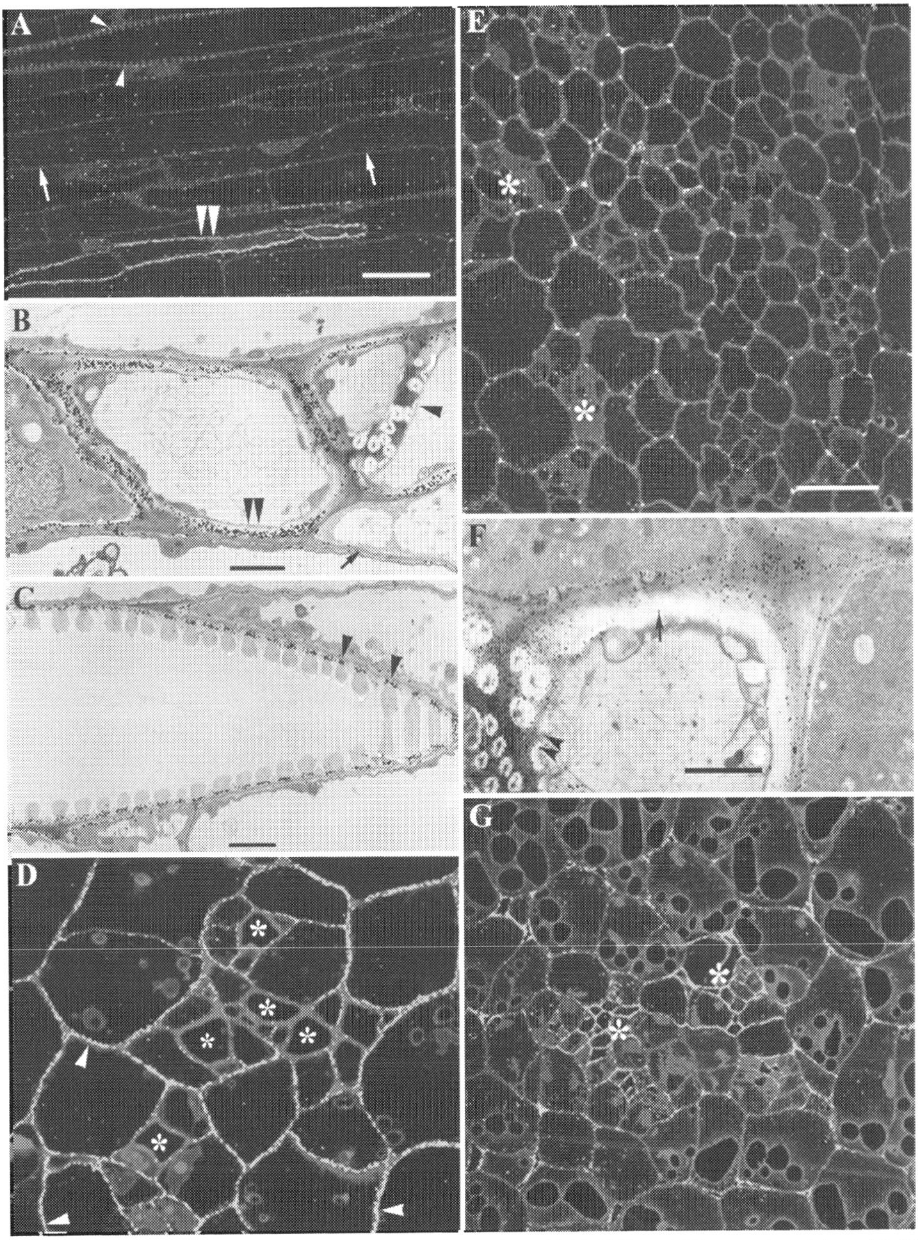

Figure 2. Sections of vascular tissue from elongating stolons (A-C, E-F) and 32mm mature tubers (D, G) immunogold labelled and silver-enhanced with monoclonal antibodies LM5 (A-D) and 2F4 (E-G) and then imaged by reflection confocal laser scanning microscopy (A, D-E, G) and transmission electron microscopy (B-C, F). Monoclonal antibody LM5 recognises β-(1,4)-galactan in RG I hairs, whilst the 2F4 epitope is a calcium-induced conformation of HGA. The walls of phloem sieve tubes in stolons label strongly with LM5 (A-B, double arrowheads), whereas parenchymal cell walls label weakly (A, B, arrows). The LM5 epitope is present at the base of xylem secondary thickenings (A, C, single arrowheads), but not in phloem sieve plates (B, single arrowhead). In mature tubers (D), the LM5 epitope is absent in vascular tissue (D, *), but more abundant in parenchymal walls (D, arrowheads); this contrasts to the situation in stolons (A). The 2F4 epitope is more abundant in vascular tissue of mature tubers (G, *) compared to stolons (E, *). The phloem sieve plate is labelled by 2F4 (F, double arrowheads), whilst the 2F4 epitope is absent from the electron-lucent inner regions of the sieve tube walls (F, arrows). There is therefore a complementary developmental pattern of galactan and HGA epitope expression. Scale bars represent 20μm in A, D, G, 40μm in E and 2μm in B-C, F.

In vascular walls and at pit fields of parenchymal walls, the LM5 epitope is greatly reduced in abundance, whilst the abundance of the LM6 arabinan epitope is reduced to a lesser extent in vascular walls and is not altered at pit fields (Bush and McCann, 1999). These labelling studies clearly show that in potato some co-localisation of the galactan and arabinan epitopes occurs. Nevertheless, the distinct distribution of these epitopes indicates the existence of different populations of RG I bearing different hairs with the implication that each type may influence the developmental characteristics of the wall.

Galactans and arabinans of other plants also show differences in developmental and spatial localisation. (McCartney *et al.*, 2000) showed that, in pea cotyledon cell walls, both HGA and (1$\rightarrow$5)-α-L-arabinan are continuously present, in contrast to a late developmental appearance of (1$\rightarrow$4)-β-D-galactan. Further, the galactan was localised in the region of the cell wall close to the plasma membrane. This localisation of galactan is also found in other species (Jones *et al.*, 1997), and in most cases might be explained by deposition at a later stage of development (Bush and McCann, 1999; Vicré *et al.*, 1998). The developmental deposition of galactan in pea correlated with an increased firmness of the cotyledons (McCartney *et al.*, 2000). In tomato and kiwi, the loss of galactan is correlated with fruit softening and is expected to be an important trigger of additional cell wall changes associated with fruit ripening (Jones *et al.*, 1997; Redgwell *et al.*, 1997). The fact that these correlations were only found for the galactan and not for arabinan suggests that the different types of hairy regions have different functions in the cell wall.

It remains difficult to assign a specific characteristic of the wall or tissue to a particular pectic structure. The major reason for this is that during tissue development, other cell wall polymers will also be modified. The ability to produce plants with a specific, tailor-made, cell wall composition will be extremely useful to analyze the functions of individual cell wall polysaccharides further. In the following section we discuss different strategies to generate modified hairy region structures in plants. Even though little information is available concerning the biosynthesis of pectin, the generation of a number of mutants with a modified hairy region composition has shown the relevance of particular structures (Table 1). Together with this mutant strategy, a limited number of genetically modified plants have also been produced (Table 1). The possibility of targeted degradation of a particular hairy region domain by the expression of fungal pectinases has proven to be extremely useful and contributed significantly to revealing the biological function of cell wall polysaccharides.

Table 1. Plants with a modified RG I composition

Mutant / Transformant	Pectin modification	Phenotype	Mutated gene	Reference
Cnr	↓ de-esterified HGA, disrupted Ara deposition	disturbed fruit ripening, reduced cell-to-cell adhesion and non-swollen cell walls in the pericarp	?	Orfila *et al.* (2001)
emb30	abnormal localisation of the pectin	seeds unable to pass through normal embryogenesis, abnormal plants	similar to Sec7p, which functions in the secretory pathway	(Shevell *et al.*, 2000)
kor	↑ HGA ↓ RG I with galactan hairs	dwarfed, cellulose ↓	membrane-bound endo-1,4-β-glucanase	(His *et al.*, 2001)
mur5-6-7	↓ Ara[1]	not observed	?	(Reiter *et al.*, 1997)
mur4	Ara ↓ 50%	not observed	UDP-D-Xyl-4-epimerase	(Burget and Reiter, 1999)
mur8	↓ Rha	not observed	?	(Reiter *et al.*, 1997)
mur10	↓ Fuc and Xyl ↑ Ara	slow growth, dark-green leaves, low seed-set	?	(Reiter *et al.*, 1997)
mur11	↓ Rha, Fuc and Xyl ↑ Man	not observed	?	(Reiter *et al.*, 1997)
nolac-H14	abnormal localization of the pectin	has non-organogenic callus with loosely attached cells	?	Iwai *et al.* (2001)
* TBG4	no decrease in galactan	delayed and reduced fruit softening	not applicable	Brummell and Harpster (2001)
* UDP-Glc-4-epimerase	no galactan ↑ during growth on galactose medium	not observed	not applicable	Dörmann and Benning (1998)
** *e*GAL	Gal ↓	not observed	not applicable	Sørensen *et al.* (2000)
** *e*ARA[A]	?	no stolons, tubers, flowers and side shoots	not applicable	Skjøt *et al.* (2002)
** *e*ARA[B]	?	degenerate stolons, no tubers	not applicable	Skjøt *et al.* unpublished
** *e*GARA	Ara ↓	not observed	not applicable	Skjøt *et al.* (2002)
** *e*RGL	↓ Ara and Gal	disturbed histology of the potato tuber cortex and periderm tissue	not applicable	Oomen *et al.* (2002)

[1]It is not clear if this reduction is an effect of a decreased RG I bound arabinan or a decrease in AGPs
[A]Arabinanase expression under control of the granule bound starch synthase promoter
[B]Arabinanase expression under control of the patatinB33 promoter
no labelling = mutants
* = anti-sense expression of an endogenous gene
** = sense expression of a heterologous gene

3. The generation, isolation and characterisation of cell wall mutants

Most plants with a modified cell wall composition have been produced in *Arabidopsis thaliana* by a random mutagenesis approach, using chemicals, X-ray irradiation or T-DNA insertion (for an overview see (Fagard *et al.*, 2000). These mutants have been the starting point for the isolation and characterization of genes involved in the biosynthesis of cell wall polysaccharides.

Some of the cell wall mutants described so far have been isolated based on altered growth (His *et al.*, 2001). Identification of cell wall mutants is also possible by analysing the monosaccharide composition of isolated cell wall material, but this is a very laborious way to screen a large mutagenised population (Reiter *et al.*, 1997). New methods, with minimal sample preparation, have been developed to screen directly for modified cell wall polysaccharides. Such methods will facilitate the identification of new mutants. Fourier transform infrared (FTIR) microspectroscopy has proven to be a powerful tool for the screening of cell wall mutants (Chen *et al.*, 1998). Further, the various anti-pectin antibodies (Willats *et al.*, 2000) can be used for the identification of mutants with a modified cell wall composition (Willats *et al.*, 2001b).

A number of *Arabidopsis* mutants having modified hairy regions as a result of mutations in genes involved in pectic biosynthesis, degradation or deposition are shown in Table 1. The *Arabidopsis mur* mutants, as described by (Reiter *et al.*, 1997), were isolated by a screening based on monosaccharide composition analysis. Among these, *mur4, 5, 6* and *7* show a reduction of arabinose, which is due to a decrease in arabinan in either hairy regions or arabinogalactan-proteins (AGPs). The 50% reduction of arabinose in *mur*4 is too high to be accounted for solely by a reduction in AGP-arabinose. Further analysis of this mutant suggested a mutation in a membrane-bound UDP-D-xylose 4-epimerase, that blocks the conversion of UDP-D-xylose to UDP-L-arabinose (Burget and Reiter, 1999). The *mur8* mutant is reduced in rhamnose and the *mur10* has a modification in fucose, xylose and arabinose. The *mur11* mutants show modifications in rhamnose, fucose, xylose and mannose. Most of the *mur* mutants showed no visible phenotype; of the *mur* mutants listed in Table 1 only *mur10* showed a reduction in growth rate and vigour.

Not all mutants with modified pectin are likely to have a mutation in a gene directly involved in the biosynthesis or degradation of pectic polymers. An example is the *Arabidopsis emb30* mutant (Shevell *et al.*, 2000) with a mutation in a gene that is likely to be part of the secretory pathway. In these mutants the pectin shows an abnormal localisation, and the seeds are unable to develop into a normal plant which is most likely due to a lack of control of the polarity of cell divisions and expansions during embryogenesis. Another example is the *Nicotiana plumbaginifolia nolac-H14* mutant (Iwai *et al.*, 2001) which is characterized by having a *n*on-*o*rganogenic callus with *l*oosely *a*ttached *c*ells. In contrast to wild-type callus, the mutant, which can only be maintained as a callus line, does not stain with ruthenium red (recognizing pectin) in the middle lamella and cell walls, and the pectin seems to be relocated to the surface of the callus and the growth medium. Sugar composition analysis, comparing the *nolac-H14* with wild-type callus, shows relative decreases in arabinose, xylose, galactose, glucose and mannose in whole cell extracts and relative increases of these sugars and galactose in the culture medium. These changes are likely to be linked to an aberrant deposition of pectin, although a modification in arabino-galactan proteins (AGPs) can not be excluded.

The *Cnr* ripening mutant of tomato also shows a disrupted deposition of several pectic polysaccharides (Orfila *et al.*, 2001). The mutant, for which the genetic basis has not yet been identified, has a pericarp with altered physical properties, including non-swollen cell walls and reduced intercellular adhesion (Thompson *et al.*, 1999). Even though no major differences were found by comparing the monosaccharide composition of the *Cnr* mutant and the wild type, antibody labelling clearly identified modifications in HGA and arabinan localization. A reduction of long de-esterified stretches of HGA possibly explains the reduced calcium-binding of the *Cnr* middle lamella, resulting in a reduced cell-to-cell adhesion. The additional disrupted deposition of (1→5)-α-arabinan did not correlate to the altered cell wall properties (Orfila *et al*,. 2001).

Modifications in pectin structure and localization can also be an indirect effect of modifications of other cell wall polymers. In the *Arabidopsis korrigan* mutant (His *et al.*, 2001), a mutation in a membrane-bound endo-1,4-β-glucanase resulted in a reduction of the cellulose content. This reduction in cellulose was accompanied with an increase in HGA and a decrease in RG I with galactan hairs, suggesting some compensatory mechanism for the cellulose-related reduction in wall strength. A similar result was observed in *Nicotiana benthamiana* after antisense inhibition of a cellulose synthase (*CesA*) gene (Burton *et al.*, 2000). These plants also showed that a decrease in cellulose content was accompanied with an increase of HGA with a decreased degree of esterification.

It is apparent from these studies that although the analysis of pectin mutants is a very useful approach to clarify the complexities of pectin biosynthesis, the generation of novel mutants remains a random process. In crop plants, the generation of mutants can be extremely difficult due to their polyploidy. An alternative here is genetic modification, which is possible in a large number of crop plants. Further, the down or up-regulation of an endogenous gene or the introduction of a heterologous gene provides a direct approach to specifically modify a known cell wall polysaccharide structure. Ideally, one would like to manipulate the biosynthetic machinery of the pectin, but unfortunately this approach is still limited since, to date, not many genes involved in wall polysaccharide biosynthesis have been cloned (Mohnen, 1999).

4. Genetic modification to generate plants with an altered cell wall composition

The biosynthetic and degradative pectic pathways can be targeted by genetic modification to generate cell walls with an altered pectin composition *in planta*. Genetic modification has already been used extensively to manipulate the processes causing fruit ripening and softening (reviewed by (Brummell and Harpster, 2001)). In these experiments, the expression of several genes (e.g. pectin methyl esterases and *endo*-polygalacturonases) responsible for modification of the pectin structure were down-regulated. These experiments are not further discussed here since the activity of these enzymes is mostly directed to HGA and not to the hairy regions. An important modification of the hairy regions during fruit ripening is the decline in pectic galactan early in the ripening process (reviewed in Brummell and Harpster, 2001). Suppression of the TBG4 gene, encoding a β-galactosidase activity (Smith and Gross, 2000) in tomato, showed a reduced fruit softening (Brummell and Harpster, 2001). This suggested that RG I galactan chains contribute to maintaining the firmness of the fruit, and that their degradation by endogenous enzymes is one of the changes leading to fruit softening during ripening.

An alternative direct approach to obtain a modified cell wall composition in plants is by manipulation of the biosynthetic machinery. One example of such a biosynthetic gene is the UDP-Glc-4-epimerase, which was used for sense and antisense expression in *Arabidopsis* (Dörmann and Benning, 1998). The enzyme, which catalyzes the reversible epimerization of UDP-galactose to UDP-glucose, is important in maintaining the pool of UDP-galactose that is the building block for the galactan hairs. The sense and antisense expression resulted in 3-fold increases and 90% reduction of the corresponding enzyme activity, respectively. Nevertheless, this did not result in a modified composition of the pectin during normal growth of the plants. Growth of the *Arabidopsis* plantlets on agar containing galactose, increased the amount of cell wall bound galactose in both wild type as well as in antisense plants with a decreased UDP-Glc-4-epimerase activity. In plants with an increased epimerase activity this effect was not found. This suggests that the UDP-Glc-4-epimerase is indeed important for maintaining the UDP-galactose pool and indirectly influences the amount of cell wall bound galactose. Nevertheless, in these transformants, enzyme levels as low as 10% of wild type are still sufficient to maintain normal UDP-galactose levels.

5. The introduction of pectin-degrading enzymes in potato

5.1. Fungal enzymes with potential to degrade hairy regions

Many enzymes degrading different pectic polymers have been described, and even though some of these enzymes are from plants (Fry, 1995), the best-characterised enzymes originate from fungi and bacteria (for an overview see Prade *et al.*, 1999).

Focussing on the removal of the branched part of the pectin, five hairy region-modifying enzymes, which can be divided in two categories, have been introduced in potato so far (Oomen *et al.*, 2001; Skjøt *et al.*, 2001a; Sørensen *et al.*, 2000). The first category encompasses an *endo*-galactanase and an *endo*-arabinanase, which are referred to as 'shaving enzymes'. These remove the galactan and arabinan side-chains of the RG I leaving the backbone polymer intact. The *endo*-galactanase (*e*GAL) from *Aspergillus aculeatus* is able to degrade (unsubstituted) 1→4-β-D-galactan to galactose and galactobiose (Christgau *et al.*, 2001; Lahaye *et al.*, 1991; Vis van de *et al.*, 1991). The *endo*-arabinanase (*e*ARA) also from *Aspergillus aculeatus* hydrolyzes α-1→5-L-arabinan and is specific for α-1→5-L-arabinofuranosidic linkages (Skjøt *et al.*, 2001b).

The second category includes the 'breaking enzymes' which degrade the RG I backbone and liberate the hairs attached to small backbone fragments. The rhamnogalacturonan hydrolase (*e*RGH) (Mutter *et al.*, 1996; Schols *et al.*, 1990) cleaves the backbone by hydrolysis at the non-reducing side of a rhamnosyl residue. In contrast, rhamnogalacturonan lyase (*e*RGL) cleaves the RG I backbone at the reducing side of the rhamnose. *In vitro* experiments have shown that de-acetylation of an RG I preparation by addition of a rhamnogalacturonan acetyl esterase (RGAE) enhances degradation by both *e*RGL and *e*RGH (Kauppinen *et al.*, 1995; Kofod *et al.*, 1994;

Schols *et al.*, 1990). This synergy can also be used *in planta* by expressing the RGAE together with the *e*RGL or *e*RGH.

5.2 INTRODUCTION OF HAIRY REGION DEGRADING ENZYMES IN POTATO

Modifications to cell walls could potentially have adverse effects on plants. These secondary effects might be limited by using tissue-specific promoters and specific targeting signals. In the majority of the experiments described below the granule-bound starch synthase (GBSS) promoter (Visser *et al.*, 1991) was used to obtain preferential expression of the genes in the potato tubers, and thus limit the effect on other parts of the plant. Since signal peptides are very similar among eukaryotes (Nielsen *et al.*, 1997), the endogenous targeting signals in these fungal enzymes were sufficient to permit targeting of the protein to the apoplast. The transformation efficiency with the *e*ARA and *e*GAL genes (Skjøt *et al.*, 2001a; Sørensen *et al.*, 2000) showed a significant decrease compared with the transformation of empty vector constructs, even though no significant differences were found for the other genes. The introduction of the genes did not have a large effect on the potato plants, which showed a clear expression of the transgene in the potato tubers.

*e*GAL, *e*ARA and *e*RGL enzyme activities were demonstrated in tissue extracts of the corresponding transformants and were not present in wild type tubers. No RG I-degrading activity was found in transformants expressing the *e*RGH or *e*RGH + *e*RGAE gene combination (unpublished results). Western and dot blot analysis using *e*RGH and *e*RGAE specific antibodies could not detect these two proteins in whole tuber extracts. Fourier transform infrared (FTIR) microspectroscopy and antibody labelling studies showed no alterations in cell wall composition in the *e*RGH or *e*RGH + *e*RGAE transformants. RNA was expressed for both *e*RGH and *e*RGAE, this together with the absence of the proteins suggests a problem during translation or post-translational modification of these two enzymes.

Both the *e*RGL and *e*ARA transformants showed phenotypic differences, compared to the wild type plants. The *e*RGL tranformants showed only phenotypic changes in the tubers (Oomen *et al.*, 2001), which had a more wrinkled morphology compared with the wild type tubers (figure 3).

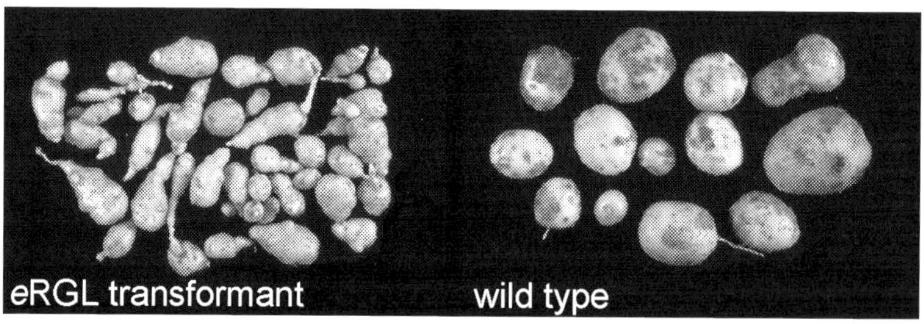

Figure 3. A selection of wild type potato tubers and tubers produced by a transformant expressing the *e*RGL gene.

The *e*ARA transformants showed a more severe phenotype. Plants expressing the arabinanase under control of the GBSS promoter produced no side shoots, flowers, stolons and tubers (Skjøt *et al.*, 2001a). Replacement of the GBSS promoter with the patatin B33 promoter (Rocha-Sosa et al., 1989) yielded plants with normal aerial parts (Skjøt et al. unpublished results). However, these plants developed degenerated stolons (Figure 4). Scanning electron microscopy (SEM) revealed major morphological differences between transgenic and wild type stolons (Figure 4). Wild type stolons viewed by SEM are long, slender diageotropic stems with a hooked tip bearing an apical meristem surrounded by leaf primordia (Fig. 4A). As the stolon tip elongates, meristematic nodes are left behind between the angle of the leaf primordia and the stolon. Leaf primordia bear trichomes on both surfaces (Fig. 4 A and B) and numerous stomata (Fig. 4C). In contrast, transgenic stolons are multi-branched stems with many secondary stolons (Fig. 4D) that develop from the leaf primordial nodes along the length of the primary stolon. Transgenic leaf primordia develop normally, but lack the numerous trichomes seen in the wild type (Fig. 4E), whilst the stomata are characteristically raised above the irregular epidermal surface (Fig. 4F).

Due to the inhibition of normal stolon and tuber development in the arabinanase-expressing potato plants, we have not been able to show, by using conventional biochemical techniques and immunogold labeling with the arabinan specific antibody LM6, that apoplastic expression of the endo-arabinanase results in a changed RGI sugar composition or epitope abundance. The carbohydrate moieties of cell wall associated proteoglycans may also be hypothetical substrates for the arabinanase when expressed to the apoplastic space. However, to the best of our knowledge, all arabinosylated hydroxyproline rich glycoproteins (HRGPs) analyzed to date except one, only contain a single terminal 1,5 linked arabinosyl residue which is not a substrate for the *A.aculeatus* endo-arabinanase (Skjøt *et al.*, 2002).

The sole described exception is a proteoglycan that has been isolated from cultured *Acacia senegal* cells. This arabinogalactan protein (AGP) carries side-chains consisting of α-1,5-linked arabinosyl groups (Mollard and Joseleau, 1994). The HRGPs are a large and heterogeneous protein family (Sommer-Knudsen *et al.*, 1998) and the finding that the *Acacia senegal* AGP carries arabinan side chains consisting of 1,5-linked arabinose does raise the possibility that non-hairy region arabinans may serve as substrate for the endo-arabinanase.

The phenotype of the *e*ARA plants clearly exemplify that a post-depositional degradation approach is not universally applicable and that other approaches must be used for *in vivo* enzymatic modification of α-1,5-linked arabinans.

Interestingly, a pre-deposition strategy where the arabinanase was targeted to the site of pectin biosynthesis, the Golgi apparatus, resulted in transformants (*eG*ARA) with a phenotype indistinguishable from WT and results in development of normal tubers (Skjøt *et al.*, 2001a). Expression of this chimeric enzyme engineered to be Golgi-membrane anchored resulted in a 69 % reduction of RG I-associated arabinosylgroups. This suggests that the severe phenotype of the apoplastic *e*ARA transformants is not an effect of the hairy regions lacking arabinan side chains. It is possible that in the *e*ARA plants normal tuber organogenesis is disturbed by secondary effects of released degradation products or by the consequences of enzymatic hydrolysis of non-hairy region associated arabinans.

Clearly, these results suggest that cell wall associated galactans and arabinans in the form of side chains of the hairy regions or as polysaccharide substituents on e.g. AGPs play different roles during plant development.

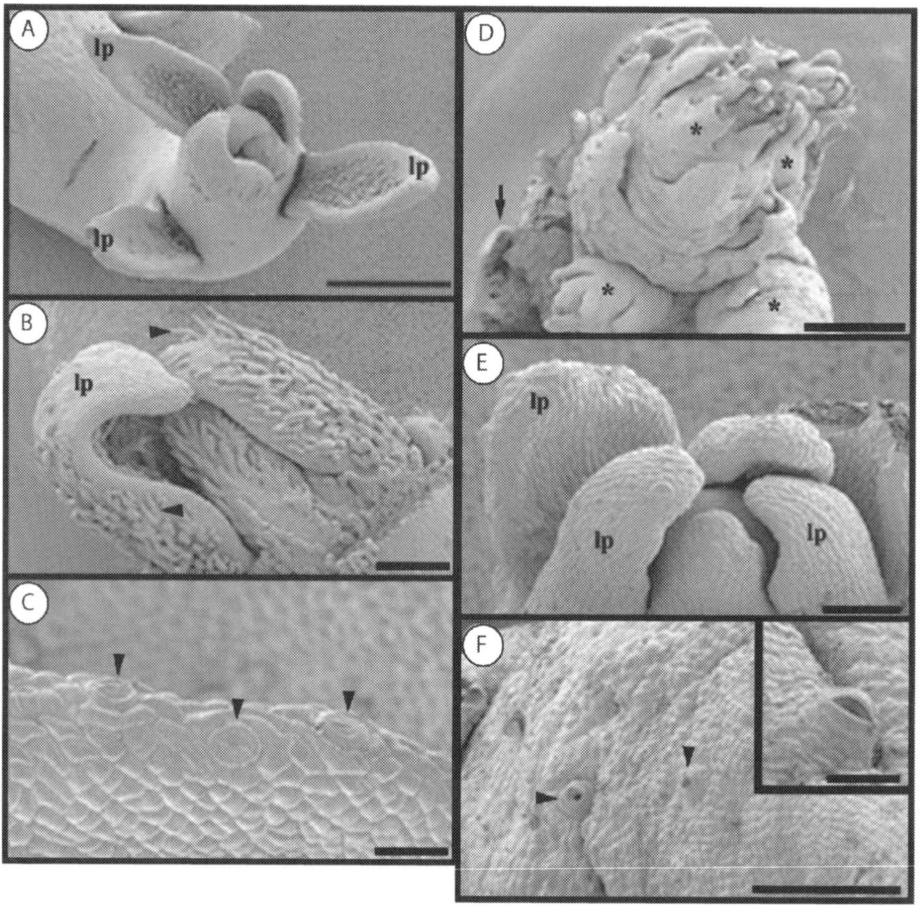

Figure 4. Scanning electron micrographs of stolons from wild type (A-C) and *e*ARA transformant with the arabinanase under control of the patatin B33 promoter (D-F). A: The tip of the wild type stolon bends through 90° to give a characteristic hook that bears leaf primordia (lp). In B the outer leaf primordia have been removed to reveal the numerous trichomes (arrowheads) that decorate both surfaces of the inner leaf primordia. C: Wild type stomata are normal in appearance (arrowheads). D: A short, brush-like arabinanase-expressing stolon bearing numerous secondary stolons (asterisks), the arrow indicates where the primary stolon was attached to the main stem of the plant. E: Transgenic leaf primordia lack the trichomes seen in the wild type plant (compare with A and B). F: The epidermal surface is very irregular and the stomata are abnormally raised (see insert). Scale bars are 1mm, D; 0.5mm, A; 200μm, B, F; 100μm, E and 50μm, C, inset in F.

5.3 EFFECT ON THE CELL WALL COMPOSITION

In addition to the predicted enzyme activities in the transgenic plants, both biochemical and microscopic analysis showed that the enzymes were able to modify the hairy region composition *in planta*.

The differences in the sugar composition of cell wall material isolated from the different transformants (Oomen *et al.*, 2001; Skjøt *et al.*, 2001a; Sørensen *et al.*, 2000) are summarized in Table 2. Both the *e*GAL and Golgi-targeted *eG*ARA transformants showed a reduction in galactose and arabinose respectively to a level corresponding to about 30% of the wild type. These reductions did not lead to obvious alterations in the other sugar residues, which suggests that these modifications are not compensated. Microscopic analysis, using the LM5 and LM6 antibodies respectively recognizing (1→4)-β-D-galactan (Jones *et al.*, 1997) and (1→5)-α-L-arabinan (Willats *et al.*, 1998), confirmed these biochemical observations. The *e*GAL and *eG*ARA transformants showed a reduced labelling with LM5 and LM6 respectively. No phenotypic changes were found in these transformants.

Table 2. Modifications in the sugar composition of the potato tubers from the *e*RGL, *e*ARA and *e*GAL transformants in comparison with tubers from wild type plants

	*e*RGL	*eG*ARA	*e*GAL
Rhamnose	–	≈	≈
Arabinose	–	–	≈
Galactose	–	≈	–
Mannose	+	not determined	≈
Xylose	+	≈	≈
Glucose	≈	≈	≈
Uronic acid	+	≈	≈

+ and – represent, respectively, increases and decreases of the particular sugar residue in comparison to wild type levels

≈ indicates no major changes in the particular sugar residue

In the *e*RGL transformants, where the RG I is degraded as opposed to 'shaved' as with the *e*GAL and *e*ARA, a greater impact on the potato tuber phenotype was observed. Sugar composition analysis showed a reduction of both galactose and arabinose to levels corresponding to respectively 20 and 57% of the wild type, this was accompanied by relative increases in mannose, xylose and uronic acid. Antibody labelling studies with LM5 and LM6 showed that both epitopes were reduced in abundance in walls, indicating that the side-chains were released from the cell wall by a fragmentation of the RG I backbone. Together with the reduced antibody labelling there was also a relocation of the LM5 galactan epitopes towards the middle lamella at cell corners and intercellular spaces. These relocations and reductions suggest that the RG I side chains are not cross-linked to other wall polymers, since the *e*RGL can only degrade the RG I backbone (Kofod *et al.*, 1994). The relocation of the hairy region components in the cell wall is only observed in the *e*RGL transformants and not in the *e*GAL and *eG*ARA plants.

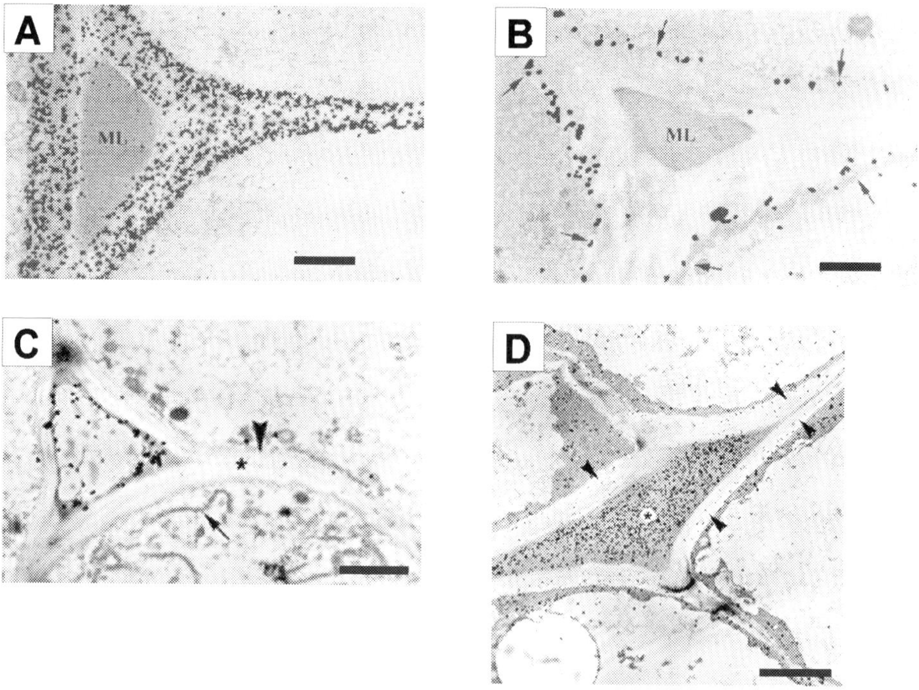

Figure 5. Transmission electron micrographs of cell walls from wild type (A), *e*GAL (B) and *e*RGL (C-D) expressing tubers immuno-labelled with LM5. ML (in A and B) and asterisk (in D) indicate the expanded middle lamella. Asterisk in C indicates the middle lamella. Arrows and arrowheads in B, C and D indicate remaining LM5 labelling. Scale bars represent 2 μm (A, B and C) and 4 μm (D).

Figure 5 shows the LM5 labelling in transmission electron micrographs for the *e*GAL, *e*RGL and wild type potato tubers. In wild type tubers, the LM5 epitopes are located in the primary cell wall and are excluded from the middle lamella and cell corners. The *e*GAL tubers show a reduction of LM5 epitopes and some remaining epitopes close to the plasma membrane. In the *e*RGL tubers, most of the epitope is present in the middle lamella at cell corners and there is a reduced epitope abundance in the primary wall. The remaining LM5 epitopes in the *e*GAL are likely to be located closer to the RG I backbone of the hairy region due to the (partial) degradation of the galactan hairs. The localization of these residual epitopes close to the plasma membrane suggests that (part of) the RG I (with enough remaining galactan to be still recognized by LM5) is located there. This RG I could either be anchored at this location or could represent newly deposited RG I with galactan hairs. However, it would be expected that newly deposited intact RG I would lie very close to the plasma membrane and that as it became incorporated into the wall, the *e*RGL would degrade it. As we did not see LM5 or LM6 labelling very close to the plasma membrane in the *e*RGL transformants, this would argue against the residual epitopes belonging to newly deposited RG I. We interpret the residual epitopes as belonging to a population of RG I backbones anchored close to the plasma membrane. The release of both galactan and arabinan after fragmentation of the RG I backbone confirms that they are only attached to the RG I

backbone. This is consistent with the results of (Foster *et al.*, 1998) and (Fenwick *et al.*, 1999) who showed that galactan side-chains in onion parenchyma walls are attached to a less mobile molecule at one end but are otherwise very mobile.
Both LM5 and LM6 are indirect antibody markers to detect RG I since they only recognize the respective side-chains and not the backbone. An antibody recognizing the RG I backbone structure could indicate if the relocated fragments contain RG I backbones or if RG I structures are still left in the cell wall. The mAb CCRC-M2 (Puhlmann *et al.*, 1994) does not cross-react with potato (Bush and McCann, 1999).
The *e*RGL potato tubers have a modified histology (Oomen *et al.*, 2001). In contrast to wild type, cell layers in the periderm and cortex tissues of transformants are severely disorganised and the number and size of cortical intercellular spaces is increased.
The comparison of the *e*GAL, *eG*ARA and *e*RGL transformants shows the importance of the complete hairy region structures, or the RG I backbone in particular, compared to its individual galactan and arabinan side-chains. The histological and morphological changes in the *e*RGL transformants suggest an important role of the RG I backbone for the integrity and function of the wall, at least in potato tubers.

6. A possible biological function for RG I and the hairs?

The expression of fungal cell wall degrading enzymes in plants was a useful approach to study pectic polysaccharide function. The specific deletion of parts of the hairy region enabled us to assign a putative function to the galactan hairs and the RG I backbone. Even though the *e*RGH and *e*RGAE were not successfully translated into a functional protein, the *e*GAL, *eG*ARA and *e*RGL enzymes modified pectin structure *in planta*. Nevertheless, tissue specific expression and subcellular localization were important determinants to confine the consequences of these modifications (Skjøt *et al.*, 2001a). Future use of constructs where the expression of the hairy region modifying enzymes is under control of inducible promoters may be beneficial.
The *e*GAL transformants (Sørensen *et al.*, 2000) are a useful tool to study the function of the RG I galactan hairs. It has been shown that galactan is present mostly in relatively older cells of a number of plant tissues (Bush *et al.*, 2001; Vicré *et al.*, 1998; Willats *et al.*, 1999). Some of these developmental changes, showing either addition (McCartney *et al.*, 2000) or degradation (Jones *et al.*, 1997; Redgwell *et al.*, 1997) of galactan, suggest that galactan is positively correlated with firmness of the plant tissues. However, in the *e*GAL transformants, with a reduction of RG I galactan hairs, there were no indications of altered tuber firmness, development or morphology. Nevertheless, the suppression of galactan degradation in tomato fruit, by antisense expression of a β-galactosidase/exo-galactanase (Brummell and Harpster, 2001), reduced fruit softening, but did not stop this process.
Combination of the results from the *e*GAL potatoes with the observations of pea and fruit ripening suggests that removal of the galactan chains does not directly reduce the strength of the cell wall. It is likely that the loss of galactan chains enables other enzymes to access regions of the cell wall to modify other cell wall components. This is in line with the observations by (Foster *et al.*, 1998) and (Fenwick *et al.*, 1999), who suggested that galactan hairs might be important for decreasing the pore size of cell walls and hence play a role in wall porosity. Nevertheless, the exact mechanism by

which galactan deposition can decrease the pore size is unclear. In contrast to RG II, which is suggested to actively determine the pore size in the cell wall by forming a pectic network with borate ester cross-linking (Fleischer *et al.*, 1999), the galactan is probably only decreasing the pore size by functioning as a filling material.
The fact that *e*GAL expression in potato tubers had no detectable effect on tuber development and morphology (Sørensen *et al.*, 2000), may be due to an absence of endogenous cell wall modifying enzymes in the tubers when the *e*GAL was expressed. This may explain why the loss of RG I galactan hairs did not have a similar impact to galactan degradation in fruit ripening processes. It seems that galactan degradation has only an impact when other wall polysaccharidases are expressed concomitantly. This is in line with the observation that a down regulated TBG4 expression always gives a reduced galactan degradation but results only in a delayed softening when the down regulation takes place early in the ripening process (Brummell and Harpster, 2001). Since potato tubers, directly after their formation, enter a dormant period, these organs are not likely to show many structural and or textural changes as a result of galactan degradation.
No particular function has been suggested to date for the RG I arabinan hairs, even though they are developmentally and spatially regulated (Bush *et al.*, 2001). The severe phenotype of the apoplastic *e*ARA transformants is likely to be caused by secondary effects or possibly by enzymatic hydrolysis of non-hairy region associated arabinans and not purely by a removal of arabinan from RG I. This is supported by the normal phenotype of transformants in which the *eG*ARA protein was targeted to the Golgi (Skjøt *et al.*, 2001a) and the arabinan content in the wall was successfully reduced. These latter *eG*ARA transformants are comparable with the *Arabidopsis mur4* mutant (Burget and Reiter, 1999) which also shows a reduction in L-arabinose (respectively 69 and 50%). These plants do not show any phenotypic defect and the reduction in L-arabinose does not seem to be compensated for by another specific monosaccharide. This suggests that a reduced level of RG I bound arabinan (at these environmental and physiological conditions) does not interfere with normal plant development although the extended phenotypes of the plants remain to be described.
The altered histology and morphology of potato tubers expressing the *e*RGL suggests an important function of the RG I backbone structure, since the removal of only the galactan and arabinan hairs did not result in cytological differences. The altered tuber morphology in the *e*RGL plants, caused by changes in the periderm and cortex, are probably the result of random divisions and abnormal expansions in these tissues. These results suggest that the RG I is necessary for normal periderm development even though the underlying reason is not yet clear. A similar effect was found in the *Arabidopsis emb30* mutants (Shevell et al., 2000). These mutants show a localization of pectic polysaccharides at cell corners and interstitial spaces together with the normal wild type localization in the cell wall. The authors hypothesize that the abnormal localization, which is likely to be caused by a failure in the deposition of the pectic polysaccharides, might interfere with normal cell division and disturb the cellular polarity. It is not known that the mutation in *emb30* directly affects the hairy region, but it cannot be excluded that the abnormalities found in the *e*RGL transformants may be the result of a similar mechanism.
Primary analysis of the first generation of genetically modified plants with an altered RG I composition was very useful in assigning a putative biological function to the

galactan hairs and the RG I backbone. The deposition of RG I with galactan hairs is likely to alter access of wall-modifying enzymes to the cell wall by decreasing the pore size. RG I backbones in the wall appear to be necessary for normal cell division and tissue development in potato tuber periderm and cortical tissues. Nevertheless, much more research on these, and future plants, is necessary. Apart from the generation of an antibody recognizing the potato RG I backbone, the generation of double transformants containing both the *e*GAL as well as the *e*GARA will help to confirm if indeed the fragmentation of the RG I backbone structure (as in the *e*RGL transformants) is causing the altered potato tuber morphology. Analysis of the *e*GAL and *e*GARA tranformants under normal growth conditions has so far not revealed any far-reaching implications for wall architecture or plant development. It should be noted here that the potato tubers were studied during the dormancy stage, in which developmental and biochemical processes are at a very low level. It is likely that e.g. an increase in the pore size of the cell walls in the *e*GAL transformants, may have an effect at other developmental stages, for instance, sprouting of the tubers. Studies of the extended phenotypes of these plants during different developmental stages and at different environmental and physiological conditions might reveal differences between the various transformants and wild type plants. It is expected that the use of these transgenic plants in developmental studies will deliver an important contribution and may result in a detailed description of the biological functions of the hairy regions.

Acknowledgements

This work was supported by the European Union project 'Remodelling pectin structure in plants' in the Fifth Framework Programme (contract CT97 2224).

References

Aldington, S. and Fry, S. C. (1993) Oligosaccharins. *Adv. Bot. Res.*, **19**, 2-101.

Beldman, G., Searle-van Leeuwen, M.J.F., De Ruiter, G.A., Siliha, H.A. and Voragen, A.G.J. (1993) Degradation of arabinans by arabinanases from *Aspergillus aculeatus* and *Aspergillus niger*. *Carbohydr. Polym.* **20**, 159-168.

Brummell, D. A. and Harpster, M. H. (2001) Cell wall metabolism in fruit softening and quality and its manipulation in transgenic plants. *Plant Mol. Biol.*, **47**, 311-340.

Burget, E. and Reiter, W.-D. (1999) The *mur4* mutant of *Arabidopsis* is partially defective in the de novo synthesis of uridine diphospho L-arabinose. *Plant Physiol.*, **121**, 383-389.

Burton, R. A., Gibeaut, D. M., Bacic, A., Findlay, K., Roberts, K., Hamilton, A., Baulcombe, D. C., and Fincher, G. B. (2000) Virus-induced silencing of a plant cellulose synthase gene. *Plant Cell*, **12**, 691-705.

Bush, M. S. and McCann, M. C. (1999) Pectic epitopes are differentially distributed in the cell walls of potato (*Solanum tuberosum*) tubers. *Physiol. Plantarum*, **107**, 201-213.

Bush, M. S., Marry, M., Huxham, I. M., Jarvis, M. C., and McCann, M. C. (2001) Developmental regulation of pectic epitopes during potato tuberisation. *Planta*, **213**, 869-880.

Carpita, N. C. and Gibeaut, D. M. (1993) Structural models of primary cell walls in flowering plants: consistencey of molecular structure with the physical properties of the walls during growth. *Plant J.*, **3**, 1-30.

Chapple, C. and Carpita, N. C. (1998) Plant cell walls as targets for biotechnology. *Curr. Opin. Plant Biol.*, **1**,
179-185.

Chen, L., Carpita, N. C., Reiter, W.-D., Wilson, R. H., Jeffries, C., and McCann, M. C. (1998) A rapid method

to screen for cell-wall mutants using discriminant analysis of Fourier transform infrared spectra. *Plant J.*, **16**, 385-392.
Christgau, S., Sandal, T., Kofod, L. V., and Dalbøge, H. (1995) Expression cloning, purification and characterization of a β-1,4-galactanase from *Aspergillus aculeatus*. *Curr. Genet.*, **27**, 135-141.
Darvill, A., Augur, C., Bergmann, C., Carlson, R. W., Cheong, J.-J., Eberhard, S., Hahn, M. G., Meyer, B., Marfa, V., and Meyer (1992) Oligosaccharins - oligosaccharides that regulate growth, development and defence responses in plants. *Glycobiology*, **2**, 181-198.
Dörmann, P. and Benning, C. (1998) The role of UDP-glucose epimerase in carbohydrate metabolism of Arabidopsis. *Plant J.*, **13**, 641-652.
Ermel, F.F., Follet-Gueye, M.-L., Cibert, C., Vian, B., Morvan, C., Catesson, A.-M., Goldberg, R. (2000) Differential localization of arabinan and galaactan side chains of rhamnogalacturonan I in cambial derivatives, *Planta*, **210**, 732-740.
Fagard, M., Höfte, H., and Vernhettes, S. (2000) Cell wall mutants. *Plant Physiol. Biochem.*, **38** , 15-25.
Fenwick, K. M., Apperley, D. C., Cosgrove, D. J., and Jarvis, M. C. (1999) Polymer mobility in cell walls of cucumber hypocotyls. *Phytochemistry*, **51**, 17-22.
Fleischer, A., O'Neill, M.A. and Ehwald, R. (1999) The pore size of non-graminaceous plant cell walls is rapidly decreased by borate ester cross-linking of the pectic polysaccharide rhamnogalacturonan II. *Plant Physiol.*, **121**, 829-838.
Foster, T. J., Ablett, S., McCann, M. C., and Gidley, M. J. (1998) Mobility-resolved ^{13}C-NMR spectroscopy of primary plant cell walls. *Biopolymers*, **39**, 51-66.
Freshour, G., Clay, R.P., Fuller, M.S., Albersheim, P., Darvill, A.G., and Hahn, M.G. (1996) Developmental and tissue-specific structural alteration of the cell-wall polysaccharides of *Arabidopsis thaliana* roots. *Plant Physiol.*, **110**, 1413-1429.
Fry, S. F. (1995) Polysaccharide-modifying enzymes in the plant cell wall. *Annu. Rev. Plant Physiol. Plant Mol. Biol.*, **46**, 497-520.
His, I., Driouich, A., Nicol, F., Jauneau, A., and Höfte, H. (2001) Altered pectin composition in primary cell walls of *korrigan*, a dwarf mutant of *Arabidopsis* deficient in membrane-bound endo-1,4- -glucanase. *Planta*, **212**, 348-358.
Iwai, H., Ishii, T. and Satoh, S. (2001) Absence of arabinan in the side chains of the pectic polysaccharides strongly associated with cell walls of *Nicotiana plumbaginifolia* non-organogenic callus with loosely attached constituent cells. *Planta,* **213**, 907-915.
Jarvis, M. C., Hall, M. A., Threlfall, D. R., and Friend, J. (1981) The polysaccharide structure of potato cell walls: chemical fractionation. *Planta*, **152**, 93-100.
Jarvis, M. C. and McCann, M.C. (2000) Macromolecular biophysics of the plants cell wall: Concepts and methodology. *Plant Physiol. Biochem.* **38**, 1-13.
Jones, L., Seymour, G. B., and Knox, J. P. (1997) Localization of pectic galactan in tomato cell walls using a monoclonal antibody specific to (1→4)- -D-Galactan. *Plant Physiol.*, **113**, 1405-1412.
Kauppinen, S., Christgau, S., Kofod, L. V., Halkier, T., Dörreich, K., and Dalbøge, H. (1995) Molecular cloning and characterization of a rhamnogalcturonan acetylesterase from *Aspergillus aculeatus*, Synergism between rhamnogalacturonan degrading enzymes. *J. Biol. Chem.*, **270**, 27172-27178.
Knox, J. P. (1997) The use of antibodies to study the architecture and developmental regulation of plant cell walls. *Int. Rev. Cytol.*, **171**, 79-120.
Kofod, L. V., Kauppinen, S., Christgau, S., Andersen, L. N., Heldt-Hansen, H. P., Dörreich, K., and Dalbøge, H. (1994) Cloning and characterization of two structurally and functionally divergent rhamnogalacturonases from *Aspergillus aculeatus*. *J. Biol. Chem.*, **269**, 29182-29189.
Lahaye, M., Vigouroux, J., and Thibault, J.-F. (1991) Endo-β-1,4-D-galactanase from *Aspergillus niger* var. *Aculeatus*: Purification and some properties. *Carbohyd. Polym.*, **15**, 431-444.
Liners, F., Gaspar, T., Van Cutsem, P. (1994) Acetyl- and methyl-esterification of pectins of friable and compact sugar-beet calli: Consequences for intercellular adhesion. *Planta*, **192**, 545-556.
McCann, M. C. and Roberts, K. (1994) Changes in cell wall architecture during elongation. *J. Exp. Bot.*, **45**, 1683-1691.
McCann, M. C., Shi, J., Roberts, K., and Carpita, N. C. (1994) Changes in pectin structure and localization during the growth of unadapted and NaCl-adapted tobacco cells. *Plant J.*, **5**, 773-785
McCann, M. C. and Roberts, K. (1996) Plant cell wall architecture: the role of pectins. In Pectins and Pectinases (Visser, J. and Voragen, A. G. J., eds). Amsterdam: Elsevier Science B.V., pp. 91-107.
McCann, M.C., Chen, L., Roberts, K., Kemsley, E.K., Sene, C., Carpita, N.C., Stacey, N.J. and Wilson, R.H. (1997) Infrared microspectroscopy: Sampling heterogeneity in plant cell wall composition and architecture. *Physiol. Plant.* **100**, 729-738.
McCartney, L., Ormerod, A. P., Gidley, M. J., and Knox, J. P. (2000) Temporal and spatial regulation of

pectic (1→4)-beta-D-galactan in cell walls of developing pea cotyledons: implications for mechanical properties. *Plant J.*, **22**, 105-113.
Mohnen, D. (1999) Biosynthesis of pectins and galactomannans. In Comprehensive natural products chemistry, volume 3: carbohydrates and their derivatives including tannins, cellulose, and related lignins (Pinto, B. M., Barton, D., Nakanishi, K., and Meth-Cohn, O., eds). Oxford: Elsevier Science Ltd, pp. 497-527.
Mollard, A. and Joseleau, J.P. (1994) *Acacia senegal* Cells cultured in suspension secrete a hydroxyproline-deficient Arabinogalactan-protein. *Plant Physiol. Biochem.* **32,** 703-709.
Mutter, M., Colquhoun, I. J., Schols, H. A., Beldman, G., and Voragen, A. G. J. (1996) Rhamnogalacturonase B from *Aspergillus aculeatus* is a rhamnogalacturonan -L-rhamnopyranosyl-(1→4)- -D-galactopyranosyluronide lyase. *Plant Physiol.*, **110**, 73-77.
Nielsen, H., Engelbrecht, J., Brunakd, S., and Heijne von, G. (1997) Identification of prokaryotic and eukaryotic signal peptides and prediction of their cleavage sites *Protein Eng.*, **10**, 1-6.
O'Neill, M., Albersheim, P., and Darvill, A. (1990) The pectic polysaccharides of primary cell walls. In Methods in Plant Biochemistry (Dey, P. M. and Harborne, J. B., eds). London: Academic Press Limited, pp. 415-441.
Oomen, R. J. F. J., Doeswijk-Voragen, C. H. L., Bush, M. S., Vincken, J.-P., Borkhardt, B., Broek van den, L.
A. M., Corsar, J., Ulvskov, P., Voragen, A. G. J., McCann, M. C., and Visser, R. G. F. (2002) *In muro* fragmentation of the rhamnogalacturonan I backbone in potato (*Solanum tuberosum* L.) results in a reduction
and altered location of the galactan and arabinan side-chains and abnormal periderm development. *Plant J.*, in press.
Orfila, C. and Knox, J. P. (2000) Spatial regulation of pectic polysaccharides in relation to pit fields in cell walls of tomato fruit pericarp. *Plant Physiol.*, **122**, 775-781.
Orfila, C., Seymour, G.B., Willats, W.G.T., Huxham, M., Jarvis, M.C., Dover, C.J., Thompson, A.J., and Knox, J.P. (2001) Altered middle lamella homogalacturonan and disrupted deposition of (1→5)-α-L-arabinan in the pericarop of *Cnr*, a ripening mutant of tomato. *Plant Physiol.*, **126**, 210-221.
Prade, R.A., Zhan, D. F., Ayoubi, P. and Mort, A. J. (1999). Pectins, pectinases and plant-microbe interactions. In Harding, S. E. (Ed), *Biotechnol. Gen. Eng. Rev*, **16**, 361-391.
Puhlmann, J., Bucheli, E., Swain, M. J., Dunning, N., Albersheim, P., Darvill, A. G., and Hahn, M. G. (1994) Generation of monoclonal antibodies against plant cell-wall polysaccharides. *Plant Physiol.*, **104**, 699-710.
Redgwell, R. J., MacRae, E., Hallet, I., Fisher, M., Perry, J., and Harker, R. (1997) *In vivo* and *in vitro* swelling of cell walls during ripeing. *Planta*, **203**, 162-173.
Reiter, W.-D., Chapple, C., and Somerville, C. R. (1997) Mutants of *Arabidopsis thaliana* with altered cell wall polysaccharide composition. *Plant J.*, **12** , 335-345.
Rocha-Sosa, M., Sonnewald, U., Frommer, W., Stratmann, M., Schell, J. and Willmitzer, L. (1989) Both developmental and metabolic signals activate the promoter of a class I patatin gene. *EMBO J.* **8,** 23-29.
Ryden, P. and Selvendran, R. R. (1990) Structural features of cell-wall polysaccharides of potato (*Solanum tuberosum*). *Carbohydr. Res.*, **195**, 257-272.
Schols, H.A., Posthumus, M.A. and Voragen, A.G.J. (1990a) Structural features of hairy regions of pectins Isolated from apple juice produced by the liquefaction process. *Carbohydr. Res.*, **206**, 117-129.
Schols, H. A., Geraeds, C. C. J. M., Searle-van Leeuwen, M. J. F., Kormelink, F. J. M., and Voragen, A. G. J. (1990b) Rhamnogalacturonase: a novel enzyme that degrades the hairy regions of pectins. *Carbohydr. Res.*, **206**, 105-115.
Schols, H. A. and Voragen, A. G. J. (1994) Occurence of pectic hairy regions in various plant cell wall materials and their degradability by rhamnogalacturonase. *Carbohydr. Res.*, **256**, 83-95.
Shevell, D. E., Kunkel, T., and Chua, N. H. (2000) Cell wall alterations in the *Arabidopsis emb30* mutant. *Plant Cell*, **12**, 2047-2059
Skjøt, M., Pauly, M., Bush, M., Borkhardt, B., McCann, M., and Ulvskov, P. (2002) Direct interference with rhamnogalacturonan I biosynthesis in Golgi vesicles. *Plant Physiol.* in press.
Skjøt, M., Kauppinen, S., Kofod, L. V., Fuglsang, C., Pauly, M., Dalbøge, H., and Andersen, L. N. (2001) Functional cloning of an endo-arabinanase from *Aspergillu aculeatus* and its heterologous expression in *A. oryzae* and tobacco. *Mol. Gen. Genet.*, **265**, 913-921.
Smith, R. C. and Gross, K. C. (2000) A family of at least seven -galactosidase genes is expressed during tomato fruit development. *Plant Physiol.*, **123** , 1173-1183.
Sommer-Knudsen, J., Bacic, A. and Clarke, A.E. (1998) Hydroxyproline-rich plant glycoproteins. *Phytochemistry* **47,** 483-497.
Sørensen, S. O., Pauly, M., Bush, M. S., Skjøt, M., McCann, M. C., Borkhardt, B., and Ulvskov, P. (2000)

Pectin engineering: Modification of potato pectin by *in vivo* expression of an endo-1,4- -D-galactanase. *Proc. Natl. Acad. Sci. USA*, **97**, 7639-7644.
Thakur, B. R., Singh, R. K., and Handa, A. K. (1997) Chemistry and uses of pectin. *Crit. Rev. Food Sci.*, **37**, 47-73.
Thompson, A.J., Tor, M., Barry, C.S., Vrebalov, J., Orfila, C., Jarvis, M.C., Giovannoni, J.J., Grierson, D., and Seymour, G.B. (1999) Molecular and genetic characterization of a novel pleiotropic tomato-ripening mutant. *Plant Physiol.* **120**, 383-389.
Vicré, M., Jauneau, A., Knox, J. P., and Driouich, A. (1998) Immunolocalization of 1 →4)- and 1→6)-D galactan epitopes in teh cell wall and Golgi stacks of developing flax root tissues. *Protoplasma*, **203**, 26-34.
Vincken, J.-P., Borkhardt, B., Bush, M., Doeswijk-Voragen, C. H. L., Dopico, B., Labrador, E., Lange, L., McCann, M., Morvan, C., Schols, H. A., Oomen, R., Peugnet, I., Rudolph, B., Schols, H., Sørensen, S., Ulvskov, P., Voragen, A., and Visser, R. (2000) Remodelling pectin structure in potato. In Conference Proceedings of Phytosfere'99 European Plant Biotechnology Network (Vries de, G. E. and Metzlaff, K., eds).
Amsterdam: Elsevier Science B.V., pp. 245-256.
Vis van de, J. W., Searle-van Leeuwen, M. J. F., Siliha, H. A., Kormelink, F. J. M., and Voragen, A. G. J. (1991) Purification and characterization of endo-1,4-beta-D-galactanases from *Aspergillus niger* and *Aspergillus aculeatus*: Use in combination with arabinanases from *Aspergillus niger* in enzymic conversion of potato arabinogalactan. *Carbohyd. Polym.*, **16**, 167-187.
Visser, R. G. F., Stolte, A., and Jacobsen, E. (1991) Expression of a chimaeric granule-bound starch synthase-GUS gene in transgenic potato plants. *Plant Mol. Biol.*, **17**, 691-699.
Voragen, A. G. J., Pilnik, W., Thibault, J.-F., Axelos, M. A. V., and Renard, C. M. G. C. (1995) Pectins. In Food Polysaccharides and Their Applications (Stephen, A. M., eds). New York: Marcel Dekker Inc, pp. 287-
339.
Vries de, J. A., Voragen, A. G. J., Rombouts, F. M., and Pilnik, W. (1981) Extraction and purification of pectins from alcohol insoluble solids from ripe and unripe aples. *Carbohyd. Polym.*, 117-127.
Willats, W. G. T., Marcus, S. E., and Knox, J. P. (1998) Generation of a monoclonal antibody specific to 1->5- -L-arabinan. *Carbohydr. Res.*, **308**, 149-152.
Willats, W. G. T., Steele-King, C. G., Marcus, S. E., and Knox, J. P. (1999) Side chains of pectic polysaccharides are regulated in relation to cell proliferation and cell differentiation. *Plant J.*, **20**, 619-628.
Willats, W. G. T., Steele-King, C. G., McCartney, L., Orfila, C., Marcus, S. E., and Knox, J. P. (2000) Making
and using antibody probes to study plant cell walls. *Plant Physiol. Biochem.*, **38**, 27-36.
Willats, W. G. T., McCartney, L., Mackie, W., and Knox, J. P. (2001a) Pectin: cell biology and prospects for functional analysis. *Plant Mol. Biol.*, **47**, 9-27.
Willats, W. G. T., McCartney, L., and Knox, J. P. (2001b) In-situ analysis of pectic polysaccharides in seed mucilage and at the root surface of *Arabidopsis thaliana*. *Planta*, **213**, 37-44.

DETERGENT-SOLUBILISATION OF A HOMOGALACTURONAN GALACTURONOSYLTRANSFERASE FROM MUNG BEAN

H. J. CROMBIE, C. SCOTT and J. S. G. REID
University of Stirling, Department of Biological Sciences, University of Stirling, Stirling
FK9 4LA E-mail: hjc1@stirling.ac.uk

Abstract

A membrane-bound enzyme [homogalacturonan 4- -galacturonosyltransferase (HGA-GalAT)] that catalyses the transfer of galacturonic acid onto the homogalacturonan (HGA) component of plant cell wall pectin has previously been identified and partially characterised in mung bean (*Vigna radiata*) seedlings (Villmez *et al.*, 1965; Villmez *et al.*, 1966; Smith, 1999). This study reports the successful detergent-solubilisation of the enzyme from the microsomal membranes of the growing hypocotyl using a range of detergents. Highest levels of radioincorporation were obtained with the detergent Zwittergent Z3-14 at a concentration of 1 % in 25 mM Tris-HCl pH 7.0 and a manganese concentration of 5 mM. Under these conditions, using polygalacturonic acid (PGA) as acceptor, activity solubilised by this detergent was comparable with that detected in intact particulate microsomal membranes. The soluble enzyme transferred a small amount of [^{14}C]-galacturonic acid from the nucleotide precursor UDP-[^{14}C]-galacturonic acid onto an endogenous acceptor and the amount of radioincorporation was enhanced by the addition of exogenous methylated or non-methylated pectins. The presence of methyl groups on the exogenous pectin did not appear to influence the activity of the enzyme. Oligogalacturonides [degree of polymerisation (DP) = 6, 10, 13, 14 or 15] also enhanced radioincorporation, the oligogalacturonide of DP 15 being slightly more effective than the others. Successful detergent-solubilisation has enabled a partial purification of HGA-GalAT by isoelectric focusing (IEF) and the results have revealed the presence of at least two isoforms of the enzyme (pI~5 and pI~6).

1. Introduction

Pectin plays an important structural role in the cell walls of higher plants. The polysaccharide is present in the primary cell wall and middle lamella and is deposited in the early stages of growth during cell wall expansion. The cell wall lends strength and support to the plant and yet it is a very dynamic structure. New material is constantly being laid down and old material degraded and removed. There are three main types of pectin, homogalacturonan (HGA), rhamnogalacturonan I (RGI) and rhamnogalaturonan II (RGII), present in plant primary cell walls (Albersheim *et al.*, 1996; O'Neill *et al.*, 1990) and others such as xylogalacturonan and apiogalacturonan that are present in

F. Voragen et al. (eds.), Advances in Pectin and Pectinase Research, 35-45.
© 2003 *Kluwer Academic Publishers. Printed in the Netherlands.*

some plants (Hart & Kindel, 1970; Kikuchi *et al.*, 1996; Schols *et al.*, 1995). The structural complexity of pectin implies the presence of many transferase enzymes for the biosynthesis of these polymers. Transferases that have already been identified in various tissues include galactosyltransferases, arabinosyltransferases, galaturonosyltransferases and methyltransferases (see Doong and Mohnen, 1998 for review). To date none of these enzymes have been purified to homogeneity nor have any of the corresponding cDNAs been sequenced.
Homogalacturonan is the major type of pectin in cell walls accounting for around 60% of the total amount (Mohnen *et al.*, 1996; O'Neill *et al.*, 1990). The polymer consists of a backbone of α-1,4-linked galacturonic acid (GalA) residues. This backbone may be methylesterified (Mort *et al.*, 1993) and/or *O*-acetylated at *O*-2 and *O*-3 (Ishii, 1995). The enzyme responsible for HGA backbone biosynthesis is homogalacturonan galacturonosyltransferase (HGA-GalAT), a membrane-bound protein thought to be located in the Golgi. A membrane-bound enzyme like HGA-GalAT usually requires that the protein be detergent-solubilised with retention of activity prior to purification. Solubilisation can often be a critical step since loss of activity may be experienced as a result of irreversible protein denaturation or loss of natural acceptor substrate. An enzyme (referred to as 4-α-galacturonosyltransferase), which catalysed the transfer of $[^{14}C]$ - labelled galacturonic acid from UDP - $[^{14}C]$ galacturonic acid onto an endogenous polysaccharide, was previously identified in mung bean seedlings (Villemez *et al.*, 1965; Villemez *et al.*, 1966). The particulate form of this enzyme [referred to as homogalacturonan synthase (HGS)] has since been further characterised by Smith (1999). A similar enzyme from tobacco cell microsomal membranes [referred to as homogalacturonan galacturonosyltransferase (HGA-GalAT)] has been identified, partially characterised (Doong *et al.*, 1995) and solubilised (Doong and Mohnen, 1998). The successful detergent-solubilisation of the enzyme required 40mM CHAPS and the presence of EDTA and manganese. The soluble form of the enzyme required the addition of an exogenous pectin or oligogalacturonide acceptor. Its action on oligogalacturonide acceptors was found to be non-processive, ie it catalysed the addition of only one galacturonic acid residue onto an oligogalacturonide of DP 15 giving a product of DP 16. We now report the detergent-solubilisation of a homogalacturonan galacturonosyltransferase from mung bean hypocotyls, some properties of the soluble enzyme and a partial purification of the enzyme by isoelectric focusing.

2. Materials and Methods

2.1. CHEMICALS

Radioactive uridine diphospho-D-[U-^{14}C]glucuronic acid was purchased from Amersham Pharmacia Biotech, Little Chalfont, Buckinghamshire, UK and converted to uridine diphospho-D-[U-^{14}C]galacturonic acid by the method described by Liljebjelke *et al.* (1995). TritonX-100, Nonidet-P40 and deoxycholate were from Sigma Chemical Company, Poole, Dorset, UK and bigChap, deoxybigChap and Zwittergent Z3-14 were from C N Biosciences, Nottingham, UK. Pectins that had been prepared by controlled alkali treatment of a highly-methylated pectin (Sigma) and the red and green tomato pectin extracts were provided by the Unilever Research Colworth Laboratory, Sharnbrook, Bedfordshire, UK. All other chemicals were of Analar grade or better.

2.2. PLANT MATERIAL

Mung beans (*Vigna radiata*) were imbibed overnight in tap water and then grown in the dark at 20°C on a bed of wet vermiculite. Hypocotyls were harvested 3-4 days after sowing.

2.3. PREPARATION OF MICROSOMAL MEMBRANES

Mung bean cotyledons were removed and the hypocotyls homogenised in an Omnimix blender in 50 mM Tris-HCl pH 7.5 / 1 mM EDTA (1 ml buffer per gram fresh weight). The extract was spun at 20,000 *g* for 15 minutes and the resultant supernatant then spun at 40,000 *g* for 60 minutes to obtain the microsomal pellet (Sorvall RC28S centrifuge, SS34 rotor). Pellets were resuspended in 50 mM Tris-HCl pH 7.0 in a hand-held glass homogeniser (0.1 ml buffer per gram fresh weight of starting tissue). The preparation was referred to as the intact microsomal membrane or particulate membrane preparation. Centrifugations were at 4°C and all other procedures were carried out on ice.

2.4. DETERGENT-SOLUBILISATION OF PROTEIN

Microsomal pellets were resuspended in 0.05 ml buffer per gram of fresh starting tissue in a hand-held glass homogeniser. An equal volume of detergent was added to give the required final concentration and the extract homogenised on ice (5 strokes every 10 min. for a total of 30 min.). The detergent-treated extract was then spun at 100,000 *g* for 1 hour and the supernatant kept (detergent-solubilised fraction).

2.5. ASSAY OF HGA-GalAT ACTIVITY

An aliquot (50 μl) of extract (particulate or soluble) was incubated with 10 μl of radiolabelled uridine diphospho-D-[$U^{14}C$]galacturonic acid (4,000 dpm, 0.074kBq, specific activity 7.34 mCi/mmol) and 40μl distilled water at 25°C. Other reaction components such as divalent cations and acceptor substrates could be added and the volume maintained by adjusting the volume of distilled water. Boiled microsomal extracts were also assayed and all samples (active and boiled) were assayed in duplicate. Incubations were carried out for up to 1 hour and terminated by boiling for 5 min. Methanol was added (final concentration 70% v/v), reactions heated to 70°C for 10 min. and then spun at 15,000 *g* for 5 min. in a bench centrifuge (Heraeus BiofugeA) to pellet polysaccharide products. The pellets were then washed twice with 70% methanol followed by centrifugation at 15,000 *g*. For liquid scintillation counting, pellets were resuspended in 50 μl distilled water, 1 ml scintillant was added and samples vortexed to mix. Liquid scintillation counting was carried out using a Packard Tri-Carb 2000CA liquid scintillation analyser (Canberra Packard, Pangbourne, Berkshire, UK). Scintillation fluid was Emulsifier-Safe™ (Packard Bioscience DV,

Ulgersmaweg 47, 9731BK Groningen, The Netherlands). HGA-GalAT activities were expressed as dpm per pellet.

2.6. PRODUCT ANALYSIS

Methanol-precipitated pellets were resuspended in 50µl of buffer and digested with the pectin-specific enzymes (0.5 units) *endo*-polygalacturonase (EPG) and pectin methylesterase (PME). EPG digestion was carried out at pH 5.0 in 50 mM ammonium acetate and PME treatment at pH 7.5 in 50 mM Tris-HCl. Products of these reactions were assayed for radioactivity as above or separated by TLC and analysed by charring and/or autoradiography.

2.7. THIN-LAYER CHROMATOGRAPHY AND AUTORADIOGRAPHY OF RADIOLABELLED PRODUCTS

Thin layer chromatography was carried out using aluminium foil-backed silica gel plates, 0.2 mm thickness (E Merck-DC-Alufolien, Kieselgel 60). Separation was by development in the solvent propan-1-ol : nitromethane : water (5 : 3 : 2 by volume). Radiolabelled products that had been separated by TLC were analysed by a Fujifilm BAS-1500 phosphorimager (Fuji Photo Film Co. Ltd., Kanagawa, Japan) and carbohydrates were detected by spraying the plates with a solution (10% v/v) of concentrated sulphuric acid in ethanol followed by charring at 120°C for 5 min.

2.8. HPAEC OF PRODUCTS

Reaction products were separated on a Carbopac PA1 column (4 x 250 mm) (Dionex, Camberley, UK) with a Carbopac PA Guard Column (3 x 25 mm). A linear gradient of 50 – 950 mM ammonium formate, pH 6.6 (flow rate 1 ml /min) was used. The gradient was produced by mixing 1 M ammonium formate buffer with deionised water. Sugars were detected using a Pulsed Amperometric Detector (PAD). Fractions were analysed by scintillation counting. 100 µl of each was mixed with 1 ml scintillation cocktail and assayed for radioactivity as before.

2.9. ISOELECTRIC FOCUSING (IEF)

Detergent-solubilised proteins were separated according to their pI by the technique of isoelectric focusing described by Edwards *et al.* (1999).

3. Results

3.1. SOLUBILISATION USING DIFFERENT DETERGENTS

In two separate experiments, microsomal membranes from mung bean hypocotyls were treated with various detergents (bigChap, deoxybigChap, deoxycholate, Nonidet-P40, tritonX-100 and Zwittergent Z3-14) at 1 % and 5 % concentrations and the resultant 100,000 *g* supernatants were checked for soluble HGA-GalAT activity. Activity was detected in all supernatants but the amount varied according to the type and concentration of the detergent. At a detergent concentration of 5 %, the highest activities of HGA-GalAT were obtained using bigChap, deoxybigChap and Zwittergent Z3-14 (Figure 1). At a detergent concentration of 1 %, highest activities were obtained using deoxybigChap and Zwittergent Z3-14 (Figure 1). Although these results are from two separate experiments and cannot be directly compared, it was found that the starting activities in the intact microsomal membrane preparations were very similar. On this basis, activities of HGA-GalAT were generally higher with 5 % detergent.

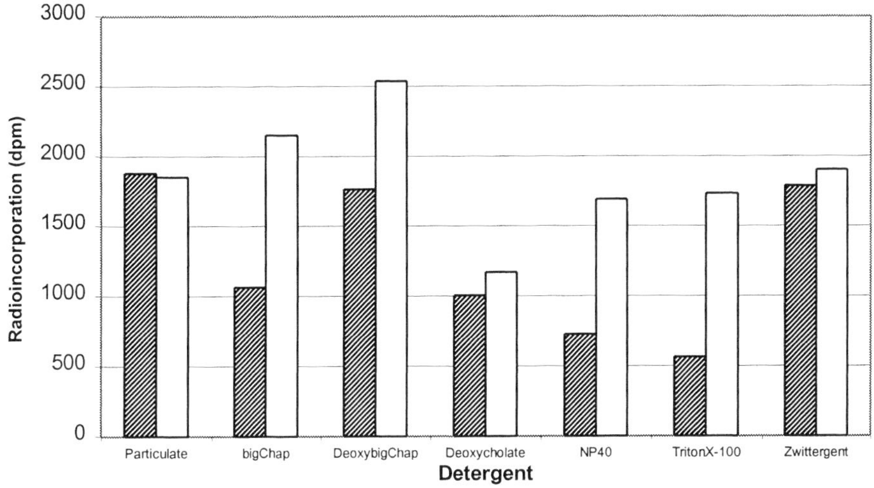

Figure 1. Radioincorporation by mung bean HGA-GalAT following detergent-solubilisation using a range of detergents. The results are from two separate experiments where the final concentration of detergent was 1 % (hatched boxes) or 5 % (blank boxes). Assays were carried out in the presence of PGA (0.1 %) and manganese (5 mM). The amount of radiolabelled substrate added was equivalent to 4,000 dpm. Incorporation of radiolabel into product by intact particulate membranes (approximately 45% of starting radiolabel) is also shown as a comparison.

3.2. EFFECT OF DETERGENT CONCENTRATION ON SOLUBLE ACTIVITY

The effect of detergent concentration on HGA-GalAT activity was examined further for the two most effective detergents deoxybigChap and Zwittergent Z3-14. HGA-GalAT activity was measured in extracts where the concentration of detergent used was 0.5, 1.0, 2.5, 5.0, 7.5 or 10.0 %. HGA-GalAT activity was similar at all concentrations of detergent tested for both deoxybigChap and Zwittergent Z3-14 (Figure 2). In this series of experiments maximal activity was detected with 1 % Zwittergent Z3-14.

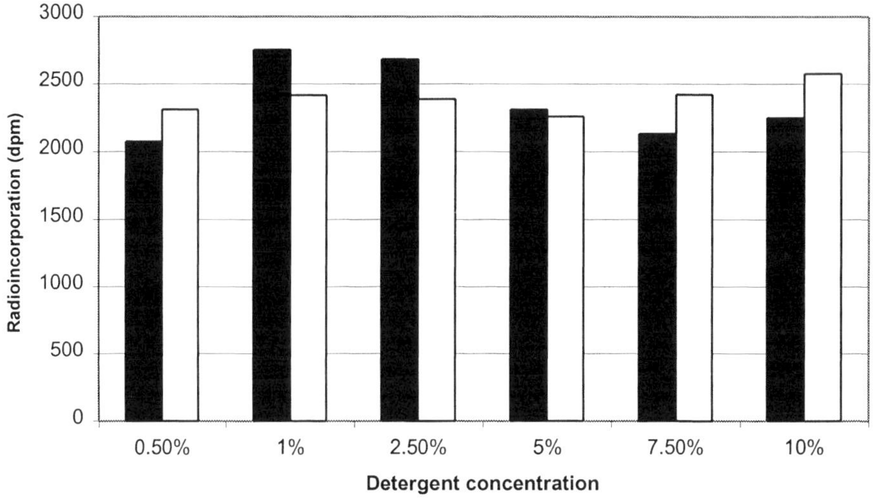

Figure 2. The effect of detergent concentration on HGA-GalAT activity. Radioincorporation by the enzyme was measured following detergent-solubilisation with deoxybigChap (blank boxes) or Zwittergent Z3-14 (filled boxes) using different detergent concentrations. PGA (0.1 %) and manganese 5 mM were included in the assay for HGA-GalAT. The amount of radiolabelled substrate added was equivalent to 4,000 dpm. The results represent the mean values of at least two measurements.

3.3. EFFECT OF EXOGENOUS PECTINS

Using 1% Zwittergent Z3-14 to solubilise HGA-GalAT activity, the effect of a range of exogenous pectin acceptor substrates was investigated. Pectins examined included PGA, those extracted from red and green tomato pericarp, a series of pectins with different degrees of methylesterification prepared by limited digestion of a high-methylated pectin (Sigma) and commercially available pectins with different degrees of methylesterification. A small amount of activity was detected in the absence of added pectin [Table 1, experiment (i)] but this was enhanced by between 2 - 5.5 fold when pectin was added. All pectins examined enhanced HGA-GalAT activity [Table 1, experiment (i)] and the degree of methylation of the pectin had no effect on the ability of the pectin to act as a substrate for HGA-GalAT. In a separate experiment, a series of oligogalacturonides of well-defined degree of polymerisation (DP 6, 10, 13, 14, 15) were also tested for their ability to act as acceptors. Each of the oligogalacturonides tested was able to act as an acceptor for [^{14}C]-GalA addition. Highest levels of radioincorporation were obtained with the oligogalacturonide DP = 15 [Table 1, experiment (ii)].

Table 1. The effect of pectins and oligogalacturonides on soluble HGA-GalAT. The amount of radiolabelled substrate added to each reaction was equivalent to 4,000 dpm.

Experiment (i) Acceptor (0.1 %)	Radioincorporation (dpm)
No addition	500
Polygalacturonic acid	1,854
Green tomato extract	1,083
Red tomato extract	1,731
DE 1 % (alkali-prepared)	2,390
DE 35 % "	2,129
DE 40 % "	1,860
DE 66 % "	2,035
DE 93 % "	1,923
DE 5 % (commercial preparation)	1,910
DE 37 % "	2,487
DE 67 % "	2,778
Experiment (ii) Acceptor (0.1 %)	**Radioincorporation (dpm)**
No addition	1,263
Polygalacturonic acid	2,604
Oligogalacturonide DP 6	2,474
" DP 10	2,539
" DP 13	2,614
" DP 14	2,609
" DP 15	2,729

3.4. EFFECT OF MANGANESE CONCENTRATION

Particulate HGA-GalAT activity is enhanced by the presence of manganese (Smith, 1999). In this study, activity of the soluble form of HGA-GalAT was found to be dependent upon Mn^{2+} concentration (Figure 3). HGA-GalAT activity is very low in the absence of Mn^{2+} and increases as the concentration of Mn^{2+} increases. Optimal HGA-GalAT activity was detected with 5 mM manganese and addition of higher concentrations did not enhance activity further.

3.5. RADIOACTIVE PRODUCT ANALYSIS

The nature of the radioactive product formed by soluble HGA-GalAT activity was examined by digestion of the polymeric product with hydrolytic enzymes specific to pectin. Untreated product remained on the base line on TLC plates when run in a solvent that separates smaller oligosaccharides of galacturonic acid. Treatment of product with *endo*-polygalacturonase released most of the radioactivity into the supernatant and products co-migrated with oligogalacturonides on TLC. Treatment of EPG-digested products with pectin methylesterase, known to be contaminated with a small amount of EPG, resulted in products that ran further in the same solvent system. Thus removal of methylesters from the product allowed further digestion by EPG and resulted in products that migrated further on TLC. EPG and PME are specific enzymes and their ability to hydrolyse the radioactive product from soluble HGA-GalAT activity confirms that the enzyme activity detected is indeed that of a homogalacturonan synthase.

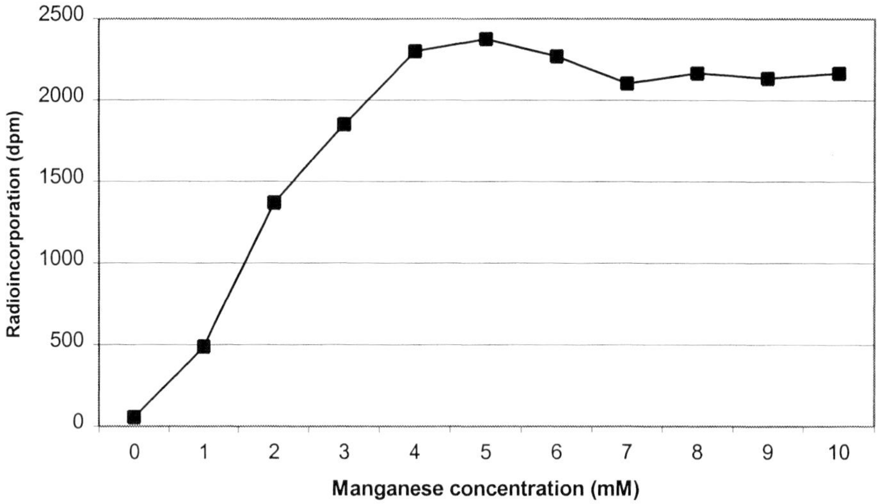

Figure 3. The effect of manganese concentration on radioincorporation by HGA-GalAT. The enzyme was solubilised with Zwittergent (1 %), and was assayed using PGA (0.1 %) as acceptor. The results are the means of three separate measurements.

3.6. APPARENT pI OF HGA-GalAT

The isoelectric point of HGA-GalAT was determined using the technique described by Edwards *et al.* (1999). A typical result from isoelectric focusing is depicted in Figure 4. A pH gradient was formed along the length of the agarose gel. Although HGA-GalAT activity was not sharply resolved, the activity was usually detected in the lower pH region of the gel. In the example given, the activity spanned the region corresponding to a pH of 4.5 – 6.7. Within this broad area of activity it was possible to distinguish at least two peaks, one at around pH 5.0 and one at pH 6.0, and this indicated that there are at least two isoforms of the HGA-GalAT enzyme in mung bean. Isoelectric focusing also fractionated mung bean detergent-solubilised proteins resulting in a substantial partial purification of HGA-GalAT. However, due to the large number of proteins present in each fraction it was not possible to correlate protein bands on gels with high HGA-GalAT activity (results not shown).

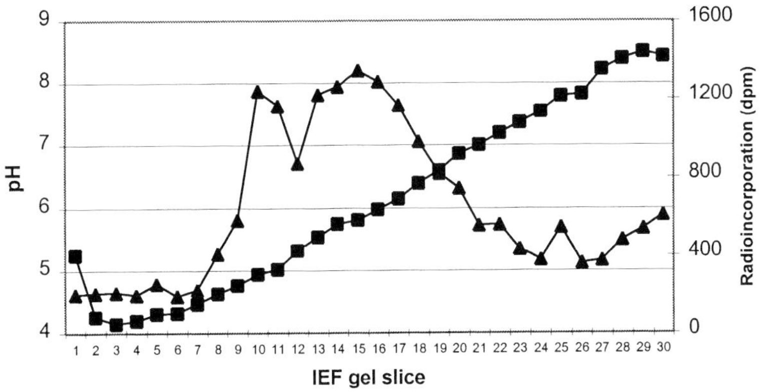

Figure 4. Separation of solubilised HGA-GalAT by isoelectric focusing in a non-denaturing agarose gel. HGA-GalAT activity (triangle) was correlated with pH (square) and two peaks of activity were found to focus at pH 5.0 and pH 6.0. The enzyme was solubilised using Zwittergent Z3-14 (1 %) and assayed in the presence of PGA (0.1 %) and manganese (5 mM).

4. Discussion

An enzyme that catalysed the transfer of galacturonic acid from UDP-galacturonic acid onto an endogenous pectin was previously shown to be present in mung bean microsomal membrane preparations and was referred to as polygalacturonate 4-α-galacturonosyltransferase (PGA-GalAT) (Villemez *et al.*, 1965; Villemez *et al.*, 1966). A similar enzyme was more recently shown to be present in microsomal membranes from tobacco cell-suspension cultures, and was referred to as homogalacturonan galacturonosyltransferase (HGA-GalAT). The mung bean HGA-GalAT activity has also recently been further characterised (Smith, 1999).

As a first step towards purification of mung bean HGA-GalAT, we have carried out the successful detergent-solubilisation of the enzyme. Detergent-solubilisation of membrane-bound proteins can be difficult and is often hindered by the loss of enzyme activity. This loss of activity can be due to several factors including, irreversible denaturation of the protein by detergent or the loss of associated acceptor substrate molecule upon membrane disruption. These problems were not encountered in the present study. Indeed, HGA-GalAT activity was solubilised with retention of activity by all six detergents investigated. These results contrast with the work carried out by Doong and Mohnen (1998) using tobacco cell-suspension cultures where an initial screen using eleven different detergents failed to solubilise HGA-GalAT with retention of activity. These authors did succeed in solubilising HGA-GalAT using 40mM CHAPS but detection of activity required the inclusion of EDTA and manganese cations. EDTA was not found to be a requirement in the present study, but mung bean HGA-GalAT activity was stimulated by divalent cations notably manganese. A minimum molar excess of 0.17 mM manganese over the EDTA concentration was required for detection of tobacco HGA-GalAT (Doong and Mohnen, 1998). In the present study, the optimum manganese concentration for soluble mung bean HGA-GalAT was 5 mM whereas the optimum concentration for the particulate form of the enzyme was 1 mM (Smith, 1999).

Of the detergents examined in this study, deoxybigChap and Zwittergent Z3-14 resulted in highest soluble activities of HGA-GalAT. Optimisation of detergent concentration showed that Zwitterent at around 1 % gave the highest HGA-GalAT activity. Solubilisation of HGA-GalAT using Zwittergent 1 % was therefore used routinely to further investigate the properties of the soluble enzyme.

Soluble HGA-GalAT activity in mung bean was enhanced by exogenously added pectin acceptors although no preference was shown for the degree of methylation of the added pectin. In tobacco, exogenous oligogalacturonides of degree of polymerisation > 10 were needed for HGA-GalAT activity and the enzyme appeared to be non-processive in its action, thus only one galacturonic acid residue was added to the oligogalacturonide DP 15 (Doong and Mohnen, 1998). Additionally, relatively low levels of radioincorporation were detected (21 % of that detected in intact microsomal membranes). In the present study, higher radioincorporation levels equivalent to those observed for particulate enzymes were obtained with all exogenously added pectins suggesting that the enzyme in mung bean may be processive. Further experiments will be needed to confirm the processivity of the mung bean HGA-GalAT.

There are clear differences between the mung bean and tobacco cell HGA-GalAT enzymes. They have different requirements for detergent-solubilisation and different acceptor preferences. This raises the question as to whether the two enzymes under study are indeed carrying out the same function in vivo. Considering the structural complexity of pectin there is a possibility that different galacturonosyltransferase enzymes might exist to carry out different biosynthetic functions. In addition to an enzyme that adds galacturonic acid residues onto the growing HGA chain there may be another transferase that incorporates galacturonic acid into RGI. Alternatively, if HGA is a pectin sidechain as suggested by Voragen (oral communication, *Pectins and Pectinases 2001*) then it might be envisaged that at least two enzymes are necessary for HGA sidechain biosynthesis: a chain initiator enzyme and a chain elongation enzyme. Indeed, the IEF separation described in this study suggests that there are at least two isoforms of the HGA-GalAT enzyme in mung bean hypocotyls.

In summary, a homogalacturonan galacturonosyltransferase from the microsomal fraction of mung bean hypocotyls has been successfully detergent-solubilised with retention of activity. Assay conditions have been optimised and the soluble enzyme has been characterised and partially purified. We are now in a position to attempt the purification of the homogalacturonan galacturonosyltransferase and the subsequent cloning of the corresponding cDNA.

5. Acknowledgements

The authors would like to thank the European Commission for funding this work (Contract No: QRLT-1999-00089 ‘Upgrading of sugar beet pectin by enzymatic modification and molecular farming’).

6. References

Albersheim, P., Darvill, A.G., O'Neill, M.A., Schols, H.A. and Voragen, A.G.J. (1996) An hypothesis: the same six polysaccharides are components of the primary cell walls of all higher plants, in J. Visser and A.G.J. Voragen (eds.), *Pectins and Pectinases*, Elselvier Sciences BV, Amsterdam, pp. 47-53.

Doong, R.L., Liljebjelke, K., Fralish, G., Kumar, A. and Mohnen, D. (1995) Cell free synthesis of pectin: identification and partial characterization of polygalacturonate 4-α-galacturonosyltransferase and its products from membrane preparations of tobacco (*Nicotiana tabacum* L. cv samsun) cell suspension cultures, *Plant Physiol.* **109**, 141-152.

Doong, R.L. and Mohnen, D. (1998) Solubilization and characterization of a galacturonosyltransferase that synthesizes the pectic polysaccharide homogalacturonan, *Plant J.* **13**, 363-374.

Edwards, M.E., Dickson, C.A., Chengappa, S., Sidebottom, C., Gidley, M.J. and Reid, J.S.G. (1999) Molecular characterisation of a membrane-bound galactosyltransferase of plant cell wall matrix polysaccharide biosynthesis, *Plant J.* **19**, 1-7.

Hart, D.A. and Kindel, P.K. (1970) Isolation and partial characterization of apiogalacturonans from the cell wall of *Lemna minor*, *Biochem J.* **116**, 569-579.

Ishii, T. (1995) Pectic polysaccharides from bamboo shoot cell-walls, *Mokuzai Gakkaishi* **41**, 669-676.

Kikuchi, A., Edashige, Y., Ishii , T. and Satoh, S. (1996) A xylogalacturonan whose level is dependent on the size of cell clusters is present in the pectin from cultured carrot cells, *Planta* **200**, 369-372.

Liljebjelke, K., Adolphson, R., Baker, K., Doong, R.L. and Mohnen, D. (1995) Enzymatic synthesis and purification of uridine diphosphate $[^{14}C]$ galacturonic acid: a substrate for pectin biosynthesis, *Anal. Biochem.* **225**, 296-304.

Mohnen, D., Doong, R.L., Liljebjelke, K., Fralish, G. and Chan, J. (1996) Cell free synthesis of the pectic polysaccharide homogalacturonan, in Visser, J. and Voragen, A.G.J. (eds.), *Pectins and Pectinases*, Elselvier Science BV, Amsterdam, pp. 109-126.

O'Neill, M., Albersheim, P. and Darvill, A. (1990) The pectic polysaccharides of primary cell walls, in Dey, P.M. (ed.), *Methods in Plant Biochemistry Vol 2*, Academic Press, London, pp. 415-441.

Schols, H.A., Vierhuis, E., Bakx, E.J. and Voragen, A.G.J. (1995) Different populations of pectic hairy regions occur in apple cell walls, *Carbohydrate Res.* **275**, 343-360.

Smith, K.J. (1999) In vitro biosynthesis of pectic polysaccharides, PhD Thesis, University of Stirling.

Villmez, C.L., Lin, T.-Y., and Hassid, W.Z. (1965) Biosynthesis of the polygalacturonic acid chain by a particulate enzyme preparation from *Phaseolus aureus* seedlings, *Proc.Natl.Acad.Sci.USA* **54**, 1626-1632.

Villmez, C.L., Swanson, A.L., and Hassid, W.Z. (1966) Properties of a polygalacturonic acid-synthesizing enzyme system from *Phaseolus aureus* seedlings, Arch. Biochem. Biophys. **116**, 446-452.

Voragen, A.G.J. (2001) Pectin – The hairy thing: A summary of the symposium, *Pectin and Pectinases 2001* An International Symposium, Rotterdam, The Netherlands.

PECTIN – THE HAIRY THING

Evidence that homogalacturonan is a side chain of rhamnogalacturonan I

JEAN-PAUL VINCKEN[1,2], HENK A. SCHOLS[2], RONALD J.F.J. OOMEN[1], GERRIT BELDMAN[2], RICHARD G.F. VISSER[1], ALPHONS G.J. VORAGEN[2*]

[1] *Laboratory of Plant Breeding, Wageningen University, Binnenhaven 5, 6709 PD Wageningen, The Netherlands*

[2] *Laboratory of Food Chemistry, Wageningen University, Bomenweg 2, 6703 HD Wageningen, The Netherlands*

Abstract

Pectin is one of the most complex biomacromolecules in nature. In fact, it is a collection of different kinds of polysaccharides, which are most likely covalently connected to each other. *In vitro* degradation studies of pectic polysaccharides with novel fungal pectinases, investigations in which these polymers were treated with dilute acid, and microscopic analysis of extracted pectins have provided clues on how these polysaccharides are linked. These new findings have prompted us to believe that pectin is not an extended backbone consisting of homogalacturonan and rhamnogalacturonan regions, but rather a rhamnogalacturonan with neutral sugar *and* homogalacturonan side chains.

1. Constituent Polymers, Primary Structure

Pectin is an important component of dicotyledonous plant cell walls, besides cellulose and hemicellulose (Carpita and Gibeaut, 1993). Pectin is probably the most complex macromolecule in nature, because it can be composed of as many as 17 different monosaccharides (Fig. 1A) (reviewed in Mohnen (1999) and Ridley *et al.* (2001)). Rather than making all possible combinations with these monosaccharides, mother nature has provided us with a number of distinct polysaccharides, which together form pectin. The structural characteristics of these are summarized in Fig. 1B. **Homogalacturonan** (HG) is an essentially unbranched molecule composed of 1,4-linked α-D-Gal*p*A residues. The GalA residues can be methyl-esterified at *C*-6, and carry acetyl groups on *O*-2 and *O*-3. The methyl-esterification, in particular, has gained a lot of attention over the years in pectin chemistry, because it determines the physical properties of pectin to a large extent. Not only the amount of methyl-esterification is important, but also their distribution. Blocks of more than 10 unesterified GalA residues

* Corresponding author. email: fons.voragen@chem.fdsci.wag-ur.nl; Tel: +31-317-482888; Fax: +31-317-484893.

F. Voragen et al. (eds.), Advances in Pectin and Pectinase Research, 47-59.
© 2003 *Kluwer Academic Publishers. Printed in the Netherlands.*

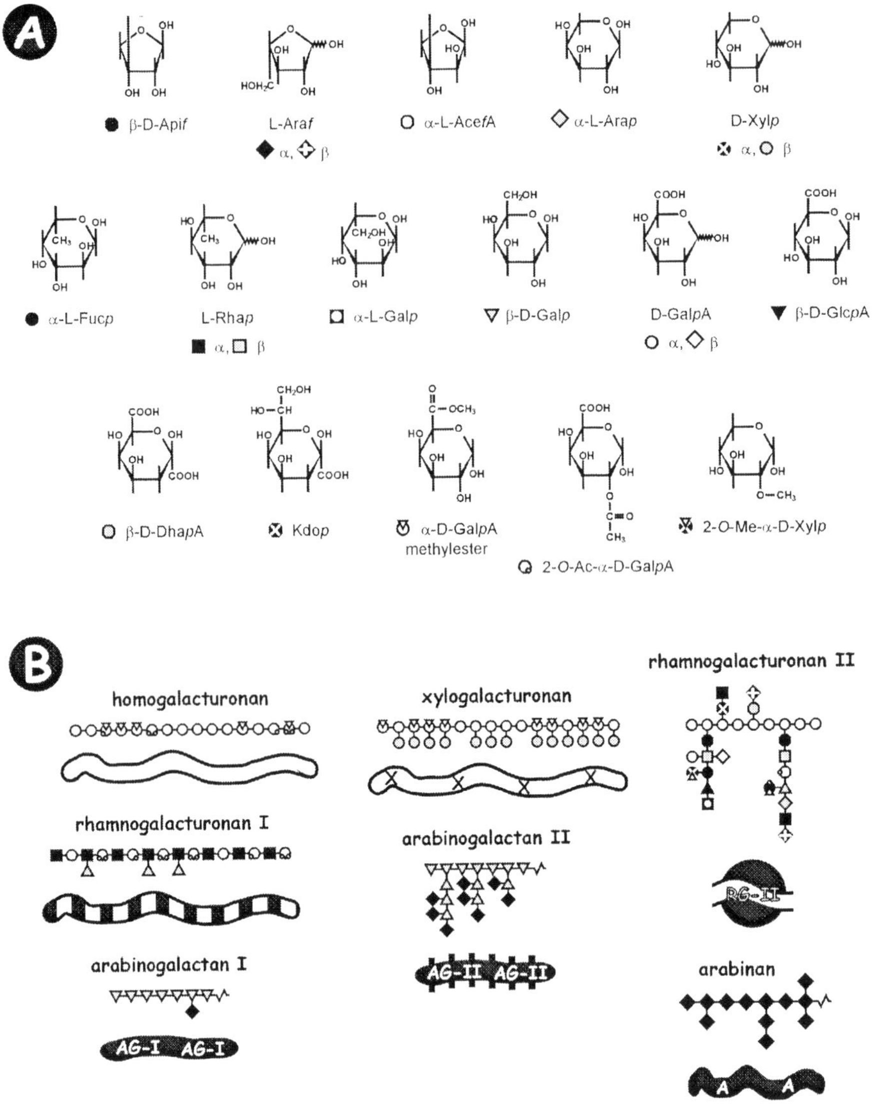

Figure 1. Summary of the structure of pectic polysaccharides. A. Monosaccharide units which are encountered in pectic molecules, including examples of esterified and etherified residues. Ferulic esters are not indicated. Symbols underneath the Haworth projections represent this monosaccharide elsewhere in this paper. B. Schematic structures of the constituent polysaccharides of pectin. Linkage types present in the various polysaccharides are detailed in the text. For symbols, see Fig. 1A. Underneath the schematic structure of a pectic constituent, the symbol for its polymeric structure is indicated. These symbols will be used in the rest of the paper.

generally yield pectin molecules, which are sensitive to Ca^{2+}-crosslinking (Daas *et al.*, 2001).

In the literature, there are a few reports describing a HG with β-D-Xylp-(1→3) side chains (Schols *et al.*, 1995a; Albersheim *et al.*, 1996; Le Goff *et al.*, 2001). This polysaccharide is referred to as **xylogalacturonan** (XGA). Sofar, its presence in plants seems to be confined to reproductive organs (fruits, seeds) (Albersheim *et al.*, 1996). Also, exudates from trees (for instance, gum tragacanth from the *Astralagus* species) can contain a XGA-like polysaccharide (Anderson and Grant, 1988). The degree of xylosylation can vary between 25% (watermelon) and 75% (apple) (Schols *et al.*, 1995a; Yu and Mort, 1996; Le Goff *et al.*, 2001). Besides the single-unit xylosyl side chains, HGs can contain clusters of 4 different side chains with very peculiar sugar residues (such as Api, AceA, Dha, and Kdo). These side chains, together with the HG fragment of ~9 galacturonosyl residues to which they are attached, are referred to as **rhamnogalacturonan II** (RG-II) (Vidal *et al.*, 2000; O'Neill *et al.*, 2001; Ridley *et al.*, 2001). The exact structure of RG-II has not yet been elucidated, and its structure is updated continuously (Vidal *et al.*, 2000; Ridley *et al.*, 2001). In particular, the distribution of the four side chains in RG-II, as well as the distribution of RG-II in the HG backbone, is unknown. The structure of RG-II seems to be highly conserved in the plant kingdom. Two molecules of RG-II can complex with boron, forming a borate-diol ester, which can cross-link two HG molecules (Ishii *et al.*, 1999; Ishii and Matsunaga, 2001). Only the apiofuranosyl residues of the 2-*O*-methyl-D-xylose-containing side chains in each of the subunits of the dimer participate in the cross-linking (Ishii *et al.*, 1999). The name RG-II is somewhat misleading, because it suggests that this structure contains a rhamnogalacturonan backbone just like RG-I (see further). However, the Rha residues are much less abundant in RG-II than in RG-I, and they are present in the side chains of RG-II instead of in the backbone. Thus, both XGA and RG-II are in principle homogalacturonan-based polymers with a special decoration pattern.

Rhamnogalacturonan I (RG-I) can have a backbone composed of as many as 100 repeats of the disaccharide [→2)-α-L-Rha*p*-(1→4)-α-D-Gal*p*A-(1→] (McNeil *et al.*, 1980; Albersheim *et al.*, 1996). The rhamnosyl residues can be substituted at *O*-4 with neutral sugars (McNeil *et al.*, 1980; Lau *et al.*, 1987). The proportion of branched Rha residues varies from ~20% to ~80% depending on the source of the polysaccharide (Albersheim *et al.*, 1996). The side chains are mainly composed of galactosyl and/or arabinosyl residues. They can be single-unit [β-D-Gal*p*-(1→4)], but also polymeric such as **arabinogalactan I** (AG-I) and **arabinan** (50 glycosyl residues or more). AG-I is composed of a 1,4-linked β-D-Gal*p* backbone; α-L-Ara*f* residues can be attached to the *O*-3 of the galactosyl residues (Carpita and Gibeaut, 1993; Mohnen, 1999; Ridley *et al.*, 2001). The arabinans consist of a 1,5-linked α-L-Ara*f* backbone, which can be substituted with α-L-Ara*f*-(1→2)-, α-L-Ara*f*-(1→3)-, and/or α-L-Ara*f*-(1→3)-α-L-Ara*f*-(1→3)- side chains (Schols *et al.*, 1990; Carpita and Gibeaut, 1993; Beldman *et al.*, 1997; Mohnen, 1999; Ridley *et al.*, 2001). Complexes of RG-I, AG-I and arabinan are often referred to as pectic hairy regions (HR), in which AG-I and arabinan are the "hairs". The abundance of hairy regions can differ from species to species. Also, the amount and nature of the hairs can differ considerably among species (Ryden and Selvendran, 1990; Renard *et al.*, 1991; Oosterveld *et al.*, 2001). Another type of arabinogalactan, **arabinogalactan II** (AG-II), is mainly associated with proteins (arabinogalactan proteins or AGPs). To this end, it is unclear whether AG-II is part of the pectin complex. The two polysaccharides often seem to co-extract, and are subsequently difficult to separate from each other. These observations fuel speculations that the two are covalently linked. AG-II is composed of 1,3-linked β-D-Gal*p* backbone, containing short side chains of α-L-Ara*f*-(1→6)-[β-D-Gal*p*-(1→6)]$_n$ (n = 1, 2, or 3) (Carpita and Gibeaut, 1993; Mohnen, 1999; Ridley *et al.*, 2001). The galactosyl residues of the side chains can be substituted with α-L-Ara*f*-(1→3) residues. The major part of AGPs (>90%) consists of polysaccharides. The protein part is rich in proline/hydroxyproline, alanine, serine and threonine (Gaspar *et al.* (2001), and references therein).

From the above, it may be concluded that the fine structure of the constituent polysaccharides of pectin is known in much detail. However, with the in-depth analysis of pectic polysaccharides from different plant sources, new structural details are added to the collection of elements already known. Two examples are discussed. Pectin derived from soybean appears to be extremely resistant against pectolytic enzymes (Huisman *et al.*, 2001a). From this and other results in our laboratory it was concluded that HG is not a structural unit of soybean pectin. Instead, soybean seems to contain XGA, possibly having side chains which are more complex than a single xylosyl residue. In addition to this, also the AG-I side chains show two new structural features: (i) 1,4-linked β-D-Gal*p* side chains terminated with a α-L-Ara*p*-(1→4) at the non-reducing end, and (ii) side chains containing [→4)-β-D-Gal*p*-(1→5)-α-L-Ara*f*-(1→4)-β-D-Gal*p*-(1→] (Huisman *et al.*, 2001b). Structural characterization of oligosaccharides released from sugar beet by treatment with dilute acid showed the presence of two fragments with a tetrasaccharide backbone [α-D-Gal*p*A-(1→2)-α-L-Rha*p*-(1→4)-α-D-Gal*p*A-(1→2)-α-L-Rha*p*], in which single-unit β-D-Glc*p*A-(1→3) side chains were attached to one of the galacturonosyl residues (Renard *et al.*, 1999).

2. Evidence That The Constituent Pectic Polysaccharides Are Covalently Linked

Although there is a rather detailed picture on the primary structure of pectic constituent polysaccharides, surprisingly little is known on how the constituent polysaccharides assemble into a macromolecular network. Selective cleavage of the galacturonosyl residues of the RG-I backbone by treatment with lithium in ethylenediamine resulted in a collection of fragments in which neutral side chains were attached to rhamnitol at the "reducing" end (Lau *et al.*, 1987). Also, oligosaccharides with small neutral sugar side chains attached to rhamnosyl residues were found in digests of pectic material obtained with either dilute acid or particular pectic enzymes (reviewed in Schols and Voragen (1996), and Ridley *et al.* (2001)). These observations suggest that the neutral sugar-containing hairs are covalently linked to RG-I. There is also evidence that HG and RG-I are (covalently) attached to each other. A number of important observations to support this are listed below. Based on these observations a putative structure for pectin was proposed which is illustrated by Fig. 2 (Schols and Voragen, 1996). This model will be discussed in more detail at the end of this section.

- Many extractants have been used to solubilize pectins from plant material. For instance, chelating agents (such as EDTA) are thought to capture Ca^{2+} ions, as a result of which HG-HG interactions (*in situ* the Ca^{2+}-pectate gel) can be broken, and HG is solubilized. However, besides to HG, also RG-I, arabinan, and galactan are solubilized, which implies that these polysaccharides are interconnected (de Vries *et al.*, 1982). Anion-exchange chromatography of pectic material extracted with EDTA solutions does not resolve the neutral polysaccharides (arabinan, AG-I), RG-I and HG in separate fractions, despite differences in their charge density.
- When EDTA-extracted pectins are treated with endo-polygalacturonase (endoPG) or endo-pectin lyase (endoPL), the forth-coming fragments containing the hairs have a

lower molecular weight than the original material, providing further evidence that the polysaccharides are covalently connected (Schols *et al.*, 1995b).

❑ Imidazole-extracted borate-containing pectin from sugar beet was treated with acid to hydrolyze the borate ester (Ishii and Matsunaga, 2001), resulting in a decrease in molecular weight of the pectin. When this mixture was incubated with boric acid and di-valent cations (conditions promoting borate cross-linking), the molecular weight distribution of the original imidazole-extracted pectin could be completely restored. Treatment of the extracted pectin with a mixture of endoPG and exoPG released oligogalacturonides, the RG-II dimer, and RG-I from the complex. These results demonstrate that the 3 polysaccharides are connected, and that borate esters between RG-II can cross-link macromolecular pectin.

A detailed structural characterization of the so-called "modified hairy regions" (MHR) has been extremely valuable in defining the various pectic subunits, and in providing clues on how the various subunits are connected to each other. MHR can be obtained by treating plant tissues with a rather crude mixture of enzymes, including pectinases (*e.g.* endoPG, galactanases), cellulases and hemicellulases. During this treatment, the native pectic polysaccharides can be partially degraded, and therefore the forth-coming material is referred to as *modified* HR. The MHR was subsequently fragmented at specific sites with novel enzymes. Based on these studies a tentative structure was put forward, which was consistent with most of observations made at that time (Schols and Voragen, 1996). Pectin was represented as one molecule containing smooth and hairy regions (see Fig. 2). From incubations of MHR with endo-rhamnogalacturonan hydrolase (endoRGH), it was concluded that these regions were assembled from 3 subunits: (i) stretches of RG-I, containing single-unit galactosyl side chains, (ii) xylogalacturonan, and (iii) a rhamnogalacturonan (with a GalA:Rha ratio of ~4) containing long arabinan and AG-I side chains. Release of the two latter subunits

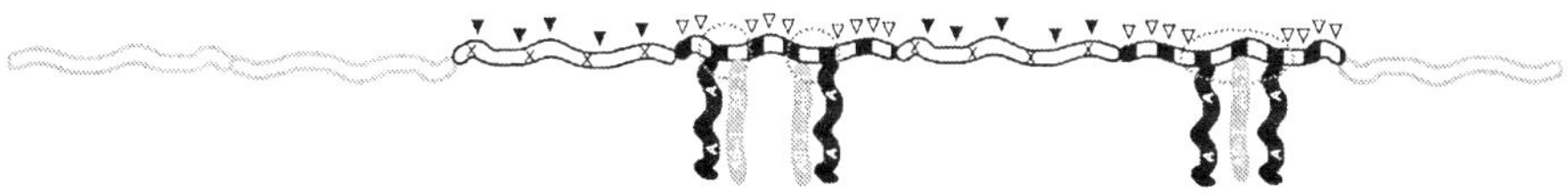

Figure 2. Schematic representation of the macromolecular structure of pectin. The "smooth" regions are mainly composed of HG. The "hairy" regions consist of RG-I, XGA, arabinan, AG-I, and (probably) AG-II. The abundance of the various polysaccharide units and their length is arbitrary. Modified hairy regions (MHR) can be obtained by incubation of a cell wall preparation with a crude mixture of enzymes. The parts of the macromolecule, which are indicated in gray, are removed during this treatment. The (modified) hairy regions can be fragmented with endoRGH (indicated with open triangles) after removal of the acetyl groups. Analysis of the forth-coming products has provided insight in the subunit structure of hairy regions. Based on these studies and the literature, the various subunits were combined in the "putative macromolecular structure" shown in the figure (adapted from Schols and Voragen (1996)). A novel pectinase, XGH (indicated with closed triangles), can degrade XGA. This has offered the opportunity to fragment MHR in an alternative way (for details, see text). Circles with dashed lines are discussed in section 3 of the text. For symbols, see Fig. 1.

upon endoRGH treatment can only be explained when they are flanked by stretches of RG-I (see open triangles in Fig. 2). Further, it was believed that xylogalacturonan could be a continuation of HG (or a smooth region), and that the "pectic backbone" consisted of regions with varying GalA:Rha ratios (*i.e.* HG or XGA, branched rhamnogalacturonan, and RG-I having GalA:Rha ratios of ∞, ~4, and 1, respectively).

3. Recent Observations ...

Since 1996, a number of important observations have been made by us and by others. Some of these challenge the model shown in Fig. 2, others provide clues on the biological significance of the various different polysaccharides of pectin. The various observations will be elaborated below. Because the recent results are mainly related to HG and RG-I, the model in Fig. 2 has been simplified in Fig. 3A, showing only HG and RG-I, without the neutral side chains and XGA. Note that also rhamnosyl residues interconnecting HGs are indicated, which is in agreement with the observed length-periodicity of HGs after treatment with dilute acid. In the next section, an alternative model for pectin will be postulated, which is more consistent with these new findings.

❑ Experiments using exhaustive treatment of citrus pectin with endoPG, followed by careful fractionation of the forth-coming fragments, were unable to demonstrate the presence of Rha-containing oligouronides (Zhan *et al.*, 1998). It was concluded that single rhamnosyl residues (as indicated in Fig. 3A) must be extremely scarce in HG, and that pectin must be a block co-polymer composed of RG-I regions, separated by HG regions containing 70-100 GalA residues.

❑ When considering the varying GalA:Rha ratios, the model depicted in Fig. 2 implies that there is a rather flexible mechanism for inserting rhamnosyl residues in the growing "pectin backbone". The frequency of Rha insertion may be regulated by the concentration of nucleotide-Rha. Over the past 5 years an increasing number of polysaccharide synthesizing complexes have been described. One example is β-(1→3),(1→4)-glucan (or mixed-linkage β-glucan), in which the regularity of the β-(1→4)-linked backbone is interrupted by the insertion of (1→3) linkages. In these glucans, each (1→3)-insertion alternates with 2 or 3 (1→4)-linkages (Buckeridge *et al.*, 1999), indicating that the interruptions follow a regular pattern. The length of RG-I interuptions in a HG backbone seem to be of a different order, whereas the HG-interconnecting rhamnosyl residues seem to be difficult to insert deliberately. Therefore, it seems likely that different synthases are involved in polymerizing the HG and RG-I backbones, and that there is not one enzyme complex synthesizing the "pectin backbone".

❑ It was evident that novel enzymes to fragment pectic hairy regions in a different way as with endoRGH were required to elucidate the structure of pectin in more detail, and to validate the model of Fig. 2. Therefore, a screening for XGA-degrading enzymes was initiated resulting in two *Aspergillus* enzymes: exoPG (Beldman *et al.*, 1996) and endo-xylogalacturonan hydrolase (endoXGH; van der Vlugt-Bergmans *et al.*, 2000). The exoPG was able to release GalA and the GalA-Xyl disaccharide from the non-reducing terminus of xylosyl-substituted galacturonans. EndoXGH is homologous to "classical endoPGs". It differs in its mode of action from these, in that its activity increases with the degree of xylosylation of (unesterified) homogalacturonan, whereas the activity of "classical endoPGs" is hindered by xylosylation (unpublished results). Apple MHR was treated with exoPG and endoXGH. ExoPG released large amounts of the GalA-Xyl disaccharide from MHR, suggesting that a large part of the xylogalacturonan is accessible for enzyme action. EndoXGH released also oligosaccharides larger than this disaccharide from MHR, but typically a similar (modest) shift in the MHR molecular weight distribution was observed as with exoPG (unpublished results). These observations strongly argue against an "internal" position of xylogalacturonan (see Fig. 2).

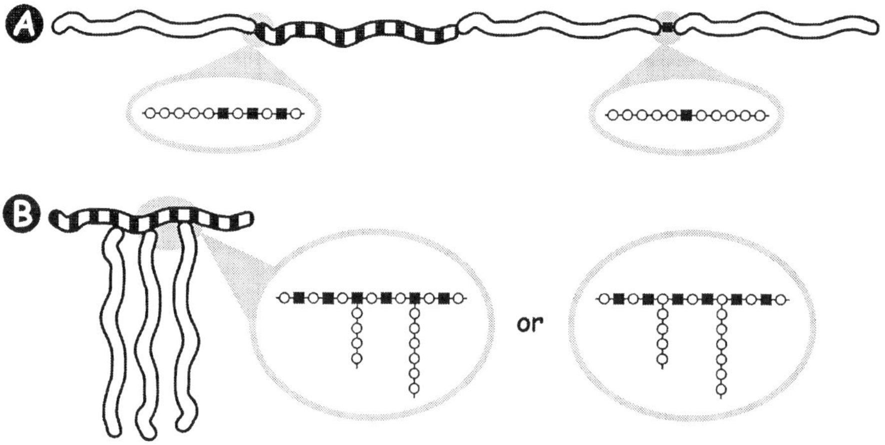

Figure 3. Schematic illustration of two putative pectin complexes, both leading to the observed length-periodicity of HG. Note that the molecules are not drawn to scale, the length of HG and RG-I is arbitrary. In the upper model (A), the galacturonans form an extended chain with HG and RG-I regions. HG regions can be interspersed with rhamnosyl residues. In the lower model (B), RG-I is decorated with HG side chains. It is unknown where these side chains are attached. They can be linked to rhamnosyl or galacturonosyl residues of RG-I; other ways of attachment can not be excluded. The distance between branch points is arbitrary. The symbols and polymeric structures are explained in Fig. 1.

❑ Besides RG-I oligosaccharides and xylogalacturonan, also the third class of products released from MHR by endoRGH, the rhamnogalacturonan with arabinan and AG-I side chains (GalA:Rha ratio of ~4; indicated with dashed circles in Fig. 2), has been subjected to a more detailed structural analysis. EndoPG can release oligogalacturonides from these rhamnogalacturonan fragments after they were treated with dilute acid to remove arabinan side chains (unpublished results). These results suggest that this subunit of MHR contains HG remainders, which could not be removed by endoPG due to the presence of (arabinan) side chains. It seems likely that these homogalacturonides are no integral part of the RG-I backbone, because we have not found oligosaccharides other than homogalacturonides. Thus, no evidence for the release of rhamnosyl-containing oligosaccharides was obtained, which rules out the possibility of [Rha-$(GalA)_n$] structural units with $n \approx 4$. Interestingly, also the RG-I released from sycamore cell walls by treatment with endoPG contains approximately twice as much GalA as Rha (McNeil *et al.*, 1980). It was suggested that part of the GalA residues were present in the side chains of RG-I.

❑ Round *et al.* (2001) investigated two pectin fractions from tomato by atomic force microscopy (AFM): a CDTA-extracted pectin containing ~15% of hairs (arabinose and galactose), and a Na_2CO_3-extracted pectin containing ~40% of hairs. For both pectins, it was found that the branches accounted for approximately 10% of the total measured polymer length, and that the number of branch points was approximately the same. It was expected that the Na_2CO_3-extracted pectin had many more or longer branches than the CDTA-extracted one. The authors hypothesized that the pectins have relatively short branches composed of neutral sugars, which are coiled around the RG-I regions of pectin, and that these are consequently very difficult to detect by AFM. The longer branches were thought to represent HG, connected to the "pectic backbone" by an undetermined linkage. Experiments in which AFM is combined with selective enzymic degradation of pectin are in progress.

❑ The distribution of pectic epitopes in the cell wall is rather well-documented (Bush and McCann, 1999; Bush *et al.*, 2001; reviewed in Willats *et al.* (2001)). Nevertheless, our picture of whether or how individual pectic components contribute to particular properties of the cell wall is very incomplete. The main reason for this is that our knowledge on pectin biosynthesis is still in its infancy (Mohnen, 1999; Vincken *et al.*, 2000). In order to obtain clues on the biological significance of the individual pectic polysaccharides fungal pectinases, endogalactanase (Oxenbøll-Sørensen *et al.*, 2000), endoarabinanase (Skjøt *et al.*, 2002), and endo-rhamnogalacturonan lyase (Oomen *et al.*, 2002), were expressed *in planta*. Typically, potato tuber cell walls seem to tolerate a low level of either galactan or arabinan hairs under the growth conditions applied, because both the tubers and the appearance of the cells look normal (Oxenbøll-Sørensen *et al.*, 2000; Skjøt *et al.*, 2002). In contrast to this, fragmentation of the RG-I backbone leads to wrinkled tubers with abnormal periderm development, and large intercellular spaces in the cortex (Oomen *et al.*, 2002). This suggests that RG-I plays a more pivotal role than the neutral hairs in plant development.

4. Pectin Macromolecular Structure Revisited ...

The observations elaborated in the previous section argue against a pectin structure, in which HG is an extension of RG or vice versa. This has prompted us to revisit pectin macromolecular structure, and formulate an alternative model which is consistent with the new insights (see Fig. 3B). Here, HG is represented as a side chain of RG-I, like arabinan and galactan.

At this point, it is important to mention a few words on the length periodicity observed for HGs. Many authors have suggested that (long) stretches of GalA can be interspersed with isolated rhamnosyl residues, the so-called Rha-kinks (see Fig. 3A). It has been estimated (using dilute acid hydrolysis) that these (so-called) Rha-kinks occur at more or less regular intervals of approximately 70-100 GalA residues (Thibault *et al.*, 1993). These results can be explained by the fact that the α-L-Rha*p*-(1→4)-α-D-Gal*p*A is more acid-labile than the aldobiuronic acid α-D-Gal*p*A-(1→2)-α-L-Rha*p*, and the α-D-Gal*p*A-(1→4)-α-D-Gal*p*A linkages. This is in line with dilute acid hydrolysis of sugar beet pulp, which generated (among others) [(1→4)-α-D-Gal*p*A-(1→2)-α-L-Rha*p*]$_n$-type of oligosaccharides (Renard *et al.*, 1998). The structures in Fig. 3A and 3B are consistent with the length periodicity observed for HG. Upon treatment with dilute acid, the RG-I backbone will be split between Rha and GalA, and the HG chains will be released. The degradation studies with dilute acid do not allow us to give preference to one of the hypothetical structures.

Taking the hypothetical structure of Fig. 3B as a starting point, treatment of pectin with endoPG is expected to remove most of the HG, without releasing Rha-containing fragments. This is in agreement with the results of Zhan *et al.* (1998). Also the existence of pectic fragments with a GalA:Rha ratio of ~4 are consistent with a subunit structure in which HG and arabinan side chains are attached to a RG-I backbone. The release of homogalacturonides by endoPG following dilute acid hydrolysis of this fraction are more easily explained. Treatment of MHR with exoPG or with endoXGH resulted in a similar, modest shift in the molecular weight distribution of MHR, whereas the XGA subunit can be released from MHR by the action of endoRGH. The most plausible explanation for these observations is that XGA is a side chain of RG-I. With this in mind, it does not seem to be unreasonable to assume that also HG is a side chain of RG-I, because HG and XGA have the same backbone structure. The alternative structure in Fig. 3B is in agreement with the structural characterization of RG-I by McNeil *et al.* (1980). It is possible that the excess of GalA residues, which were suggested to be present in the side chains of RG-I, actually represent remainders of HG after endoPG treatment. Interestingly, the unexpected, putative HG side chains of pectin observed with AFM can be explained better with the new model. We interprete these as HG hairs connected to a RG-I backbone.

Several lines of evidence supporting the idea of HG side chains attached RG-I were summarized. Although this concept is still speculative, it seems to accommodate a number of observations which are more difficult to explain with the old model. Therefore, we favour a macromolecular structure of pectin, in which RG-I carries neutral AG-I/arabinan (AG-II) hairs *and* HG/XGA hairs. One important question remains: why has a crosslink between HG and RG-I not been demonstrated until now? The scarcity of the HG/RG-I crosslink, in combination with the difficulty to find selective methods to enrich a particular fraction in the crosslink, are beyond doubt important reasons for this. It is expected that endoRGH will not be able to cleave an RG-I backbone that is branched with galacturonan side chains (Fig. 3B). This means that endoRGH is probably not a suitable tool to generate diagnostic fragments, which are appropriate for a detailed spectroscopic analysis. Treatment with dilute acid may destroy the crosslink. The HG side chains can either be attached to the rhamnosyl residues of RG-I, or to the GalA residues of RG-I. Some support for the former may be found in the identification of an α-D-Gal*p*A-(1→4)-α-L-Rha*p* disaccharide in acid hydrolysates of *Artemisia princeps* (Yamada *et al.*, 1987), and in RG-I preparations released from sycamore cell wall material by endoPG treatment (McNeil *et al.*, 1980). The latter may be in accordance with the observation that GalA residies of RG-I can be branched with GlcA, albeit at low frequencies (Renard *et al.*, 1999). The presence of branched GalA residues has also been suggested a long time ago (Talmadge *et al.*, 1973), but this observation was dismissed as a possible artefact due to undermethylation. It is obvious that the exact crosslink between HG and RG-I remains to be established. We are currently trying to obtain RG-I oligosaccharides with short homogalacturonosyl side chains.

5. Implications Of The Alternative Macromolecular Pectin Structure

The previous sections have shown that evidence for an alternative pectin structure is accumulating. It seems that pectin is much more hairy than thought before, *i.e.* besides neutral side chains such as AG-I, AG-II and arabinan, also charged galacturonans may be attached to a RG-I backbone. The literature suggests that RG-I molecules can differ in their "hairstyle". Arabinan and AG-I epitopes appear to be developmentally regulated in a tissue-specific manner (Bush and McCann, 1999; Bush *et al.*, 2001; Willats *et al.*, 2001). For instance, arabinans seem to be the predominant neutral hair species in the younger tissues, whereas AG-I is generally deposited later. In Fig. 4 two different pectin molecules are shown. In pectin A different kinds of hairs are distributed in an *at random* fashion in the RG-I backbone. Pectin B only contains two kinds of

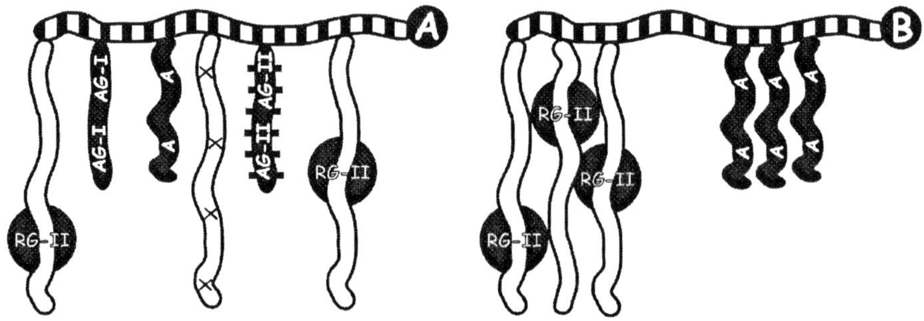

Figure 4. Two new tentative structures for pectin. Note that HG is a side chain of RG-I in both cases. It is possible that the side chains are randomly distributed in the RG-I backbone (A). However, it is also possible that various populations of RG-I molecules exist, each having their own set of side chains, which may be organized in clusters (B). The existence of RG-I molecules with only one type side chain can not be excluded. For symbols, see Fig. 1.

hairs, which are attached to the RG-I backbone in a cluster-like fashion. It goes without saying that other configurations can not be excluded. The distribution of hairs in the RG-I backbone will probably be one of the major challenges of pectin research in the near future, besides establishing how the various hairs are attached to RG-I. The HG, arabinan, and galactan hairs are very different in nature (charge, length, conformation). In principle, an unprecedented number of pectin molecules can be made by combining the different hairs in various fashions. In this way, different micro-environments may be created within the cell wall, which can adapt the wall to its specific functions at a given time point or location. A predominant role for RG-I in cell wall integrity is anticipated, which is in line with our observations (Oomen *et al.*, 2002). In a separate paper (Vincken *et al.*, 2002), the implications of the alternative pectin structure for cell wall assembly will be detailed.

Pectins are used as gelling agents in the food industry. Usually they are obtained from plant material (for instance, apple pomace, or citrus pulp) after treatment with dilute acid. During this treatment a relatively pure HG is obtained; neutral side chains, arabinans in particular, are removed. Industrial pectins always contain a small percentage of rhamnose and other neutral sugars. At the moment, much research is directed towards modulating the physical properties of pectins in a specific manner, mainly by using different pectin methylesterases. According to our new model, pectins are comb-like molecules locking into each other. In principle, this interaction is different from the other situation where pectin gels are simply a stack of homogalacturonans. The comb-like structures can acquire an increasingly intimate interaction when the degree of methylation decreases, in particular when this is done in a block-wise fashion. Our new tentative structure for pectin may influence our ideas on how industrial pectin molecules engage in gel formation.

Acknowledgements

The EC (CT97 2224) is gratefully acknowledged for financial support of part of the work on pectin biosynthesis. The authors would like to thank dr. Miranda Huisman and dr. Lex Oosterveld for stimulating discussions, and their careful reading of the manuscript.

References

Albersheim, P., Darvill, A.G., O'Neill, M.A., Schols, H.A., and Voragen, A.G.J. (1996) An hypothesis: the same six polysaccharides are components of the primary cell walls of all higher plants. *Progress in Biotechnology 14: Pectins and Pectinases* (Visser, J., Voragen, A.G.J., eds.), pp. 47-53, Elsevier

Anderson, D.M.W. and Grant, D.A.D. (1988) The chemical characterization of some *Astralagus* gum exudates. *Food Hydrocoll.* 2, 417-423

Beldman, G., van den Broek, L.A.M., Schols, H.A., Searle-van Leeuwen, M.J.F., van Laere, K.M.J., and Voragen, A.G.J. (1996) An exogalacturonase from *Aspergillus aculeatus* able to degrade xylogalacturonan. *Biotechnol. Lett.* 18, 707-712

Beldman, G., Schols, H.A., Pitson, S.M., Searle-van Leeuwen, M.J.F., and Voragen, A.G.J. (1997) Arabinans and arabinan degrading enzymes. *Advances in Macromolecular Research 1* (Sturgeon, R.J., ed.), pp. 1-64, JAI Press Inc., London

Buckeridge, M.S., Vergara, C.E., and Carpita, N.C. (1999) The mechanism of synthesis of a mixed-linkage (1→3),(1→4)β-D-glucan in maize. Evidence for multiple sites of glucosyl transfer in the synthase complex. *Plant Physiol.* 120, 1105-1116

Bush, M.S. and McCann, M.C. (1999) Pectic epitopes are differentially distributed in the cell walls of potato (*Solanum tuberosum*) tubers. *Physiol. Plant.* 107, 201-213

Bush, M.S., Marry, M., Huxham, I.M., Jarvis, M.C., and McCann, M.C. (2001) Developmental regulation of pectic epitopes during potato tuberisation. *Planta* 213, 869-880

Carpita, N.C. and Gibeaut, D.M. (1993) Structural models of primary cell walls in flowering plants: consistency of molecular structure with the physical properties of the walls during growth. *Plant J.* 3, 1-30

Daas, P.J.H., Boxma, B., Hopman, A.M.C.P., Voragen, A.G.J., and Schols, H.A. (2001) Nonesterified galacturonic acid sequence homology of pectins. *Biopolymers* 58, 1-8

Gaspar, Y., Johnson, K.L., McKenna, J.A., Bacic, A., and Schultz, C.J. (2001) The complex structures of arabinogalactan-proteins and the journey towards understanding function. *Plant Mol. Biol.* 47, 161-176

Huisman, M.M.H., Fransen, C.T.M., Kamerling, J.P., Vliegenthart, J.F.G., Schols, H.A., and Voragen, A.G.J. (2001a) The CDTA-soluble pectic substances from soybean meal are composed of rhamnogalacturonan and xylogalacturonan but not homogalacturonan. *Biopolymers* 58, 279-294

Huisman, M.M.H., Brüll, L.P., Thomas-Oates, J.E., Haverkamp, J., Schols, H.A., and Voragen, A.G.J. (2001b) The occurrence of internal (1→5)-linked arabinofuranose and arabinopyranose residues in arabinogalactan side chains from soybean pectic substances. *Carbohydr. Res.* 330, 103-114

Ishii, T., Matsunaga, T., Pellerin, P., O'Neill, M.A., Darvill, A., and Albersheim, P. (1999) The plant cell wall polysaccharide rhamnogalacturonan II self-assembles into a covalently cross-linked dimer. *J. Biol. Chem.* 274, 13098-13104

Ishii, T., and Matsunaga, T. (2001) Pectic polysaccharide rhamnogalacturonan II is covalently linked to homogalacturonan. *Phytochem.* 57, 969-974

Lau, J.M., McNeil, M., Darvill, A.G., and Albersheim, P. (1987) Treatment of rhamnogalacturonan I with lithium in ethylenediamine. *Carbohydr. Res.* 168, 245-274

Le Goff, A., Renard, C.M.G.C., Bonnin, E., and Thibault, J.-F. (2001) Extraction, purification and chemical characterisation of xylogalacturonans from pea hulls. *Carbohydr. Polym.* 45, 325-334

McNeil, M., Darvill, A.G., and Albersheim, P. (1980) Structure of plant cell walls. X. Rhamnogalacturonan I, a structurally complex pectic polysaccharide in the walls of suspension-cultured sycamore cells. *Plant Physiol.* 66, 1128-1134

Mohnen, D. (1999) Biosynthesis of pectins and galactomannans. In *Comprehensive Natural Products Chemistry* (Barton, D., Nakanishi, K., Meth-Cohn, O., eds.), Vol 3, pp. 497-527, Elsevier

O'Neill, M.A., Eberhard, S., Albersheim, P., and Darvill, A.G. (2001) Requirement of borate cross-linking of cell wall rhamnogalacturonan II for *Arabidopsis* growth. *Science* 294, 846-849

Oomen, R.J.F.J., Doeswijk-Voragen, C.H.L., Bush, M.S., Vincken, J.-P., Borkhardt, B., van den Broek, L.A.M., Corsar, J., Ulvskov, P., Voragen, A.G.J., McCann, M.C., and Visser, R.G.F. (2002) *In muro* fragmentation of the rhamnogalacturonan I backbone in potato (*Solanum tuberosum* L.) results in a reduction and altered location of the galactan and arabinan side-chains and abnormal periderm development. *Plant J.*, in the press

Oosterveld, A., Beldman, G., and Voragen, A.G.J. (2000) Characterization of arabinose and ferulic acid rich polysaccharides and hemicelluloses from sugar beet pulp. *Carbohydr. Res.* 328, 185-197

Oxenbøll-Sørensen, S., Pauly, M., Bush, M., Skjøt, M., McCann, M.C., Borkhardt, B., and Ulvskov, P. (2000) Pectin engineering: modification of potato pectin by *in vivo* expression of an endo-1,4-β-D-galactanase. *Proc. Natl. Acad. Sci. U.S.A.* 97, 7639-7644

Renard, C.M.G.C., Voragen, A.G.J., Thibault, J.-F., and Pilnik, W. (1991) Studies on apple protopectin V: structural studies on enzymatically extracted pectins. *Carbohydr. Polym.* 16, 137-154

Renard, C.M.G.C., Lahaye, M., Mutter, M., Voragen, A.G.J., and Thibault, J.-F. (1997) Isolation and structural characterisation of rhamnogalacturonan oligomers generated by controlled acid hydrolysis of sugar-beet pulp. *Carbohydr. Res.* 305, 271-280

Renard, C.M.G.C., Crépeau, M.-J., and Thibault, J.-F. (1999) Glucuronic acid directly linked to galacturonic acid in the rhamnogalacturonan backbone of beet pectins. *Eur. J. Biochem.* 266, 566-574

Ridley, B.L., O'Neill, M.A., and Mohnen, D. (2001) Pectins: structure, biosynthesis, and oligogalacturonide-related signaling. *Phytochem.* 57, 929-967

Round, A.N., Rigby, N.M., MacDougall, A.J., Ring, S.G., and Morris, V.J. (2001) Investigating the nature of branching in pectin by atomic force microscopy and carbohydrate analysis. *Carbohydr. Res.* 331, 337-342

Ryden, P. and Selvendran, R.R. (1990) Structural features of cell wall polysaccharides of potato (*Solanum tuberosum*). *Carbohydr. Res.* 195, 257-272

Schols, H.A., Posthumus, M.A., and Voragen, A.G.J. (1990) Structural features of hairy regions of pectins isolated from apple juice produced by the liquefaction process. *Carbohydr. Res.* 206, 117-129

Schols, H.A., Bakx, E.J., Schipper, D., and Voragen, A.G.J. (1995a) A xylogalacturonan subunit present in the modified hairy regions of apple pectin. *Carbohydr. Res.* 279, 265-279

Schols, H.A., Vierhuis, E., Bakx, E.J., and Voragen, A.G.J. (1995b) Different populations of pectic hairy regions occur in apple cell walls. *Carbohydr. Res.* 275, 343-360

Schols, H.A. and Voragen, A.G.J. (1996) Complex pectins: structure elucidation using enzymes. *Progress in Biotechnology 14: Pectins and Pectinases* (Visser, J., Voragen, A.G.J., eds.), pp. 3-19, Elsevier

Skjøt, M., Pauly, M., Bush, M., Borkhardt, B., McCann, M., and Ulvskov, P. (2002) Direct interference with rhamnogalacturonan I biosynthesis in Golgi vesicles. *Plant Physiol.*, in the press

Talmadge, K.W., Keegstra, K., Bauer, W.D., and Albersheim, P. (1973) The structure of plant cell walls. I. The macromolecular components of the walls of suspension-cultured sycamore cells with a detailed analysis of the pectic polysaccharides. *Plant Physiol.* 51, 158-173

Thibault, J.-F., Renard, C.M.G.C., Axelos, M.A.V., Roger, P., and Crépeau, M.-J. (1993) Studies of the length of homogalacturonic regions in pectins by acid hydrolysis. *Carbohydr. Res.* 238, 271-286

Vidal, S., Doco, T., Williams, P., Pellerin, P., York, W.S., O'Neill, M.A., Glushka, J., Darvill, A.G., and Albersheim, P. (2000) Structural characterization of the pectic polysaccharide rhamnogalacturonan II: evidence for the backbone location of the aceric acid-containing oligoglycosyl side chain. *Carbohydr. Res.* 326, 227-294

Vincken, J.-P., Borkhardt, B., Bush, M., Doeswijk-Voragen, C., Dopico, B., Labrador, E., Lange, L., McCann, M., Morvan, C., Muñoz, F., Oomen, R., Peugnet, I., Rudolph, B., Schols, H., Sørensen, S., Ulvskov, P., Voragen, A., Visser, R. (2000) Remodelling pectin structure in potato. In *Phytosfere '99 – Highlights in European Plant Biotechnology* (de Vries, G.E., Metzlaff, K., eds.), pp. 245-256, Elsevier

Vincken, J.-P., Schols, H.A., Oomen, R.J.F.J., McCann, M.C., Ulvskov, P., Visser, R.G.F., and Voragen, A.G.J. (2002) If homogalacturonan were a side chain of rhamnogalacturonan I...... Implications for cell wall architecture. *Trends Plant Sci.*, submitted for publication

van der Vlugt-Bergmans, C.J.B., Meeuwsen, P.J.A., Voragen, A.G.J., and van Ooyen, A.J.J. (2000) Endo-xylogalacturonan hydrolase, a novel pectinolytic enzyme. *Appl. Environm. Microbiol.* 66, 36-41

de Vries, J.A., Rombouts, F.M., Voragen, A.G.J., and Pilnik, W. (1982) Enzymic degradation of apple pectins. *Carbohydr. Polym.* 2, 25-33

Willats, W.G.T., McCartney, L., MacKie, W., and Knox, J.P. (2001) Pectin: cell biology and prospects for functional analysis. *Plant Mol. Biol.* 47, 9-27

Yamada, H., Kiyohara, H., and Otsuka, Y. (1987) Further structural studies of an anti-complementary acidic heteroglycan from the leaves of *Artemisia princeps*. *Carbohydr. Res.* 170, 181-191

Yu, L., and Mort, A.J. (1996) Partial characterization of xylogalacturonans from cell walls of ripe watermelon fruit: inhibition of endopolygalacturonase activity by xylosylation. *Progress in Biotechnology 14: Pectins and Pectinases* (Visser, J., Voragen, A.G.J., eds.), pp. 79-88, Elsevier

Zhan, D., Janssen, P., and Mort, A.J. (1998) Scarcity or complete lack of single rhamnose residues interspersed within the homogalacturonan regions of citrus pectin. *Carbohydr. Res.* 308, 373-380

COVALENT CROSS-LINKING OF PRIMARY CELL WALL PECTIC POLYSACCHARIDES IS REQUIRED FOR NORMAL PLANT GROWTH

M. A. O'NEILL[1*], S. EBERHARD[1], B. REUHS[1], W.-D. REITER[2], T. ISHII[3], T. FUJIWARA[4], P. ALBERSHEIM[1], AND A. DARVILL[1]
[1]*Complex Carbohydrate Research Center and Dept. of Biochemistry and Molecular Biology, University of Georgia, Athens, GA 30602 USA.* [2] *Dept of Molecular and Cell Biology, University of Connecticut, Storrs, CT 06269, USA.* [3] *Forestry and Forestry Products Research Institute, Ibaraki 305-8687, Japan,* [4] *Dept of Applied Biological Chemistry, University of Tokyo, Tokyo113, Japan.*

*corresponding author. e-mail: mao@ccrc.uga.edu

Abstract

Rhamnogalacturonan II (RG-II) is a structurally complex pectic polysaccharide that is present in the primary walls of all higher plant cells. Recent research has revealed much about the structure and function of RG-II. RG-II exists in the wall predominantly as a dimer that is cross-linked by a 1:2 borate diol ester. The formation of the RG-II dimer *in muro* is proposed to generate a covalently cross-linked pectic network that contributes to the physical and biochemical properties of the wall. We have investigated the function of this pectic network using a dwarf *Arabidopsis* mutant (*mur1*) that synthesizes RG-II with an altered glycosyl residue composition and a dwarf *Arabidopsis* mutant (*bor1*) that is defective in root-to-shoot translocation of boron. The walls of these mutants contain reduced amounts of the RG-II dimer. The amounts of borate cross-linked RG-II in the walls and morphology of *mur1* and *bor1* plants sprayed with boric acid are comparable to wild-type plants. Our study demonstrates a major role for borate cross-linking of primary wall pectic polysaccharides in plant growth.

1. Introduction

Boron was shown to be an element that is essential for the normal growth and development of higher plants in the 1920s [1, 2]. Subsequent studies have established that there is a relationship between boron and cell wall pectin [3 – 7] and that a plants boron requirement is correlated with the pectin content of its cell walls [8, 9]. The chemical nature of this relationship was established when the structurally conserved pectic polysaccharide rhamnogalacturonan II (RG-II, see Fig. 1) was shown to be present in the primary wall predominantly as a dimer that is cross-linked by a 1:2 borate-diol ester [10 – 13, see Fig. 2A].

Dimer formation *in muro* results in a cross-linked pectic network since RG-II itself is covalently inserted within homogalacturonan chains [14, 15, see Fig. 2B]. Borate

F. Voragen et al. (eds.), Advances in Pectin and Pectinase Research, 61-73.
© 2003 *Kluwer Academic Publishers. Printed in the Netherlands.*

cross-linking of pectin contributes to the physical and biochemical properties of the wall because the initial phenotype of boron deficiency is structurally abnormal walls [14, 16, 17]. Many researchers now agree that a physiologically important function of boron is to cross-link pectins in the walls of growing plant cells [12, 18, 19]. However, there has been no direct demonstration that borate ester cross-linking of RG-II is required for normal plant growth.

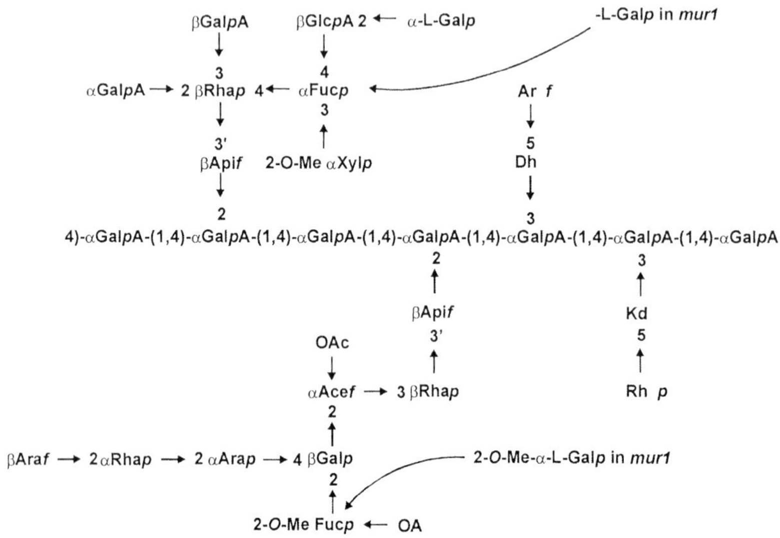

Figure1. A. The glycosyl sequence of rhamnogalacturonan II. In *mur1* RG-II the L-fucosyl and 2-*O*-Me L-fucosyl residues are replaced by L-galactosyl and 2-*O*-Me L-galactosyl residues, respectively.

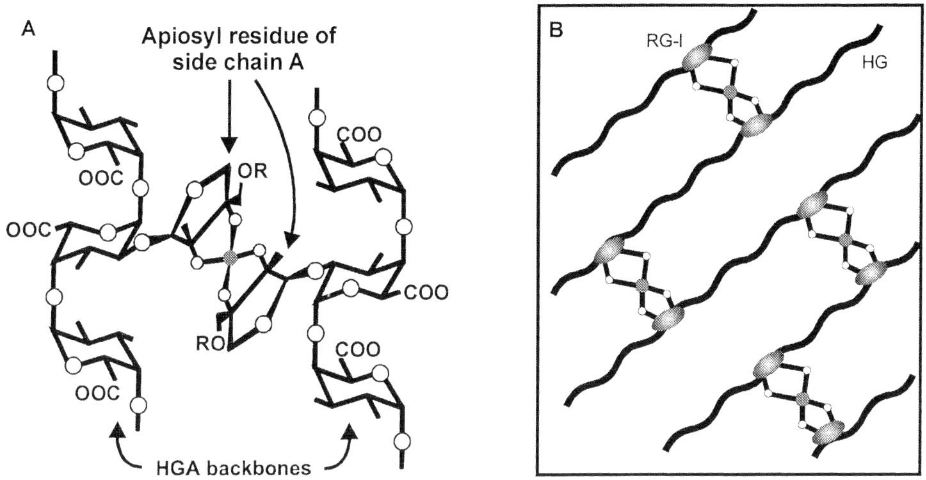

Figure 2. A. Two molecules of RG-II are cross-linked together by a 1:2 borate-diol ester. The cross link is formed between each apiosyl residue of the 2-*O*-Me xylosyl-containing side chain (side chain A). B. Borate cross-linking of RG-II is likely to generate a covalently cross-linked pectic network *in muro* because RG-II itself is covalently inserted within homogalacturonan chains

The aerial portions of *Arabidopsis thaliana mur1-1* and *mur1-2* mutant plants contain less than 2% of the amount of L-fucose (6-deoxy-L-galactose) present in wild type plants [20, 21]. These mutants are dwarfed and have brittle stems [20]. The altered gene in *mur1* plants has been shown to encode an isoform of GDP-mannose 4,6-dehydratase [22], a enzyme required for the biosynthesis of L-fucose (see Fig. 3). L-fucosyl residues are present in several Arabidopsis wall polysaccharides including RG-II, rhamnogalacturonan I, and xyloglucan [25], and in the N-linked oligosaccharides of glycoproteins [26]. Thus, the absence of L-fucosyl residues in one or more of these complex glycans may result in the dwarf phenotype of *mur1* plants [27]. The absence of fucosyl residues on xyloglucan is not likely to be responsible for the dwarf phenotype of *mur1* plants since Arabidopsis *mur2* plants, which are deficient in a xyloglucan- specific fucosyl transferase [28, 29], synthesize xyloglucan that contains ~2% of the normal amounts of L-Fuc but grow normally [29]. The absence of L-Fuc residues on the N-linked complex glycan side chains of glycoproteins synthesized by *mur1* plants [26] is also unlikely to result in the dwarf phenotype since an Arabidopsis mutant (*cgl1*) which synthesizes glycoproteins that lack L-Fuc grows normally [30]. In this chapter we outline the data showing that *mur1-1* and *mur1-2* rosette leaf cell walls contain normal amounts of RG-II but only half exists as a borate cross-linked dimer [31]. Plants carrying the *mur1* mutation synthesize RG-II in which the L-fucosyl (L-Fuc) and 2-*O*-methyl L-fucosyl (2-*O*-Me L-Fuc) residues are replaced by L-galactosyl (L-Gal) and 2-*O*-methyl L-galactosyl (2-*O*-Me L-Gal) residues, respectively [31, 32].

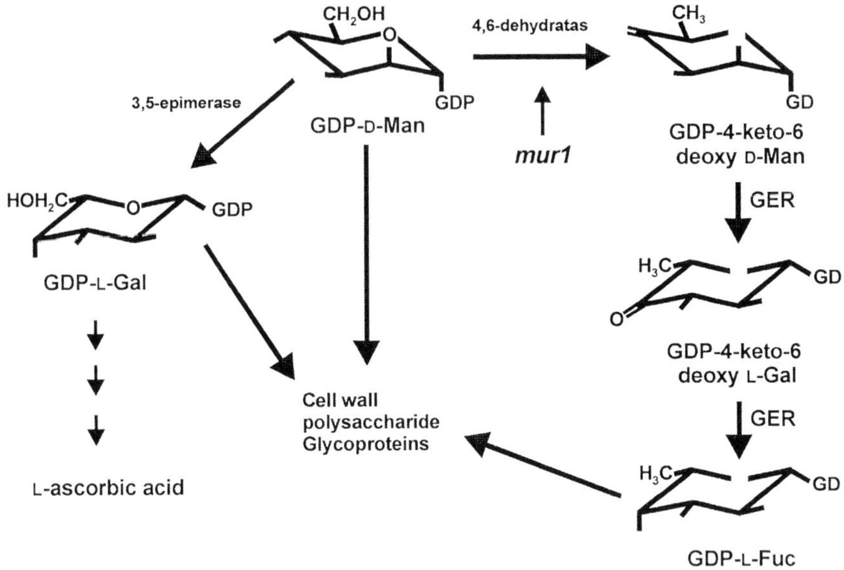

Figure 3. In Arabidopsis two proteins (MUR1 and GER1) are required to catalyze the formation of GDP-L-fucose from GDP-D-Man [22, 23]. The altered gene in *mur1* plants encodes for an isoform of GDP-mannose-4,6-dehydratase. GDP-L-Gal is generated from GDP-mannose by a 3,5-epimerase. GDP-L-galactose is incorporated into cell wall polysaccharides and the N-linked oligosaccharides of glycoproteins synthesized by *mur1* plants. GDP-L-galactose is known to be a precursor of ascorbate in higher plants [24].

The altered structure of *mur1* RG-II reduces the rate of formation and stability of the borate cross link. Data is presented showing that in *mur1* plants exogenous aqueous borate results in an increase in the extent of borate cross-linking of RG-II and that the morphology of borate-treated *mur1* and wild type plants are comparable. We also show that less than half of the RG-II is cross-linked by borate in the rosette leaf walls of the dwarf Arabidopsis mutant *bor1-1*. This mutant requires higher concentrations of B than wild-type plants for normal growth [33, 34]. Exogenous aqueous borate rescues the growth of soil-grown *bor1* plants. Thus, reduced cross-linking of RG-II is likely to be responsible for the dwarf phenotypes of *mur1* and *bor1* plants. We conclude that the covalent cross-linking and the organization of wall pectic polysaccharides has a major role in controlling normal plant growth.

2. *mur1-1, mur1-2,* and *bor1-1* plants are dwarfed and only half of the RG-II in their rosette leaf cell walls is cross-linked by borate

mur1 (Fig. 4) and *bor1-1* (Fig. 5) plants have smaller rosette leaves than wild-type (WT) plants. These physiological symptoms are characteristic of B deficiency [35] even though the plants were fertilized with normal amounts of borate.

The RG-II dimer accounts for ~95% of the RG-II in WT plants but for only ~50% of the RG-II in the mutants (Fig. 6 and Table 1). The L-Fuc and 2-*O*-Me-L- Fuc residues of *mur1* RG-II are replaced by L-Gal and 2-*O*-Me L-Gal residues, respectively [31, 32, see Fig. 1]. Wild type and *mur1* RG-II contain comparable amounts of galacturonic, glucuronic, 3-deoxyheptulosaric (Dha), and 3-deoxyoctulosonic acid (Kdo) [32]. However, the 2-*O*-methyl xylose (2-*O*-Me-Xyl) content of *mur1* RG-II

Table 1. The percentage of RG-II present as dimer in the walls of WT, *mur1-2*, and *bor1-1* plants.

Plant	Dimer (% of total RG-II)
WT control	95±2
+ 2.5 mM borate	94±3
+ 10 mM Fuc	96±2
mur1-2 control	50±5
+ 2.5 mM borate	74±3
+ 10 mM Fuc	81±7
bor1-1 control	45
+ 0.1 mM borate	86
+ 1 mM borate	98

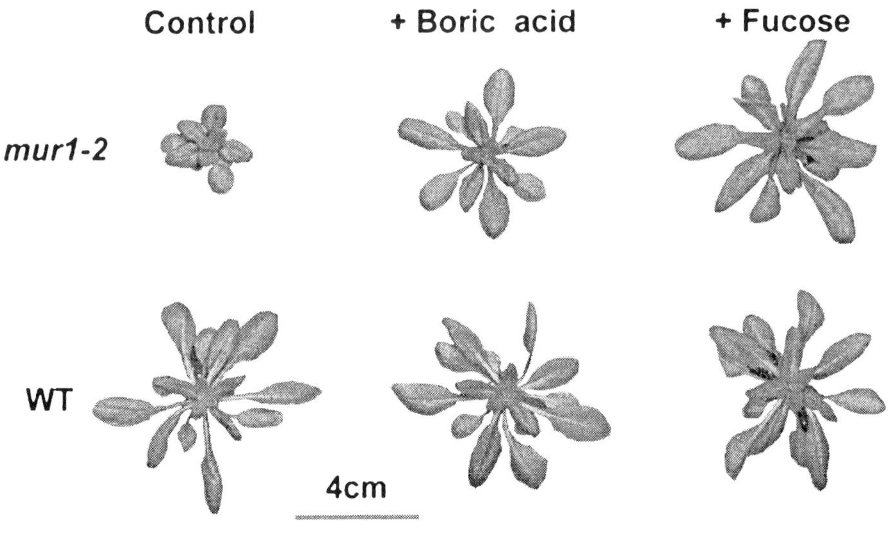

Figure 4. Boric acid rescues *mur1-2* plants. Control plants were treated every seven days with fertilizer containing boric acid (15 μM). Borate-treated plants were sprayed every seven days with

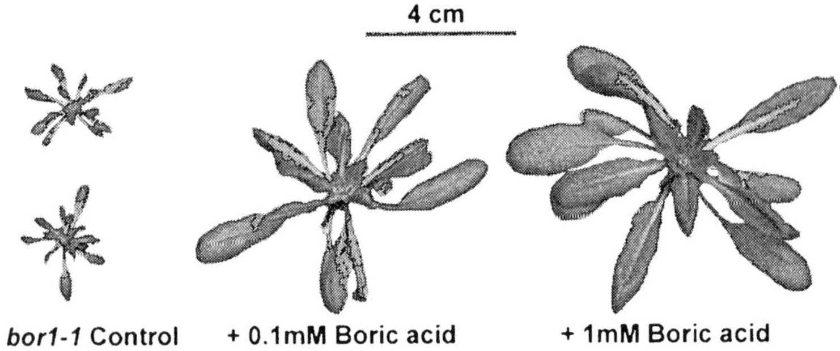

Figure 5 Boric acid rescues *bor1-1* plants. *bor1-1* plants were treated every seven days with fertilizer containing either 0.1 or 1 mM boric acid. Control plants were treated with fertilizer containing boric acid (15 μM)

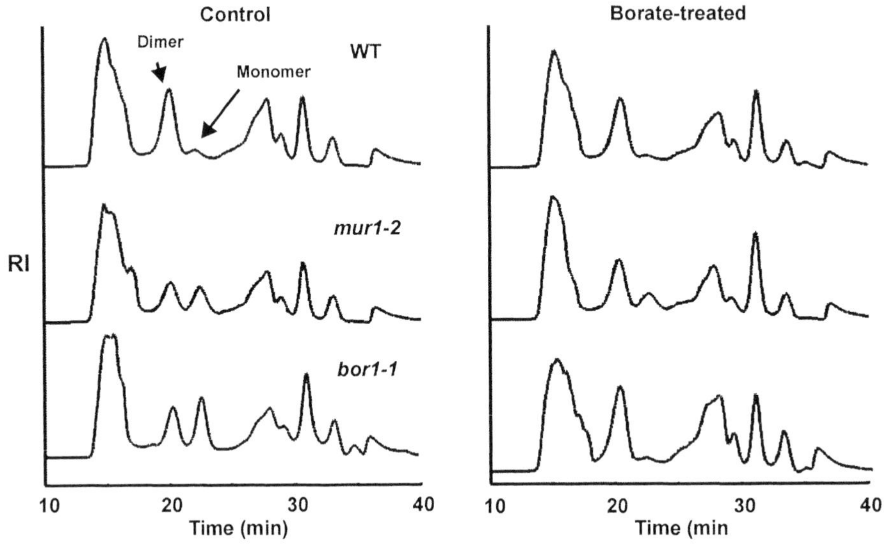

Figure 6. Superdex-75 size-exclusion chromatography of the material released by endopolygalacturonase treatment of the walls of wild type (WT), *mur1-2*, and *bor1-1* plants grown in the presence or absence of added borate. Control plants were sprayed every seven days with commercial fertilizer (15 μM boric acid). Wild type and *mur1-2* plants were sprayed every seven days with fertilizer solution containing added boric acid (2.5 mM). *bor1-1* plants were sprayed every seven days with fertilizer solution containing added boric acid (1 mM). The elution position of the RG-II dimer and monomer are shown.

is reduced by ~50% (Table 2). In contrast, WT and *bor1-1* RG-IIs have virtually identical glycosyl residue compositions (Table 2). Thus, it is unlikely that the altered gene in *bor1-1* plants encodes for a enzyme involved in the biosynthesis of RG-II. Indeed, recent studies indicate that the *bor1* mutation results in reduced uptake and/or reduced root-to-shoot translocation of B [34].

Exogenous aqueous borate rescues the growth of *mur1-2* and *bor1-1* plants (Figs. 4 and 5) and most of the RG-II exists as the dimer (Table 1). The reduced cross-linking of RG-II in *bor1-1* plants most likely results from a reduction in the availability of leaf B [34]. In contrast, reduced cross-linking of RG-II in *mur1* plants results from replacing L-Fuc and 2-*O*-Me L-Fuc with L-Gal and 2-*O*-Me L-Gal, respectively [31]. Indeed L-Fuc also rescues the growth of *mur1* plants (Fig. 4) and in these plants most of the RG-II exists as the dimer and this RG-II contains L-Fuc and 2-*O*-Me L-Fuc (Table2). The 2-*O*-Me Xyl content of *mur1-2* RG-II is also restored to normal levels (Table 2) by L-Fuc treatment which suggests that L-Fuc is more effective than L-Gal as a acceptor for the attachment of the xylosyl residue.

Table 2. The neutral glycosyl residue compositions of the RG-II isolated from WT, *mur1-2* and *bor1-1* plants grown in the absence or presence of added boric acid or L-Fuc.

					Mol %				
Plant	MeFuc	Rha	Fuc	MeXyl	Ara	Api	Ace	MeGal	Gal
WT	5	21	4	6	27	8	3	Nd	26
+ B[1]	5	21	4	6	28	8	3	Nd	25
+ Fuc[2]	6	22	4	6	28	9	3	Nd	22
mur1-2	Tr	23	Tr	3	27	9	3	5	30
+ B[1]	Tr	22	Tr	2	28	9	2	5	32
+ Fuc[2]	5	23	4	5	28	9	2	2	22
bor1-1	5	20	5	6	30	8	3	Nd	23
+ B[3]	5	19	5	6	27	10	3	Nd	25

[1] + 2.5 mM borate. [2] + 10 mM l-Fuc. [3] + 1 mM borate. Tr = trace, less than 1% of WT amounts. Nd = not detected.

3. The *mur1* RG-II dimer is less stable and forms less readily than the wild-type RG-II dimer

Mur1 plants require a higher concentration of B for normal growth than their WT counterparts (Fig. 4). This suggested that essential structural and conformational features of RG-II are altered in *mur1* plants and that such changes affect the formation or stability of the borate cross-link. Thus, we compared the *in vitro* pH stability and rate of formation of WT and *mur1* RG-II dimers (Fig. 7). The RG-II dimer from *mur1* plants grown in the presence or absence of boric acid is less stable at low pH than the dimer obtained from either WT plants or L-Fuc-treated *mur1* plants (Fig. 7A and Table 3).

The *mur1* RG-II dimer forms less rapidly than the WT dimer irrespective of whether the reaction is performed in the presence or absence of divalent cations (Fig. 7B). Thus, the replacement of L-Fuc and 2-*O*-Me L-Fuc with L-Gal and 2- *O*-Me L-Gal, respectively, together with a reduction in amount of 2-*O*-Me Xyl reduces the ability of *mur1* RG-II to form a dimer and this dimer once formed is less stable than its WT counterpart [31].

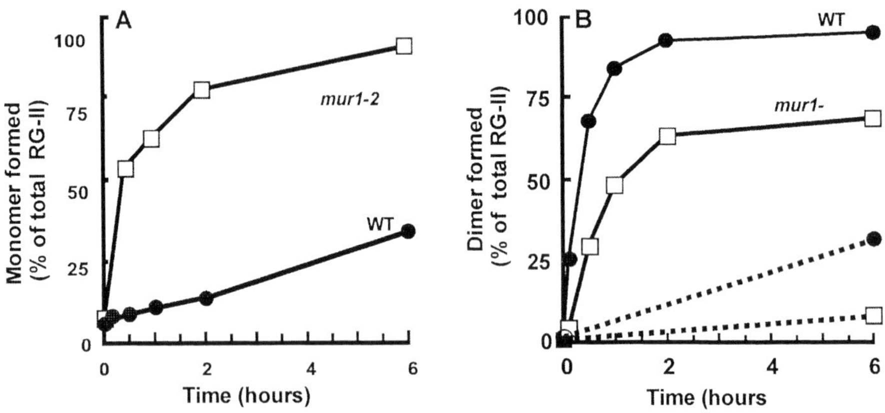

Fig. 7. The *in vitro* formation and stability of WT and *mur1-2* RG-II dimers. A. The amount of monomer formed when the WT and *mur1-2* RG-II dimers were kept at pH 2.5. B. The amount of RG-II dimer formed when WT and *mur1-2* RG-II monomers were reacted at pH 3.6 with boric acid (1 mM) in the absence (dotted lines) and presence (solid lines) of $SrCl_2$ (1mM).

Table 3. The stability at pH 2.5 of the RG-II dimers from WT and *mur1-2* plants grown in the presence and absence of boric acid and L-fucose

Plant	RG-II monomer formed (% of total RG-II)[1]
WT control	3
+ 2.5 mM borate	5
+ 10 mM L-Fuc	6
mur1-2 control	58
+ 2.5 mM borate	55
+ 10 mM L-Fuc	14

[1] The amount of RG-II monomer formed when the EPG-soluble, RG-II-containing fractions of the walls of WT and *mur1-2* plants grown in the presence or absence of added boric acid or L-Fuc were treated for 1h at pH 2.5.

4. Discussion

The extent of borate cross-linking of RG-II is reduced by ~50% in the rosette leaf walls of two genotypically different dwarf Arabidopsis mutants (*mur1* and *bor1*). The reduction of RG-II cross-linking in *mur1* plants is a consequence of replacing L-fucosyl residues with L-galactosyl residues. This modest change in RG-II structure reduces the ability of RG-II to form a borate cross-linked dimer and the dimer once formed is less stable than the WT dimer. In contrast, reduced cross-linking of RG-II in *bor1* plants does not result from altering RG-II structure but is a consequence of a decrease in the availability of leaf borate. Irrespective of the nature of the altered gene in *mur1* and *bor1* plants the reduction in RG-II cross-linking is responsible for abnormal plant growth (Fig. 8) since exogenous borate treatment increases the extent of borate cross-linking of RG-II and rescues the phenotypes of both mutants.

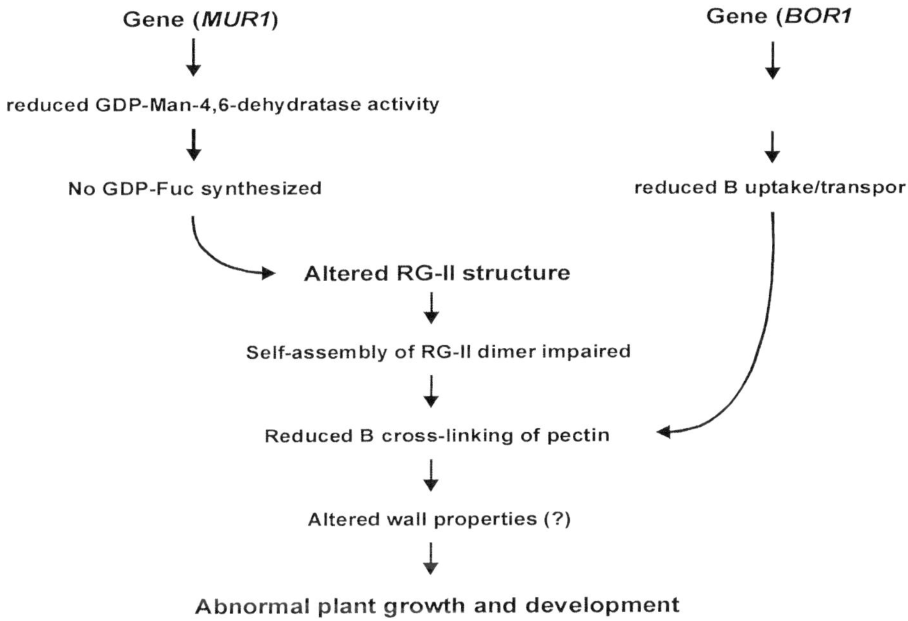

Figure 8. The relationship between a mutated gene, altered polysaccharide structure and interaction with B, and abnormal plant growth.

Our data together with the results of several other studies suggest that covalent cross-linking of pectin has a role in determining the mechanical and biochemical properties of the primary wall [14, 17]. The cumulative results of several studies suggest that boron's primary role is in turgor-driven wall expansion rather than in cell division [31, 34, 35]. Nevertheless, the mechanism(s) by which reduced cross-linking of RG-II affects plant growth is not understood. It is unlikely that borate cross-links are load-bearing bonds since reducing the number of such bonds would be expected to result in "wall loosening" together with uncontrolled wall expansion. This is in direct contrast to our experimental findings showing that reducing borate cross-linking results in a decrease in cell expansion. We suggest that covalent cross-linking of pectin generates a structurally well-defined three dimensional network that facilitates restructuring of the cellulose-hemicellulose network that itself is required for wall expansion. However, the possibility cannot be discounted that the borate cross-link is dynamic and its controlled hydrolysis and reformation is required for wall expansion to occur.

The results of several studies have indicated that connections exist between the plant cell wall and the plasma membrane [36]. A putative linkage between wall-associated kinases (WAKS) and pectin [37, 38] has been reported, although the nature of the connections have not been characterized. Arabidopsis plants expressing anti-sense RNAs to WAK genes exhibit reduced cell expansion and the plants are dwarfed [38, 39]. These results have led to the suggestion that the covalent linkage of WAKS to pectins has a role in plant cell expansion [38]. We extend this hypothesis by suggesting that plant cells perceive changes in the extent of covalent cross-linking of pectin through wall-membrane protein connections (Fig. 9). Such interactions may involve WAKS or mechanosensory transduction of signals either directly or via the cytoskeleton. There is, as yet, little experimental evidence to support this form of signal transduction in plants. However, there is accumulating evidence that animals cells do respond to mechanical signals [40, 41].

Our study has shown that a reduction in covalent cross-linking of pectin that results from either a modest change in the structure of RG-II or a decrease in the availability of leaf borate affects plant growth. Our results also illustrate how a mutation in a gene that encodes for an isoform of GDP-D-mannose-4,6- dehydratase may have unexpected biochemical and functional consequences.

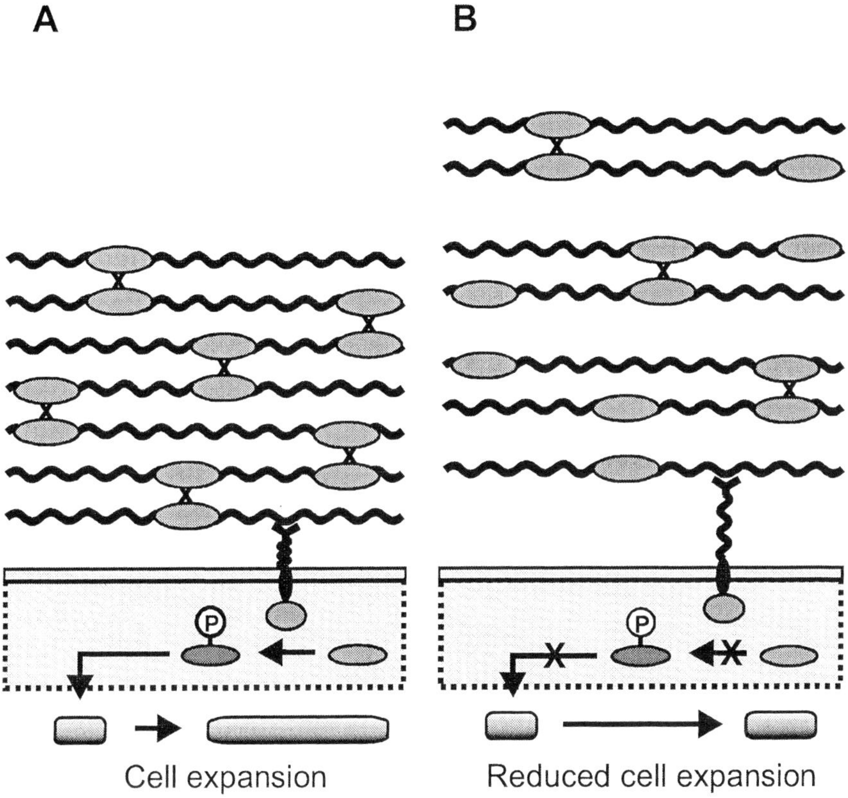

Figure 9. A hypothetical mechanism whereby altering the extent of covalent cross-linking of pectin in the primary wall is perceived by plant cells. A. Most (>95%) of the RG-II in the primary wall is cross-linked by borate. The physical/mechanical properties of wall pectin are transmitted to the cell by pectin-wall-membrane interactions, either via wall associated kinase (e.g. WAKS) or via wall-associated mechanical-force-sensing membrane proteins. Under boron-sufficient conditions such signals facilitate normal cell expansion. B. Under boron-deficient conditions the extent of covalent cross-linking of pectin is reduced. Such conditions are known to result in a increase in wall porosity [14] and swelling of the primary cell wall [17]. Changes that alter the physical/mechanical properties of the wall may be perceived by wall-membrane interactions. Such changes in the wall then lead to the transduction of signals that ultimately reduce the ability of the cell to expand normally.

5. Acknowledgments

This research was supported by US Dept of Energy Grants DE-FG02-96ER20220 (to AD), DE-FG05-93ER20097 (to AD and PA), DE-FG02-95ER20203 (to W.-D.R) and by grants from the Japanese Ministries of Agriculture (BDP-01-II-4, to TI) and Education, Culture, Sports, Science and technology (to TF).

6. References

1. Warrington, K.: The effect of boric acid and borax on the broad bean and certain other plants, *Ann. Bot.* **37** (1923), 629 – 672.
2. Sommer, A.L., and Lipman, C.B.: Evidence on the indispensable nature of zinc and boron for higher green plants, *Plant Physiol.* **1** (1926), 231 – 249.
3. Schmucker, T.: Zur Blütenbiologie tropischer nymphea-arten. II. Bor als entscheidender Faktor,*Planta* **18** (1933), 641 – 650.
4. Smith, M.E.: The role of boron in plant metabolism. 1. Boron in relation to the absorption and solubility of calcium, *Aust J. Exp. Biol. Med. Sci.* **22** (1944), 257 – 263.
5. Yamauchi, T., Hara, T., and Sonoda, Y.: Distribution of calcium and boron in the pectic fraction of tomato leaf cell wall, *Plant Cell Physiol.* **27** (1986), 729 – 732.
6. Loomis, W.D., and Durst, R.W.: Chemistry and biology of boron, *BioFactors* **3** (1992), 229 – 239.
7. Hu, H., and Brown, P.H.: Localization of boron in cell walls of squash and tobacco and its association with pectin. Evidence for a structural role of boron in the cell wall, *Plant Physiol.* **105** (1994), 681 – 689.
8. Hu, H., Brown, P.H., and Labavitch, J.M.: Species variability in boron requirement is correlated with cell wall pectin, *J. Exp. Bot.* **47** (1996), 227 – 232.
9. Matoh, T., Kawaguchi, S., and Kobayashi, M.: Ubiquity of a borate-rhamnogalacturonan II complex in the cell walls of higher plants, *Plant Cell Physiol.* **37** (1996), 636 – 640.
10. Kobayashi, M., Matoh, T., and Azuma, J.: Structure and glycosyl composition of the boron-polysaccharide complex of radish roots, *Plant Cell Physiol.* **36S** (1995), 139.
11. Kobayashi, M., Matoh, T., and Azuma, J.: Two chains of rhamnogalacturonan II are cross-linked by borate-ester bonds in higher plant cell walls, *Plant Physiol.* **110** (1996), 1017 – 1020.
12. O'Neill. M.A., Warrenfeltz, D., Kates, K., Pellerin, P., Doco, T., Darvill, A.G., and Albersheim, P.: Rhamnogalacturonan-II, a pectic polysaccharide in the walls of growing plant cells, forms a dimer that is covalently cross-linked by a borate ester, *J. Biol. Chem.* **217** (1996), 22923 – 22930.
13. Ishii, T., Matsunaga, T., Pellerin, P., O'Neill, M.A., Darvill, A.G., and Albersheim, P.: The plant cell wall polysaccharide rhamnogalacturonan II self-assembles into a covalently cross-linked dimer, *J. Biol. Chem.* **274** (1999) 13098 – 13104.
14. Fleischer, A., O'Neill, M.A., and Ehwald, R.: The pore size of non-graminaceous plants is rapidly decreased by borate ester cross-linking of the pectic polysaccharide rhamnogalacturonan II, *Plant Physiol.* **121** (1999), 829 – 838.
15. Ishii, T. and Matsunaga, T.: Pectic polysaccharide rhamnogalacturonan II is covalently linked to homogalacturonan, *Phytochemistry* **57** (2001), 969 – 974.
16. Fleischer, A., Titel, A., and Ehwald, R.: The boron requirement and cell wall properties of growing- and stationary-phase suspension-cultured *Chenopodium album* L. cells, *Plant Physiol* **117** (1997), 1401 – 1410.
17. Ishii, T., Matsunaga, T., and Hayashi, N.: Formation of rhamnogalacturonan II-borate dimer in pectin determines cell wall thickness of pumpkin tissue, *Plant Physiol.* **126** (2001), 1698 –1705.
18. Brown, P.H., and Hu, H.: Does boron play only a structural role in the growing tissues of higher plants? *Plant Soil* **196** (1997), 211 – 215.
19. Matoh, T., and Kobayashi, M.: Boron and calcium, essential inorganic constituents of pectic polysaccharides in higher plant cell walls, *J. Plant Res.* **111** (1998), 179 – 190.
20. Reiter, W.D., Chapple, C.C.S., and Somerville, C.R.: Altered growth and cell walls in a fucose-deficient mutant of *Arabidopsis*, *Science* **261** (1993), 1032 – 1035.
21. Reiter, W.D., Chapple, C.C.S., and Somerville, C.R.: Mutants of *Arabidopsis thaliana* with altered cell wall polysaccharide composition, *Plant J.* **12** (1997), 335 – 345.
22. Bonin C.P., Potter, I., Vanzin, G.F., and Reiter, W.D.: The MUR1 gene of Arabidopsis thaliana encodes an isoform of GDP-mannose-4,6-dehydratase, catalyzing the first step in the de novo synthesis of GDP-L-fucose, Proc. Natl. Acad. Sci. USA, **94** (1997), 2085 – 2090.
23. Bonin, C.P. and Reiter, W.D.: A bifunctional epimerase-reductase acts downstream of the MUR1 gene product and completes the de novo synthesis of GDP-L-fucose in Arabidopsis, Plant J. *21, (2000), 445 – 454.*
24. Smirnoff, N., Conklin, P.L., and Loewus, F.A.: Biosynthesis of ascorbic acid in plants. A renaissance, Ann. Rev. Plant Physiol. Plant Mol. Biol. **52** (2001), 437- 467.
25. Zablackis, E., Huang, J., Muller, B., Darvill, A.G., and Albersheim, P.: Characterization of the cell-wall polysaccharides of *Arabidopsis thaliana* leaves, *Plant Physiol.* **107** (1995), 1129 – 1136.

26. Rayon, C., Cabanes-Macheteau, M., Loutelier-Bourhis, C., Salliot-Maire, I., Lemoine, J., Reiter, W.D., Lerouge, P., and Faye, L.: Characterization of N-glycans from *Arabidopsis*. Application to a fucose-deficient mutant, *Plant Physiol.* **119** (1999), 725 – 734.
27. Reiter W.D.: *Arabidopsis thaliana* as a model system to study synthesis, structure, and function of the plant cell wall, *Plant Physiol. Biochem*. **36** (1998) 167 – 176.
28. Perrin, R.M., DeRocher, A.E., Bar-Peled, M., Zeng, W.Q., Norambuena, L., Orellana, A., Raikhel, N.V., and Keegstra, K.: Xyloglucan fucosyltransferase, an enzyme involved in plant cell wall biosynthesis, *Science* **284** (1999), 1976 – 1979.
29. Vanzin, G.F., Madson, M., Carpita, N.C., Raikhel, N.V., Keegstra, K., and Reiter, W.D.: The *mur2* mutant of *Arabidopsis thaliana* lacks fucosylated xyloglucan due to a lesion in fucosyltransferase AtFT1. Submitted for publication.
30. von Schaewen, A., Sturm, A. O'Neill, J., Chrispeels, M.J. Isolation of a mutant *Arabidopsis* plant that lacks N-acetyl glucosaminyl transferase I and is unable to synthesize Golgi-modified complex N-linked glycans, *Plant Physiol.* **102**, (1993), 1109 - 1118.
31. O'Neill, M.A., Eberhard, S., Albersheim, P., and Darvill, A.G.: Arabidopsis growth requires borate cross-linking of the cell wall pectic polysaccharide rhamnogalacturonan II, *Science* **294** (2001), 846 - 849.
32. Reuhs, B., Stephens, S., Kim, J., O'Neill, M.A., Albersheim, P., and Darvill, A.G.: L-galactosyl residues replace L-fucosyl residues in the rhamnogalacturonan II synthesized by the Arabidopsis *mur1* mutant. Manuscript in preparation.
33. Noguchi, K., Yasumori, M., Imai, T., Naito, S., Matsunaga, T., Oda, H., Hayashi, H., Chino, M., and Fujiwara, T, *bor1-1*, an *Arabidopsis thaliana* mutant that requires a high level of boron, *Plant Physiol.* **115** (1997), 901 – 906.
34. Takano, J., Yamagami, M., Noguchi, K., Hayashi, H., and Fujiwara, T, Preferential translocation of boron to young leaves in *Arabidopsis thaliana* regulated by the *BOR1* gene, Soil Sci. Plant Nutr. 47 (2001), 345 – 357.
35. Dell, B and Huang, L.: Physiological response of plants to low boron, *Plant Soil* **193** (1997), 103 – 120.
36. Kohorn, B.: Plasma membrane-cell wall contacts, *Plant Physiol.* **124** (2000), 31 – 38.
37. He, Z.H., Fujiki, M., and Kohorn, B.: A cell-wall associated, receptor-like kinase, *J. Biol. Chem.* **271** (1996), 19789 – 19793.
38. Wagner, T.A., and Kohorn, B.: Wall-associated kinases are expressed throughout plant development and are required for cell expansion, *Plant Cell* **13** (2001), 303 – 318.
39. Lally, D., Ingmire, P., Tong, H.Y., and He, Z.H.: Antisense expression of a cell wall-associated protein kinase, WAK4, inhibits cell elongation and alters morphology, *Plant Cell* **13** (2001), 1317 – 1331.
40. Chen, C.S., Mrksich, M., Huang, S., Whitesides, G.M., and Ingber, D.E.: Geometric control of cell life and death, *Science* **276** (1997), 1425 – 1428.
41. Walker, R.G., Willingham, A.T., and Zuker, C.S.: A Drosophila mechanosensory transduction channel, *Science* **287** (2000), 2229 – 2234.

DIFFERENCES IN THE METHYL ESTER DISTRIBUTION OF PECTINS

H.A. SCHOLS, M.M.H. HUISMAN, E.J. BAKX AND A.G.J. VORAGEN
Wageningen University, Agrotechnology and Food Sciences, Laboratory of Food Chemistry, Bomenweg 2, 6703 HD Wageningen, The Netherlands. E-mail: Henk.Schols@chem.fdsci.wag-ur.nl

Abstract

Both commercial pectin preparations as well as pectins as present in plant cell walls are very complex and heterogeneous molecules. Although the relative amount of the rhamnogalacturonan structural element may very significant for pectins from different origine, most pectins have in common that homogalacturonans forms the major part of the molecule. These homogalacturonan segments partly determine the functional properties of the pectin like its thickening and gelling properties. Especially the level of methyl esters and their distribution over the galacturonan backbone is of major importance. Many effort has been directed to the reveal the pattern of methyl ester distribution.
We now present a strategy in which endo-polygalacturonase is used to degrade (methyl esterified) homogalacturonans, followed by the identification and quantification of degradation products released. The term 'degree of blockiness' is introduced as a measure for the number of galacturonic acid residues present in non-esterified, enzyme degradable segments. Next to some other indicators, the possibility of using a distance tree to obtain a visual presentation of the characteristics of many different pectins is discussed.

1. Introduction

Pectins or pectic substances are collective names for a group of closely associated polysaccharides present in plant cell walls where they contribute to complex physiological processes like cell growth and cell differentiation and so determine the integrity and rigidity of plant tissue. They also play an important role in the defense mechanisms against plant pathogens and wounding. As constituents of plant cell walls and due to their anionic nature, pectic polysaccharides are considered to be involved in the regulation of ion transport, they determine the porosity of the walls and in this way they may control the permeability of the walls for enzymes. They also determine the water holding capacity of the wall[1, 2, 3].
The texture of fruits and vegetables during growth, ripening and storage is strongly influenced by the amount and composition of the pectic molecules present. Consequently, the structure of pectins present in fruit and vegetables depends on enzymatic and chemical modifications occurring during these processes. Endogenous as well as exogenous enzymes (processing aids) play an important role in determining the pectic structures present in a given plant material or food product at a given time.

F. Voragen et al. (eds.), Advances in Pectin and Pectinase Research, 75-90.
© 2003 *Kluwer Academic Publishers. Printed in the Netherlands.*

Pectins used in the food industry as natural ingredients for their gelling, thickening and stabilizing properties are extracted from suitable agro-byproducts like citrus peel and apple pomace[2,4]. Commercially produced pectins usually contain more than 65 % of galacturonic acid by weight, since the extraction conditions used remove the greatest part of the neutral side chains present in native pectin rather specifically[2,4,5].

A general representation of native pectins is given in figure 1. This figure shows that the major part of the galacturonosyl residues is present in homogalacturonan segments (smooth regions of pectin), whereas only a relatively small part of the uronic acids is present in so-called hairy regions[6].

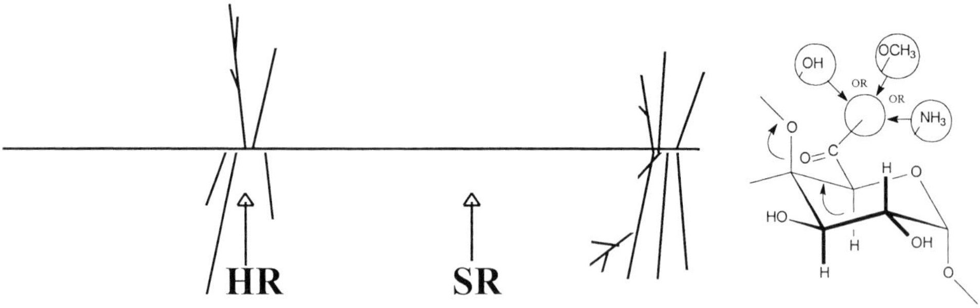

Figure 1. Schematic structure of pectin including the homogalacturonan (smooth) region (SR) and the rhamnogalacturonan (hairy region, HR). alfa-D-Galacturonic acid is the most dominant building block. The location of methyl esterification as found in native pectins is indicated. Furthermore, the location of the amide group as present in industrially modified amidated pectins is pointed out. Arrows indicate the potential for degradation by beta-elimination in the ester form. Acetyl groups may be present at O-2 and/or O-3 [4].

The physical properties of particularly acid extracted pectins used as food ingredient are determined to a large extent by the homogalacturonan segments within the pectin molecule. Important characteristics of pectins are the GalA content, the neutral sugar (linkage) composition, amount and distribution of methyl esters (and acetyl groups), and molecular weight. Within this homogalacturonan, the methyl ester distribution is extremely important and even relative small variations in methyl ester amount and/or distribution results in large differences in the physical behavior of such pectins. Therefore, the inter- and intramolecular distribution of pectins have been studied for many years.

2. The use of enzymes in revealing the methyl ester distribution

The methyl ester distribution of pectin has been the subject of study for many years. This distribution is very complex due to the fact that the methyl ester distribution should be revealed on a intramolecular level (within one molecule) and on a intermolecular level (distribution of methyl esters over various pectin molecules within a mixture[7,8]. Methods reported until 1982 have been reviewed by Taylor[9].

In 1982 Tuerena et al. developed a method that allowed for the study of the intramolecular distribution of free (non-esterified) carboxyl groups in pectin with enzymes[10,11]. The non-esterified carboxyl groups of pectin were glycolated with ethylene oxide and a mixture of pectic enzymes was used to degrade the polymer. The non- degraded glycolated oligomers were separated and the molecular weight distribution was determined by SEC. Computer studies were performed to determine the distributions expected for random methyl esterified samples of varying DM[10,11]. The experimental results were in reasonable agreement with the calculated data, although the methyl ester distribution could not be resolved in great detail.

The methyl ester distribution of carefully extracted apple pectin was meticulously studied by De Vries et al. with both highly purified pectin lyase and endo-pectate lyase[7]. Comparison of the SEC patterns of the 'native' apple pectin digests with those of enzyme digests of a 'transesterified' (an assumed random esterified) apple pectin of similar DM, revealed clear dissimilarities. The 'native' and transesterified pectins were degraded to a different extent by the enzymes used. Fractionation of the pectin digests by anion-exchange chromatography enabled the separation and quantification of the oligomers produced according to the number of free carboxyl groups[12]. Here, only minor differences were observed, suggesting a similar intramolecular methyl ester distribution[7]. Ion-exchange fractionation of the undigested 'native' apple pectin did reveal the presence of an non-homogeneous intermolecular methyl ester distribution. Similar results were found for carefully extracted lemon (albedo) pectin[8]. Using pectin lyase, de Vries also investigated the methyl ester distribution of commercially extracted lemon, lime, orange peel, and apple pectin[12]. Here, data suggesting the occurrence of quite large sequences of methyl esterified GalA in the commercially extracted pectins was obtained. A non-homogeneous intermolecular methyl ester distribution was reported for such pectins as well[13]. Kiyohara et al. used purified endo-PG to characterize the methyl-ester distribution of complementary activating pectic poly-saccharides extracted from the root of Angelica acutiloba[14] and fractionated by anion-exchange chromatography. Meticulous analysis and more detailed fractionation demonstrated that the homo-galacturonan regions of the various fractions differed in the number of homogeneous sequences of non-esterified GalA.

The methyl ester distribution pattern of commercially extracted lemon and apple pectins of high DM was also investigated by Kravtchenko et al.[15]. Highly purified endo-PG was used for the study of the intramolecular distribution of non-esterified GalA. The intermolecular distribution of these pectins was independently determined by both anion-exchange and size exclusion chromatography[16,17]. The latter revealed a non-homogeneous intermolecular methyl ester distribution, which mostly resembled that of a mixture of pectic polymers with blocks of non-methyl esterified GalA of various sizes. High-performance anion-exchange chromatography (HPAEC) of the endo-PG degradation products of the commercial lemon and apple pectins enabled the separation and identification of oligomers ranging from DP 1 to 17. Because of the high pH employed during HPAEC separation, the information regarding the methyl ester content of the oligomers produced was lost[15]. Computer simulation studies revealed that the amount of DP 1-3 oligomers produced after endo-PG degradation of the commercial pectins was much higher than the amount expected for random methyl esterified pectins of identical DM. The amounts of DP 4-17 oligomers found were lower than the amounts expected for random esterified pectins[15]. These results again confirmed the non-random methyl-ester distribution of commercially extracted lemon and apple pectin. In general, it was concluded that the commercial lemon and apple pectins investigated contained at least part of the non-esterified carboxylic acids grouped in blocks: sequences of non-esterified GalA residues[18]. Though very exact data was obtained on the size of the oligomers produced, the unavoidable loss of methyl groups during HPAEC analysis at high pH and the insufficient possibility to quantify higher oligouronides, resulted in the concomitant loss of important information.

3. New method for the determination of the methyl ester distribution

Detection of methyl esterified galacturonides

One important breakthrough of the last years was the development of HPAEC at high pH followed by PAD-detection[19] for the analysis of GalA oligomers[15,19,20]. This separation technique has been adapted to elution conditions at lower pH (5-7) to increase the number of oligomers separated[21] or to improve the separation between pectin- and pectate lyase degradation products[22]. This procedure was adapted by us for the examination of the composition of methyl esterified oligomers[23]. Post-column alkali addition was used to enable PAD-detection. When this method was used to separate an endo-PG digest of a low methoxyl (DM = 30%) pectin at pH 5, the complex elution pattern of Figure 2 (line a) was obtained. In addition to a series of non-esterified GalA oligomers (DP 1, 2, 3, and some 4) other compounds were also eluted in a regular fashion. The latter were identified as methyl-esterified oligomers after fractionation by HPAEC (in absence of post-column detection and alkali addition) and analysis with Matrix Assisted Laser Desorption / Ionisation Time of Flight Mass Spectroscopy (Maldi-Tof MS)[20]. An example of the Maldi Tof mass spectrum of the PG-digest of a DM 30 pectin is shown in figure 3. The various methyl esterified oligomers can easily be recognised.

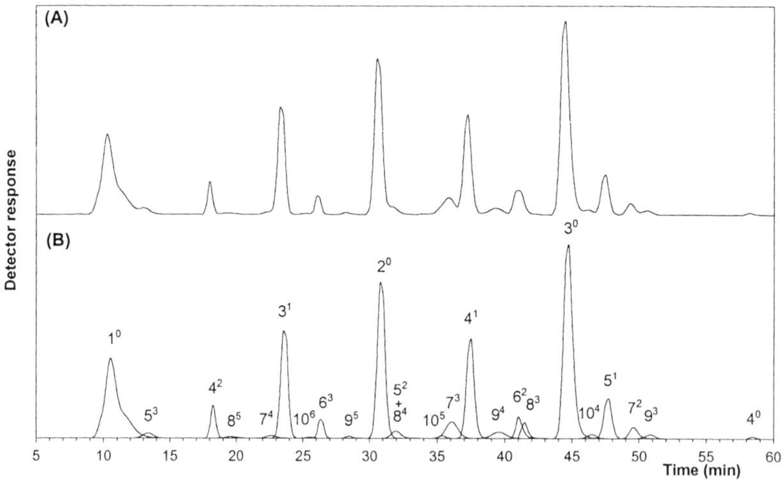

Figure 2. Separation of an endo-PG digest of a DM 30 pectin with HPAEC at pH 5. Elution profiles before (A) and after integration (B) are shown. The arabic number indicates the DP. The number in superscript denote the number of methyl esters.

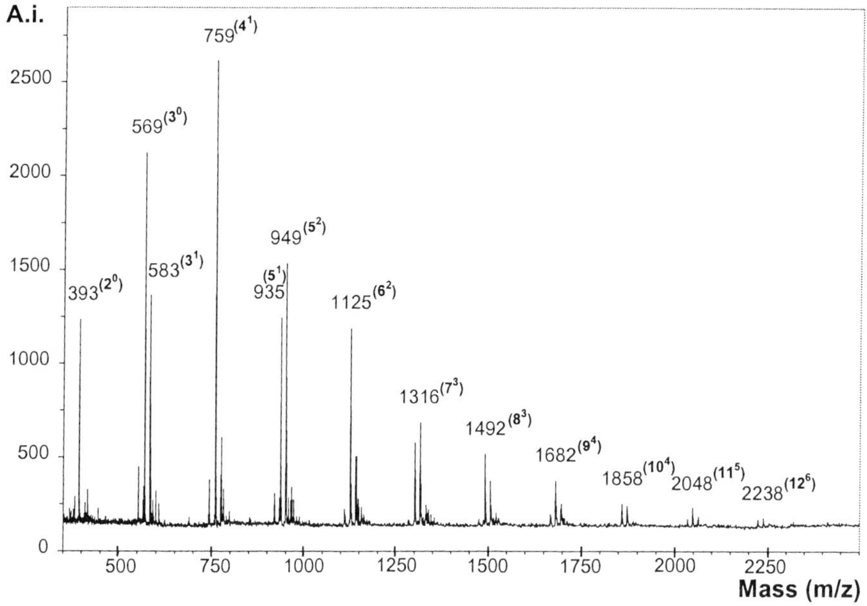

Figure 3. Positive-ion matrix assisted laser desorption/ionisation time of flight (MALDI-TOF) mass spectrum of the endo-polygalacturonase digest of the DM 30 pectin with 2,5-dihydroxy-benzoic acid as the matrix. The masses of a selected number of peaks are shown. The numbers in parenthesis display the composition of the corresponding monosodiated galacturonic acid oligomers. The arabic number indicates the DP of the oligomer, whereas the number in superscript denotes the amount of methyl esters present. A.i., accumulated intensity.

The peaks corresponding to the compounds detected are listed in the resolved elution profile of Figure 2 (line b). Almost all peaks could be assigned to specific galacturonides[23,24]. With the two techniques, methyl-esterified GalA oligomers can be easily detected even when present in complex mixtures. Maldi-TOF-MS can also be used for the identification of the oligomers produced by PL and PAL[25,26].

Unfortunately, quantification of these components is not yet possible with the MS-technique and forms a topic for future research. With HPAEC, the oligomers can be quantified after determination of the PAD-response factors[24]. This enabled the exact determination of specific structural features related to the methyl-ester distribution of pectin by using enzymes. Recently, techniques have become available that are able to differentiate the various isomers of partially methyl-esterified GalA oligomers[25,26,27]. In this way, the exact location of the methyl-ester(s) within a GalA oligomer can be determined. The use of Maldi-TOF MS post-source decay fragmentation for the precise determination of the position of the methyl ester has been demonstrated[26,28].

Non-esterified regions in pectin

For the study of the methyl ester distribution, commercially extracted lemon pectins of varying DMs (17-74%) as well as PE de-esterified (DMs 56-71%), highly esterified

(DM 85 and 93%), and random esterified (DMs 30-70%) pectins and polygalacturonic acid were used. Highly esterified pectins were prepared by methanol/sulfuric acid treatment, whereas the random esterified pectins were prepared by chemical de-esterification[29]. Endo-PG of Kluyveromyces fragilis was used to extensively degrade the pectic polymers. HPAEC analysis at pH 5 was employed to identify the components formed. With the exception of the DM 93% pectin digest, non-esterified mono-, di-, and tri-GalA were always formed as indicated by HPAEC analysis at pH 5. Non-esterified tetra-GalA was never observed in DM ~35% digests and only in small amounts in pectin digests of lower DM.

Upon analysis of pectins that were sequentially de-esterified with tomato PE, we found that large amounts of non-esterified mono-, di-, and tri-GalA were released. The more methyl esters removed with PE, the more non-esterified residues liberated[29]. Random methyl esterified pectins liberated the lowest amounts of non-esterified GalA, even when the DM was relatively low. These observations can easily be explained through the known mode of action of the endo- PG used as shown in figure 4. In general it could be concluded that the absolute amount of non-esterified mono-, di-, and tri-GalA residues produced after extended endo-PG degradation of pectin indicates the occurrence of sequences of non-esterified GalA (so-called blocks) in pectin. The more of these blocks present and/or the larger the average size of these blocks, the more non-esterified residues formed.

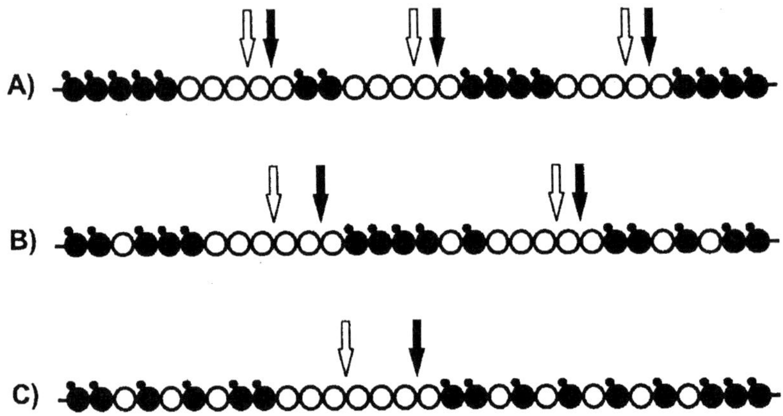

Figure 4. Schematic representation of endo-PG action on three DM 50 pectins with different methyl ester distributions. Methyl- and non-esterified GalA are represented by black and open circles. White and black arrows indicate the first and second linkage split, respectively. Here, it is assumed that endo-PG needs 4 adjacent none-esterified GalA residues to act.

In Figure 5, the HPAEC elution profiles of endo-PG digests of three DM ~67% pectins are shown that release different amounts of non-esterified GalA; indicating a clearly different methyl ester distribution.

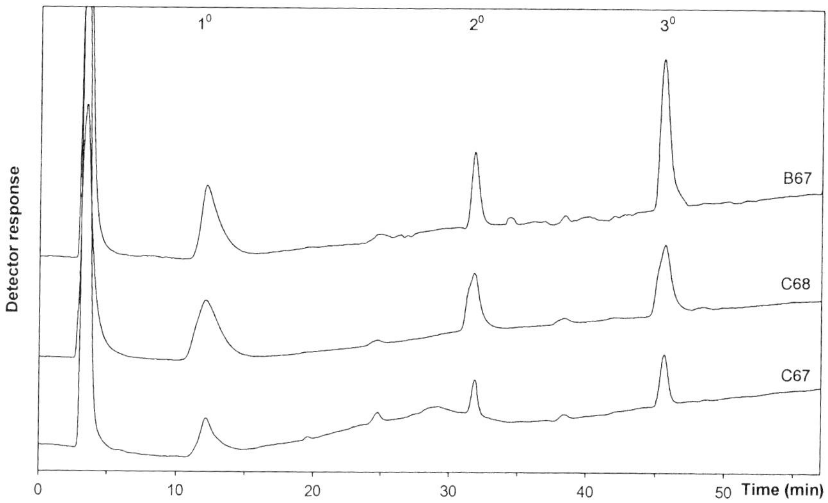

Figure 5. Three non-identical pH 5 HPAEC elution patterns of pectin endo-PG digests of nearly identical DM (~67%). The elution positions of non-esterified mono-, di-, and tri-GalA are shown. B, blockwise (PE) de-esterified; C, commercially extracted.

For the total amount of non-esterified GalA liberated, expressed as the percentage of the total number of non-esterified GalA residues present in pectin, the term "degree of blockiness" (DB) was introduced[29]. In Figure 6, the DBs found for pectins of various DM are shown. In this figure, the results of similar series of pectins are connected with lines to better illustrate the information obtained. Pectins having the same DM, but a different DB have a non-identical methyl-ester distribution. The pectin of high DB will have more and/or larger regions of non-esterified residues present. With random methyl-esterified pectins as a reference, the DB readily reveals to what extent the overall methyl-ester distribution differs from a random one.

The DB, however, does not discriminate between the methyl-ester distribution of two pectins (of similar DM) of which one contains a few large and the other more but smaller non-esterified blocks if both release the same amount of non-esterified GalA residues upon enzyme degradation.

Size of the non-esterified blocks

Even if both the DM and DB were nearly identical, some pectin digests gave dissimilar elution patterns. In Figure 7, the chromatograms of the endo-PG digest of two blockwise de-esterified pectins of DM 56.4 and 57.1% are shown. Both digests contain large amounts of non-esterified mono-, di-, and tri-GalA. Besides the DM, the total amount of GalA residues liberated is also nearly identical; resulting in DBs of 32.7 and 34.4%, respectively.

However, from Figure 7, it is clear that the ratio at which the non-esterified mono-, di-, and tri-GalA molecules were produced differed greatly. Varying ratios of mono-, di-, and tri-GalA were also discerned in a lot of the other pectin digests.

When the individual amounts of mono-, di-, and tri-GalA were expressed as the percentage of the total number of non-esterified molecules released clear differences were observed. These differences can best be viewed when plotted in a ternary graph (Figure 8). In this figure, the data for the random esterified samples (R70-R32 and polygalacturonic acid) are connected with lines to better illustrate the information revealed. Overall, a gradual increase in the relative occurrence of mono-, di-, and tri-GalA can be observed for random esterified pectins of decreasing DM. The higher the methyl ester content, the larger the relative percentage of mono- and di-GalA released by the enzyme. For random esterified pectins, it can be calculated that the average size and occurrence of non-esterified blocks decreases with increasing DM39. As a result, endo-PG will predominantly have to degrade blocks close to the limit of its action in high DM, random esterified pectins (e.g. R70). For the latter this will result in the release of (relative) a lot of small (di- and mono-)GalA molecules. During the degradation of polygalacturonic acid, endo-PG is not hindered by methyl esters and produces mono-, di-, and tri-GalA in a ratio strictly determined by the mode of action of the enzyme.

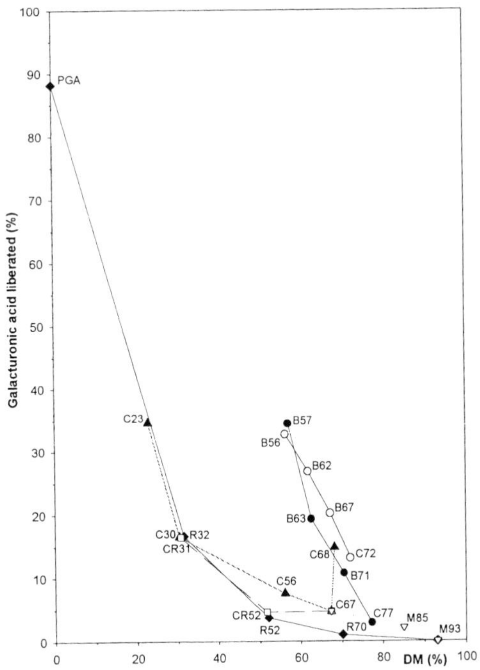

Figure 6. Percentage of non-esterified GalA residues liberated by endo-PG versus the DM of pectin. Random esterified (M93,R70,R32 and PGA; ◆), alkaline de-esterified (C67,CR52, CR32; --◊--), commercial extracted (C68- C23; ▲), highly esterified (C67,M85, M93; ··∇··), and PE de-esterified (C77,B71,B63,B57; ●) and C72 (C72,B67,B62,B56; O) pectins are shown. CR, commercial random de-esterified; M, additionally esterified; PGA, polygalacturonic acid; R, random esterified.

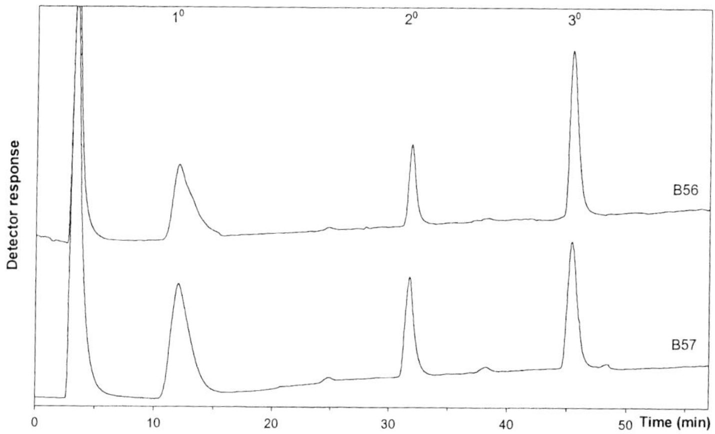

Figure 7 HPAEC pH 5 elution patterns of B56 and B57 pectin endo-PG digests.

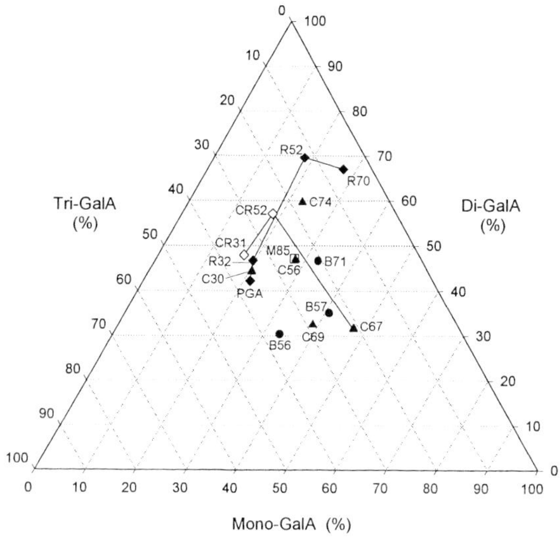

Figure 8. Ternary graph of the normalized percentages of non-esterified mono-, di-, and tri-GalA observed in pectin endo-PG digests. Commercial (C69-C30; ▲), PE de-esterified (B71-B56; ●) and highly esterified M85 (□) pectins are shown. The random esterified (R70-R32, PGA; ◆) and commercial de-esterified (C67, CR52-CR32; --◊--) pectin series are connected with lines (see text).

In Figure 8, it can be observed that this "preferred" ratio greatly differs from the ratio observed in most of the pectin digests and this ratio seems to be indicative for the presence of short blocks of non-esterified galA sequences in (HM) pectins. The smaller the average size of the "endo-PG degradable" blocks, the smaller the GalA molecules produced. Only during the degradation of DM 30 pectins (C30, CR31, and R32) endo-PG is apparently not hindered by any of the substituents.

Distribution of non-esterified blocks

Endo-PG action on pectins of similar DM, DB, and non-esterified mono-, di-, and tri-GalA ratio could still result in non-identical elution profiles. This is illustrated in the chromatograms of the R32 and C30 pectin digests of Figure 9. Here, for example, clear differences in the areas of the peaks corresponding to mono-esterified trimer and mono-esterified tetramer can be observed. In the C30 digest higher amounts of these components are present. Because the methyl-esterified GalA oligomers were not available in sufficient quantities, no PAD-response factors could be determined. Therefore, no exact quantification of the oligomers was possible so peak areas were used instead. As a means to express the (relative) amount of all esterified components liberated, the area of all peaks corresponding to methyl-esterified oligomers were summarized and divided by the total area of the non-esterified GalA peaks[29]. In Figure 10, the ratios obtained are plotted versus the DM of the corresponding pectins. For all digests, the methyl- and non-esterified peak area ratio was highest for random esterified pectins. The information indicated by the methyl- and non-esterified peak area ratio is again best explained by the results of the random esterified pectin series. As expected, the methyl- vs. non-esterified area ratio will be high for pectins releasing large amounts of methyl-esterified oligomers and small numbers of non-esterified GalAs. For random esterified pectins the occurrence of "endo-PG degradable" blocks will increase with decreasing DM whereas, simultaneously, the chance of liberating a methyl esterified oligomer after endo-PG action gradually decreases. As a result, starting from a highly, random esterified pectin, the ratio of methyl- vs. non- esterified GalA production will increase up to a specific DM and then gradually decrease. From Figure 10, it can be concluded that this transition occurs at a DM of around 50%.

With the HPAEC method employed, esterified oligomers up to DP 10 can be detected[24]. As a result of this and the composition of the active site of the enzyme, "endo-PG degradable" blocks located at distances of more than a certain number of GalA residues (>10) will not release any detectable methyl-esterified oligomers. Endo-PG digests of pectins with a lot of closely located blocks (so-called clustered blocks), in contrast, will contain large amounts of detectable esterified oligomers. Hence, a high ratio of the total methyl-esterified peak area versus the total non-esterified peak area is indicative for the occurrence of blocks located at distance closer than ~6-10 GalA residues[29]. The data in Figure 11 demonstrates that all of the commercial lemon HM pectins measured contain almost no clustered blocks. Random esterified (HM) pectins contain considerable amounts of such blocks.

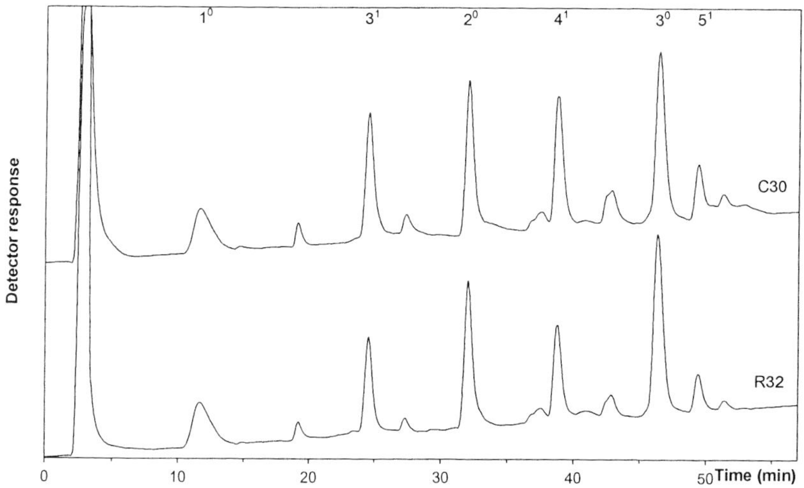

Figure 9. Non-identical pH 5 HPAEC chromatograms of C30 and R32 pectin endo-PG digests. Only the peaks corresponding to non- and mono-methyl-esterified GalA are indicated.

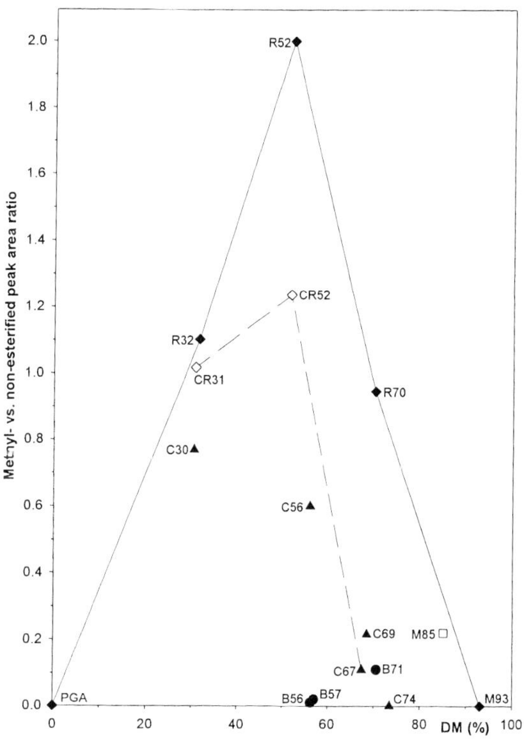

Figure 10. Ratio of the methyl- versus the non-esterified GalA peak areas of pectin endo-PG digests plotted versus the DM. Commercial (C69-C30; ▲), PE de-esterified (B71-B56; ●) and highly esterified M85 (□) pectins are shown. The random esterified (R70-R32, PGA; ◆) and alkaline de-esterified (C67, CR52-CR32; --◊--) pectin series are connected with lines as explained in the text..

Combined results

With the enzymatic approach developed, three characteristics indicative for the distribution of non-esterified GalA residues in pectin are obtained:

i) the degree of blockiness[24],
ii) the normalized percentages of non-esterified mono-, di-, and tri-GalA liberated[29],
iii) the ratio of the total methyl-esterified peak versus the non-esterified peak area[29].

These three characteristics all indicate a typical aspect of the non-esterified GalA distribution. By combining these aspects -including thc DM- a better and more complete overview of the GalA distribution in pectins can be obtained. Computational techniques commonly employed in the determination of the sequence similarity of DNA and proteins were used to discriminate the various type of distributions found for pectins and construct a distance tree[30]. In Figure 11, the distance tree obtained for the non-esterified sequence similarity data is shown. In the tree, the more closely the GalA distribution resemble one another, the smaller the distance between the pectins. In general, three types of GalA distributions can be discerned: A) random, B) high esterified, and C) blockwise esterified. Group A is almost solely composed of chemically modified pectins. These pectins all have DMs of 30 to 70 and quite high methyl vs. non-methyl esterified peak area ratios (0.6), indicative for a random or close to random methyl ester distribution. Group B is solely composed of highly esterified pectins of DMs 74. A typical feature of these polymers is the presence of very small non-esterified blocks. Group C pectins have, from left to right, increasing amounts and/or sizes of non-esterified blocks. This group also includes polygalacturonic acid. Preliminairy results indicate that within these groups even some subgroups might be distinguished. Since the distribution of the methyl esters of pectin is very important for its the functional properties, it is highly conceivable that the three general types of distributions found could be correlated to one or more of the typical applications of pectin[1,4,30].

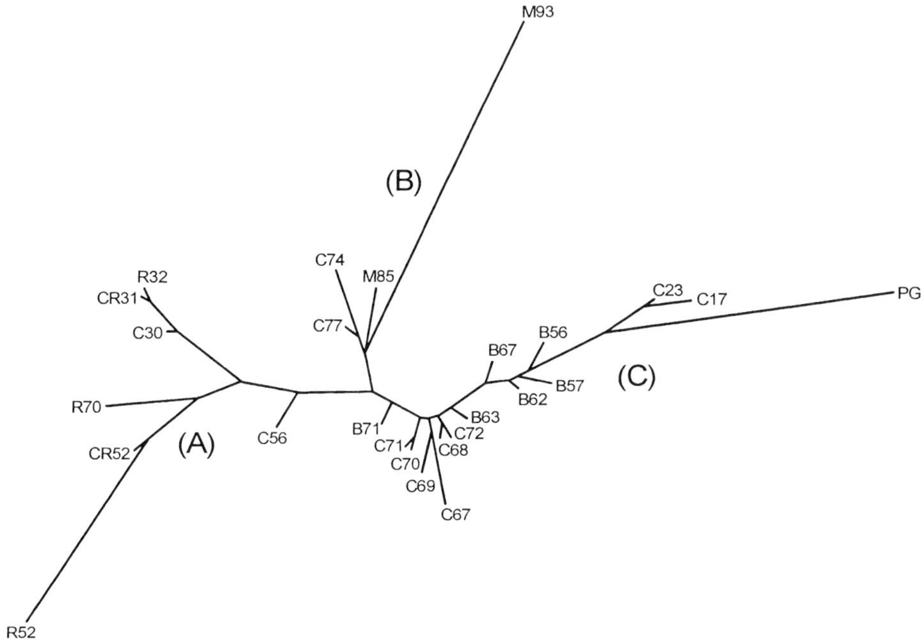

Figure 11. Non-rooted tree showing the similarity of the methyl ester distribution in pectin. The letters in parenthesis denote the three groups discerned (see text).

4. Application of the new pectin characteristics in real life tomato samples

The pectins in some commercial tomato samples were characterised using the analyses described above. These new characteristics enabled us to discriminate the pectins in the samples. The water-soluble tomato pectins (WE) and chelator-soluble tomato pectins (ChSS) were extracted from the tomato products and analysed.

The ripeness of the tomatoes picked to make tomato paste appeared to be of importance for the measured pectin characteristics.

Table 1. Changes in tomato pectin characteristics during ripening.

Sample	DM (%)	DB (%)	Ratio methyl-esterified peak area:non-esterified peak area
WE green paste	62	0	-
ChSS green paste	56	6	0
WE red paste	61	n.d.	n.d.
ChSS red paste	40	35	0.07

During ripening, the DM of the water-soluble pectins does not change, but the DM of the chelator-soluble tomato pectins decreases. This decrease in DM is accompanied by an increase in the DB from 6 to 35%. This can be explained by the blockwise release of methyl groups by endogenous PME in the ripening tomato.

To see if processing of the tomato paste changes the characteristics of the tomato pectin, the original paste, the end product, and two intermediates were fractionated and analysed. The results are shown in Table 2.

Table 2. Changes in chelator-soluble tomato pectin characteristics during processing.

Sample	DM (%)	DB (%)	Ratio monoGalA:diGalA:triGalA	Ratio methyl-esterified peak area:non-esterified peak area
Tomato paste	30	36	8:16:76	0.15
Intermediate 1	31	31	6:15:79	0.14
Intermediate 2	24	34	8:14:78	0.09
End product	23	43	9:15:76	0.14

During processing of the paste, the DM decreases from 30 to 23%. So during processing methyl groups are released from the pectin. This is accompanied by an increase of the DB from 31% in intermediate 1 to 43% in the end product. It can be concluded that the way of removing the methyl groups between intermediate 1 and 2 is different from the way between intermediate 2 and the end product. The size of the non-esterified blocks and their distribution over the pectin is not affected by processing. With the new pectin characteristics we are able to address these differences in the way methyl groups are released. With measuring only DM this would never have been noticed.

5. Conclusions

It can be concluded that as a result of the development of two techniques capable of detecting methyl-esterified GalA oligomers, great progress has been made in the study of the chemical fine structure of pectin. As a result of this, a vast amount of information has become available with respect to the precise distribution of methyl esters over the homogalacturonan backbone of pectin. In has been shown that the same techniques and approaches developed can also be applied to study the physicochemical and catalytic properties of pectic enzymes resolving their mode of action in much more detail. The other way around, the increased availability of (mutants of) pectic enzymes (PG, PL, etc), having a modified but known mode of action, will even strengthen enzymatic approaches to reveal complex methyl ester distributions of pectins. Such knowledge is essential for understanding the relation between chemical structure of pectins and their functional properties. However, it is also realised that the examples on methyl ester distributions given above is based on complex mixtures of pectin molecules and we have not yet dealt with the inter- and intramolecular distribution of methyl esters. At the moment, part of our research is directed to address such questions. As mentioned before, anion-exchange chromatography prior to determination of the degree of blockiness and other parameters could be quite helpful. Recently, the use of capillary electrophoresis as a tool for the characterisation has been demonstrated nicely [31,32,33,34,35]. An advantage of CE could be that, due to the great resolution of this technique, partially methyl-esterified oligo-uronide as described above could be separated in the same run as the high-molecular weigth pectins which are not being degraded by the enzyme[34,35]. Additional information may also be obtained through a chemical approach (with or without a pre-fractionation step) like conversion of methyl esterified galacturonic acids to galactose followed by reduction with sodium borohydride and cleavage of the glycosidic linkages of the resulting galactose residues specifically with liquid HF solvolysis[36]. More recently, a specific degradation of pectins through a carbodiimide-mediated Lossen rearrangement of methyl esterified galacturonic acid residues have been reported by Needs et al[37]. Combination of all strategies mentioned in this contribution, together with new methods to be developed should enable us to obtain more insights in the complex methyl ester distribution of pectins and their effect on pectins functionality in the near future.

References

1. Bacic, P.J. Harris, and B.A. Stone. In: *The Biochemistry of Plants* ***14***, *Carbohydrates.* J. Preiss (Ed.), Academic Press, London, 1988, 297-371.
2. A.G.J. Voragen, W. Pilnik, J-F. Thibault. M.A.V. Axelos, and C.M.G.C. Renard. In: *Food polysaccharides and their applications.* A.M. Stephen (Ed.), Marcel Dekker Inc., New York, 1995, 287-339.
3. B.L. Ridley, M.A. O'Neill, D. Mohnen. *Phytochemistry 57* (2001) 927-967.
4. C.D. May, In: *Handbook of hydrocolloids.* G.O. Phillips and P.A .Williams (Ed.) Woodhead Publishing Limited, Cambridge, 2000, 169-188.
5. Thibault
6. H.A. Schols and A.G.J. Voragen. In: *Progress in Biotechnology* ***14****: Pectins and pectinases.* J. Visser and A.G.J. Voragen (Ed.), Elsevier, Amsterdam, 1996, 3-19.
7. J.A. De Vries, F.M. Rombouts, A.G.J. Voragen, and W. Pilnik. *Carbohydr. Polym.* **3** (1983) 245-258.
8. J.A. De Vries, F.M. Rombouts, A.G.J. Voragen, and W. Pilnik. *Carbohydr. Polym.* **4** (1983) 89-101.
9. A.J. Taylor *Carbohydr. Polym.* **2** (1982) 9-17.
10. C.E. Tuerena, A.J. Taylor, and J.R. Mitchell. *Carbohydr. Polym.* **2** (1981) 193-203.
11. C.E. Tuerena, A.J. Taylor, and J.R. Mitchell. *J. Sci. Food Agric.* **35** (1984) 797-804.
12. J.A. De Vries, M. Hansen, J. Spderberg, P.E. Glahn, and J.K. Pedersen. *Carbohydr. Polym.* **6** (1986) 165-176.
13. Glahn and C. Rolin. In: *Gums and Stabilizers for the Food Industry* ***8***. G.O. Philips, D.J. Wedlock, and P.A. Williams (Ed.), IRL Press, Oxford, 1996, 393-402.
14. H. Kiyohara, J.-C. Cyong, and H. Yamada. *Carbohydr. Res.* **182** (1988) 259-275.
15. T.P. Kravtchenko, M. Penci, A.G.J. Voragen, and W. Pilnik. *Carbohydr. Polym.* **20** (1993) 195-205.
16. T.P. Kravtchenko, A.G.J. Voragen, and W. Pilnik. *Carbohydr. Polym.* **19** (1992) 115-124.
17. T.P. Kravtchenko, A.G.J. Voragen and W. Pilnik. *Carbohydr. Polym.* **18** (1992) 253-263.
18. T.P. Kravtchenko. Studies on the structure of industrial high methoxyl pectins. Thesis Wageningen Agricultural University, 1992.
19. Y.C.Lee. *J. Chromatogr. A* , **720** (1996) 137-149.
20. P.J.H. Daas, J.W.M. van Alebeek, A.G.J. Voragen and H.A. Schols. In : *Gums and stabelisers for the food industry* ***10***. P.A. Williams and G.O. Phillips (Ed.), Royal Society of Chemistry, Cambridge, 2000, 3-18.

21. A.T. Hotchkiss, Jr. and K.B. Hicks. *Carbohydr. Res.* **247** (1993) 1-7.
22. H.P. Lieker, K.Thielecke, K. Buchholz, and P.J. Reilly. *Carbohydr. Res.* **238** (1993) 307-311.
23. P.J.H. Daas, P.W. Arisz, H.A. Schols, G.A. de Ruiter, and A.G.J. Voragen. *Anal. Biochem.* **257** (1998) 195-202.
24. P.J.H. Daas, K. Meyer-Hansen, H.A. Schols, G.A. De Ruiter, and A.G.J. Voragen. *Carbohydr. Res.*, **318** (1999) 135-145.
25. R. Körner, G. Limberg, J. Dalgaard Mikkelsen, and P. Roepstorff. *J. Mass. Spectrom.* **33** (1998) 836-842.
26. G.J.W.M. van Alebeek, O. Zabotina, G. Beldman, H.A. Schols, and A.G.J. Voragen. *Carbohydr. Polym.* (2000) **43**, 39-46.
27. R. Körner, G. Limberg, T.M.I.E. Christensen, J. Dalgaard Mikkelsen, and P. Roepstorff. *Anal. Chem.* **71** (1999) 1421-1427.
28. G.J.W.M. van Alebeek, O. Zabotina, G. Beldman, H.A. Schols and A.G.J Voragen. *J. Mass Spec.* **35** (2000) 831-840.
29. P.J.H. Daas, A.G.J. Voragen and H.A Schols. *Carbohydr. Res.* **326**, (2000) 120-129.
30. Daas, P.J.H., Boxma, B., Hopman, A.M.C.P., Voragen, A.G.J., Schols, H.A. *Biopolymers,* **58** (2001) 1-8.
31. H.J. Zhong, M.A.K. Williams, R.D. Keenan, D.M. Goodall and C. Rollin. *Carbohydr. Polym.* **32** (1997) 27-32.
32. H.J. Zhong, M.A.K. Williams, D.M. Goodall and M.E. Hansen. *Carbohydr. Res.* **308** (1998) 1-8.
33. M.A.K. Williams, G.M.C. Buffet, T.J. Foster and I.T. Norton. *Carbohydr. Res.* **334** (2001) 243-250.
34. M.A.K. Williams, G.M.C. Buffet and T.J. Foster. In: *Gums and Stabilizers for the Food Industry* **11**. P.A. Williams and G.O. Phillips (Ed.), Royal Society of Chemistry, Cambridge, 2002a, 14-26.
35. M.A.K. Williams, T.J. Foster, H.A. Schols and J.A.E. Benen. In press (2002b)
36. A.J. Mort, F. Qiu and N.O. Maness. *Carbohydr. Res.* 247 (1993) 21-35
37. P.W. Needs, N.M. Rigby, S.G. Ring and A.J. MacDougall. *Carbohydr. Res.* **333** (2001) 47-58

PHYSICO-CHEMICAL PROPERTIES OF PECTINS IN THE CELL WALLS AND AFTER EXTRACTION

J.-F. THIBAULT, M.-C. RALET
Unité de Recherche sur les Polysaccharides, leurs Organisations et Interactions
Institut National de la Recherche Agronomique B.P. 71627
F-44316 Nantes Cedex 03 France
E-mail: thibault@nantes.inra.fr

Abstract

Pectins are present as structural polysaccharides in the middle lamella and the primary cell-walls of higher plants. They are part of our diet either as additives in some food products or as constituants of the raw materials used in the food products. Their physico-chemical properties, within the cell wall matrix or after extraction, are of prime importance both from a functional and a nutritional point of view.
Industrial pectins are extracted from by-products of the fruit juice industry (apple pomace or citrus peels). They are extracted in acidic conditions and chemically modified to give High Methoxy (HM) pectins and Low Methoxy (LM) pectins (some may be also amidated). The properties of these two main types of differ widely. The HM pectins form gels in acidic conditions and in presence of sugars whereas the LM pectins needs calcium for gelation. The gelation mechanisms are directly linked to the conformation of the molecules, to their polyelectrolyte nature, and to the amount and distribution of the substituents (methyl, amid ...). The characteristics of the surrounding solutions (pH, ionic strength, nature of the ions, presence of cosolutes, temperature..) may also play key roles. The application areas of these gels are mainly in the food industries: jams, marmelades, jellies, confectionery...Their stabilizing properties are now more and more used, for instance in acid milk products.
Within the cell wall, the pectins may have another range of important properties in relation to ion-binding and to hydration which depend on numerous parameters. Furthermore, the physical structure of the matrix is also very important and particle size, surface area, porosity have to be considered in the dietary fibre context.

1. Introduction

Pectins are heterogeneous branched polysaccharides arising from the primary cell walls and intercellular regions of higher plants. Pectins are therefore natural constituents of terrestrial plants and form part of our diet:

(i) as components of fruits or vegetables consumed in their traditional presentation;

F. Voragen et al. (eds.), Advances in Pectin and Pectinase Research, 91-105.
© 2003 *Kluwer Academic Publishers. Printed in the Netherlands.*

(ii) as ingredients in food products supplemented in pectin-rich dietary fibre;
(iii) as additives in gel food manufactured products such as jams and jellies.

This last point is of peculiar importance and the remarkable gelling properties of pectins make them key additives for food industries. Pectins and pectin-rich material also exhibit some health benefit. Pectins are resistant to hydrolysis by human digestive enzymes and belong therefore to the dietary fibre class. The physico-chemical properties of pectins are significant from both a functional and a nutritional point of view. Physico-chemical properties that have to be taken into account differ however if we consider pectin as isolated molecules or pectin molecules within the cell-wall matrix. As isolated molecules, pectins exhibit physico-chemical properties (viscosity, gelation, ion-exchange) mainly related to their chemical structural characteristics (Voragen *et al.*, 1995). Pectin-rich cell-walls exhibit peculiar hydration, ion-exchange and rheological properties. Pectins chemical structure surely plays a key role in these physico-chemical properties but the physical structure (surface area, pore size, particle size) of the cell-wall material has also to be taken into account. In this paper, the origin, extraction, production, physico-chemical and functional properties of isolated pectic molecules considered as food additives will be first presented. The physico-chemical properties of pectin-rich cell-walls will then be discussed.

2. Pectins as food additives

The origins of pectin production go back to the practice by preserve manufacturers of making a pectin-rich extract from apple residues, from the juice and cider industries, in order to supplement the gelling power of "difficult" fruits (May, 1990).

2.1. RAW MATERIALS

Two by-products are at present used as raw materials for the production of commercial pectins:

(i) the press cake of apple juice manufacture (apple pomace which contains around 10-15% of dry matter of extractable pectins);
(ii) the peels and rags obtained after the extraction of citrus juice (citrus pomace which contains around 20-30% of dry matter of extractable pectins). Among the citrus, lime and lemon are the most widely used (Rolin *et al.*, 1998).

Citrus peel and apple pomace can be used directly for pectin manufacture, but the major part is dried so that pectin can be produced outside the fruit-harvesting season and/or at another location. Washing is required to eliminate sugars and to minimise thereby caramelisation during drying. Apple pomace must be rapidly dried to inhibit the development of various fungi, producing pectolytic enzymes. Citrus (lime, lemon, orange or grapefruit) peels have to be blanched to avoid de-esterification reactions due to particularly active endogenous pectin methyl-esterases.
Other interesting sources considered for pectin production are sugar-beet pulp, sunflower heads, potato fibre, onion skins, tobacco leaves, wastes from the processing

of tropical fruits and more particularly papaya, mango, coffee and cocoa (Voragen *et al.*, 1995). Sugar-beet pectins, although they exhibit poor gelling properties in the usual conditions (calcium or sugars/acid), are extracted by some producers and currently marketed for their stabilising properties (Rolin *et al.*, 1998).

2.2. EXTRACTION

An overview of pectin production process is shown in Figure 1. Pectins are extracted from citrus peels and apple pomace by water acidified with nitric acid. The extraction conditions (pH, temperature, time) must be optimised in order to provide good yields of material that also has the desired gelling capacity and DM. Typical conditions are in the range of pH 1.5 to 3.0, 60 to 100°C and 0.5 to 6h. The solid/liquid ratio has to be well defined for an efficient liquid/solid separation. Ratios of 1/17 for apple and 1/35 for citrus are often used. The separation of the viscous aqueous pectin extract from the strongly swollen and partly disintegrated plant material can be achieved by a combination of centrifugation and filtration, but this still remains a key problem in pectin manufacture. Water-insoluble materials such as diatomaceous earth may be added as filter aids. Rotary drum vacuum filtration is commonly used (Joye and Luzio, 2000). Efficient filtration requires quite low viscosity and the extract should not contain more than 0.6 to 1% pectin (Rolin *et al.*, 1998). Pectin production is therefore quite energy-consuming.

Extracts are rapidly brought to pH 3 to 4 and temperature is lowered to avoid pectin de-methylation and de-polymerisation. The clarified extract is concentrated to 3-4% pectin content. For the recovery of pectin from the clarified extracts, the aluminium precipitation has been largely replaced by the "environment friendlier" alcoholic (ethanol or isopropanol) precipitation. The precipitate is separated by filtration, washed to remove contaminants (heavy metals, acid, sugars, polyphenols, pigments ...) and finally dried and milled to the desired particle size. During processing, the pectin may undergo an ion-exchange step to put it in the sodium form for ease of use in foodstuffs applications (Joye and Luzio, 2000).

The suspension in alcohol is a very suitable form for changing the characteristics of the extracted pectins, typically the degree of methylation (DM). By acid or alkali treatment in alcohol and at temperature not exceeding 50°C, they can be de-esterified to the desired DM. This treatment can yield HM pectins with DM values in the range of 55-75% or LM pectins with DM values in the range of 20-45%. LM pectins can also be obtained by treatment with ammonia in the alcoholic suspensions. By choosing proper conditions, pectins with various proportions of amide, carboxyl and methyl functional groups can be produced (Figure 2).

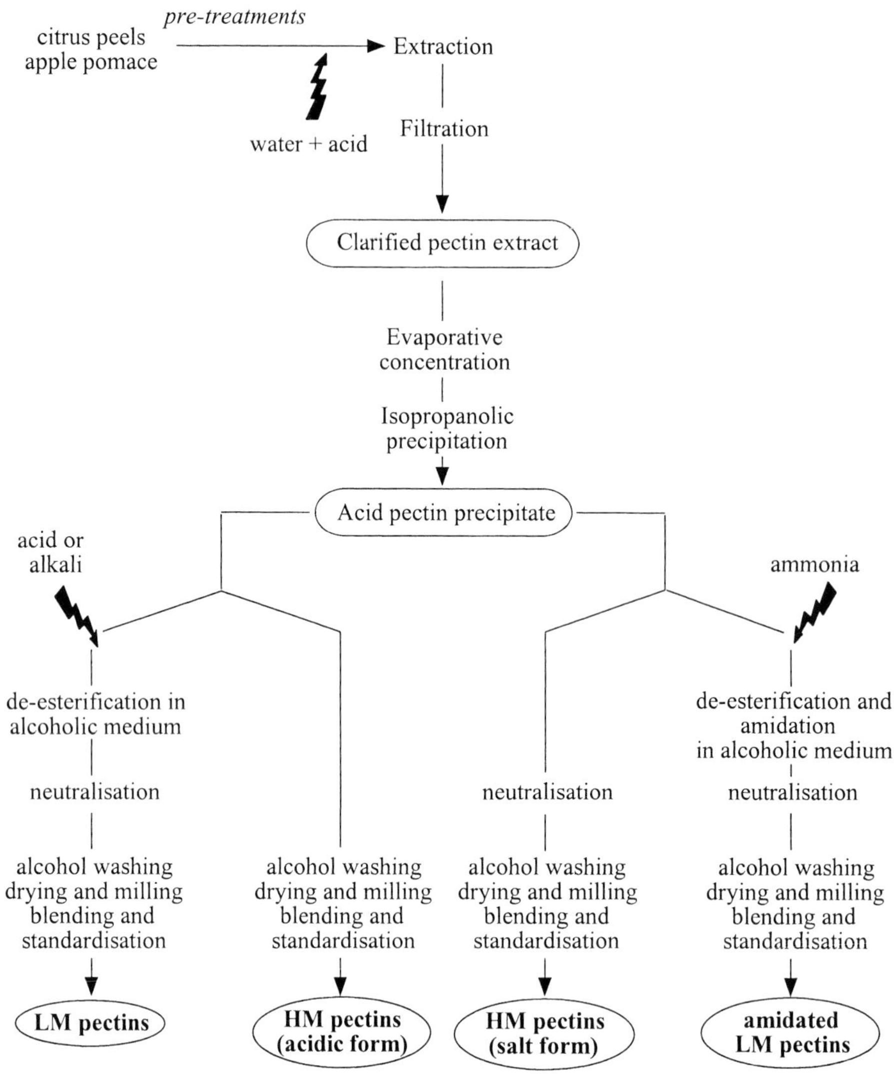

Figure 1. Industrial production of pectins
(HM: high methoxyl; LM: low methoxyl)

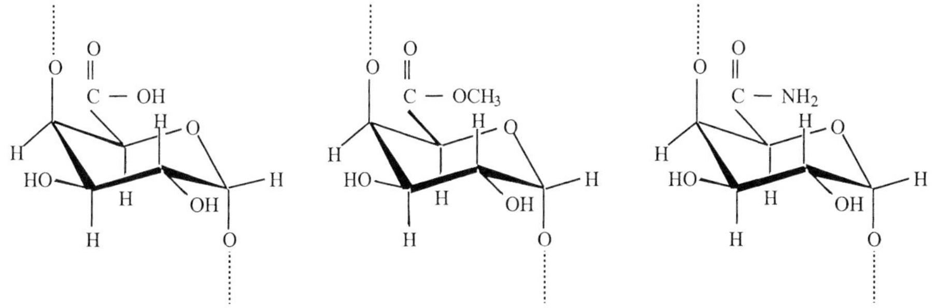

carboxyl methyl amide

Figure 2. Functional groups in pectins

There are large variations in the raw materials (i) because of agro-environmental conditions, which to plants have been exposed, and (ii) because of peel pre-treatment conditions. Pectins are therefore standardised in order to achieve constant and reproducible gel strength. HM pectins are typically standardised by dilution with sucrose. The average composition of the extracted pectins before standardisation is given in Table 1.

Table 1. Typical features of commercial pectins

Galacturonic acid	> 65% (typically 75-80%)
Degree of methylation	30-75
Degree of acetylation	< 5 (except for e.g. sugar-beet pectin)
Degree of amidation	0-25
Neutral sugars	< 15%
Proteins (N x 6.25)	< 5%
Molar mass	100,000-200,000

2.3. PECTIN GELLING PROPERTIES AND APPLICATIONS

The most important physico-chemical property of pectins is their gelling ability. A good solubilisation of pectins is a pre-requisite of the gelation process. Pectins solubility in aqueous media depends on multiple parameters, the main ones being the counterion nature, the ionic strength, the pH and the temperature. Once pectins are solubilised, they may gel through inter-chain associations. The numerous hydroxyl groups and free, methylated or amidated carboxyl functions, can lead to the setting up of hydrogen bonds, hydrophobic interactions or ionic bonds between molecules. The DM value is an important criteria because it governs the gelation conditions: HM pectins (DM > 55%) will gel with at least 55% of soluble solids and at pH 3.5; conventional and amidated LM pectins (DM < 45%) will gel in presence of calcium ions.

2.3.1. *HM pectins*

Gelation of HM pectins requires an acidic medium (pH < 3.5) and a high soluble solids content (> 55%). The high sugar concentration creates conditions of low water activity which promote chain-chain interactions rather than chain-solvent interactions (Rees, 1972). The acid reduces the carboxylic dissociation, thus diminishing electrostatic chain repulsion. The mechanism for HM pectin gelation is thought to be an alignment of molecular helices which are associated to each other by hydrogen bonds between non-dissociated carboxyl and secondary alcohol groups (Morris *et al.*, 1982) and by hydrophobic interactions between methoxyl groups (Oakenfull and Scott, 1984).

The analytical parameter that allows prediction of gelling behaviour and particularly of gel setting time is the DM. This fact has led to a further subdivision of HM pectins based on the setting time and temperature with DM ranging from 77 (for ultra-rapid-set pectins; URS) to 58 (for slow-set pectins; SS) (Table 2). For a given pH, the setting time increases and the setting temperature decreases when decreasing DM.

Table 2. HM pectins classification, gelation conditions and principal uses

DM	Gelation conditions			Commercial name	Principal uses
	General conditions	pH	Setting time (min)		
74-77	Brix > 55	3.1-3.4	1-3	Ultra rapid set	Jams with whole fruits, marmalade, acid milk products
71-74	"	3.0-3.3	3-7	Rapid set	"Classical" jams
66-69	"	2.8-3.1	15-25	Medium rapid set	Acid jams and jellies (raspberry), confectionery
58-65	"	2.6-2.9	30-120	Slow set	Acid to very acid jams and jellies (blackcurrant)

HM pectins require 55-85% sugar and pH 2.5 to 3.8 in order to gel. These requirements limit the possible uses of HM pectins as a gelling agent to sweetened fruit products and about 80% of the world production of HM pectins are used in the manufacture of jams and jellies. Depending on the pH conditions and on the type of pectin, different setting times or setting temperatures can be achieved. For example, the production of jams with whole fruits requires URS pectin in order to ensure a uniform distribution of fruit particles in the continuous jelly phase (Table 2). On the opposite, the production of jams from very acid fruits such as blackcurrant requires the use of SS pectins to avoid pre-gel formation (Table 2). HM pectins are also used within the confectionery industry for making fruit jellies and jelly centres; in fruit juices and fruit drink concentrates as a stabiliser and/or to provide a "natural mouthfeel".

Pectins are increasingly used as stabilisers in acidified milk systems including sour milk products and mixtures of fruit juice and milk (Rolin *et al.*, 1998). Pectins react with caseins and prevent coagulation at acidic pH. At milk's pH (6.8), the casein particles exhibit a net negative charge and they repel each other. When pH is lowered, electrostatic repulsion decreases and casein particles tend to aggregate. Thus, in the absence of stabiliser, high viscosity, whey exsudation and sandy mouthfeel are likely to occur in low-pH milk drinks (Rolin *et al.*, 1998). HM pectins can minimise protein-protein interactions, thereby reducing protein coagulation. This property is explained by the remaining carboxylic functions that are sufficiently dissociated at the pH of acidified milk products (4.0). Pectins will then electrostatically stick to the positive areas of casein particles producing a highly hydrated layer next to the surface of the casein which prevents aggregation by "steric stabilisation" (Parker *et al.*, 1994). Such pectin-stabilised particles are often seen as particles with "hair" or as "fuzzy golfballs" (Figure 3).

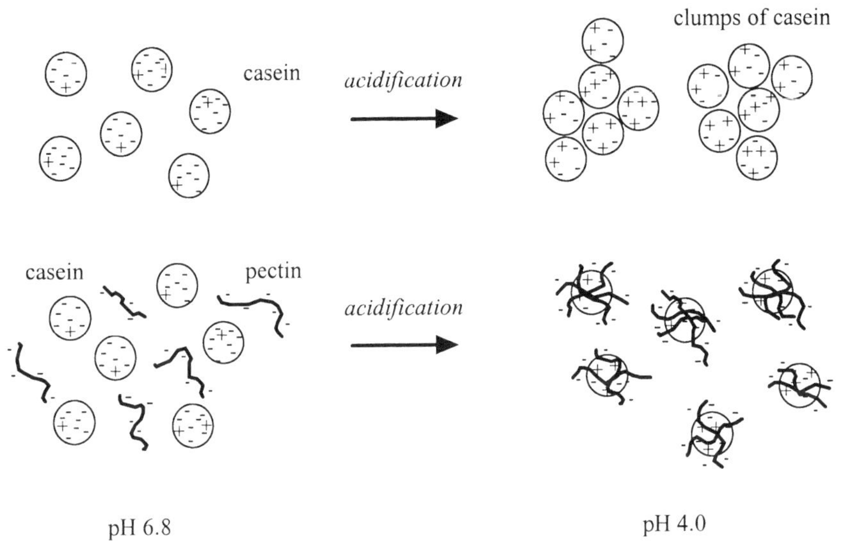

Figure 3. Schematic representation of pectins stabilising functions in acidified milk products

In fact, the stabilisation of casein with pectin is only effective in the pH interval 3.2-4.5 (Rolin *et al.*, 1998). Below pH 3.5, pectin carboxylic functions are not sufficiently dissociated and do not grapple efficiently to casein particles. Above pH 4.5, casein does not possess enough positively charged areas to induce interactions with pectins. The pectin content affects protein stabilisation in a very characteristic way (Leskauskaitė *et al.*, 1998; Rolin *et al.*, 1998). The charge distribution on the pectin molecules could also play a significant role.

2.3.2. LM pectins

LM pectins form gels in the presence of divalent ions (calcium for food purposes (Axelos and Thibault, 1991; Garnier *et al.*, 1994). To achieve gelation, the pH range required is much larger than for HM pectins (2.0 to 6.0), and sugar addition is not necessary. Gelation is due to the formation of junction zones between "smooth" regions of different chains through calcium bridges. Dissociated carboxyl groups as well as hydroxyl groups are believed to play a role in calcium-pectin interaction. The most well known model for Ca^{2+}-LM pectins gelation is the "egg-box" model (Gidley *et al.*, 1979). Other cross-linking mechanisms are however involved. pH influences the texture of the gel; when pH is below 3.5 (pK_a of galacturonic acid), there is a predominance of non-dissociated acid groups which leads to more hydrogen bonding in the gel network. This gives rise to a more rigid, non-shear reversible gel network. When the pH is above 3.5, there is a predominance of ionised acid groups, which favours calcium cross-linking. This leads to the formation of a more spreadable, shear reversible gel network. Soluble solids content also affects the gelling performances of LM pectin; with a given LM pectin, as the soluble solids increases, the calcium requirement decreases and the calcium "bandwidth" becomes more narrow. LM pectin gels are very sensitive to structural parameters which can modify the environment of the carboxyl groups, such as the nature, the amounts and the distribution of substituents along the homogalacturonic backbone (Kohn *et al.*, 1983, Thibault and Rinaudo, 1985; Renard and Jarvis, 1999; Ralet *et al.*, 2001). The gel forming ability increases with decreasing DM; furthermore, LM pectins with a blockwise distribution of free carboxyl groups are very sensitive to calcium (Kohn *et al.*, 1983, Thibault and Rinaudo, 1985; Ralet *et al.*, 2001) (Table 3).

Table 3. Fraction of "free" calcium (determined as calcium transport parameter) for lime pectins with a random or a blockwise distribution of the charges
[*Data from Ralet et al. (2001)*]

DM (± 2)	Calcium transport parameter *Distribution of free carboxyl groups*	
	Random	*Blockwise*
75	0.58	0.54
65	0.47	0.44
58	0.40	0.34
40	0.28	0.21
30	0.25	0.18
20	0.15	0.15

Amidation increases or improves the gelling ability of LM pectins. Amidated pectins need less calcium to gel and are less prone to precipitation by high calcium level (May, 1990). Amide groups are thought to be distributed in a blockwise manner along the pectic chain, and it has been suggested that the gelation of amidated pectins might be reinforced by hydrogen bonding within these blocks (Racape *et al.*, 1989).

The applications of LM pectins are directly based on their peculiar gelling properties. As LM pectins do not require sugars for gelation, they are traditionally used in jams with soluble solids below 55% (low/reduced-calorie jams, jellies preserves and conserves), which is the limit for the use of HM pectin. As for HM pectins, different types of LM pectins can be considered, according to their DM (slow set, medium set and rapid set for DM of 50, 40 and 30, respectively). The type of LM pectin must be carefully selected according to the soluble solids/pH conditions in the application medium. As LM pectins do not require low pH for gelation, they can be used in food products that have to be gelled in non acidic conditions, for instance to preserve aroma. The heat reversibility of LM pectin gels may be utilised in bakery jams and jellies for glazing purposes. LM pectins also find applications in the production of fruit preparations for yoghurt. And finally, because of their calcium reactivity, LM pectins - particularly amidated ones - are well adapted for gelled milk products and in fruit/milk desserts.

3. Pectin-rich cell-walls as dietary fibres

Dietary fibre is defined as the part of foodstuffs which is not digested by the secretions of the human gastrointestinal tract (Eastwood, 1992). Dietary fibre is almost entirely plant-derived and cell-walls constitute the major part of it. Dietary fibres from cereals, wheat bran in particular, are more frequently used than those from fruits and vegetables (Larrauri, 1999). However, fruits and vegetables dietary fibres, and among these those rich in pectins, have some special functional and nutritional interest. The physical characteristics and the hydration, ion-exchange and adsorption properties of the fibres are of peculiar importance:

(i) for their optimal use in food products and in the potential development of a wide range of fibre-enriched products (Endress and Fischer, 2001). Fibre-enriched ingredients can indeed be used only if they can be easily incorporated in the formula and if the final product exhibits good sensory characteristics, regardless of the nutritional benefits of the fibres (Thébaudin *et al.*, 1997);

(ii) for their nutritional impact as physico-chemical characteristics of pectin-rich fibres are related to some nutritional benefits, mainly connected with fermentability (Guillon *et al.*, 1998; Ryden and Robertson, 1998).

3.1. PHYSICAL CHARACTERISTICS

Chemical composition alone cannot adequately reflect the complexity and heterogeneity of the fibre matrix. The three-dimensional network that forms the structure of the cell walls imparts a number of physical characteristics among which the available surface area, the particle size and the pore size and distribution are probably of paramount importance (Chesson, 1998).

Information about surface features is derived from the amount of gas that physically adsorbs, as a monolayer, onto the surface of the cell-wall sample. Surface area values are very scarce. The cell walls of cereal grains generally exhibit a quite low total surface area (2-10 m^2/g) while that of fruits and vegetables may be an order of magnitude greater (Chesson, 1998). Results are however very contradictory ranging from 0.2 to 0.6 m^2/g for citrus and apple commercial fibres (Cloutour, 1995; Robertson *et al.*, 2000) to more than 30 m^2/g for carrot cell walls (Chesson, 1998). Such variations might be ascribed to different modes of preparation of the cell-wall material.

More data can be found concerning particle size of dry and hydrated cell-wall materials. The interest in mean particle size and its distribution arises from the recognition that this physical characteristic plays an important role:

(i) in food processing (Thébaudin *et al.*, 1997) where fibres of small particle size are usually preferred to avoid "granular mouthfeel";

(ii) and in controlling a number of events that occur in the digestive tract. An increased fermentation when decreasing particle size has been demonstrated on compact lignocellulosic materials (Gama et al., 1994; Cloutour, 1995). Data on pectin-rich cell-walls are more contradictory (Cloutour, 1995; Guillon *et al.*, 1998).

In a nutritional context, it is more relevant to measure the particle size of hydrated fibres (Robertson *et al.*, 2000).

Finally, the pore size affects also significantly the physiological properties of dietary fibre in the digestive tract. Plant cell-walls exhibit a distribution of pore size ranging from 0.5 to 5 nm radii (Chesson, 1998). Pore size and distribution of cereal cell-walls is little affected by cooking while that of pectin-rich cell-walls is significantly affected (Chesson, 1998). Several studies have highlighted the role of pectin in controlling pore size distribution in primary cell-walls (Carpita *et al.*, 1979; Baron-Epel *et al.*, 1988; Carpita and Gibeaut, 1993; Guillon *et al.*, 1998). Moreover, the importance of matrix porosity in controlling the fermentation behaviour of fibre has been demonstrated (Guillon *et al.*, 1998).

3.2. HYDRATION PROPERTIES

It is crucial to consider fibre-associated water when examining fibre as an ingredient in a commercial formulation, or when investigating the effect of fibre along the gut. Furthermore, hydration data on fibre could be helpful to understand the behaviour of plant cell-walls as an hydrated complex polymer network.

Table 4. Hydration properties of some cell-wall-rich samples

		Swelling mL/g	Water retention capacity (mL/g)	Water absorption (mL/g)
Pea hull (1)	ϕ 950 μm	9.9 (10.3) [a]	4.3	1.9
	ϕ 490 μm	8.7 (7.9)	4.8	3.0
	ϕ 300 μm	7.8 (6.8)	6.2	2.8
Wheat bran (1)	ϕ 540 μm	9.8 (8.6)	4.3	0.9
Apple fibre (2)	ϕ ~ 80 μm	6.8	6.4 6.3 (pH 6-7)	3.4
Apple pulp (3)	ϕ ~ 133 μm	7.4	5.4	-
Apple cell-wall (2)	ϕ ~ 800 μm	48.4	14.2 26.1 (pH 6-7)	8.2
Micronised apple cell-wall (2)	ϕ ~ 80 μm	21.0	8.1	-
Sugar-beet fibre (1)	ϕ 540 μm	21.5 (19.8)	24.2	8.5
	ϕ 385 μm	21.4 (19.3)	22.6	8.8
	ϕ 205 μm	15.9 (16.3)	19.2	7.3
Sugar-beet pulp	Native (4)	11.0 (10.0)	-	-
	H^+ (5)	17.8 (13.4)	-	-
	Na^+ (5)	32.0 (15.3)	-	-
Sugar-beet fibre (6)	Native	-	11.7	14.9
	Dried at 100°C	-	11.1	12.7
	Depectinised	-	14.6	15.2
	Depectinised and dried at 100°C	-	6.4	6.1

[a] in presence of supporting salts
(1) Auffret *et al.* (1994)
(2) Renard and Thibault (1991)
(3) Robertson *et al.* (2000)
(4) Renard *et al.* (1994)
(5) Dronnet *et al.* (1997)
(6) Guillon *et al.* (1998)

Hydration properties vary widely depending on:

(i) intrinsic factors and mainly chemical composition and particle size. Pectin-rich cell-walls usually exhibit much higher hydration properties than cereal brans cell-walls (Auffret *et al.*, 1994; Robertson *et al.*, 2000). This can be attributed to the high hydrophilicity and the charge of pectic substances (Tibbits *et al.*, 1998) but other characteristics such as the ratio of hydrophilic to hydrophobic regions (Gieringer *et al.*, 1995) and/or cristalline to amorphous regions, have surely to be considered too. A decrease in particle size generally induces lower hydration properties but this effect is very fibre-specific (Auffret *et al.*, 1994). Particle size reduction was associated with decreased hydration properties in wheat bran and pectin-rich fibres, probably because of alterations in regions of fibres that previously bound and held water. On the opposite, cellulose-rich pea hulls exhibited unmodified hydration properties when decreasing particle size.

(ii) The physico-chemical environment. pH, ionic strength of the hydrating solution and its ion composition can influence the hydration properties of fibres containing polyelectrolytes (Auffret *et al.*, 1994; Renard *et al.*, 1994; Dronnet *et al.*, 1997, 1998).

(iii) The fibre preparation, especially the drying process and the chemical modifications applied (Renard and Thibault, 1991; Cloutour, 1995; Guillon *et al.*, 1998).

(iv) The method used to measure hydration properties (Auffret *et al.*, 1994; Robertson *et al.*, 2000). Within the recent "Profibre" EU concerted action group, clear definitions and standards for measurements of hydration properties were major considerations (Robertson *et al.*, 2000). Definitions arising from "Profibre" were: ***swelling***, "the volume occupied by a known weight of fibre" (measured as settled bed volume); ***water retention capacity (WRC)***, "the amount of water retained by a known weight of fibre" (measured by centrifugation); ***water absorption***, the kinetics of water movement under defined conditions (measured using a Baumann apparatus or osmotic pressure/dialysis techniques).

The effects of the fibre-source (pectin-rich cell-walls, cereal brans cell-walls, cellulose-rich pea hulls cell-walls), the particle size, the physico-chemical environment, the "history" of the fibre and the hydration measurement method used on hydration properties are illustrated in Table 4.

3.3. CATION-EXCHANGE CAPACITY

Pectins exhibit a strong affinity for divalent cations. This tough interaction between pectins and divalent cations, Ca^{2+} in particular, is the basis of LM pectins gelation mechanism. The binding of multivalent cations by extracted pectic molecules is quite different from that of pectins *in muro*, when they are embedded in the cell-wall matrix. Pectin-rich cell-walls present indeed all the characteristics of a Donnan system, the Donnan effect producing an excess of ions within the cell-wall compared to the external solution. The swelling properties of pectin-rich cell-walls could be under the strong control of the Donnan effect (Tibbits *et al.*, 1998).

Pectin-rich fibre behave as weak monofunctional cation-exchange resins with a maximum cation-exchange capacity of around 0.5 meq/g. The ion-binding capacity is essentially due to the presence of non-methylesterified galacturonic residues, the other charged groups (other uronic acids, proteins ...) playing only a very minor role. There is indeed a very close agreement between experimental data and theoretical values (Table 5).

Table 5. Experimental and theoretical cation-exchange capacity of pectin-rich fibres

	CEC_{exp} [a]	CEC_{theo} [b]	GalA (mg/g)	DM
Apple cell-wall (1)	0.35	0.34	230	71
Apple fibre (1)	0.38	0.38	148	55
Depectinised apple pomace (1)	0.31	0.27	99	51
Sugar-beet pulp (2)	0.55	0.55	200	52
Saponified sugar-beet pulp (3)	1.13	1.15	205	< 2

(1) Renard and Thibault (1991)
(2) Dronnet *et al.* (1997)
(3) Dronnet *et al.* (1998a)

[a] by titrimetry
[b] calculated as: GalA mmol/g x (100 - DM)/100

As pectin-rich fibre is able to bind metal cations in solution, studies have been carried out on the potential use of pectin-rich by-products, sugar-beet pulp in particular, as natural sorbent for the decontamination of industrial effluents contaminated with toxic metals (Dronnet *et al.*, 1997; 1998a, b). Cross-linked or saponified sugar-beet pulp exhibit an interesting order of selectivity (Cu^{2+} $Pb^{2+} >> Zn^{2+} \sim Cd^{2+} > Ni^{2+} > Ca^{2+}$), a very rapid and a very strong binding of heavy metals.
On a nutritional point of view, fibres, and more particularly pectin-rich fibres, have been suspected of impairing mineral absorption. Although cation-exchange capacity studies of dietary fibre show clear evidence of binding, extrapolation of these effects *in vivo* has not been consistently demonstrated (Cho *et al.*, 1997).

4. Conclusion

The physico-chemical properties of isolated pectins or of pectins within the cell wall relate primarily with gelation, viscosity, hydration and ion-exchange. Both chemical and physical structure have to be taken into account in order to study these properties. The polyelectrolyte behaviour of pectins, and more particularly their peculiar affinity for divalent cations such as Ca^{2+}, is a clue to explain their functional properties. Pectin networks within the cell-wall are exposed to the localised electrical environment and cation fluxes. Ionic interactions within the pectic polysaccharide matrix could affect:

(i) the growth and the mechanical properties of the cell-walls. The amount and length of non-esterified galacturonic acid stretches are differently distributed and therefore, the extent of cross-linking changes within the cell wall. Moreover, the extracellular Ca^{2+} concentration varies widely in plant tissues. Meristems are particularly low in Ca^{2+}, and the amount increases markedly as cells elongate and differentiate (Carpita and Gibeaut, 1993).
(ii) the swelling properties of the cell-wall.
(iii) the porosity of the cell-wall.

The ion binding capacity of pectin-rich cell wall also allows to consider some agro-industrial by-products as natural cation-exchange resins.

5. References

Auffret A., Ralet, M.-C., Guillon, F., Barry, J.-L. and Thibault, J.-F. (1994) Effect of grinding and experimental conditions on the measurement of hydration properties of dietary fibres, *Lebensm. Wiss. und Technol.* **27**, 166-172.

Axelos, M.A.V. and Thibault, J.-F. (1991) The chemistry of low-methoxyl pectin gelation, in: *The Chemistry and Technology of Pectin* (Walter, R.H. Ed.), pp. 109-117. New York: Academic Press.

Baron-Epel, O., Gharyl, P.K. and Schindler, M. (1988) Pectins as mediators of wall porosity in soybean cells, *Planta* **175**, 389-395.

Carpita, N.C. and Gibeaut, D.M. (1993) Structural models of primary cell walls in flowering plants: consistency of molecular structure with the physical properties of the walls during growth, *Plant J.* **3**, 1-30.

Carpita, N.C., Sabularse, D., Montezinos, D. and Delmer, D.P. (1979) Determination of the pore size of cell walls of living plant cells, *Science* **205**, 1144-1147.

Chesson, A. (1998) Cell wall porosity and available surface area: their measurement and significance, in: *Plant Polysaccharides in Human Nutrition: Structure, Function, Digestive Fate and Metabolic Effects* (Guillon, F., Abraham, G., Amado, R. *et al.* Eds.), pp. 2-5. Nantes: Imprimerie Parenthèses, INRA.

Cho, S., DeVries, J.W. and Prosky, L. (1997) Dietary fiber content of foods, in: *Dietary Fiber Analysis and Applications* (gaithersburg, M.D. Ed.), pp. 119-138. AOAC International.

Cloutour, F. (1995) Caractéristiques des fibres alimentaires: influence sur leur fermentation *in vitro* par la flore digestive de l'homme, PhD thesis, University of Nantes, France.

Dronnet, V.M., Renard, C.M.G.C., Axelos, M.A.V. and Thibault, J.-F. (1997) Binding of divalent metal cations by sugar-beet pulp, *Carbohydr. Polym.* **34**, 73-82.

Dronnet, V.M., Axelos, M.A.V, Renard, C.M.G.C. and Thibault, J.-F. (1998a) Improvement of the binding capacity of divalent metal cations by sugar-beet pulp. 1. Impact of cross-linking treatments on composition, hydration and binding properties, *Carbohydr. Polym.* **35**, 29-37.

Dronnet, V.M., Axelos, M.A.V, Renard, C.M.G.C. and Thibault, J.-F. (1998b) Improvement of the binding capacity of divalent metal cations by sugar-beet pulp. 2. Binding of divalent metal cations by modified sugar-beet pulp, *Carbohydr. Polym.* **35**, 239-247.

Eastwood, M.A. (1992) The physiological effects of dietary fibre: an update, *Ann. Rev. Nutr.* **12**, 19-35.

Endress, H.-N. and Fischer, J. (2001) Fibres and fibre blends for individual needs: a physiological and technical approach, in: *Advanced Dietary Fibre Technology* (McCleary, B.V. and Prosky, L. Eds.), pp. 283-297. Oxford: Blackwell Science.

Gama, F.M., Teixeira, J.A. and Mota, M. (1994) Cellulose morphology and enzymatic reactivity: a modified solute exclusion technique, *Biotechnol. Bioeng.* **43**, 381-387.

Garnier, C., Axelos, M.A.V. and Thibault, J.-F. (1994) Selectivity and cooperativity in the binding of calcium ions by pectins, *Carbohydr. Res.* **256**, 71-81.

Gidley, M.J., Morris, E.R., Murray, E.J., Powell, D.A. and Rees, D.A. (1979) Spectroscopic and stoichiometric characterization of the calcium-mediated association of pectate chains in gels and in the solid state, *J. Chem. Soc. Chem. Commun.* **22**, 990-992.

Gieringer, R., Steinert, P., Buttersack, C. and Buchholz, K. (1995) Anisotropic swelling of cell walls of sugar beet tissue: influence of ion-exchange and sucrose, *J. Sci. Food Agric.* **68**, 439-449.

Guillon, F., Auffret, A., Robertson, J.A., Thibault, J.-F. and Barry, J.-L. (1998) Relationships between physical characteristics of sugar-beet fibre and its fermentability by human feacal flora, *Carbohydr. Polym.* **37**, 185-197.

Joye, D.D. and Luzio, G.A. (2000) Process for selective extraction of pectins from plant material by different pH, *Carbohydr. Polym.* **43**, 337-342.

Kohn, R., Markovic, O. and Machova, E. (1983) Deesterification mode of pectin by pectin esterases of *Aspergillus foetidus*, tomatoes and alfalfa, *Collec. Czech. Chem. Commun.* **48**, 790-797.

Larrauri, J.A. (1999) New approaches in the preparation of high dietary fibre powders from fruit by-products, *Trends in Food Sci. and Technol.* **10**, 3-8.

Leskauskaitė, D., Liutkevichius, A. and Valantinaitė, A. (1998) Influence of the level of pectin on the process of protein stabilization in an acidified milk system, *Milchwiss.* **53**, 149-152.

May, C. (1990) Industrial pectins: sources, production and applications, *Carbohydr. Polym.* **12**, 79-99.

Morris, E.R., Powell, D.A., Gidley, M. and Rees, D.A. (1982) Conformations and interactions of pectins I. Polymorphism between gel and solid states of calcium polygalacturonate, *J. Mol. Biol.* **155**, 507-516.

Oakenfull, D. and Scott, A. (1984) Hydrophobic interaction in the gelation of high methoxyl pectins, *J. Food Sci.* **49**, 1093-1098.

Parker, A., Boulenguer, P. and Kravtchenko, T.P. (1994) Effect of the addition of high methoxy pectin on the rheology and colloidal stability of acid milk drinks, in: *Food Hydrocolloids: Structure, Properties and Functions* (Nishinari, K. and Doi, E. Eds.), pp. 307-312. New-York: Plenum Press.

Racapé, E., Thibault, J.-F., Reitsma, J.C.E. and Pilnik, W. (1989) Properties of amidated pectins. II. Polyelectrolyte behavior and calcium binding of amidated pectins and amidated pectic acids, *Biopolymers* **28**, 1435-1448.

Ralet, M.-C., Dronnet, V., Buchholt, H.C. and Thibault, J.-F. (2001) Enzymatically and chemically de-esterified lime pectins: characterisation, polyelectrolyte behaviour and calcium binding properties, *Carbohydr. Res.* **336**, 117-125.

Renard, C.M.G.C., Crépeau, M.-J. and Thibault, J.-F. (1994) Influence of ionic strength, pH and dielectric constant on hydration properties of native and modified fibres from sugar-beet and wheat bran, *Indust. Crops Prod.* **3**, 75-84.

Renard, C.M.G.C. and Jarvis, M.C. (1999) Acetylation and methylation of homogalacturonans. Part II. Effect on ion-binding properties and conformations, *Carbohydr. Polym.* **39**, 209-216.

Renard, C.M.G.C. and Thibault, J.-F. (1991) Composition and physico-chemical properties of apple fibres from fresh fruits and industrial products, *Lebensm. Wiss. und Technol.* **24**, 523-527.

Robertson, J.A., de Monredon, F.D., Dysseler, P., Guillon, F., Amado, R. and Thibault, J.-F. (2000) Hydration properties of dietary fibre and resistant starch: a European collaborative study, *Lebensm. Wiss. und Technol.* **33**, 72-79.

Rolin, C., Nielsen, B.U. and Glahn, P.-E. (1998) Pectin, in: *Polysaccharides, structural diversity and functional versatility* (Dimitriu, S. Ed.), pp. 377-431. London: Marcel Dekker.

Ryden, P. and Robertson, J.A. (1998) Fermentability of defined fibres and the effect of fermentation on their binding properties for MeIQx, in: *Plant Polysaccharides in Human Nutrition: Structure, Function, Digestive Fate and Metabolic Effects* (Guillon, F., Abraham, G., Amado, R. *et al.* Eds.), pp. 20-23. Nantes: Imprimerie Parenthèses, INRA.

Thébaudin, J.Y., Lefebvre, A.C., Harrington, M. and Bourgeois, C.M. (1997) Dietary fibres: nutritional and technological interest, *Trends in Food Sci. and Technol.* **8**, 41-48.

Thibault, J.F. and Rinaudo, M. (1985) Interactions of mono- and divalent counterions with alkali- and enzyme-deesterified pectins in salt-free solutions, *Biopolymers* **24**, 2131-2143.

Tibbits, C.W., MacDougall, A.J. and Ring, S.G. (1998) Calcium binding and swelling behaviour of a high methoxyl pectin gel, *Carbohydr. Res.* **310**, 101-107.

Voragen, A.G.J., Pilnik, W., Thibault, J.-F., Axelos, M.A.V. and Renard, C.M.G.C. (1995) Pectins, in: *Food Polysaccharides* (Stephen, A.M. Ed.), pp. 287-339. London: Marcel Dekker.

A COMPARISON OF LIME AND ORANGE PECTIN WHICH WERE RAPIDLY EXTRACTED FROM ALBEDO.

M.L. FISHMAN. H.K. CHAU, D.R. COFFIN, AND A.T. HOTCHKISS JR.
U.S. Dept. Agriculture, Agricultural Research Service, Eastern Regional Research Center, 600 East Mermaid Lane, Wyndmoor, PA 19038.
mfishman@arserrc.gov

Abstract

Pectin was acid extracted from lime albedo by microwave heating under pressure. Extraction times ranged from 1 to 10 minutes. Solubilized pectin was characterized for molar mass (M), rms (root mean square) radius of gyration (R_g) and intrinsic viscosity (0) by HPSEC with online light scattering and viscosity detection. M, R_g and 0 were relatively constant up to an extraction time of 3 minutes, but decreased rapidly thereafter. At heating times of 3.0 minutes or less, M, R_g and 0 were significantly higher than for a commercial citrus pectin. At a heating time of 2.5 minutes, weight average molar mass (M_w) was 335 H 10^3, Z-average rms radius of gyration (R_{gz}) was 42 nm and weight average intrinsic viscosity (0_w) was 10.5 dL/g. These values were comparable to values for orange albedo pectin which were obtained previously by microwave heating under pressure (Fishman et al., Carbohydr. Res. **323,** 126, 2000). Furthermore gel strength measurements revealed that the gel strength order was lime > orange > commercial citrus pectin.

1. Introduction

In the late eighties, Manabe et al. [1], reported that irradiating mandarin orange pulp immersed in acid solution with microwaves extracted 5% more pectin in 15 minutes than could be extracted by conventional methods in 60 minutes. Furthermore after 10 minutes, when 95% of the pectin has been extracted, microwave extracted pectin had higher relative viscosity, galacturonic acid content and degree of esterification than pectin extracted by conventional heating. The extraction was performed in an open vessel. Subsequently, Kratchanova et al. [2,3]reported that appropriate microwave irradiation of fresh navel orange peels prior to extraction by conventional means gave pectin with somewhat higher galacturonic acid content, degree of esterification and gel strength but more than twice the yield of soluble pectin than peels with no microwave pretreatment. Intrinsic viscosity and viscosity average molar mass were found to be comparable or slightly lower. Recently, we have shown that acid extraction of orange albedo by heating with microwaves in a closed microwave transparent cell permits rapid extraction of pectin under conditions of controlled temperature and pressure [4]. Optimum heating time was between 2.5 and 3.0 minutes. The solvent to albedo

F. Voragen et al. (eds.), Advances in Pectin and Pectinase Research, 107-122.
© 2003 *Kluwer Academic Publishers. Printed in the Netherlands.*

extraction ratio was 25:1 (v/w). At those heating times, weight average molar mass (M_w) was about 1.2 times higher and weight average intrinsic viscosity about 1.5 to 2 times higher than commercial citrus pectin. Degree of esterification of microwave heated pectin was in the range of 91 to 93% whereas galacturonic acid content was 93 to 96%.

Among the common citrus fruits from which pectin is extracted by conventional heating methods, orange is considered the least suitable [5]. In this paper, we compare the molecular properties of orange and lime pectin which have been rapidly extracted with microwave heating to determine the efficacy of the method.

2. Materials and Methods

2.1. ALBEDO

Fresh albedo was obtained from Florida early Valencia oranges [4] and Florida tropical seedless limes. Upon arrival, the flavedo was stripped from the skin with a potato peeler, followed by removal of the albedo with a paring knife. After cutting the albedo into small pieces it was stored at -20EC in sealed polyethylene bags until extraction.

2.2. EXTRACTION

The method of extraction has been described previously [4]. Briefly, microwave heating was performed in a CEM, model MDS-2000 microwave sample preparation system (CEM Corp., Mathews, NC). Samples were irradiated with 630 watts of microwave power at a frequency of 2450 MHz. For each experiment, six equally spaced cells were placed in the sample holder, a rotating carousel. One vessel was equipped with temperature and pressure sensing devices which measured and controlled the temperature and pressure within the cell. Time of irradiation varied between 1 and 10 minutes followed by rapid cooling in a cold water bath to room temperature. The maximum allowed pressure level within the cell was set at 52 ∀ 2 psi and the maximum temperature within the cell was set at 195EC. Experiments were run with HCl at pH 2.0 prior to addition of albedo. Cells were loaded with 1 g of albedo dispersed in 25 mL of acid solution. In one set of experiments for each kind of albedo, the same residue was extracted three times to determine the effect of repeated extraction on pectin structure and yield. Solubilized pectin was precipitated with 70% isopropyl alcohol (IPA), washed once with 70% IPA and once with 100% IPA. Finally, the sample was vacuum dried at room temperature and prepared for chromatography. Percentage of galacturonic acid content [6] and degree of methyl esterification were determined [7] for selected pectin samples.

2.3 CHROMATOGRAPHY

Chromatography has been described elsewhere [4]. A brief description follows. Depending on molar mass, dried sample at a final concentration of 1 or 2 mg/ml was dissolved in 0.05 M $NaNO_3$, stirred until dissolved, centrifuged at 50,000 g for 10 minutes and filtered with a 0.22 or 0.45 :m Millex HV filter (MilliporeCorp., Bedford,

MA). Injection volume was 200 :L. Samples were run in triplicate. The flow rate was 0.70 mL/min. The solvent delivery system consisted of a model 1100 series degasser, pump and auto sampler (Hewlett-Packard Corp.). Pectins were separated by two PL-Aquagel OH-60 and one PL Aquagel OH-40 size exclusion columns in series. Column effluents were detected by a Dawn DSP multi-angle laser light scattering photometer (LS) (Wyatt Technology), model 100 differential pressure viscometer (DPV) (Viscotek Corp.) and an Optilab DSP interferometric refractometer (RI) (Wyatt Technology) in series. The electronic output from the detectors were connected to two personal computers in a manner which permitted data to be collected and processed by ASTRA™ (Wyatt Technology) and TRISEC (Viscotek Corp.) software simultaneously [4]. The dn/dc values in $NaNO_3$ for pectin from lime and orange albedo were measured offline using the HPLC pump, the auto injector and the Optilab refractometer [8]. The value found for lime albedo pectin was 0.130 whereas the value for orange albedo pectin was 0.132.

2.4 GEL STRENGTH DETERMINATION

Gels were prepared by dissolving 0.7 grams of the pectin to be tested in 102.2 grams of a 0.01 M citrate buffer. The pH of the buffer was controlled by the ratio of citric acid and sodium citrate. The beaker containing the pectin solution was placed in a 95-98 EC water bath, and 189.0 grams of sucrose was added with stirring. The mixture was heated for 10 minutes, and then poured into three 45 mL weighing bottles (40 mm o.d. H 50 mm h.) to which tape had been placed around the top to allow the gelling solution to be poured to above the rim of the bottle. The solutions were allowed to cool for 24 hours. The tape was then removed, and the part of the gel above the rim of the bottle was cut off using a cheese cutter.

Gel strengths were determined with a Stable Micro Systems TA-XT2 Texture Analyzer. Probe speed was 1.0 mm/sec and the 0.5 inch diameter Delrin probe was allowed to penetrate to a depth of 15 mm. . The maximum in the force-penetration curve was taken as the gel strength at break. Three replicates were run for each sample.

3. Results and Discussion

Figure 1. contains plots of temperature and pressure above ambient against time in the microwave cell for 1g of fresh lime albedo dispersed in 25 mL of acid solution. The plateau region commences after about four minutes. In this region the temperature and pressure within the cell have reached the maximum pressure set point. When this occurs, the instrument is under the control of the pressure stat which turns the microwave radiation on and off to maintain constant pressure and temperature. If the maximum temperature set point were less than about 140EC, the temperature stat would control the on/off cycles of the microwave radiation. A comparable temperature/pressure/time curve was obtained for orange albedo [4].

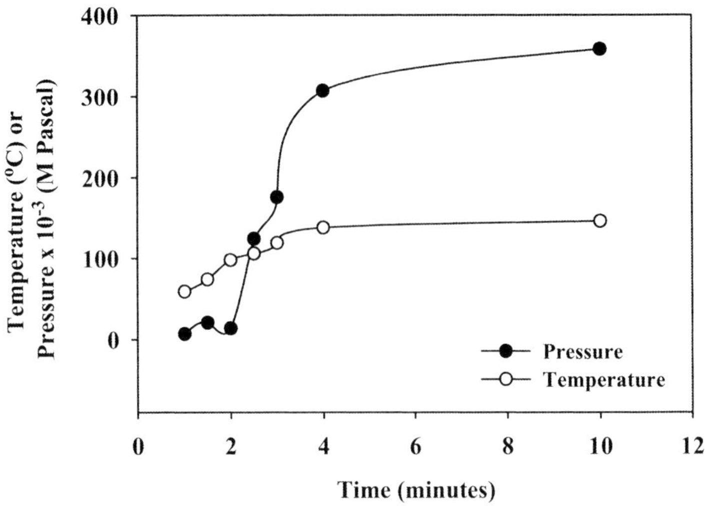

Figure 1. Temperature and pressure as a function of heating time during extraction.

For lime pectin, global values of M_w, R_{gz} and $[0]_w$ are plotted against time of heating in Figure 2 for samples irradiated with microwaves over the range of 1 to 10 minutes. Over the first 2.0 minutes there are relatively small changes in M_w, R_{gz} and $[0]_w$. The value of $[0]_w$ decreases somewhat at 2.5 minutes whereas M_w, and R_{gz} remain relatively constant. At 4 minutes and beyond, all three values have diminished, greatly. Table 1 contains molar mass, size and intrinsic viscosity data for lime and orange [4] albedo pectins. Comparison reveals that values for molecular properties for lime and orange pectins are comparable at 2.5 minutes. For orange pectin at 3 minutes, M_w does not undergo a significant change from 2.5 minutes but R_{gz} and $[0]_w$ decrease. Comparison of lime and orange albedo heated for 3 minutes revealed that M_w of orange pectin was slightly higher but that $[0]_w$ was somewhat lower than lime pectin. At heating times of 4 minutes, values of M_w, R_{gz} and $[0]_w$ are larger for orange pectin than lime pectin. Heating orange albedo for 2 minutes produced insufficient pectin for analysis. Data in Table 2 reveals that lime albedo generally gives greater yields than orange albedo when heated for 2.5 to 4 minutes. It would appear that lime pectin can be extracted more quickly from its albedo than orange pectin can be extracted from its albedo. Thus under certain circumstances, lime pectin could be extracted with a larger value of $[0]_w$ than orange pectin because it could be extracted more rapidly. Nevertheless, orange albedo heated for 3 minutes by microwaves in a closed cell yields pectin with higher value of M_w and $[0]_w$ than commercial citrus pectin (see data in Table 1).

Previously [4], we found that sequential extraction of orange pectin with 2.5 minute heating times and removal of pectin at each step gave pectin with higher values of M_w and $[0]_w$ than pectin from a 6 minute single extraction (c.f. Tables 1 and 3) and improved yield (c.f. Tables 2 and 4). In those experiments, the temperature was allowed to rise to about 110EC and the pressure to rise to about 0.12 M (Pascal) above ambient. We concluded from these experiments that pectin contained in its albedo matrix was more heat stable than pectin dissolved in extracting solvent. Here, we sequentially extracted lime albedo 3 times in 3 minute steps and set the maximum temperature of extraction at 80EC which is lower than the 110EC that would be reached in the cell after 3 minutes if the temperature of the cell were allowed to rise without restraint. In this case (see Table 3), M_w increased with each step and R_{gz} decreased whereas $[0]_w$ remained constant for the first two extraction steps and decreased after the third step.

TABLE 1. Comparison of Molecular Properties of Pectin from Lime and Orange Albedo

	Lime				Orange[a]			
Time[b] min.	M_w $x10^{-3}$	R_{gz} nm	0_w dL/g	a[c]	M_w $x10^{-3}$	R_{gz} nm	0_w dL/g	a[c]
1	335(2)[d]	42(4)	15.2(0.1)	0.78				
1.5	323(1)	43(1)	14.1(0.3)	0.77				
2	372((4)	42(1)	14.3(0.1)	0.73				
2.5	335(8)	42(1)	10.5(0.1)	0.71	394(22)	38(1)	10.8(0.1)	0.71
3	311(10)	38(4)	9.5(0.1)	0.74	373(11)	33(1)	7.7(0.5)	0.63
CCP[e]	254(4)	35(2)	6.1(0.1)	0.64				
4	97(1)	25(1)	3.2(0.1)	0.94	132(11)	35(5)	4.8(0.1)	0.75
5					58(7)	19(1)[f]	1.8(0.2)	0.98
6					61(5)	15(1)[f]	1.4(0.1)	0.99
10	17(1)	7.7(0.2)[f]	0.55(0.02)	1.18				

[a] Data taken from reference 4.
[b] Extraction time
[c] Mark-Houwink exponent.
[d] Standard deviation of triplicate analysis.
[e] Commercial pectin.
[f] Determined by LS/Viscometry method.

TABLE 2. Percentage Recovery of Pectin in the Albedo[a]

Extraction Time (min)	Lime	Orange[b]
1	4.1	
1.5	6.4	
2	10.5	
2.5	12	4.7
3	16.1	12.4
4	12.7	9.1
5		12.4
6		12.4
10	6.8	

[a] Gram pectin extracted per 100 g albedo.
[b] Data taken from reference 4.

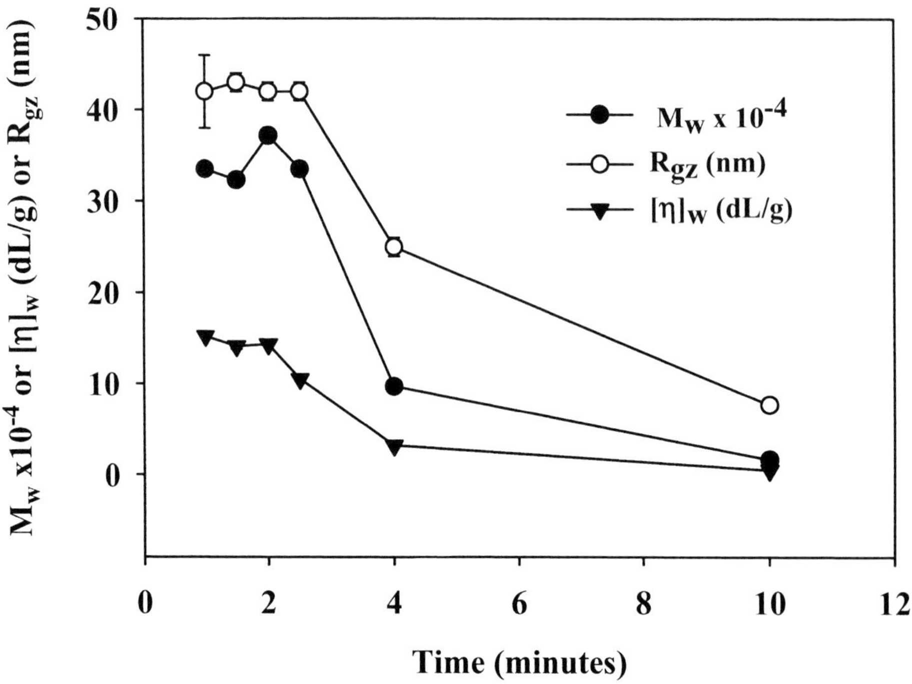

Figure 2. Effect of heating on lime pectin properties.

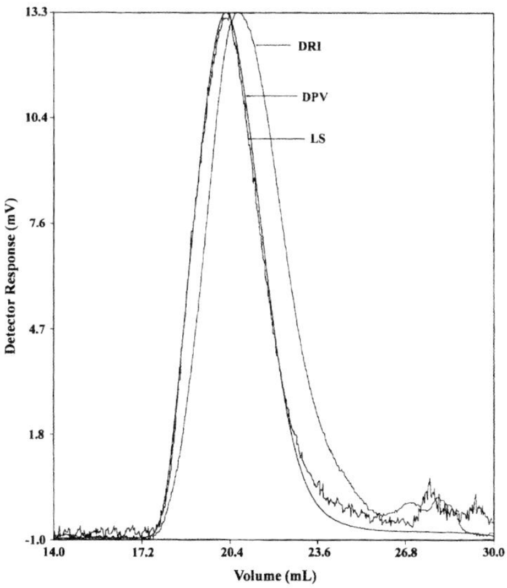

Figure 3. Chromatograms for lime pectin extracted after 3 minutes of microwave heating. DRI is trace from differential refractometer detector, DPV is trace from differential pressure viscometer detector, and LS is trace from light scattering detector.

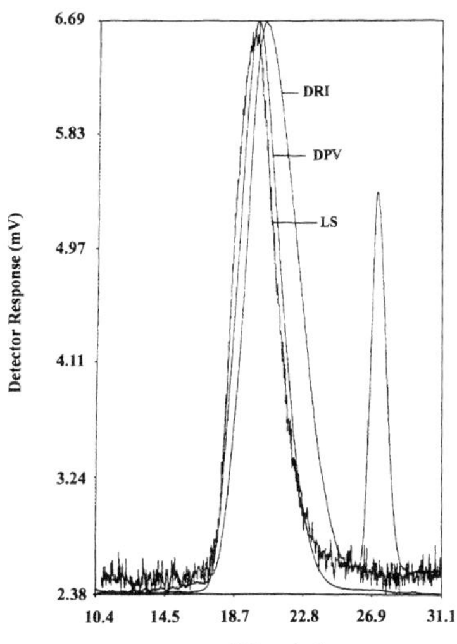

Figure 4. Chromatograms for orange pectin extracted after 2.5 minutes of microwave heating. DRI is trace from differential refractometer detector, DPV is trace from differential pressure viscometer detector, and LS is trace from light scattering detector. Taken from reference 4.

TABLE 3. Molecular Properties of Sequentially Extracted Pectins

Lime[a]					Orange[a,b]			
Ext. No.[c]	M_w $x10^{-3}$	R_{gz} nm	0_w dL/g	a[d]	M_w $x10^{-3}$	R_{gz} nm	0_w dL/g	a[d]
1	336(2)[e]	46(1)	14.0(0.2)	0.80	285(22)	37(4)	9.9(0.1)	0.71
2	494(6)	43(1)	14.3(0.3)	0.67	274(11)	37(1)	8.3(0.2)	0.54
3	573(7)	39(1)	13.1(0.2)	0.49	296(22)	34(4)	6.7(0.4)	0.43

[a] Lime albedo heated 3 minutes, orange albedo heated 2.5 minutes.
[b] Data taken from reference 4.
[c] Extraction sequence.
[d] Mark-Houwink exponent.
[e] Standard deviation of triplicate analysis.

TABLE 4. Percentage Recovery from Sequential Extractions

Ext. No[a]	Lime	Orange[b]
1	4.1	7.9
2	6.6	3.1
3	4.1	5.8
Total	14.8	16.8

[a] Extraction sequence.
[b] Data taken from reference 4.

In the case of the sequential extraction of orange pectin, M_w and R_{gz} remained constant throughout the 3 steps of extraction whereas $[0]_w$ decreased. The M_w, R_{gz} and $[0]_w$ from both single extractions as a function of time (see Table 1) and from sequential extractions (see Table 4) appear to act independently of each other. For a homologous series of single macromolecules with the same shape but changing molar mass, radius or intrinsic viscosity, the relationship of $[0]_w$, R_{gz} and $[0]_w$ is governed by equation 1, the Einstein relationship.

(1)

In that case, an increase of $[0]_w$ must be followed by a proportionate increase in R_{gz} and M_w in order for K to remain a constant. One explanation for the failure of equation 1 to hold is a change in shape in pectin with time of extraction or extraction step. This in turn could be rationalized if pectin existed in different states of aggregation as reported previously [4, 9-11].

Figure 3 contains superimposed chromatograms produced by 90E light scattering (LS), differential pressure (viscosity) (DP) and refractive index (RI) traces for lime pectin heated for 3 minutes. Comparable chromatograms for orange albedo heated for 2.5 minutes are shown in Figure 4. As indicated in Table 1, lime pectin heated for 1 minute gave the highest values for [0], 15 dL/g whereas lime heated for 2 minutes gave the highest M_w value, 372 x 10^3. All samples heated for 3 minutes or less gave R_{gz} values of about 42 nm. The polydispersity of sample 1 as indicated by the ratios of M_w/M_n and M_z/M_n (see Table 5) were 1.12 and 1.24, respectively. These values indicate a relatively narrow molar mass distribution when the sample was heated for 1 minute. As time of heating increased, polydispersity indices passed through a maximum. An interpretation consistent with this data is that pectin in the albedo is organized as a series of loosely held aggregated networks. Upon release into solution, these networks start to break down and the distribution broadens. With continued heating, the component structures within the networks are released and tend to dominate. Since these components are of comparable size, as more and more are released into solution, the distribution tends to narrow. Thus polydispersity indices pass through a maximum with heating time. This broadening and narrowing of molar mass distributions with heating time for lime pectin is demonstrated in Figure 5.

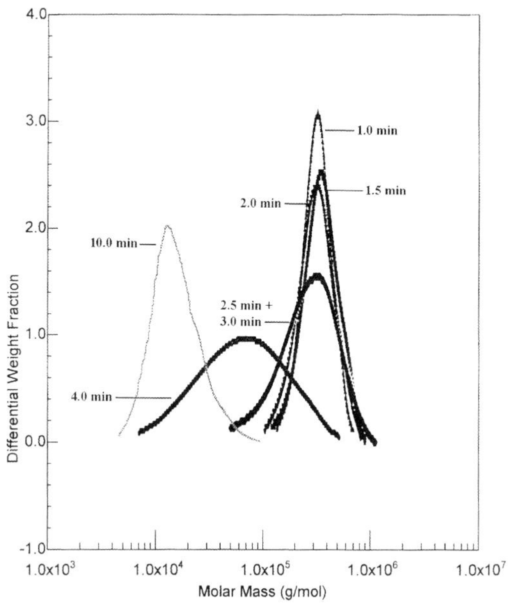

Figure 5. Overlaid differential weight fractions against molar mass for lime pectins heated during extraction from 1 to 10 minutes.

Figure 6 contains Mark-Houwink plots [12] of lime pectin heated during extraction for lengths of time ranging between 1 and 10 minutes. For molar masses covering more than two orders of magnitude, the curves are concave down which may indicate the presence of pectin aggregates with two or more shapes. Figure 7 contains Mark-Houwink plots of orange pectin heated during extraction for lengths of time ranging between 2.5 and 6 minutes. For molar masses covering more than two orders of magnitude, the curves for orange pectin are concave down with more concavity, i.e, greater change in slope than lime pectin. These curves may indicate even bigger changes in shape with shorter length of heating time during extraction. Table 1 contains Mark-Houwink exponents which are obtained by determining slopes from plots of log [0] against log M_w. The slopes obtained are those of the best linear least squares lines of the plots found in Figures 6 and 7. In Figure 8 are plotted Mark-Houwink exponents against molar mass. At 2.5 minutes, orange and lime pectins have about the same molar masses and Mark-Houwink exponents. At all other extraction times, the Mark-Houwink exponent for orange pectin is lower than that for lime pectin when both pectins have the same molar mass. This indicates that as molar mass decreased due to increased length of heating during extraction, orange pectin structures were always more compact than lime pectin structures and possibly more branched.

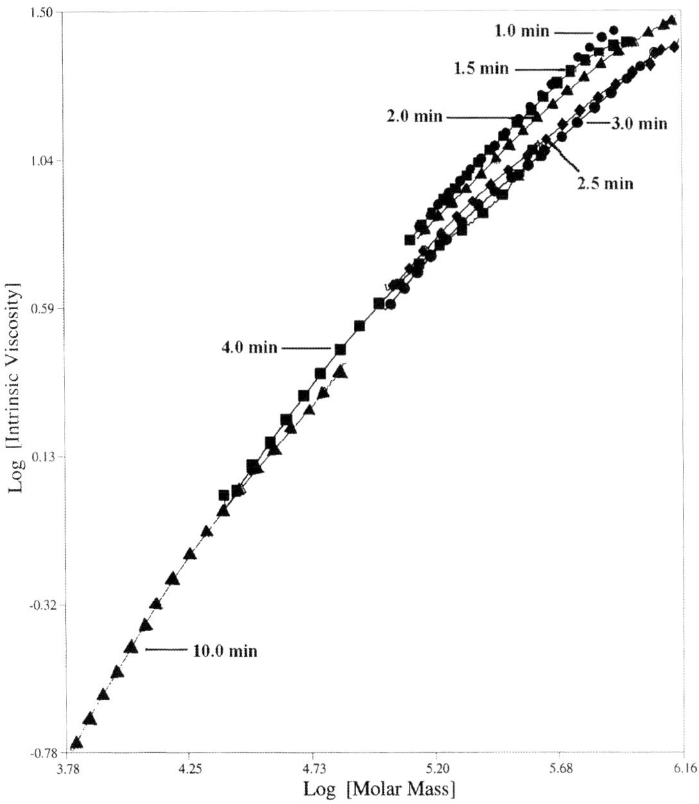

Figure 6. Overlaid Mark-Houwink plots for lime pectins heated during extraction from 1 to 10 minutes

Interestingly, the data in Table 3 indicated that when sequentially extracted, lime and orange pectin had decreasing Mark-Houwink exponents with increasing extraction step. In the case of lime pectin, its molar mass increased with increased extraction step whereas for orange pectin molar mass remained about the same. Since for both pectins, molar mass did not decrease with increased extraction step as might happen if degradation due to heating had occurred, one might conclude that with each extraction step a more compact or branched and possibly different population of pectin molecules were being extracted.

As indicated by Figure 8, the Mark-Houwink exponent for lime and orange albedo pectin may pass through a minimum with increased length of heating time during extraction. In the case of lime pectin the minimum is either much more shallow or does not occur. Nevertheless, both pectins ultimately exhibit an increase in Mark-Houwink exponents with increased length of heating time.

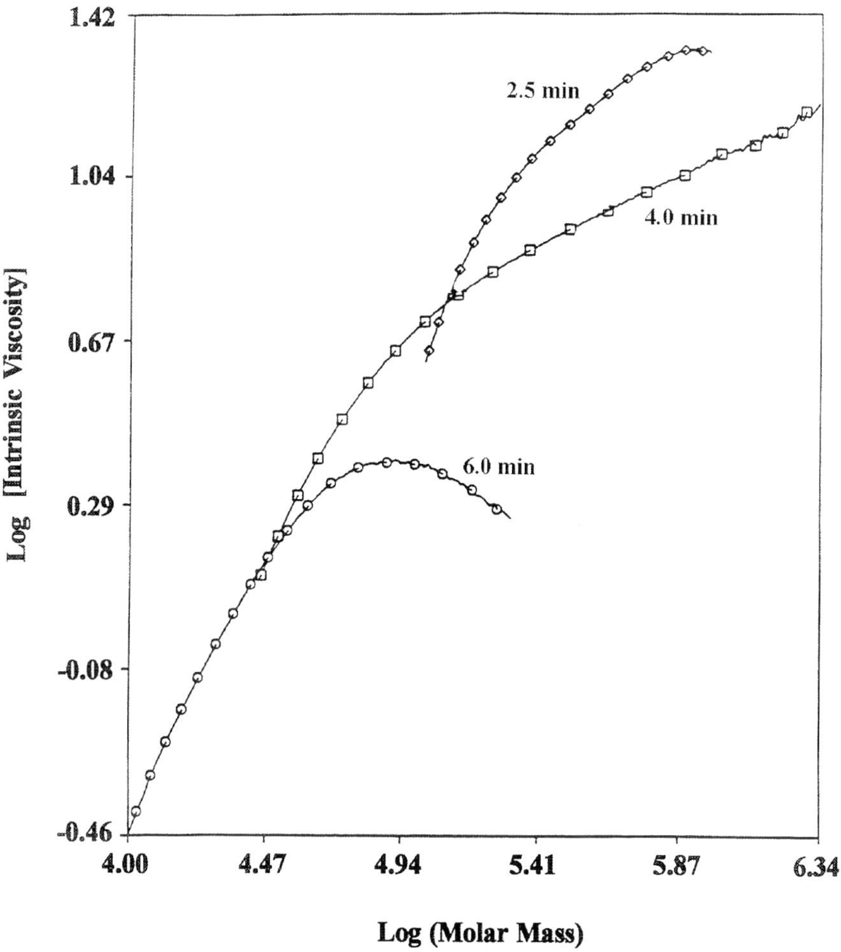

Figure 7. Overlaid Mark-Houwink plots for orange pectins heated during extraction from 2.5 to 6 minutes

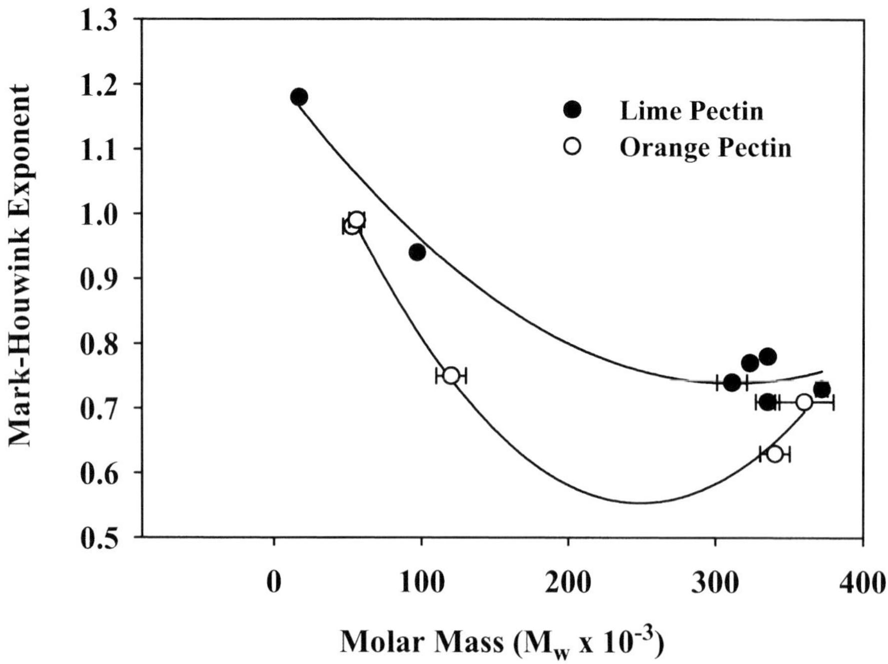

Figure 8. Change in shape as function of time of heating as related to molar mass changes.

Previously [9,13], we have used electron microscopy with rotary shadowing to image pectin molecules dried from dilute solution. These images enabled us to observe the progressive dissociation of pectin network structures induced by adding NaCl or glycerol to dilute aqueous pectin solutions. Both NaCl and glycerol could induce the dissociation of pectin by breaking hydrogen bonds. Comparable images are shown in Figure 9. In Figure 9A is a circular microgel network which was observed in neat water in which segmented rods (SR), kinked rods (K) and rods (R) are visible. In Figure 9B are branched structures of partially dissociated networks found in 0.005 M NaCl in which SR, K and R are visible, also. In Figures 9C and 9D are fully dissociated pectin networks consisting of rods (R), segmented rods (SR), and kinked rods (K) which were found to be even more prominent in 0.05 M NaCl and 50% glycerol respectively. Mark-Houwink plots of distributions of each of these sets of structures could be expected to give exponents ("a") with increasing values in the order: image A < B < C = D. In addition values of M_w and $[0]_w$ would decrease in the order A >B > C = D. These are the same trends that we have found for values of M_w and $[0]_w$ when we increased the time of microwave heating. Our results suggest that pectin network structures being extracted from albedo may break down in the same fashion with increased heating time.

In Figure 10 we have plotted gel strengths at breaking point for commercial citrus pectin; and for lime and orange pectin extracted from albedo by heating for 3 minutes with a microwave energy source. Three minute heating times gave the highest yields. Ehrlich [4], measured gel strength against pH of the gelling solution for a series of pectins with differing degrees of methyl esterification (DE). Pectins with the same DE gave gel strengths that were arrayed almost symmetrically around the maximum gel strength. Thus as an empirical starting point, gel strength against pH data was fitted to a Gaussian distribution which is symmetrical. As indicated by the R^2 values in Table 5, 90 to 98% of the error in predicted values as compared to experimental values in the distributions was accounted for by the model. Table 5 contains the pH and gel strength at peak maximum whereas Table 6 contains the galacturonic acid content and DE of pectins whose gel strengths were measured. The orange albedo pectin (OAP) and the commercial citrus pectin (CCP) had larger than expected standard deviations in DE. Nevertheless, in the cases of CCP and standard pectin, average values of DE found in this report, agreed with supplier values within experimental error. Microwaved albedo pectins had greater gel strengths at optimum pH than commercial citrus pectin. This result is consistent with the data in Table 1 which shows that M_w and $[0]_w$ at 3 minutes heating time are larger for lime and orange pectin than for commercial citrus pectin. Interestingly, when heated for 3 minutes, M_w for lime is smaller than M_w for orange whereas $[0]_w$ and "a" for lime is larger than $[0]_w$ and "a" for orange. This would indicate that at 3 minutes heating time, orange pectin is more compact and possibly more branched than lime pectin. Since the gel strength at break point for lime is greater than that for orange pectin, one might conclude that pectin main chain molar mass is more highly correlated with gel strength than side chain molar mass for partially dissociated network structures. Furthermore, partially dissociated network structures may be the precursors of gel networks formed upon the addition of sugar to high methoxy pectin at acid pH's.

TABLE 5. Maximum Gel Strength

Sample	pH	Gel Strength(g)	R^{2a}
Lime Albedo Pectin	3.53	117(9)[b]	0.98
Orange Albedo Pectin	3.23	107(11)	0.96
Commercial Citrus Pectin	3.00	87(6)	0.90

[a]coefficient of determination.

[b]Number in parentheses is standard error.

TABLE 6. Galacturonate Composition of Pectins

Sample	% galacturonic acid content	% Degree Methyl Esterification
Lime Albedo Pectin	76(2)[a]	88(2)
Orange Albedo Pectin	76(6)	78(13)
Commercial Citrus Pectin[b]	81(6)	77(12)
Commercial Citrus Pectin[c]	74(2)	72(3)
Standard Pectin[d]	88(8)	60(5)
Standard Pectin[e]	85	64

[a]standard deviation of triplicate analysis.
[b]sample gift from pectin manufacturer.
[c]values supplied by manufacturer.
[d]sample purchased from Sigma Chemical Co.
[e]values supplied by Sigma Chemical CO.

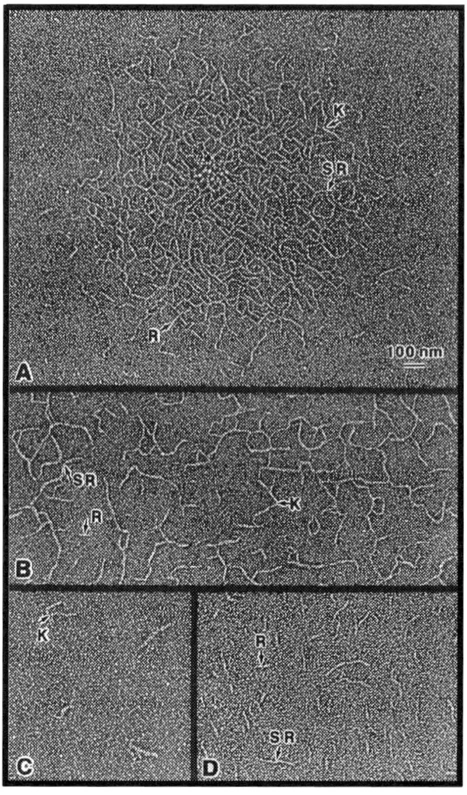

Figure 9. Electron micrographs of rotary shadowed images of pectin aggregates and dissociated aggregates. Abbreviations: (R),rods; (SR), segmented rods; (K), kinked rods.

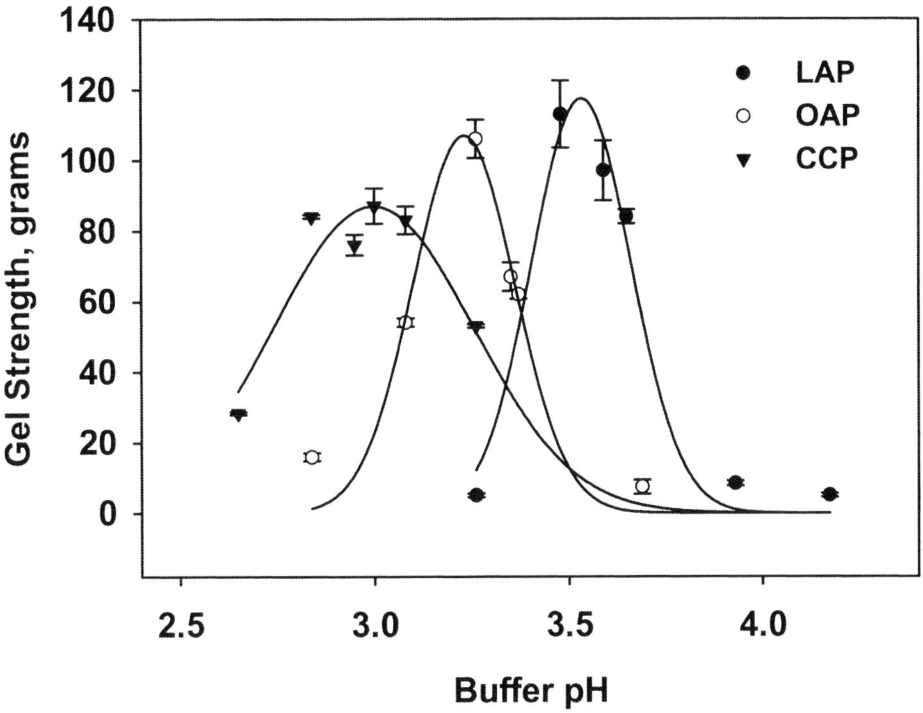

Figure 10. Effect of pH on gel strength of pectins from various sources. Abbreviations: (LAP), lime albedo pectin; (OAP), orange albedo pectin; (CCP), commercial citrus pectin.

4. Conclusion

Microwave flash extraction can produce lime and orange pectin with molecular properties which are comparable or better than commercial citrus pectin. Considering molecular properties and yield, lime pectin is somewhat better than orange pectin when both are microwave flash extracted from albedo. Lime pectin is more readily extracted than orange pectin which may account to some extent for the better molecular properties and yield of lime as compared to orange pectin. Concave down Mark-Houwink plots and failure of the Einstein equation to correctly predict relationships between M, [0] and R_g are given as evidence that molecules of pectin change shape due to degradation of aggregates with prolonged heating times. A possible explanation for these behaviors is that initially extracted network structures of pectin are progressively broken down to their more asymmetric components upon prolonged heating. Finally, we have shown that the gel forming properties of high methoxy orange and lime pectin prepared by rapid microwave heating are better than commercial pectin prepared by conventional methods of heating.

5. References

1. Manabe, M., Naohara, J., Sato,T., and Okada, J.: The extraction of pectin by microwave heating, *Nippon Shokuhin Kogyo Gakkaishi* **35** (1988), 497-501.
2. Kratchanova, M., Panchev, I., Pavlova, E., and Shtereva, L.: Extraction of pectin material pretreated in an electromagnetic field of super high frequency, *Carbohydr. Polymers* **25** (1994), 141-144.
3. Kratchanova, M., Pavlova, E., Panchev, I., and Kratchanov, Chr.: Influence of microwave pretreatment of fresh orange peels on pectin extraction, in Visser, I. and Voragen, A.G.J. (eds.), *Pectins and Pectinases.* Elsevier, Amsterdam, 1996, pp. 941-946.
4. Fishman, M.L., Chau, H.K., Hoagland, P., and Ayyad, K.: Characterization of pectin flash extracted from orange albedo by microwave heating under pressure, *Carbohydr. Res.* **323** (2000), 126-138.
5. Kertez, Z.I.: *The Pectic Substances*, Interscience Publishers, Inc. New York, 1951, pp. 457.
6. Blumenkrantz, N., and Asboe-Hansen,G.: New method for quantitative determination of uronic acids, *Anal. Biochem.* **54** (1973), 484-489.
7. Voragen, A.G.J., Schols, H.A., Pilnik, W.: Determination of the degree of methylation and acetylation of pectins by h.p.l.c., (1986), 65-70.
8. Anonymous.: *Optilab DSP Interferometric Refractometer Hardware Manual,* Wyatt Technology Corporation, Santa Barbara CA. 1998, Section 4.
9. Fishman, M.L., Cooke, P., Hotchkiss, A., and Damert, W.: Progressive dissociation of pectin, *Carbohydr. Res.* **248** (1993), 303-316.
10. Fishman, M.L., Pepper, L., and Pfeffer, P.E.: Dilute solution properties of pectin, in Glass, J.E., (ed.)., *Water Soluble Polymers: Beauty with Performance*, Advances in Chemistry Series 213, American Chemical Society Publications, Washington, DC, 1986, pp 57-70.
11. Fishman, M.L., Gillespie, D.T., Sondey, S.M., Barford, R.A.: Characterization of pectin by size exclusion chromatography in conjunction with viscosity detection, *J. Agr. Food Chem* **37** (1989), 584-591.
12. Tanford, C.: *Physical Chemistry of Macromolecules,* John Wiley and Sons, Inc. New York, 1961, pp. 411-412.
13. Fishman, M.L., Cooke, P., Levaj, B., Gillespie, D.T., Sondey, S.M., and Scorza, R.: Pectin microgels and their subunit structure, *Arch. Biochem. Biophys* **294** (1992), 253-260.
14. Crandall, P.G. and Wicker, L.: Pectin internal gel strength: theory, measurement, and methodology, in Fishman, M..L. and Jen, J.J. (eds.), *Chemistry and Function of Pectins,* ACS Symposium Series 310, American Chemical Society, Washington, D.C., 1986, pp. 88-102.

Acknowledgement.

We thank Andre White for technical assistance in the compositional analysis of pectin and Shobhanna Upadhyay for advice on the measurement of gel strength.

THE HYDRATION BEHAVIOUR OF PECTIN NETWORKS AND PLANT CELL WALLS

A.J. MacDOUGALL, and S.G. RING,
Division of Food Materials Science,
Institute of Food Research,
Norwich Research Park,
Colney, Norwich,
NR4 7UA
e-mail: alistair.macdougall@bbsrc.ac.uk

Abstract

The functional properties of pectins in food systems and in plant cell walls arise particularly from their ability to form crosslinked polymer networks that are highly hydrated. This article describes the physical basis of pectin hydration, setting it in the context of the hydration behaviour of polymers in general and polyelectrolytes in particular. The interplay between network expansion (driven by water uptake) and elastic resistance to deformation is discussed. Special emphasis is placed on the opportunities that exist to make use of mathematical relationships describing the thermodynamics of polymer hydration to build predictive models for the behaviour of pectin networks in different aqueous environments.

1. Introduction

The role of pectins in the plant primary cell wall is a subject of continuing interest. These polymers account for approximately 50% of the dry weight of the wall, and they form a hydrated polymer matrix within which the cellulose microfibrils and associated hemicelluloses are dispersed [10,37].

1.1. THE PECTIN NETWORK IN PLANT CELL WALLS

There has previously been some doubt about whether or not the pectin network plays a mechanical role in the plant cell wall. Pectic polysaccharides extracted from plant cell walls under non-degradative conditions have generally shown a high degree of methylesterification (dm) which is not conducive to gel formation with calcium ions [37]. More recently it has been shown that native pectins which form gels in the presence of calcium ions can be extracted from tomato cell walls [15]. However it has also been shown that at higher pectin concentrations, typical of the plant cell wall, ionic interaction and crosslinking occurs more readily [32]. Debates about the relative ability of different pectins to form calcium gels at polymer concentrations of 1 to 2% are therefore not very meaningful for the functional role of pectins *in vivo*.

F. Voragen et al. (eds.), Advances in Pectin and Pectinase Research, 123-135.
© 2003 *Kluwer Academic Publishers. Printed in the Netherlands.*

The absence of a mechanical role for pectins in the plant cell wall has been suggested from experiments examining the behaviour of plant tissues under applied loads [29]. However it is possible to conceive of complementary mechanical roles for the cellulose and pectin networks [14]. Consistent with this view is the observation from FT-IR studies that the cellulose-xyloglucan network and the pectin network are mechanically independent [39].

Ionic association through interaction with calcium ions is likely to form the main mechanism for crosslinking the pectin network. However the exact form of interaction of pectins with calcium ions *in vivo* remains unclear. Association of pectins in a 2_1 or a 3_1 helical conformation with calcium ions has been proposed from studies of isolated pectins [20,38]. Evidence for the presence of both chain conformations *in vivo*, together with intermediates between them has been found in an NMR study [10]. Recent molecular modelling studies have further explored the influence of pectin conformation on calcium binding [2].

Besides calcium, other forms of ionic association may also contribute to formation of a stable pectin network. Basic amino acid sequences present in plant cell wall structural proteins have been shown to be able to crosslink pectins [17]. Junction zone formation in concentrated pectins may also be possible through interaction with magnesium ions [16].

In addition pectins can be crosslinked covalently. Neutral sugar sidechains containing apiose have been found to be crosslinked via boron [24], and in certain plant families dimerisation of ferulic acid substituents on the pectin sidechains appears to play a role in pectin crosslinking [25,33]. Evidence has also been obtained for covalent crosslinking between pectins and structural proteins [28].

1.2. PLANT CELL WALL MODELS

Attempts to provide a coherent overview of these data have led to the production of a series of pictorial models of the plant cell wall [3,13]. Whilst this can be a useful exercise, these models of the plant cell wall can also be misleading. In picturing the plant cell wall polymer network two important features are necessarily discarded - namely the molecular motion of the polymers and their interaction with water. These properties crucially underlie the behaviour of the cell wall, and without them cell wall models are critically weakened in their ability to provide insight into the effect of biochemical and physiological events on wall behaviour.

The purpose of this article is to show how the hydration characteristics of pectic polysaccharides can be related to the hydration of polymer networks in general, and consequently to show that this approach provides a basis on which to build a different form of cell wall model which predicts the effect of physiological changes on the behaviour of the pectin network in the plant cell wall.

2. The Thermodynamic Basis of Polymer Hydration

Polymer networks immersed in a solvent are subject to a hydration or swelling force which has the same thermodynamic basis as the free energy of mixing of individual polymer molecules in solution [7]. Swelling of the plant cell wall polymer network is clearly observed in studies carried out on cell separation in ripening fruit [30] and in studies on the effect of cooking on the cell walls of vegetables [12]. However no real attempt has previously been made to explain these changes in hydration in terms of the physico-chemical behaviour of a polymer network immersed in an aqueous solvent.

2.1. NEUTRAL POLYMERS

As with all changes in free energy, entropic and energetic factors need to be considered. For neutral polymers the entropic gain on mixing is classically derived from notional division of the system into a lattice of compartments which can hold either a solvent molecule or a section of the polymer chain. The entropy of mixing is then determined from the number of different ways that a linear polymer molecule can be arranged within this lattice. Energetic interactions between the polymer and the solvent are accounted for by introduction of a dimensionless interaction parameter χ, although this term may also include an allowance for the effect of interactions with the polymer chain on the entropy of the adjacent water molecules. The resultant expression for water activity is as follows [7]:

$$\frac{p_1}{p_1^0} \approx (1 - v_2)e^{[\chi v_2^2 + v_2(1 - 1/x)]} \tag{1}$$

where p_1 is the vapour pressure of the system of interest, p_1^0 is the vapour pressure of pure water, v_2 is the volume fraction of polymer, and x is the number of segments in the polymer. A segment is defined as that part of the polymer molecule which is equivalent in volume to a solvent molecule, and is obtained from the ratio of the molar volume of the polymer to that of the solvent, water. The osmotic pressure generated by these forces can be calculated from the relationship:

$$\pi \approx \frac{RT}{\overline{V}_w} \ln \frac{p_1}{p_1^0} \tag{2}$$

where $\overline{V}_w$ is the partial molar volume of water, and R and T have their usual meanings. The effect of this hydration force is seen for example in the swelling of crosslinked dextran beads which are used in gel-filtration chromatography.

2.2. POLYELECTROLYTES

For polyelectrolytes an additional contribution to network hydration comes from electrostatic repulsion between fixed charges on the network, which follows as a consequence of the entropic dispersion of mobile counterions into the surrounding medium [7]. The forces arising as a consequence of this effect are reduced by the tendency of the mobile ions to congregate in the vicinity of the fixed charges, setting up a Donnan equilibrium. Addition of salts to the surrounding medium can further contribute to the shielding of the charge on the polymer network and reduce the swelling force. Conversely depletion of salts in the surrounding medium can lead to a very large hydration force, as it maximises the impact of the translational entropy of the mobile counterions. The hydration behaviour of polyelectrolyte networks is a subject of continuing research, and different equations have been derived to estimate this hydration force depending on the relative concentrations of the ionisable groups on the polymer and the salts in the bathing medium.

For high levels of added salt (relative to the concentration of ionisable groups on the polyelectrolyte) the appropriate form of the equation is [7]:

$$\pi \approx RT(ic_2)^2 / 2wvc_s^* \tag{3}$$

where i is the degree of ionisation multiplied by the valency charge on the polymer, c_2 is the concentration of polymer charge, expressed as moles of repeating unit (for pectin this is the anhydrogalacturonic acid residue), w is the valency factor of the electrolyte, v is the number of cations and anions into which the electrolyte dissociates and c_s^* is the concentration of strong electrolyte external to the polymer network.

For intermediate levels of added salt the relationship can be expressed as follows [5]:

$$\pi \approx RT\,c^2 / A(c + 4Ac_s) \tag{4}$$

where c and c_s are the molar concentrations of polymeric monomer and salt, and A is the number of monomers between effective charges.

3. Relevance to the Hydration of Biopolymers

The applicability of these concepts of polymer hydration has been well appreciated in studies on the biomechanical behaviour of mammalian tissues. Particular examples of this are cartilage [19,27] where hydration behaviour is an important contributor to the tensile properties, and the cornea [6] where hydration affects both mechanical and optical properties. In plant science the relevance of this approach for the hydration of the plant cell wall has been overlooked largely because of the historical development of the understanding of the water relations of plant tissues. Early discussions of the role of water in plant tissues leant heavily on the terminology and descriptions of the behaviour of water in soils [23]. The assumption was made that the construction of the plant cell wall was not significantly different from a clay soil in which an aqueous phase is found in the pores between solid (impermeable) particles that bear a surface charge. For highly lignified tissues this assumption appears justified, and the forces holding water in these tissues can essentially be reduced to capillary forces occurring at air:water interfaces and osmotic forces associated with the presence of low molecular weight solutes. But for the plant primary cell wall this view is currently being challenged by observations on cell wall hydration and swelling which can only be adequately explained in terms of the theory of polymer hydration outlined above [11,16,40].

3.1. HYDRATION OF PECTINS

For pectins it is instructive to compare the relative size of the hydration force that arises from the entropy of dispersion of the polymer chain and solvent affinity (Equation 1 and 2), with that arising from a polyelectrolyte effect (Equations 3 and 4). In Figure 1 the theoretical hydration force that would arise from each of these two considerations is given for a solution containing polygalacturonic acid which has a dm of 50%, and in which all the carboxylic acid groups are ionised.

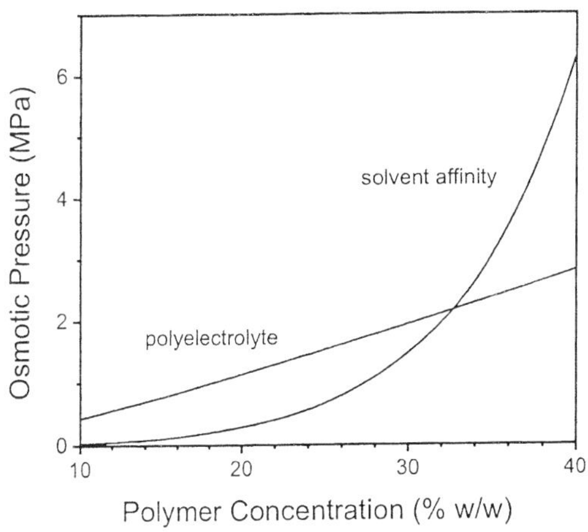

Figure 1. Theoretical contribution of different hydration forces to the osmotic swelling pressure of a pectin network. Calculations were carried out for a 50% dm pectin using Equations 1,2 and 4.

Figure 1 indicates that at polymer concentrations found in the plant cell wall (20 to 40% w/w) the pectin network is potentially subject to a large hydration force, which includes contributions from the entropy of dispersion of the polymer chains, the affinity of the solvent for the polymer, and mutual electrostatic repulsion of the fixed negative charges. From the data presented in Figure 1 it can be seen that the approximate range over which it is appropriate to study the hydration behaviour of pectic polysaccharides is 0 to 4 MPa. This is equivalent to a relative humidity range from 100 to 99.7%. Studies of the hydration of polymers in this range can be carried out in different ways, but one of the most convenient is to look at the hydration and swelling behaviour of polymer films when placed in concentrated solutions of a high molecular weight polymer [26]. For instance, solutions of 12 and 45% polyethylene glycol (PEG) of 20,000 MWt provide osmotic stresses of 0.12 and 4.7 MPa respectively.

The data presented on pectin hydration in this article were mainly obtained using a chelator extracted pectin from unripe tomato fruit. This material is of high intrinsic viscosity (810 ml/g). The sugar composition of the pectin (in mole percent) was determined as: GalA 83, Gal 12, Ara 3.2, Rha 0.8, Glc 0.6, Xyl 0.1, Fuc 0.1, Man < 0.1, with a dm of 65% [15]. Native pectins extracted with minimal degradation, rather than commercial pectins have been used by choice because the principle focus of the work is to understand the behaviour of pectins *in vivo*. Nevertheless these materials illustrate the general principles that underlie the hydration behaviour of all pectins.

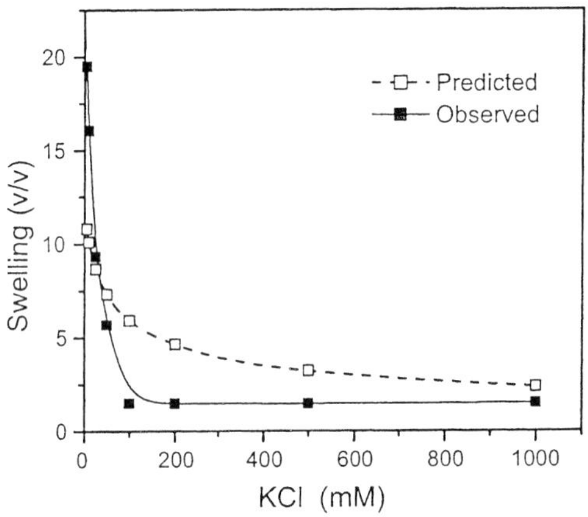

Figure 2. Salt dependence of tomato pectin swelling. All samples were subject to an external osmotic pressure of 0.5 MPa. The predicted data was calculated according to Equation 4. Reproduced with permission from *Biomacromolecules* **2**, 450-455. Copyright 2001 Am. Chem. Soc.

To obtain the data presented in Figure 2, air-dried films of tomato pectin were initially equilibrating in solutions of PEG which had an osmotic pressure of 0.5 MPa, and in addition contained 1 M KCl. Samples were then transferred to solutions of similar PEG but lower KCl concentrations. The data show that as the KCl concentration is lowered the pectin films show a marked increase in swelling. This behaviour is typical of a polyelectrolyte based swelling force, with the hydration force acting on the network being at its greatest when the salt concentration is low. The predicted behaviour of the pectin assuming full ionisation of the carboxyl residues is given for comparison. The plot has the same general form as the experimental data although the extent of swelling is overestimated at high polymer concentrations, and underestimated at low polymer concentrations.

3.2. FACTORS OPPOSING PECTIN HYDRATION

Since the predicted data is only calculated on the basis of polyelectrolyte hydration of the network the underestimate of swelling could be due to the presence of an additional hydration force associated with polymer dispersion and solvent affinity, although this would be more likely to be a factor at higher polymer concentrations (Figure 1).

Several factors can explain failure of the network to swell as much as predicted. For highly charged polyelectrolytes, such as polygalacturonic acid, it is anticipated that the phenomenon of counterion condensation may reduce the effective ionisation of the network, and increase the average distance between effective charges [18]. To estimate whether counterion condensation is likely to occur a dimensionless parameter ξ can be calculated from the relationship [18]:

$$\xi = q^2 / \varepsilon\varepsilon_0 4\pi kTb \tag{5}$$

where ε is the dielectric constant of the medium, ε_0 is the dielectric permittivity of free space, q is the elementary charge, k is Boltzmann's constant, and b is the spacing between charges. For many polyelectrolytes, counterion condensation is observed above a threshold value of $\xi = 1$. For water at 25 °C $\xi \approx (0.71/b)$, if the charge spacing is expressed in nm, indicating a minimum charge spacing that will be supported of 0.71 nm. For polygalacturonic acid the spacing of ionisable groups is approximately 0.43 nm. Regions of the pectin which contain contiguous stretches of unesterified galacturonic acid residues can therefore be expected to undergo counterion condensation until the charge spacing increases to 0.71 nm - giving a reduction in the effective charge on the polymer in these regions of approximately 40%.

In addition to this non-specific interaction with the mobile cations, the charge on the pectin backbone can also be reduced as the result of specific complexation with inorganic and organic cations. Complexation with calcium has been studied in dilute solution [8], and in semidilute solution [34], where pectin gelation has been observed to lead to the formation of complexes of markedly increased stability. Circular dichroism studies also suggest that there is a specific interaction between basic amino acid sequences present in plant cell wall hydroxy-proline rich glycoproteins and pectins [17]. Experimentally basic peptides have been shown to reduce the swelling observed in calcium pectate gels [17].

Some of these ionic interactions lead to crosslinking of the polymer network. Polymer networks immersed in a solvent reach a swelling equilibrium in which the forces driving hydration and expansion of the network are balanced by an elastic resistance to deformation. Ionic complexes which serve to crosslink polyelectrolytes therefore reduce the swelling of the network both by reducing the ionisation of the polyelectrolyte, and by increasing the elastic resistance to expansion. Overall there can be a marked reduction in swelling as a result. This is seen for calcium in Figure 3 where a low level of added calcium has a dramatic effect on the hydration of tomato pectin.

An additional observation arising from these studies, which is of particular relevance for understanding network formation in pectins and the role of pectins in the plant cell wall, is the marked effect of polymer concentration on the interaction of pectins with cations. Whereas gelation of semi-dilute solutions of pectins with calcium ions occurs only when extended blocks of contiguous unesterified galacturonic acid residues are present, in more concentrated pectin matrices network formation occurs with pectins of lower charge densities, and can be induced by a wider range of cations [16]. For instance under 0.5 MPa of external osmotic pressure the volumetric swelling ratio for tomato pectin in the presence of 50 mM $MgCl_2$ was 1.7, closely resembling the behaviour seen in the presence of 6 mM calcium (Figure 3). In the extreme, molar concentrations of monovalent salts can lead to stabilisation of pectin matrices at low levels of hydration. The swelling data for tomato pectin in the presence of 1 M KCl are almost identical to that obtained for 6 mM $CaCl_2$ (Figure 3) down to 0.25 MPa of externally applied osmotic pressure [16].

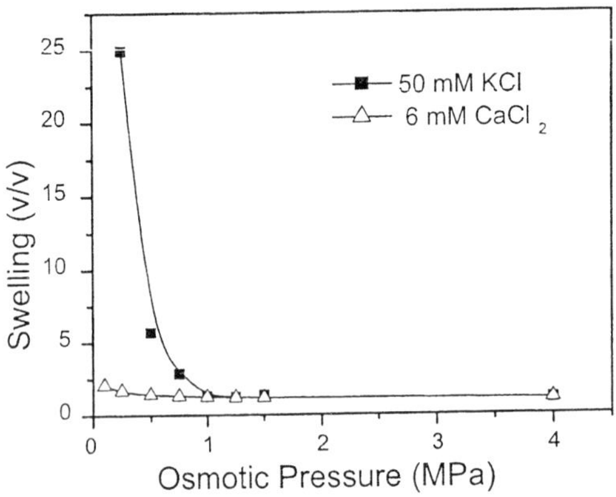

Figure 3. Variation in tomato pectin swelling with osmotic pressure.

3.3. THE DUAL ROLE OF CARBOXYLIC ACID RESIDUES

For pectic polysaccharides it is not easy to determine the relative importance of the role of ionisable groups in promoting swelling of the network, and in crosslinking. However estimates of the level of dissociation of the ionisable groups on the network can be obtained from the shear modulus. The shear modulus can be used as an approximation to the volumetric swelling modulus, which at swelling equilibrium is directly equivalent to the osmotic swelling pressure being experienced by the network [31].

For calcium pectate gels (1.1% w/w tomato pectin) washed in salt solutions to minimise the level of calcium in the bathing medium, it has been estimated that ionisation of about 20% of the ionisable groups is necessary to explain the hydration behaviour of this material [34].

Further evidence for the dual role of the ionisable groups is seen in the hydration behaviour of films prepared from citrus pectin of different levels of random esterification (Table 1) [35]. The highest extent of swelling is shown by films at intermediate levels of esterification. Films prepared from pectins of high dm fail to swell as much because of the low level of charge carried by the polymer network. Films

TABLE 1. Swelling of randomly esterified citrus pectin in 50 mM KCl or 6 mM $CaCl_2$ at 0.5 MPa of external osmotic stress provided by PEG.

dm (%)	Swollen volume/dry volume	
	50 mM KCl	6 mM $CaCl_2$
13	1.1	1.0
26	Dispersed	1.3
57	Dispersed	Unmeasurable
76	Dispersed	1.9
87	1.0	1.2

prepared from pectins of low dm fail to swell because the high charge density on the polymer network leads to effective crosslinking through the ionic interactions discussed above.

4. The Distribution of Unesterified Residues in Pectins

An appreciation of the importance of charge density in determining the properties of pectins has led to a number of attempts to improve the structural analysis of pectins and to quantify the distribution of the unesterified galacturonic acid residues. The techniques developed have included preferential hydrolysis with HF of neutral (galactose) residues in pectins reduced with sodium borohydride [21], and the use of NMR [1]. Recently selective enzymic degradation with endoPG and exoPG has been used to liberate the galacturonic acid originating from extended blocks [4]. An additional development of potential interest has been the selective degradation of the methylesterified residues of pectin through their initial conversion to hydroxamic acids and further coupling to a carbodiimide, followed by alkaline hydrolysis of the resulting isocyanates [22]. An anion-exchange hplc trace for the tomato pectin fragmented by this method is presented in Figure 4. Peak markers indicate the number of contiguous unesterified galacturonic acid residues present in the oligomeric series which form the main products of this degradation. In addition to the galacturonic acid residues, oligomers contain a terminal arabitol residue resulting from the degradation of a methylesterified galacturonic acid residue. Following correction for variation in response factors the data show a decrease in relative abundance for each successive oligomer in the range 34 to 40%. This is very close to the figure of 35% which is the predicted relative decrease in abundance for the products liberated from a randomly esterified pectin of 65% dm.

In contrast to these data the transport parameters for counterions associated with this pectin were found to be consistent with a block distribution of the unesterified galacturonic acid residues (V Dronnet and J-F Thibault, personal communication). The

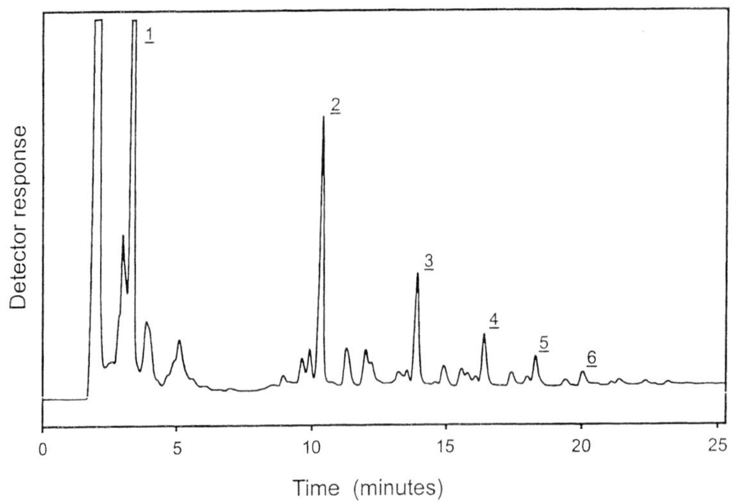

Figure 4. Anion-exchange hplc chromatogram of the products of a tomato pectin degradation procedure designed to liberate the contiguous unesterified sequences of galacturonic acid. A CarboPac PA1 column was used with a gradient of Na acetate (pH 6) from 50 mM to 1 M over 30 min, followed by post-column addition of base and pulsed electrochemical detection.

observed gelation behaviour also indicates that blocks of unesterified residues must clearly be present.

It therefore appears that this pectin contains two distinct distributions for the unesterified residues. To some extent this may be explained as the mixture of two different populations of pectins which are structurally distinct. Following calcium induced gelation of this pectin nearly 30% of the total material did not take part in network formation and could be leached from the gels [15]. The leached material had the same dm as the unfractionated tomato pectin, and presumably does not have a block distribution of unesterified residues. It remains probable, however, that individual molecules contain both a random and block distribution of the unesterified residues.

5. Cell Wall Hydration

Studies of the salt dependence of the hydration of cell walls isolated from mature green tomatoes indicate that polyelectrolyte swelling of the pectin network is a significant determinant of tomato cell wall hydration [16]. The dependence of hydration on osmotic pressure was also investigated and the data is presented in Figure 5. *In vivo* the plant cell wall is subject to osmotic stress generated by the solutes contained within the protoplast. This osmotic stress can be directly measured as the hydrostatic pressure acting on a fine capillary introduced into the vacuole of a single cell [36]. For most plant tissues this expansive force (called the turgor pressure) lies in the range 0.1 to 1 MPa. In unripe tomato pericarp cells it has been estimated at 0.13 MPa [16]. The result of an osmotic stress of this magnitude acting on the cell wall would be to maintain the level of hydration at approximately 3 g/g dry weight (Figure 5). During ripening of tomato fruit the turgor pressure of the cells is reported to fall to 0.03 MPa [16]. As a direct consequence of this one would expect to see a marked swelling of the cell wall to a hydration of 6 g/g dry weight (Figure 5). This in turn is likely to affect the mobility of enzymes, and accessibility of the substrate. It is also likely to affect the stability of ionic complexes which crosslink the pectin network.

These studies characterising the basis of pectin hydration have therefore enabled us to identify the driving force that lies behind the physiological changes that occur during cell separation of ripening tomato fruit. Control of cell wall hydration through modification of pectin structure and the ionic environment of the cell wall is likely to be a significant general mechanism of physiological control of the mechanical properties of the plant cell wall. In this respect the mode of action of pectinmethylesterase and regulation of pectinmethylesterase activity are likely to be topics of particular relevance.

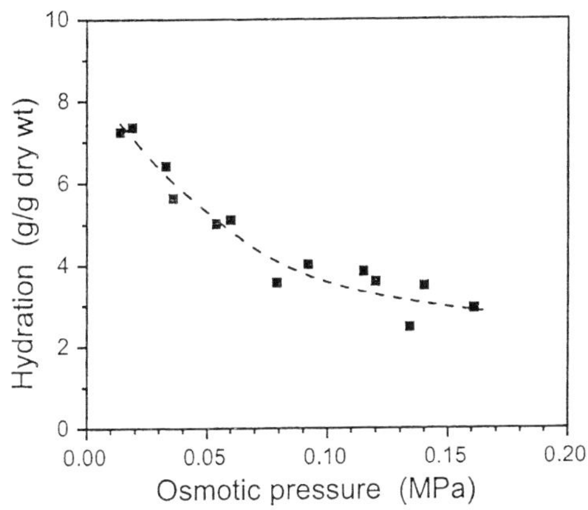

Figure 5. The effect of external osmotic pressure on the hydration of cell walls isolated from unripe tomato fruit. Reproduced with permission from *Biomacromolecules* **2**, 450-455. Copyright 2001 Am. Chem. Soc.

6. Conclusion: Predictive Models of Plant Cell Wall Behaviour

We have described the physical basis of the hydration of pectin networks, and argued that pectin plays an important role in determining the extent of hydration of the plant cell wall *in vivo*. This being so, it is possible to begin to predict the effect of physiological change on the hydration of the pectin network and consequently on the properties of the plant cell wall. Pictorial illustrations of the assembly of polymers in the plant cell wall have their uses, but in order to seriously address questions about the functional behaviour of plant cell walls more sophisticated approaches to cell wall modelling of the type we have described are needed.

Acknowledgements:

The authors would like to thank the BBSRC for funding this work through the Core Strategic Grant, and Syngenta for their support in a CASE studentship.

References:

1. Andersen A.K., Larsen B., Grasdalen H. (1995) Sequential structure by ^{1}H NMR as a direct assay for pectinesterase activity. *Carbohydr. Res.* **273**, 93-98.
2. Braccini I., Grasso R.P., Perez S. (1999) Conformational and configurational features of acidic polysaccharides and their interactions with calcium ions: a molecular modelling investigation. *Carbohydr. Res.* **317**, 119-130.
3. Carpita N.C., Gibeaut D.M. (1993) Structural models of primary cell walls in flowering plants: consistency of molecular structure with the physical properties of the walls during growth. *Plant J.* **3**, 1-30
4. Daas P.J.H., Meyer-Hansen K., Schols H.A., De Ruiter G.A., Voragen A.G.J. (1999) Investigation of the non-esterified galacturonic acid distribution in pectin with endopolygalacturonase. *Carbohydr. Res.* **318**, 135-145.
5. Dobrynin A.V., Colby R.H., Rubinstein M. (1995) Scaling theory of polyelectrolyte solutions. *Macromolecules* **28**, 1859-1871.
6. Elliott G.F., Hodson S.A. (1998) Cornea, and the swelling of polyelectrolyte gels of biological interest. *Rep. Prog. Phys.* **61**, 1325-1365.
7. Flory P.J. *Principles of Polymer Chemistry*, Cornell University Press, Ithaca, New York, 1953.
8. Garnier C., Axelos M.A.V., Thibault J-F. (1994) Selectivity and cooperativity in the binding of calcium ions by pectins. *Carbohydr. Res.* **256**, 71-81.
9. Jarvis M.C. (1984) Structure and properties of pectin gels in plant cell walls. *Plant Cell Environ.* **7**, 153-164.
10. Jarvis M.C., Apperley D.C., (1995) Chain conformation in concentrated pectic gels: evidence from ^{13}C NMR. *Carbohydr. Res.* **275**, 131-145
11. Ktitorova I.N., Skobeleva O.V. (1999) Changes in elastic properties of cell walls and some parameters of plant water relations in response to acidification of the medium. *Russ. J. Plant Physiol.* **46**, 201-206.
12. Lecain S., Ng A., Parker M.L., Smith A.C., Waldron K.W. (1999) Modification of cell wall polymers of onion waste - Part I. Effect of pressure-cooking. *Carbohydr. Polym.* **38**, 59-67.
13. McCann M.C., Roberts K. (1991) Architecture of the primary cell wall. In: Lloyd C.W., ed. *The Cytoskeletal Basis of Plant Growth and Form.* Academic Press, pp 109-129.
14. McCann M.C., Roberts K. (1994) Changes in cell wall architecture during cell elongation. *J. Exp. Bot.* **45**, 1683-1691.
15. MacDougall A.J., Needs P.W., Rigby N.M., Ring S.G. (1996) Calcium gelation of pectic polysaccharides isolated from unripe tomato fruit. *Carbohydr. Res.* **293**, 235-249.
16. MacDougall A.J., Rigby N.M., Ryden P., Tibbits C.W., Ring S.G. (2001) Swelling behavior of the tomato cell wall network. *Biomacromolecules* **2**, 450-455.
17. MacDougall A.J., Brett G., Morris V.J., Rigby N.M., Ridout M.J., Ring S.G. The effect of peptide interaction on the gelation behaviour of a plant cell wall pectin. *Carbohydr. Res.* (in press).
18. Manning G.S. (1996) The critical onset of counterion condensation: a survey of its experimental and theoretical basis. *Berenges Phys. Chem.* **100**, 909-922.
19. Maroudas A., Bannon C. (1981) Measurement of swelling pressure in cartilage and comparison with the osmotic pressure of constituent proteoglycans. *Biorheology* **18**, 619-632.
20. Morris E.R., Powell D.A., Gidley M.J., Rees D.A. (1982) Conformations and interactions of pectins. I. Polymorphism between gel and solid states of calcium polygalacturonate. *J. Mol. Biol.* **155**, 507-516.
21. Mort A.J., Qui F., Maness N.O. (1993) Determination of the pattern of methyl esterification in pectin. Distribution of contiguous nonesterified residues. *Carbohydr. Res.* **247**, 21-35.

22. Needs P.W., Rigby N.M., Ring S.G., MacDougall A.J. (2001) Specific degradation of pectins via a carbodiimide-mediated Lossen rearrangement of methyl esterified galacturonic acid residues. *Carbohydr. Res.* **333:** 47 - 58.

23. Nobel P.S. *Physicochemical and Environmental Plant Physiology*, 2nd edition, Academic Press, London, 1999.

24. O'Neill M.A., Warrenfeltz D., Kates K., Pellerin P., Doco T., Darvill A.G., Albersheim P. (1996) Rhamnogalacturonan-II, a pectic polysaccharide in the walls of growing plant cells, forms a dimer that is covalently cross-linked by a borate ester - *In vitro* conditions for the formation and hydrolysis of the dimer. *J. Biol. Chem.* **271**, 22923-22930.

25. Parr A.J., Waldron K.W., Ng A., Parker M.L. (1996) The wall-bound phenolics of Chinese water chestnut (Eleocharis dulcis). *J. Sci. Food Agric.* **71**, 501-507.

26. Parsegian V.A., Rand R.P., Fuller N.L., Rau D.C. (1986) Osmotic stress for the direct measurement of intermolecular forces. *Methods Enzymol.* **127**, 400-416.

27. Peitzsch R.M., Reed W.F. (1992) High osmotic stress behaviour of hyaluronate and heparin. *Biopolymers* **32**, 219-238.

28. Qi X.Y., Berhens B.X., West P.R., Mort A.J. (1995) Solubilization and partial characterization of extensin fragments from cell-walls of cotton suspension-cultures - evidence for a covalent cross-link between extensin and pectin. *Plant Physiol.* **108**, 1691-1701.

29. Rayle D.L. (1989) Calcium bridges are not load-bearing cell-wall bonds in *Avena* coleoptiles. *Planta* **178**, 92-95.

30. Redgwell R.J., MacRae E., Hallett I., Fischer M., Perry J., Harker R. (1997) *In vivo* and *in vitro* swelling of cell walls during fruit ripening. *Planta* **203**, 162-173.

31. Rubinstein M., Colby R.H., Dobrynin A.V., Joanny J-F. (1996) Elastic modulus and equilibrium swelling of polyelectrolyte gels. *Macromolecules* **29**, 398-406.

32. Ryden P., MacDougall A.J., Tibbits C.W., Ring S.G. (2000) Hydration of pectic polysaccharides. *Biopolymers* **54:** 398-405.

33. Saulnier L., Thibault J-F. (1999) Ferulic acid and diferulic acids as components of sugar-beet pectins and maize bran heteroxylans. *J. Sci. Food Agric.* **79**, 396-402.

34. Tibbits W., MaDougall A.J., Ring S.G. (1998) Calcium binding and swelling behaviour of a high methoxyl pectin gel. *Carbohydr. Res.* **310:** 101-107.

35. Tibbits C.W. (2000) Studies on the pectin network of the tomato fruit cell wall. Ph.D. Thesis, University of East Anglia.

36. Tomos A.D., Leigh R.A. (1999) The pressure probe: a versatile tool in plant cell physiology. *Annu. Rev. Plant Physiol. Plant Mol. Biol.* **50**, 447-472.

37. Voragen A.G.J., Pilnik W., Thibault J-F., Axelos M.A.V., Renard C.M.G.C. (1995) Pectins. In: Stephen A.M., ed. *Food Polysaccharides and their Applications*, Marcel Dekker, New York, pp 287-339.

38. Walkinshaw M.D., Arnott S. (1981) Conformations and interactions of pectins. II. Models for junction zones in pectinic acid and calcium pectate gels. *J.Mol. Biol.* **153**, 1055-1073.

39. Wilson R.H., Smith A.C., Kacurakova M., Saunders P.K., Wellner N., Waldron K.W. (2000) The mechanical properties and molecular dynamics of plant cell wall polysaccharides studied by Fourier-transform infrared spectroscopy. *Plant Physiol.* **124**, 397-405

40. Zwieniecki M.A., Melcher P.J., Holbrook N.M. (2001) Hydrogel control of xylem hydraulic resistance in plants. *Science* **291**, 1059-1062.

Structural features of pectic substances from hemicellulose extracts of apples

M.V. LUTZ, R. OECHSLIN, R. AMADÒ
Swiss Federal Institute of Technology Zurich
Institute of Food Science
ETH Zentrum
Schmelzbergstrasse 9
CH-8092 Zurich, Switzerland
renato.amado@ilw.agrl.ethz.ch

Abstract

Fractions rich in hemicelluloses were extracted from the depectinated residue of apples (var. Glockenapfel) at three different stages of ripeness. The fractions extracted with 4M NaOH were characterised by HPSEC, uronic acid determination and methylation analysis. Substantial amounts of galacturonic acid could be detected indicating the presence of pectic substances. An enrichment of the pectic substances was achieved by degradation of the xyloglucan with a specific *endo*-β-1,4-glucanohydrolase. A mixture rich in pectic polymers and xyloglucan oligomers was obtained. The pectic polymers were separated from the oligomers by ultrafiltration and characterised by methylation analysis and HPSEC. The results confirm the presence of pectic hairy regions in hemicellulose fractions of apples. The side chains of these pectic polymers consist mainly of highly ramified arabinans and linear galactans. The existence of a linkage between xyloglucans and pectins can still not be proven.

1. Introduction

Fruit ripening is associated with modifications of the cell wall polysaccharides which are partly responsible for changes in the texture of the fruit [1]. In order to understand the textural changes during ripening the elucidation of the fine structure of pectic polymers and their interactions with other cell wall polysaccharides is of prominent importance. Investigations of fractions at different stages help to understand the ripening process [2].

A widely accepted model of the primary cell walls of dicots proposes three distinct but interacting polymer networks: a cellulose-xyloglucan network, a network of pectic polysaccharides and a network consisting of structural proteins [3]. No interaction such as hydrogen-bonds, ionic interactions or covalent bonds between these three networks was considered. However, there is evidence for a covalent crosslink between xyloglucans and pectic polymers. Xyloglucan isolated from enzymatically depectinised apple pomace had a lower apparent molar mass than that from total pomace, either due

F. Voragen et al. (eds.), Advances in Pectin and Pectinase Research, 137-145.
© 2003 *Kluwer Academic Publishers. Printed in the Netherlands.*

to the presence of a cellulolytic activity in the commercial pectinase preparation used, or to degradation of the pectic part of a pectin-xyloglucan complex [4]. Evidence for a crosslink was also found in a fraction isolated by ion-exchange chromatography at high ionic strength in minor amounts. This fraction contained fucogalactoxyloglucans and pectins which could not be separated from each other [5]. The existence of this pectin-xyloglucan crosslink remains uncertain and has to be confirmed.
The aim of the present work was the elucidation of the structural features of pectic polymers found in hemicellulose fractions of apples.

2. Materials and Methods

2.1. MATERIALS AND SAMPLING

Unripe apples (*Malus malus* L., Rosaceae, var. Glockenapfel) were harvested three weeks after the end of the cell division phase. Ripe apples were picked at optimum maturity 14 weeks after the end of the cell division phase. Ripe apples were stored for one week, and stored apples for 19 weeks at 4°C and 95% relative humidity. Sampling and preparation of the alcohol-insoluble residues (AIR) has been performed by Fischer [6].

2.2. EXTRACTIONS

Extraction of pectic substances and hemicelluloses was carried out according to the method developed by Selvendran *et al.* [7]. The AIR was extracted by CDTA followed by two extraction steps with Na_2CO_3 under different conditions as described by Fischer [6]. The resulting depectinated residue (DR) was further extracted twice with NaOH at different concentrations to yield the fractions 1M and 4M rich in hemicelluloses as shown in Figure 1.

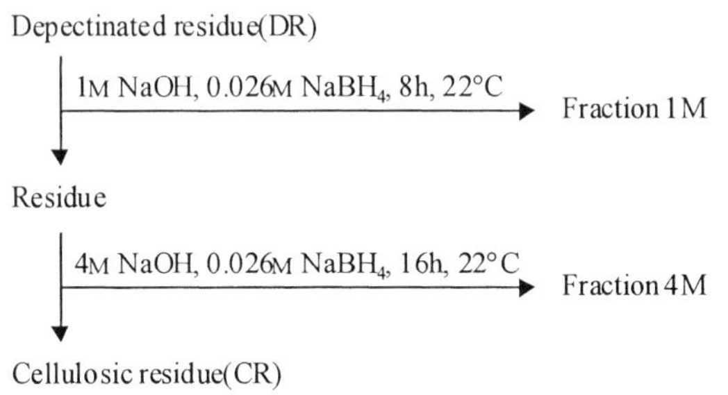

Figure 1. Alkaline extraction of the depectinated residue (DR).

2.3. ANALYTICAL METHODS

2.3.1. Reduction of uronic acids

Uronic acids were reduced to the corresponding deuterated neutral sugars by carbodiimide activation of the carboxyl groups followed by a reduction with $NaBD_4$ according to the method of Kim and Carpita [8]. The procedure was repeated once to assure complete reduction of the uronic acids.

2.3.2. Methylation analysis

Permethylation of the sample material was carried out twice according to Kvernheim [9]. Extraction, hydrolysis, reduction and acetylation was performed according to Harris *et al.* [10]. Peaks were identified by GLC-MS and quantification was done by GLC (FID) using calculated relative response factors on an effective carbon response (e.c.r.) basis [11]. For capillary GLC a DB-225 column (30 m x 0.25 mm id., 0.25 μm film thickness, J&W Scientific, Folsom, CA, USA) and a Hewlett Packard HP5890A (Agilent Laboratories, Palo Alto, CA, USA) were used. The temperature was held at 160°C for 1 min, then raised to 220°C at 2°C/min and held at 220°C for 19 min. GLC-MS was performed using a Fisons GC 8065 gas chromatograph (Carlo Erba, Milano, Italy) coupled to a Finnigan MAT SSQ 710 mass spectrometer (Finnigan MAT, San Jose, CA, USA).

2.3.3. High Performance Size Exclusion Chromatography (HPSEC)

A solution of 1 mg sample material in 350 μL sodium acetate buffer (0.05M, pH 4.5) was prepared. Insoluble material was removed by filtration (Titan Syringe Filters, PVDF-membrane, pore size 0.45 μm, Scientific Resources INC, Eatontown, NJ, USA). HPSEC was performed on a Merck Lachrom HPLC (Merck KGaA, Darmstadt, Germany) equipped with a Ultrahydrogel Linear column (Waters Corporation, Milford, MA, USA). Elution was carried out at 40°C with 0.05M sodium acetate buffer (pH 4.5) at 1.0 mL/min. The eluate was monitored using a Refractive Index detector.

2.4. ENZYMATIC DEGRADATION

To a solution (5 ml) of fraction 4M (30–35 mg in 0.05M acetate buffer, pH 3.6) 0.05 units of a xyloglucan-specific *endo*-β-1,4-glucanohydrolase were added [12]. The reaction mixture was incubated for 24 h at 37°C and fractionated by ultrafiltration after inactivation of the enzyme at 100°C for 10 min.

2.5. ULTRAFILTRATION

Ultrafiltration was carried out using Macrosep centrifugal concentrators (Pall Filtron, Northborough, MA, USA) with molecular weight cut-offs of 50 kDa and 3 kDa in a Sorvall RC-5B centrifuge (Kendro Laboratory Products, Newtown, CT, USA) at 5000 *g*. Retentates and filtrates were freeze-dried and stored at room temperature until analysed.

Results and Discussion

The results of the extraction of the depectinated residue (DR) are summarised in Table 1. Yields for the 1M fraction decreased from 18% for unripe apples to 10% for ripe apples and remained practically unchanged (8%) during storage. Yields for all 4M fractions were around 15%. The total amount of extracted hemicelluloses constituted 34% for unripe apples, decreased to 24% for ripe apples and remained practically constant (23%) for stored apples. These values are high compared to those found in other studies (14% and 20%) for different apple varieties [13-15].

TABLE 1. Yields of the hemicellulose fractions, in g/100g AIR.

	Stage of ripeness		
	unripe	ripe	stored
1M	18	10	8
4M	16	14	15
Total	34	24	23

Chromatographic separation of the pectins from xyloglucans was only partly possible. Therefore a degradation of the 4M fractions with a specific *endo*-β-1,4-glucanohydrolase was chosen to enrich the pectic polymers. The *endo*-β-1,4-glucanohydrolase was shown to hydrolyse structurally diverse xyloglucans from various sources but it did not hydrolyse any other cell wall component and was therefore considered to be xyloglucan-specific [12].

Fractions before and after enzymatic degradation were analysed using high performance size exclusion chromatography (HPSEC). Figure 2 shows the elution profile of fractions obtained with the stored apples.

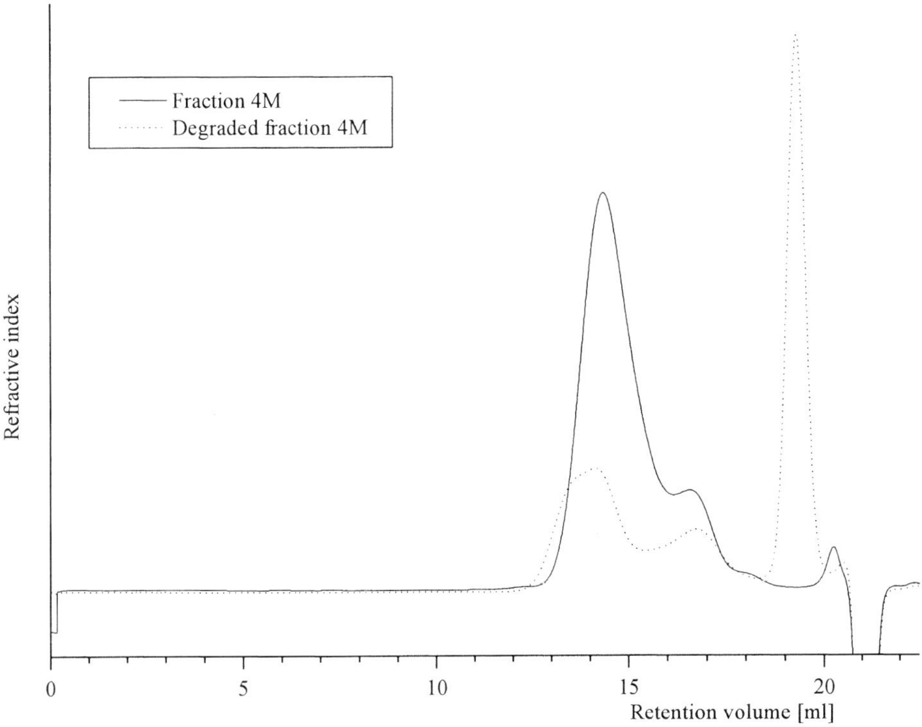

Figure 2. HPSEC of hemicellulose fractions 4M of the stored apple before and after enzymatic degradation.

The 4M fraction contained two distinct populations with different molecular weight distributions. The pattern for the degraded fraction still showed these two populations and in addition a new third low-molecular weight population which was expected due to the formation of xyloglucan-oligomers. A comparison of the chromatograms of the degraded fractions at different stages of ripeness is shown in Figure 3.
The three stages of ripeness showed a similar molecular weight distribution except for the highest molecular weight peak of the unripe apples which eluted at a lower retention volume. The shift within the first peak to lower molecular weight during ripening indicated a partial depolymerisation of the pectic polymers. Only little differences were observed during storage.
Ultrafiltration was carried out in centrifugal concentrators to enrich the pectic polymers and to remove the xyloglucan oligomers. A membrane with a MWCO of 50 kDa for proteins was used. The retentate (R50K) was freeze-dried and the filtrate (F50K) subjected again to ultrafiltration using a membrane with a MWCO of 3 kDa. The retentate (R3K) and filtrate (F3K) were freeze-dried and analysed by HPSEC (Figure 4). Two different populations of polymers were successfully separated from the oligomers by ultrafiltration. Yields are shown in Table 2. For R50K the yields increased during ripening from 25% for unripe apples to 37% for ripe and to 40% for stored apples.

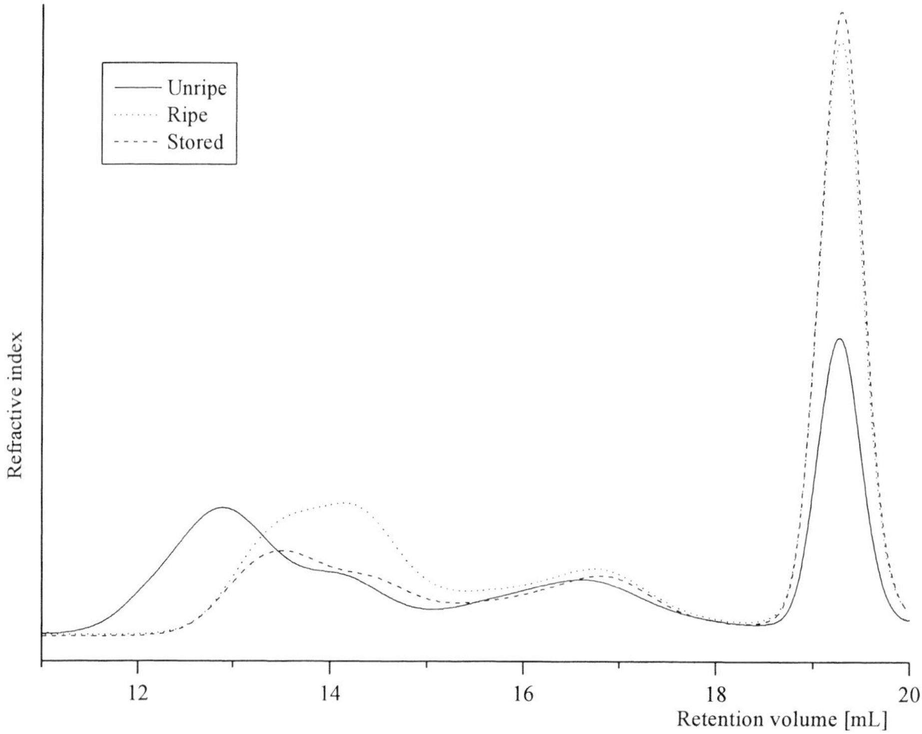

Figure 3. HPSEC of enzymatically degraded fractions 4M at different stages of ripening.

Based on the amount of AIR an overall increase of pectic polymers extracted in the 4M fractions was found. About 3% were found for R3K at all stages of ripening. Due to the small amount isolated no further characterisation of the R3K fractions was possible.

TABLE 2. Ultrafiltration of the degraded hemicellulose fractions. Yields in g/100g 4M fraction.

	Stage of ripeness		
Fraction	unripe	ripe	stored
R50K	25	37	40
R3K	4	3	3
Total	29	40	43

TABLE 3. Glycosyl linkage composition (mol%) of the pectic polymers (R50K) of the 4M fractions.

Glycosidic linkage composition		Stage of ripeness		
Sugar residue		unripe	ripe	stored
Arabinose	T-Ara*f*	14.8	15.8	16.5
	T-Ara*p*	1.0	1.3	0.9
	1,2-Ara*f*	0.2	0.2	0.5
	1,3-Ara*f*	3.0	3.9	3.5
	1,5-Ara*f*	9.5	11.0	12.9
	1,2,5-Ara*f*	0.4	1.6	1.3
	1,3,5-Ara*f*	6.7	9.3	10.5
	1,2,3,5-Ara*f*	4.0	3.6	4.1
		39.7	**46.6**	**50.2**
Fucose	T-Fuc*p*	0.1	0.1	0.2
		0.1	**0.1**	**0.2**
Galactose	T-Gal*p*	1.2	2.9	2.5
	1,3-Gal*p*	0.5	1.4	1.7
	1,4-Gal*p*	31.6	20.3	5.3
	1,6-Gal*p*	0.2	0.4	0.3
	1,2,4-Gal*p*	0.0	0.0	0.2
	1,3,4-Gal*p*	0.2	0.1	0.0
		33.8	**25.1**	**10.0**
Rhamnose	T-Rha*p*	0.1	0.1	0.2
	1,2-Rha*p*	1.3	1.6	2.2
	1,2,4-Rha*p*	2.2	3.3	3.5
		3.7	**5.0**	**5.9**
Galacturonic acid	T-GalA*p*	0.2	0.2	0.4
	1,4-GalA*p*	10.0	13.0	19.9
	1,3,4-GalA*p*	0.4	0.7	1.4
		10.6	**14.0**	**21.7**
Glucose	T-Glc*p*	0.8	0.6	0.8
	1,4-Glc*p*	5.8	2.1	2.6
	1,2,4-Glc*p*	0.3	0.2	0.4
	1,4,6-Glc*p*	0.3	0.2	0.1
		7.2	**3.1**	**3.9**
Xylose	T-Xyl*p*	1.2	2.4	2.7
	1,2-Xyl*p*	tr	tr	tr
	1,4-Xyl*p*	1.9	2.8	4.0
		3.0	**5.2**	**6.7**
Mannose	1,4-Man*p*	2.0	0.9	1.4
		2.0	**0.9**	**1.4**

tr: traces

Methylation analyses were performed on R50K at all stages of ripeness. The ratio of branching indicated highly ramified polymers for all stages of ripeness with a ratio of branching of 3.6 for the unripe apples, 2.5 for the ripe apples and 2.1 for the stored apples. The glycosyl linkage composition is shown in Table 3. The overall increase of ramifications is mainly due to an increase of the highly ramified arabinans which account for 40% in the unripe apples and increased to 50% in the stored apples.
The galactose found is mainly 1,4-linked and its amount decreased dramatically during ripening and storage from 34% to 10%. A loss of 1,4-linked galactose during ripening was also found by Wechsler *et al.* [16] for other pectic fractions from the primary cell wall and the middle lamella of apples. Linear, 1,4-linked galactans have been isolated from pectic material of different sources [16]. Arabinogalactan type I has also a backbone of 1,4-linked β-D-galactopyanosyl residues but contains short chains of 1,5-linked α-L-arabinofuranosyl residues linked to position O-3. The low amount of 1,3,4-branched galactose residues indicate that the decrease in galactose probably was related to a degradation of an unsubstituted galactan.
The detected rhamnose and galacturonic acid residues were typical for the backbone of pectins. Some of the galacturonic acid residues were branched at O-3, and could be part of a xylogalacturonan that carries side chains of single xylose units as described by De Vries *et al.* [17].
The results of the methylation analyses confirm the presence of pectic hairy regions in the hemicellulose fractions of apples. The side chains of these pectic polymers consist mainly of highly ramified arabinans and linear galactans. Traces of the fucogalactoxyloglucan residues (1,4-Glc*p*, 1,4,6-Glc*p*, T-Xyl*p*, 1,2-Xyl*p*, T-Gal*p*, 1,2-Gal*p*, T-Fuc*p*) were found in the enzymatically degraded hemicellulose fractions. They could indicate a covalent linkage between the pectic hairy regions and the xyloglucans. Nevertheless, a covalent linkage between xyloglucans and pectic hairy regions remains to be proven. Work is in progress to further characterise these pectic fractions.

Acknowledgements

The authors wish to thank Barbara Eder Aebersold for the HPSEC work and Novozymes A/S, Bagsvaerd, Denmark for providing the xyloglucan-specific *endo*-β-1,4-glucanase.

References

[1] Van Buren, J.P. (1979) The chemistry of texture in fruits and vegetables, *J. Texture Stud.* **10**, 1-23.
[2] Kunzek, K., Kabbert, R., and Gloyna, D. (1999) Aspects of material science in food processing: changes in plant cell walls of fruits and vegetables, *Z. Lebensm.-Unters. Forsch.* **208**, 233-250.
[3] Carpita, N.C. and Gibeaut, D.M. (1993) Structural models of primary cell walls in flowering plants: consistency of molecular structure with the physical properties of the wall during growth, *Plant J* **3**(1), 1-30.
[4] Watt, D.K., Brasch, D.J., Larsen, D.S., and Melton, L.D. (1999) Isolation, characterisation, and NMR study of xyloglucan from enzymatically depectinised and non-depectinised apple pomace, *Carbohydr. Polym.* **39**, 165-180.

[5] Renard, C.M.G.C., Voragen, A.G.J., Thibault, J.F., and Pilnik, W. (1991) Studies on apple protopectin. IV: apple xyloglucans and influence of pectin extraction treatments on their solubility, *Carbohydr. Polym.* **15**, 387-403.
[6] Fischer, M. (1993) Changes in the pectic substances during ripening of apples, Dissertation No 10336, ETH, Zürich.
[7] Selvendran, R.R., Stevens, B.J.H., and O'Neill, M.A. (1985) Developments in the isolation and analysis of cell walls from edible plants, in C.T. Brett and J.R. Hillman (eds.), *Biochemistry of Plant Cell Walls*, Cambridge University Press, Cambridge, pp. 39-78.
[8] Kim, J.B. and Carpita, N.C. (1992) Changes in esterification of uronic acid groups of cell wall polysaccharides during elongation of maize coleoptiles, *Plant Physiol.* **98**, 646-653.
[9] Kvernheim, A.L. (1987) Methylation analysis of polysaccharides with butyllithium in dimethyl sulfoxide, *Acta Chem. Scand.* **Ser. B41**, 150-152.
[10] Harris, P.J., Henry, R.J., Blakeney, A.B., and Stone, B.A. (1984) An improved procedure for the methylation analysis of oligosaccharides and polysaccharides, *Carbohydr. Res.* **127**, 59-73.
[11] Sweet, D.P., Shapiro, R.H., and Albersheim, P. (1975) Quantitative analysis by various G.L.C response-factor theories for partially methylated and partially ethylated acetates, *Carbohydr. Res.* **40**, 217-225.
[12] Pauly, M., Andersen, L.N., Kauppinen, S., Kofod, L.V., York, W.S., Albersheim, P., and Darvill, A. (1999) A xyloglucan-specific *endo*-b-1,4-glucanase from *Aspergillus aculeatus*: expression cloning in yeast, purification and characterization of the recombinant enzyme, *Glycobiology* **9**(1), 93-100.
[13] Knee, M. (1973) Polysaccharides and glycoproteins of apple fruit cell walls, *Phytochemistry* **12**, 637-653.
[14] Aspinall, G.O. and Fanous, H.K. (1984) Structural investigation on the non-starchy polysaccharides of apples, *Carbohydr. Polym.* **4**, 193-214.
[15] Voragen, A.G.J., Schols, H.A., and Pilnik, W. (1986) Structural features of the hemicellulose polymers of apples, *Z. Lebensm.-Unters. Forsch.* **183**, 105-110.
[16] Wechsler, D.E., Strasser, G.R., and Amadò, R. Structural features of pectic substances during growth and ripening of apples. in J. Visser and A.G.J. Voragen (eds.), *Pectins and Pectinases - Proceedings of an International Symposium*. 1995. Wageningen, The Netherlands, Elsevier Sciences B.V., pp. 651-656.
[17] De Vries, J.A., den Uijl, C.H., Voragen, A.G.J., Rombouts, F.M., and Pilnik, W. (1983) Structural features of the neutral sugar side chains of apple pectic substances, *Carbohydr. Polym.* **3**, 193-205.

during biosynthesis and by extensive modification *in muro*. The range and precise nature of the functions of the pectic network are far from clear (Willats *et al.*, 2001b).

The recent elucidation of the structure of the *Arabidopsis* genome and its analysis have emphasised the massive investment that this plant makes, in terms of genes and proteins, towards the modification of the pectic network once it is in the cell wall (Arabidopsis Genome Initiative, 2000). For example, there are over seventy genes that encode putative pectin esterases that have the capacity to modify HG. Moreover, the range of structural features found in pectic polysaccharides indicates at least 53 glycosyltransferases and associated activities that are required for their synthesis (Ridley *et al.*, 2001). It is a salutary point that none of these biosynthetic enzymes have yet been identified or characterised and few of the putative pectin-modifying genes in *Arabidopsis* are characterised in terms of precise functions in the context of plant physiology, growth, development or defence. Some modifications of the pectic cell wall matrix in response to mutations are known (Fagard *et al.*, 2000; Willats *et al.*, 2001b; His *et al.*, 2001). However, as far as we are aware, mutations have not yet been identified for any of the genes directly involved in the synthesis or modification of pectic polymers.

2. Immunochemistry and beyond: adding value to antibodies

Knowledge of pectic polysaccharide structure is generally derived from analytical techniques that typically require large amounts of material and, critically, the homogenisation of complex multicellular structures. In such cases, all information about the specific or restricted location of a pectic domain within organs, tissues or regions of a cell wall is lost. Analytical techniques, complementary to sugar composition and linkage analyses, are required to determine this contextual information. Currently, antibody-based approaches are the most useful way of identifying specific pectic domains in the context of plant cell walls and are the basis of this chapter. Other developing approaches are FT-IR and related spectroscopies (McCann *et al.*, 2001).

Anti-pectin monoclonal antibodies have been generated and used for over ten years and the early history of these has been covered elsewhere (Knox, 1997). In recent years the development of an extended range of probes to various components and forms of all three major pectic polymers has increased awareness of the complexity of pectic network within primary cell walls (Willats *et al.*, 2000b; 2001b).

There are two major aspects of anti-pectin antibodies to consider. Firstly, what is the specificity of a particular antibody? What antigen does it bind to and how far can the epitope be defined? This is not always a straightforward matter as the directed generation of antibodies to complex carbohydrates is not always easy or possible. Serendipity still can play a role during antibody selection and lead to the isolation of antibodies with (initially unlooked for) specificities of considerable interest. Secondly, antibodies can be used in a range of approaches and techniques that extend beyond localisation of particular epitopes in a section of plant material. For example, we are developing techniques to use anti-pectin probes to identify and characterise mutants and we are also developing immunomodulation strategies with the aim of disrupting the functions of specific pectic domains within growing plants.

In this chapter, we focus on the generation and use of monoclonal antibodies to the HG and the RG-I components of pectin that have been the concern of our laboratory over recent years. Table 1 summarises the characteristics of the anti-pectin monoclonal antibodies discussed in this chapter and provides source references.

TABLE 1. Specificities of anti-HG and anti-RG-I monoclonal antibodies discussed in this chapter

Antibody	Antigen / Epitope	References
JIM7	HG / high methyl-ester	Knox *et al.*, 1990; Willats *et al.*, 2000a
JIM5	HG / low methyl-ester	Knox *et al.*, 1990; Willats *et al.*, 2000a
PAM1	HG / blockwise pattern of de-esterification	Willats *et al.*, 1999a; 2000a
LM7	HG / non-blockwise pattern of de-esterification	Willats *et al.*, 2001c
LM5	RG-I / $(1\rightarrow4)$-β-D-galactan	Jones *et al.*, 1997; Willats *et al.*, 1999b
LM6	RG-I / $(1\rightarrow5)$-α-L-arabinan	Willats *et al.*, 1998; 1999b

3. Homogalacturonan: roles for substitution and *in muro* modification

HG is the most abundant pectic polymer and appears to occur throughout all primary cell walls and some secondary cell walls. However, although HG is abundant and occurs widely in the primary cell wall matrix, the final form of the polymer appears to be highly variable. This variability has two origins. The first source of variation is during synthesis within the endomembrane system. HG is thought to be synthesized in a largely methyl-esterified form (although it is not known if the extent or pattern of methyl-esterification varies along HG chains at this stage). In addition to methyl-esterification, GalA residues can be acetylated at C2 and/or C3 hydroxyl groups. A current view is that acetylated-HG is restricted to certain organs and species (e.g. sugar beet roots) but it is possible that HG acetylation may occur to some extent in some cells of all plants. Similarly, analysis of some plant materials has led to the identification of a xylose-substituted form of HG, known as xylogalacturonan (e.g. pea testa, watermelon and apple) and an apiose-substituted form of HG known as apiogalacturonan has been identified in duckweed (Mohnen, 1999). Furthermore, as RG-II has an HG backbone it is also likely to be linked to HG in biosynthetic terms and can perhaps best be viewed as a highly modified form of HG. The second source of variation is the *in muro* modification of HG by enzymes. Pectin methyl-esterases are an important group of enzymes (see below) and their action influences the subsequent cleavage of HG chains by polygalacturonases and pectate lyases. The concerted action of these enzymes will influence the potential of HG chains for association with calcium and the formation of cross-links and gels. Such *in muro* modifications will influence the local properties of the pectic network (for example porosity and pH) which in turn have important implications for cell wall properties in general.

Immunolabelling of a section of plant material with a monoclonal antibody (JIM7) that binds to highly methyl-esterified HG (Knox *et al.*, 1990; Willats *et al.* 2000a) indicates that HG occurs throughout cell walls as a major component of the matrix. However, it is the selective removal of methyl-esters that appears to be a highly variable factor and one that is responsive to local requirements in specific cells or micro-domains of cell walls. Evidence that the removal of methyl-esters is finely controlled and dynamic has been obtained from observations with two recently developed monoclonal antibodies (see below) that have different binding properties with respect to the extent and pattern of HG methyl-esterification. An indication of variation in occurrence of HG epitopes within cell walls around parenchyma intercellular spaces and an immunodot-assay indicating binding of anti-HG monoclonal antibodies to a range of model pectins with defined levels of methyl-esterification, are shown in Figure 1.

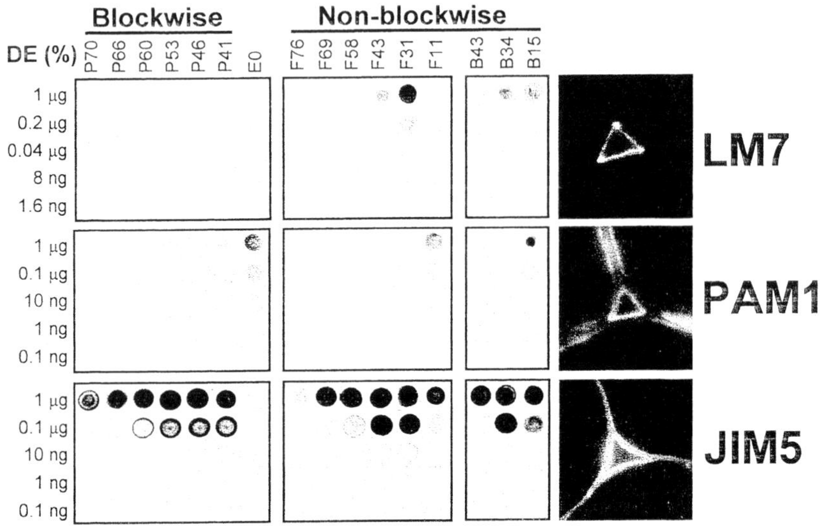

Figure 1. Immuno-dot-assays of model pectins with a range of degrees and patterns of methyl-esterification of HG components and immunofluorescent micrographs of cell walls at regions of an intercellular space as seen in transverse sections of pea stem parenchyma with the anti-HG antibodies LM7, PAM1 and JIM5. The immunoassay was performed on nitrocellulose. The degree of methyl-esterification (DE%) is indicated on the top row. Blockwise de-esterification was obtained with a plant pectin methyl-esterase (P) and non-blockwise de-esterification with a fungal pectin methyl-esterase (F) or base catalysis (B). For full details see Willats *et al.*, 2001c.

PAM1 is a phage display monoclonal antibody that appears to require at least 30 contiguous de-esterified GalA residues in a HG chain for binding (Willats *et al.*, 1999a). This would appear to be an unreasonably large epitope to be involved in direct contact between antigen and antibody. The requirement for such a large structure is likely to indicate that PAM1 recognises a length-dependent conformation of HG, i.e. above a certain number of residues the polymer takes on a specific conformation that cannot be achieved by smaller oligomers and it is this conformation that is recognised by PAM1. Precedents of length-dependent and conformational epitopes of polysaccharides have been reported (Brisson *et al.*, 1997; Zou *et al.*, 1999). The specificity of PAM1 makes it a highly selective probe for long un-esterified regions of HG and such block-structures are thought to be a major factor in the cross-linking of HG chains by calcium ions and therefore gel formation by HG domains.

The distribution pattern of methyl-ester groups also appears to be important in influencing the properties of calcium cross-linked HG. The monoclonal antibody LM7 binds to an epitope of HG that is partially methyl-esterified and is present in greatest abundance where the distribution of methyl groups is non-blockwise. As plant esterases have been generally thought to act in a blockwise manner, i.e. removing methyl-esters from contiguous residues, it was not expected that LM7 would bind to plant material in any substantial way. However, the epitope bound by LM7 is abundant and ubiquitous and is the first HG epitope identified to have a specific and universal pattern of occurrence within plants (Willats *et al.*, 2001c). LM7 binds to regions of the intercellular matrix that line developing intercellular spaces and is often particularly abundant at the corners of intercellular spaces (Willats *et al.*, 2001c). This location at the outer face of cell walls has been observed in all parenchyma systems so far examined. This indicates a specific role for a particular pattern of de-esterification of HG in regions where adhered cells are being separated during intercellular space formation. However, a very important and related observation is that the LM7 epitope is also located in restricted regions of coherent outer epidermal cell walls that are in line with the anticlinal radial adhered walls (Willats *et al.*, 2001c). In this location, of course, there is no cell separation and in aerial shoot tissue in particular, it is thought to be a critical point of cell adhesion.

Observations with LM7, PAM1 and JIM5 indicate that specific non-blockwise and blockwise patterns of HG de-esterification are components of cell wall architecture and that they are likely to have specific functions at these locations. These properties have been explored in an series of *in vitro* assays in which gels have been made from model pectins with structural features that reflect the epitopes recognised by LM7 or PAM1 (Willats *et al.*, 2001c). The results have indicated that pectins rich in non-blockwise distributions of methyl ester groups can form calcium-mediated gels with distinctive properties in terms of elasticity and porosity (Willats *et al.*, 2001c). In general terms, the highly methyl-esterified HG that is first inserted into cell walls can be viewed as a raw material from which finely-tuned functionalities can be created by the action of pectin methyl-esterases. Such fine-tuning or local remodelling of HG chains may provide an explanation for the large array of pectin methyl-esterase genes and proteins predicted from the analysis of genomes and of the different properties and activities of pectin methyl-esterases isolated from mung bean and other species (Catoire *et al.*, 1998; Goldberg *et al.*, 2001).

The extent and pattern of acetylation, and its modulation by acetyl-esterases is also likely to be another aspect of the functional fine-tuning of HG. Abundant acetylation of pectin appears to be restricted to certain organs such as sugar beet roots and potato tubers. As yet, no antibody has been developed to acetylated HG. However, anti-HG antibodies have been used in conjunction with a pectin methyl-esterase in a subtractive method to identify acetylated regions and this has indicated that acetylation was associated with decreased intercellular adhesion of sugar beet suspension-cultured cells (Liners *et al.*, 1994). As discussed above, whether HG acetylation is a feature of pectin that is restricted to certain cells or regions of cell walls in a range of plants or is taxonomically restricted is currently unclear. An indication of the potential of antibodies to answer such questions has been obtained with preliminary observations made with a monoclonal antibody to an epitope of a xylogalacturonan (XG). Using a sample of pea testa XG (Le Goff *et al.*, 2001) we have generated a monoclonal antibody that binds to XG polymers. Examination of radial sections of pea testae has indicated that the epitope is only found in cell walls of loosely attached cells at the inner face of the testa (Willats *et al.*, 2001b). Significantly, this antibody also binds to loosely attached cells in root caps in a range of dicotyledon and monocotyledon species (unpublished observations).

The potential for variation in HG structure is also emphasized by examination of seed mucilage. Development of an *in situ* immunolocalisation technique has indicated that the HG component of *Arabidopsis* seed mucilage is highly methyl-esterified: JIM7 binds abundantly throughout the mucilage and PAM1 only binds after the mucilage has been treated with base that removes methyl-ester groups (Willats *et al.*, 2001a). Seed mucilage provides an interesting system for the analysis of gel structure and HG interaction with other polymers such as cellulose (Willats *et al.*, 2001a). The interaction of pectic polymers with proteins is also likely to be of critical importance for cell wall functions (Carpin *et al.*, 2001; Mollet *et al.*, 2001).

Our current view of the functions of HG is that it contributes to the mechanical properties of cell walls that are likely to have impact upon processes such as cell expansion. HG also provides a substrate for the generation of signalling molecules (Ridley *et al.*, 2001). It is clear that HG plays an important role in regulating local properties such as porosity and pH. However, the sheer extent to which HG is modified and the functional possibilities created indicates that we have much still to learn about the biological roles of HG.

4. Rhamnogalacturonan-I: a question of side chains

Structural and compositional analysis of RG-I polymers has indicated that they are highly heterogeneous (Ridley *et al.*, 2001; Willats *et al.*, 2001b). The functions of molecules currently designated RG-I are unknown. RG-I is currently viewed as a series of polymers that have an alternating rhamnogalacturonan backbone that carries side chains that are often rich in galactose and arabinose, but may also contain other sugars. Two monoclonal antibodies have been developed to two of the major structures that occur in these side chains: $(1\rightarrow4)$-β-D-galactan and $(1\rightarrow5)$-α-L-arabinan. These monoclonal antibodies, LM5 and LM6 respectively, are well defined as they were generated subsequent to immunisation with neoglycoproteins (Jones *et al.*, 1997; Willats *et al.*, 1998). LM5 and LM6 have been used for studies in a range of systems and the observations made have supported the structural observations on RG-I variability and have also provided new insights into the occurrence of these structural features within plant tissues. In short, they have allowed aspects of RG-I structure to be put into cellular contexts (Jones *et al.*, 1997, Vicré et al., 1998; Willats *et al.*, 1998; Bush and McCann, 1999; Willats *et al.*, 1999b; McCartney *et al.*, 2000; Orfila and Knox, 2000; Orfila *et al.*, 2001). The use of LM5 and LM6 indicates that pectic galactan and arabinan components can occur in different cells of a developing system at different times. For example, the arabinan epitope is abundant in the walls of proliferating cells and the galactan epitope appears during cell differentiation at the carrot root apex (Willats *et al.*, 1999b). The extent or similarity of epitope association with cell processes in a range of species is not yet certain. Secondly, the use of these antibodies has indicated that their epitopes are not always evenly distributed within cell walls. Figure 2 shows the occurrence of the LM5 and LM6 epitopes in cell walls around an intercellular space in a legume cotyledon. In this case the galactan epitope is restricted to an inner region of the cell wall.

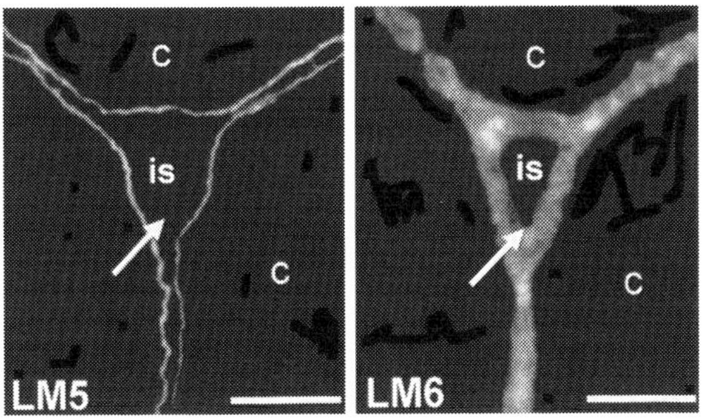

Figure 2. Immunofluorescent labelling of a resin-section through parenchyma cells in a mature aduki bean cotyledon with the monoclonal antibodies LM5 and LM6. The binding of LM5, specific to (1→4)-β-D-galactan, is restricted to an inner region of the parenchyma cell walls immediately adjacent to the plasma membrane. The binding of LM6, specific to (1→5)-α-L-arabinan, occurs throughout the parenchyma cell walls and the cell wall thickenings surrounding intercellular spaces. c, cotyledon cell; is, intercellular space. Arrows indicate a corner of the intercellular space. Scale bars = 10 μm.

A consistent feature that has been established in several systems is that the LM5 galactan epitope is absent from pit field regions of cell walls that surround plasmodesmata (Orfila and Knox, 2000). Clearly, this is likely to alter cell wall properties in some way, but it is uncertain what properties the presence (or indeed absence) of (1→4)-β-D-galactan provides. Some indication that pectic galactan may influence mechanical properties has been derived from the analysis of developing pea cotyledons (McCartney *et al.*, 2000). The galactan epitope appears relatively late in cotyledon development in an inner region close to the plasma membrane (equivalent to that shown in Figure 2) and this correlates with an increase in tissue firmness as determined by compressive testing. Figure 3 shows the force / displacement profiles with representative immunolabelled regions of cotyledon. It will be of interest to assay the mechanical properties of potato tuber in which the pectic galactan has been depleted by expression of a microbial galactanase (Sørenson *et al.*, 2000).

It is now becoming clear that the pectic (1→5)-α-L-arabinan component is likely to have roles in the cell wall that are distinct from those of (1→4)-β-D-galactan. Some insight comes from the observation that the deposition of the pectic arabinan is disrupted during development and ripening of the *Cnr* ripening mutant of tomato (Orfila *et al.*, 2001). Two of the major aspects of the phenotype of this pleiotropic mutation are reduced cell adhesion and firmer, non-swelling cell walls in the pericarp (Thompson *et al.*, 1999; Orfila *et al.*, 2001). It is currently not possible to say which of these factors the disrupted deposition of arabinan impacts on most (or indeed whether both or neither). However, the recent report that the absence of pectic arabinan side chains is associated with loosely attached cells in tobacco cell suspension cultures (Iwai *et al.*, 2001) is suggestive of a role for pectic arabinan in intercellular adhesion. In this context, it may also of particular interest that the pectic arabinan occurs abundantly in

His, I., Driouich, A., Nicol, F., Jauneau, A., Höfte, H. (2001) Altered pectin composition in primary cell walls of *korrigan*, a dwarf mutant of *Arabidopsis* deficient in a membrane-bound endo-1,4,-β-glucanase. *Planta* **212**, 348-358.

Iwai, H., Ishii, T., Satoh, S. (2001) Absence of arabinan in the side chains of the pectic polysaccharides strongly associated with cell walls of *Nicotiana plumbaginifolia* non-organogenic callus with loosely attached constituent cells. *Planta* in press (DOI 10.1007/s004250100559).

Jones, L., Seymour, G.B., Knox, J.P. (1997) Localization of pectic galactan in tomato cell walls using a monoclonal antibody specific to (1→4)-β-D-galactan. *Plant Physiol.* **113**, 1405-1412.

Knox, J.P. (1997) The use of antibodies to study the architecture and developmental regulation of plant cell walls. *Int. Rev. Cytol.* **171**, 79-120.

Knox, J.P., Linstead, P.J., King, J., Cooper, C., Roberts, K. (1990) Pectin esterification is spatially regulated both within cell walls and between developing tissues of root apices. *Planta* **181**, 512-521.

Le Goff, A., Renard, C.M.G.C., Bonnin, E., Thibault, J.-F. (2000) Extraction, purification and chemical characterization of xylogalacturonans from pea hulls. *Carbohydr. Polym.* **45**, 325-334.

Liners, F., Gaspar, T., van Cutsem, P. (1994) Acetyl- and methyl-esterification of pectins of friable and compact sugar-beet calli: consequences for intercellular adhesion. *Planta* **192**, 545-556.

McCann, M.C., Bush, M., Milioni, D., Sado, P., Stacey, N.J., Catchpole, G., Defernez, M., Carpita, N.C., Höfte, H., Ulvskov, P., Wilson, R.H., Roberts, K. (2001) Approaches to understanding the functional architecture of the plant cell wall. *Phytochem.* **57**, 811-821.

McCartney, L., Ormerod, A.P., Gidley, M.J., Knox, J.P. (2000) Temporal and spatial regulation of pectic (1→4)-β-D-galactan in cell walls of developing pea cotyledons: implications for mechanical properties. *Plant J.* **22**, 105-113.

Mohnen, D. (1999) Biosynthesis of pectins and galactomannans. *In* D Barton, K Nakanishi, O Meth-Cohn, eds, Comprehensive natural products chemistry, Vol 3. Elsevier Science B.V., Amsterdam, pp 497-527.

Mollet, J.-C., Park, S.-Y., Nothagel, E.A., Lord, E.M. (2000) A lily stylar pectin is necessary for pollen tube adhesion to an in vitro stylar matrix. *Plant Cell* **12**, 1737-1749.

O'Neill, M.A., Albersheim, P., Darvill, A. (1990) The pectic polysaccharides of primary cell walls. *In* P. M. Dey, ed, Methods in Plant Biochemistry, Vol. 2. Academic Press, London, pp 415-441.

Orfila, C., Knox, J.P. (2000) Spatial regulation of pectic polysaccharides in relation to pit fields in cell walls of tomato fruit pericarp. *Plant Physiol.* **122**, 775-781.

Orfila, C., Seymour, G.B., Willats, W.G.T., Huxham, I.M., Jarvis, M.C., Dover, C.J., Thompson, A.J., Knox, J.P. (2001) Altered middle lamella homogalacturonan and disrupted deposition of (1→5)-α-L-arabinan in the pericarp of *Cnr*, a ripening mutant of tomato. *Plant Physiol.* **126**, 210-221.

Ridley, B.L., O'Neill, M.A., Mohnen, D. (2001) Pectins: structure, biosynthesis, and oligogalacturonide related signalling. *Phytochem.* **57**, 929-967.

Sørensen, S.O., Pauly, M., Bush. M., Skjøt, M., McCann, M.C., Borkhardt, B., Ulvskov, P. (2000) Pectin engineering: Modification of potato pectin by *in vivo* expression of an endo-1,4-β-D-galactanase. *PNAS* **97**, 7639-7644.

Thompson, A.J., Tor, M., Barry, C.S., Vrebalov, J., Orfila, C., Jarvis, M.C., Giovannoni, J.J., Grierson, D., Seymour, G.B. (1999) Molecular and genetic characterization of a novel pleiotropic tomato-ripening mutant. *Plant Physiol.* **120**, 383-389.

Vicré, M., Jauneau, A., Knox, J. P., Driouich A. (1998) Immunolocalization of β(1→4)- and β(1→6)-D-galactan epitopes in the cell wall and Golgi stacks of developing flax root tissues. *Protoplasma* **203**, 26-34.

Willats, W.G.T., Gilmartin P.M., Mikkelsen, J.D, Knox, J.P. (1999a) Cell wall antibodies without immunization: Generation and use of de-esterified homogalacturonan block-specific antibodies from a naive phage display library. *Plant J.* **18**, 57-65.

Willats, W.G.T., Limberg, G., Bucholt, H.C., van Alebeek, G.-J., Benen, J., Christensen, T. M.I.E., Visser, J., Voragen, A., Mikkelsen, J.D., Knox, J.P. (2000a) Analysis of pectic epitopes recognised by hybridoma and phage display monoclonal antibodies using defined oligosaccharides, polysaccharides and enzymatic degradation. *Carbohydr. Res.* **327**, 309-320.

Willats, W.G.T., Marcus, S.E., Knox J.P. (1998) Generation of a monoclonal antibody specific to (1→5)-α-L-arabinan. *Carbohydr. Res.* **308**, 149-152.

Willats, W.G.T., McCartney, L., Knox, J.P. (2001a) In-situ analysis of pectic polysaccharides in seed mucilage and at the root surface of *Arabidopsis thaliana*. *Planta* **213**, 37-44.

Willats, W.G.T., McCartney, L., Mackie W., Knox, J.P. (2001b) Pectin: cell biology and prospects for functional analysis. *Plant Mol. Biol.* 47, 9-27.

Willats, W.G.T., Orfila, C., Limberg, G., Buchholt, H.C., van Alebeek, G.-J.W.M., Voragen, A.G.J., Marcus, S.E., Christensen, T.M.I.E., Mikkelsen, J.D., Murray, B.S. and Knox, J.P. (2001c) Modulation of the degree and pattern of methyl-esterification of pectic homogalacturonan in plant cell walls: implications for pectin methyl-esterase action, matrix properties and cell adhesion. *J. Biol. Chem.* 276, 19404-19413.

Willats, W.G.T., Steele-King, C.G., Marcus, S.E., Knox, J.P. (1999b) Side chains of pectic polysaccharides are regulated in relation to cell proliferation and cell differentiation. *Plant J.* **20**, 610-628.

Willats, W.G.T., Steele-King, C.G., McCartney, L., Orfila, C., Marcus S.E., Knox, J.P. (2000b) Making and using antibody probes to study plant cell walls. *Plant Physiol. Biochem.* **38**, 27-36.

Zou, W., Mackenzie, R., Therien, L., Hirama, T., Yang, Q.L., Gidney, M.A., Jennings, H.J. (1999) Conformational epitope of the type III group B Streptococcus capsular polysaccharide. *J. Immunol.* **163**, 820-825.

PECTIC SUBSTANCES FROM SOYBEAN CELL WALLS DISTINGUISH THEMSELVES FROM OTHER PLANT CELL WALLS PECTINS

M.M.H. HUISMAN, H.A. SCHOLS AND A.G.J. VORAGEN
Wageningen University, Agrotechnology and Food Sciences, Laboratory of Food Chemistry, P.O. Box 8129, 6700 EV Wageningen, The Netherlands. E-mail: Henk.Schols@chem.fdsci.wag-ur.nl

Abstract

In soybean cell wall material, pectic substances are the major non-starch polysaccharide. These pectic substances distinguish themselves from pectic substances of cell wall material from other plants in the absence of homogalacturonan, the presence of fucose residues in the xylogalacturonan, and two uncommon structural features of the pectic arabinogalactan side chains, namely the presence of internal (1,5)-linked arabinofuranose and terminal arabinopyranose. Therefore, these pectic substances are rather resistant to degradation by both established (like polygalacturonase) and novel (like RG-hydrolase) pectic enzymes. The hemicellulosic polysaccharides in the soybean cell wall appeared to be predominantly xyloglucans, composed of XXXG-type building units like most legume xyloglucans.

The uncommon structural features of soybean cell wall pectic substances explain their resistance to degradation by enzymes generally used to degrade this kind of polymers, and indicates that a search for new enzymes is required to enable enzymatic modification of these polysaccharides.

1. Introduction

In general, pectic polysaccharides are major components of the primary cell wall of dicotyledons (e.g. sycamore, citrus and legumes) and gymnosperms (e.g. Douglas fir)[i]. This is also true for soybean. The ChSS fraction (38% of the WUS), the DASS fraction (7% of the WUS), and about 90% of the 1 MASS fraction (16% of the WUS) consist of pectic substances. This implies that at least 60% of soybean WUS consist of pectic polysaccharides.

F. Voragen et al. (eds.), Advances in Pectin and Pectinase Research, 159-168.
© 2003 *Kluwer Academic Publishers. Printed in the Netherlands.*

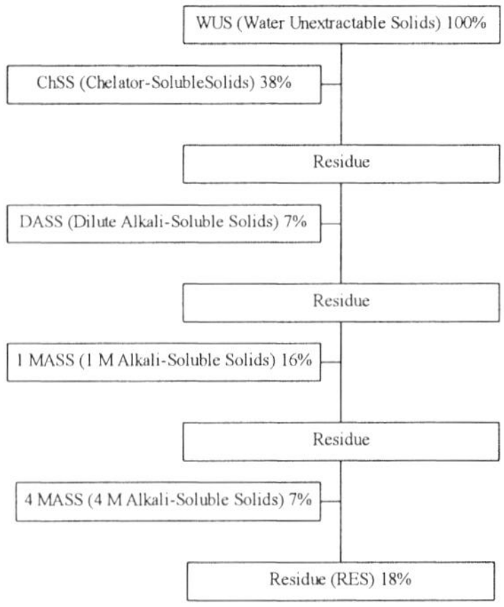

Figure 1. Sequential extraction of soybean WUS and the yield on polysaccharide basis (%)[ii].

If we let go of the classification of pectic polysaccharides in smooth and hairy regions as suggested by Schols *et al.*[iii], and choose for a classification based on the structure of the backbone, we end up with two categories of pectic polysaccharides.

The first group contains those parts of the pectin with exclusively (1,4)-linked galacturonic acid in the backbone. According to this definition, homogalacturonan, rhamnogalacturonan II, and xylogalacturonan belong to this group. The biosynthesis of the backbone of this category of pectic polymers requires a polygalacturonate 4-α-galacturonosyltransferase capable of transferring UDP-galacturonic acid residue to the non-reducing end of a pre-existing chain of galacturonic acid residues. The homogalacturonan is first synthesised as an unesterified chain and subsequently at least partially methylesterified by an enzyme that transfers methyl groups to the carboxyl groups of polygalacturonic acid[iv,v,vi]. It seems likely that other substituents (*e.g.* acetyl esters and neutral sugar residues, like xylose) can be attached to the backbone in like manner.

The second group contains those parts of the pectin with a rhamnogalacturonan backbone. The biosynthesis of the backbone of this category of pectic polymers requires two glycosyl transferases, a galacturonic acid transferase capable of adding galacturonic acid residues to rhamnose and a rhamnose transferase capable of adding rhamnose residues to galacturonic acid in a rhamnogalacturonan chain[iv]. Possibly followed by the attachment of arabinan, galactan, or arabinogalactan side chains. A rhamnogalacturonan I galactosyltransferase able to transfer a galactose residue to short pectic (1,4)-β-D-galactan chains was recently isolated[vii].

2. The absence of homogalacturonan

Homogalacturonans are chains of (1,4)-linked α-galacturonic acid, which may be partly methyl esterified. So far, pectic substances from plant cell walls were always found to consist of smooth galacturonan regions interrupted by blocks of ramified rhamnogalacturonan regions, so-called hairy regions (Figure 2). These smooth galacturonan regions were predominantly present in the fractions extracted with aqueous solvents, often containing chelating agents. It is shown that chelator-extractable pectins from soybean (38% of the WUS) do not contain homogalacturonans[viii].

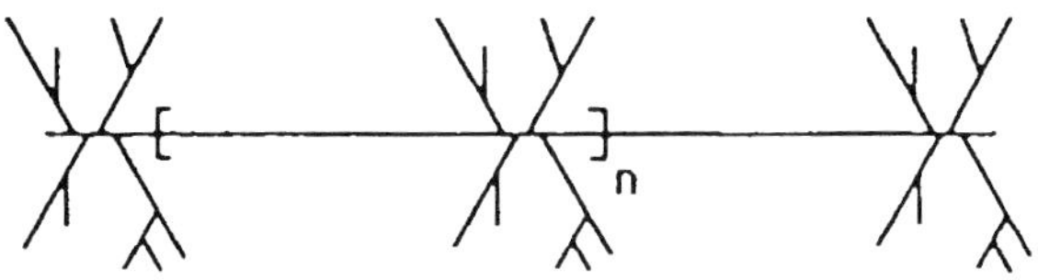

Figure 2. Schematic structure of apple pectin[ix].

In an additional experiment, monoclonal antibodies to low ester pectin (JIM5[x]), high ester pectin (JIM7[x]), and de-esterified homogalacturonan (PAM1[xi]) were used to characterise the chelator-extractable pectic substances from soybean meal. In immuno-dot assays, JIM5 was found to bind to both native and saponified pectic substances in the soybean ChSS extract. JIM 7 is only able to bind to the native pectins in the ChSS extract, being specific for high ester pectin. The epitopes recognised by these two anti-homogalacturonan monoclonal antibodies are however not defined[xi]. The epitopes recognised by PAM1 are, however, well defined. PAM1 binds specifically to de-esterified and unsubstituted homogalacturonan, optimal binding requires in the region of 30 de-esterified galacturonic acid residues[xi]. PAM1 does not bind to saponified ChSS, again indicating that blocks of pectic homogalacturonan are absent. The absence of homogalacturonan explains the resistance of (saponified) chelator-extractable pectic substances towards degradation with polygalacturonase, but makes their extraction by CDTA remarkable.

The absence of homogalacturonan in chelator-extractable pectic substances has a number of consequences. A first step in the elucidation of the structure of the pectic substances usually is the solibilisation of the pectic polysaccharides in the cell wall by using highly purified polygalacturonase[xii,xiii,xiv,xv], or a fractionation of isolated pectic substances after digestion with highly purified homogalacturonan-degrading enzymes[xvi,xvii,xviii]. Rhamnogalacturonan I and II are the major polysaccharides solubilised from various plant tissues using this approach. Due to the absence of homogalacturonan, this first step is not applicable. This implies that other routes had to be followed in this research[viii,xix].

A second consequence of the absence of homogalacturonan is the inability of soybean pectic substances to participate in calcium cross-linking and gel formation. This is not only important in the cell wall itself, but also indicates that extraction of soybean meal yields pectins unable to form the typical gels as known for other pectins (*e.g.* from apples and citrus). Jam and jelly manufacture is one of the main uses of industrially extracted pectins[xx] and soybean pectic substances are most likely unsuitable for this purpose.

Another consequence is that the absence of homogalacturonan effects the physiology of the cell wall, since pectins are mediators of wall porosity. A mild treatment with pectinase can enlarge the channels (for macromolecular transport across plant cell walls), without adversely affecting cell viability, enabling significantly larger molecules to pass through the wall[xxi].

3. Rhamnogalacturonan II

Rhamnogalacturonan II is structurally very different from the later discussed rhamnogalacturonans. It is covalently linked in the primary cell wall through homogalacturonans. This sub-unit has a highly conserved structure. The backbone is composed of about nine (1,4)-linked α-D-galacturonic acid residues. Four different, complex side chains are attached to O2 or O3 of four of the backbone residues. These side chains contain rhamnose and several rare characteristic monosaccharides such as apiose, 2-*O*-methyl-L-fucose, 2-*O*-methyl-D-xylose, aceric acid (3-*C*-carboxy-5-deoxy-L-xylose), KDO (3-deoxy-D-manno-octulosonic acid), and DHA (3-deoxy-D-lyxo-heptulosaric acid)[xxii].

The functions of rhamnogalacturonan II are not known. However, the discovery that the rhamnogalacturonan II occurs as dimers covalently cross-linked by borate diesters[xxiii,xxiv] suggest that these parts of the pectin are of importance in the interconnection of the pectic polysaccharides[xxv]. It is suggested that these boron cross-links are the "load-bearing", acid-labile linkages that are hydrolysed by a decrease in cell wall pH during auxin-induced cell expansion[xxvi].

The ChSS fraction was analysed for its apiose content, a glycosyl residue diagnostic of rhamnogalacturonan II. The chelator-extractable pectic substances appeared to be free of apiose, and consequently of rhamnogalacturonan II.

4. Soybean xylogalacturonan contains fucose residues

The presence of xylogalacturonan regions, in which terminal xylose is linked directly to the galacturonosyl residues, has been suggested before in soybean[xvi,xxvii]. Xylogalacturonans also occur in apple[xxviii], cotton suspension-cultured cell walls[xxix], watermelon[xxix], and pea hulls[xxx,xxxi]. The xylogalacturonans take up 12 to 18% of the soybean chelator-extractable pectic substances. The xylogalacturonans from different sources vary in the degree of xylose substitution, the degree of methyl esterification, and the distribution of the substituents over the backbone[iii].

So far, two enzymes are known, which are able to hydrolyse xylogalacturonan. Firstly, an exo-galacturonase, which is not hindered in its action by single branches of xylose. It releases galacturonic acid and the β-Xyl*p*-(1,3)-GalA*p* dimer from apple hairy regions[xxxii] and from soybean pectic substances[viii]. Secondly, xylogalacturonan hydrolase (XGH), a polygalacturonase which is specific for xylose-substituted galacturonan[xxxiii]. This enzyme is able to degrade xylogalacturonan derived from acid-hydrolysed gum tragacanth and apple hairy regions. Soybean pectic substances are resistant to degradation by XGH even after acid hydrolysis. The structure of soybean xylogalacturonan is obviously not identical to apple and gum tragacanth xylogalacturonan. The resistance to XGH can be caused by a different degree of xylose substitution or by an altered distribution of the substituents over the backbone. A third possibility is further substitution of the xylose substituents, as suggested by the presence of (1,2)-, (1,4)-, and (1,2,3)-linked xylose[viii].

The most obvious possibility is further substitution of the xylose residues with fucose. The fucose content of the polymeric pectins after enzymatic removal of the arabinogalactan side chains (polymeric residue P)[viii] is relatively high (7 mol%), compared to pectic substances originating from other sources, and has not been reported before except by Beldman *et al.*[xxxii] a soluble pectic polysaccharide also from soy (11 mol% of fucose). In addition, the disaccharide Fuc*p*-(1,2)-Xyl was found to be present in the digest obtained after partial hydrolysis of soybean cotyledon polysaccharides[xxvii]. This dimer cannot originate from other polysaccharides, because in the only hemicellulose containing both xylose and fucose (xyloglucans) they are not covalently linked to each other, but occur in the trimeric side chain α-L-Fuc*p*-(1→2)-α-L-Gal*p*-(1→2)-α-D-Xyl*p*-(1→. However, this disaccharide might be present as a side chain of polygalacturonic acid, since the linkage analysis composition of the ChSS fraction and polymeric residue P[viii] shows the presence of (1,2)-linked xylopyranose residues. The NMR analyses[viii] do not provide any information about the sugar residue to which the terminal fucose residue is attached.

To investigate the substitution of xylose residues with fucose, (saponified) soybean polymeric residue P was incubated with α-fucosidase[xxxiv]. This enzyme released 5.5% of the total amount of fucose in the fraction, and this increased the susceptibility of polymeric residue P for exo-galacturonase, releasing 3.6% of the galacturonic acid residues from saponified polymeric residue P and 6.4% after pre-incubation with α-fucosidase. Due to the lack of a proper standard, the content of the characteristic dimer (Xyl-GalA) could not be quantified properly. However, it is clear that the amount released from saponified polymeric residue P increased after pre-incubation with α-fucosidase. This proves that a part of the xylose substituents in the soybean xylogalacturonan are substituted with fucose residues, which would explain the resistance of this xylogalacturonan to XGH. However, XGH is still unable to degrade the saponified polymeric residue P after pre-incubation with α-fucosidase. Substitution of the xylose residues with other sugar residues, like xylose[xxxi], galactose or arabinose cannot be excluded. The structure of soybean xylogalacturonan is more complex than just a polygalacturonic acid chain substituted with single terminal xylose residues.

The α-fucosidase releases only 5.5% of the fucose from saponified polymeric residue P. The amount of α-fucosidase added to the incubation mixture was abundant, so the remaining fucose can not be released by this enzyme. These fucose residues might be unavailable for the enzymes due to (1) small alterations in the structure of the fucosyl-containing side chain, (2) steric hindrance of other parts of the molecule, or (3) the fucose residues might be present in the rhamnogalacturonan regions, as in sycamore RG-I[i,xxxv]. It was not determined if the α-fucosidase would be able to release these fucose residues.

The presence of xylogalacturonan could not be confirmed by the immuno-dot assay using antibodies to xylogalacturonan. Two xylogalacturonan monoclonal antibodies (7G3 and 5A10), developed using a pea xylogalacturonan, were used. The antibodies are likely to recognise more densely xylose-substituted homogalacturonan. No defined oligosaccharide inhibitor of antibody binding has been found (personal communication Dr. Bill Willats). Neither of the antibodies is able to bind to the native or saponified pectins in the ChSS extract and the polymeric residue P. Based on the release of galacturonic acid and the characteristic dimer (Xyl-GalA) by exo-galacturonase we are convinced of the presence of xylogalacturonan in the chelator-extractable soybean pectic substances. It is plausible that the distribution of the xylose residues over the galacturonan chain or the presence of fucose residues in the xylogalacturonan hinders binding of the antibodies.

5. Rhamnogalacturonan and the occurrence of internal (1,5)-linked arabinofuranose and arabinopyranose residues in its arabinogalactan side chains

The rhamnogalacturonan regions are substituted with neutral polymers like arabinans, galactans, and arabinogalactans[i,xx,xxxvi]. The rhamnose to galacturonic acid ratio of these pectic substances varies between 0.05 and 1[iii]. All the arabinose and galactose present in chelator- and dilute alkali-extractable pectin is present in rhamnogalacturonan side chains[ii]. The rhamnogalacturonans (including the (arabino)galactan side chains) represent over 60% of the chelator-extractable pectic substances. The arabinosyl- and galactosyl-rich side chains are attached to O4 of the rhamnosyl residues. The ratio of (1,2)- to (1,2,4)-linked rhamnose in the ChSS extract is 1:2[viii]. It can be concluded that the average length of the (arabino)galactan side chains would be in the range of 45 to 50 residues. This was confirmed by the length of the arabinan/galactan/arabinogalactan oligomers released from the ChSS extract by short incubation with endo-galactanase, exo-galactanase, endo-arabinanase, and arabinofuranosidase B. HPAEC analysis of this digest showed the presence of galactan and arabinogalactan oligomers with a degree of polymerisation up to about 15 and arabinan oligomers with a degree of polymerisation up to 30-35. The length of the (arabino)galactan side chains in soybean pectic substances exceeds the length of those in sycamore RG-I, ranging in size from one to fourteen glycosyl residues[xxxv].

RG hydrolase is unable to degrade the rhamnogalacturonans in the ChSS extract. The enzyme is probably hindered by these long (arabino)galactan side chains. RG hydrolase is only able to hydrolyse the glycosidic linkage between a galacturonic acid and a rhamnose if the rhamnose residue is unbranched or substituted through O4 with only a single galactose residue[xxxvii]. Only few unbranched rhamnose residue are available in the ChSS extract[viii] and based on the high arabinose and galactose content it is expected that none or only very few of the (1,2,4)-linked rhamnose residues will carry a single galactose residue. Besides, the long neutral side chains will sterically prevent the enzyme to approach the potential glycosidic bond to be hydrolysed. Enzymatic removal of a large number of the neutral side chains and reduction of the length of the remaining side chains increased the susceptibility of the rhamnogalacturonan for RG hydrolase, but only after saponification[viii].

A large part of the arabinogalactan side chains can be released by incubation with arabinogalactan-degrading enzymes.[xix] The remaining polymeric material was hydrolysed with acid and fractionated by size-exclusion chromatography. NMR analysis showed that terminal and 4-linked galactose were attached to the rhamnose residues in the rhamnogalacturonan of the remaining weak acid-resistant polymers.[viii] In sycamore RG-I, terminal and 4-linked galactose were also found to be attached to rhamnose. In addition, terminal arabinosyl and 3-, 6-, 2,6-, and 3,6-linked galactosyl residues were attached to O4 of the 2-linked rhamnosyl residues in sycamore[xxxviii]. These results showed that at least seven different side chains, most commencing with a galactosyl residue, were attached to the RG-I backbone in sycamore. In the soybean ChSS fraction, 3-, 6-, 2,6- and 3,6-linked galactosyl residues do not occur[viii]. However, side chains beginning with an arabinosyl residue might be present in soybean. NMR analysis of the remaining weak acid-resistant polymers[viii] did not show the presence of such side chains, but they might have been removed during acid hydrolysis. In rhamnogalacturonan I solubilised from walls of suspension-cultured sycamore cells, arabinosyl residues glycosidically linked to rhamnose were observed, and always further substituted with arabinosyl residues. Maize RG-I also contains arabinogalactan side chains, which consist of a single arabinosyl residue attached to a rhamnosyl residue, which is substituted with one or more galactosyl residues[xxxix].

So far, soybean pectic side chains were described as galactan backbones with arabinose side chains of one or two residues. The results of our studies show that soybean pectin contains rather long arabinan side chains in addition to (arabino)galactan side chains. The arabinan side chains are presumably of considerable length (up to dp 30-35). And more important arabinose was found to occur interspersed in the galactan chain as an *internal* (1,5)-linked arabinofuranose residue or as a terminal arabino*pyranose* residue[xl]. In this view it distinguishes itself from other arabinogalactans, described so far. Internal arabinose was detected once before, but from a very different source, namely arabinogalactan type II from larch[xli,xlii,xliii]. The presence of an internal arabinofuranose residue in a pectic arabinogalactan chain in cell wall polysaccharides has not been reported previously, not in soybean, nor in other fruit or vegetable cell walls. Both ring forms of arabinose also occur in arabinogalactan type II, in which L-arabinofuranosyl, and to a lesser extent L-arabinopyranosyl residues, terminate some of the side chains[xliv]. However, in soybean arabinogalactan we are dealing with linear (1,4)-linked β-D-galactan, which is type I arabinogalactan. The existence of arabinopyranose residues in pectic arabinogalactan is rarely reported. Only pectic substances isolated from the roots of *Angelica acutiloba* Kitagawa were shown to contain a small proportion of arabinopyranose[xlv].

The presence of internal (1,5)-linked arabinofuranose and terminal arabinopyranose residues in the galactan chains has far-reaching consequences for amongst others the enzymatic degradation of these arabinogalactan side chains. It (partly) explains their resistance to enzymes like endo-galactanase. It is shown[xix] that a combination of endo-galactanase, exo-galactanase, endo-arabinanase, and arabinofuranosidase B is required for optimal removal of the (arabino)galactan side chains. However, after incubation with this combination of enzymes 8% of the arabinose and galactose residues present in the ChSS extract still remained attached to the pectin, probably due to the complexity of the remaining part of the side chains. Arabinofuranosidase B is unable to release the non-reducing terminal arabinopyranose residue due to its ring form; an arabinopyranosidase would bring help here. An enzyme able to hydrolyse the glycosidic linkage between a (terminal or (1,4)-linked) galactose residue and a (1,5)-linked arabinofuranose residue is yet unknown. It is also unknown how the presence of an interspersed arabinose residue in a galactan chain effects the activity of the endo-galactanase. This could not be deduced from the structure of the degradation products, because complete degradation was not established and it is not known whether the resulting arabinogalacto-oligosaccharides can be degraded further by endo-galactanase. However, it is likely that the endo-galactanase is unable to degrade arabinogalactan chains in the vicinity of an internal arabinose residue.

So far, the incomplete removal of the arabinogalactan side chains from pectic polymers by enzymes remained unexplained. Now we have some potential causes in the presence of (1,5)-linked arabinofuranose and terminal arabinopyranose. With this knowledge and the constantly improving techniques we can focus further research on the complete removal of the pectic arabinogalactan side chains.

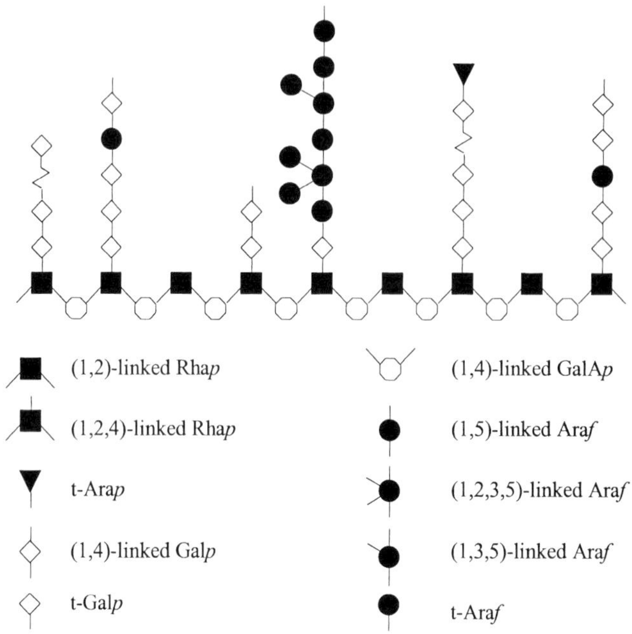

Figure 5.10. Hypothetical structure of the rhamnogalacturonan regions of chelator-extractable soybean pectic substances. The distribution of the acetyl groups is not shown. A strictly alternating sequence of Rha and GalA, termed RG-I[i], is presented here, but the structure of the rhamnogalacturonan backbone is as yet unknown. Only short (arabino)galactan side chains are presented in this figure, but the true average length of the side chains is 45 to 50 residues[xl].

6. Conclusions

We now know that the pectins from soybean differ significantly from all pectins described so far. Not only the absence of homogalacturonan, but also the presence of internal (1,5)-linked arabinofuranose and terminal arabinopyranose in the pectic arabinogalactan side chains has never before been described for parenchymatous tissues of plants.

Soybean pectic substances appeared to be rather resistant to degradation by enzymes, which are generally used in our group to degrade this kind of polymers. Their structural features, as we know them now, explain this phenomenon. Their complexity appeared, however, to be of no concern during their degradation in the gastrointestinal tract of livestock[xlvi,xlvii]. So it is very likely that enzymes able to degrade these polysaccharides do exist.

Our approach to investigation of the chemical structure of cell wall polysaccharides from soy bean meal turned out to be successful. Extraction of the polysaccharides from their original source always has to be the first step in their characterisation. Chemical characterisation and enzymatic degradation studies already reveal the main lines of their chemical structure. The use of MALDI-TOF MS, MS/MS, and NMR spectroscopy provided valuable and detailed information on the structure of the polysaccharide fragments obtained after enzymatic or chemical degradation. Although a lot of pure and well-characterised enzymes are available nowadays, this research showed the need for additional enzymes, like for example arabinopyranosidase, RG hydrolases active on side chain-containing rhamnogalacturonan, *etc.*

In general, carbohydrate researchers try to translate their findings into universal models. Our investigations show that such models are limited in their value, knowledge about the structure of plant cell wall polysaccharides is only valid for that particular tissue of that plant species.

7. Acknowledgement

We thank J.P. Knox and W.G.T. Willats for the antibodies and their assistance with the immuno-dot assays.

8. References

1. O'Neill, M.A.; Albersheim, P.; Darvill, A.G. In: Dey (Ed.) Methods in Plant Biochemistry, Vol 2, Carbohydrates; Academic Press: London, 1990; 415-441.
2. Huisman, M.M.H.; Schols, H.A.; Voragen, A.G.J. Carbohydr Res 1998, 37, 87-95.
3. Schols, H.A.; Voragen, A.G.J. In: Visser, J.; Voragen, A.G.J. (Eds.) Progress in Biotechnology, Vol 14, Pectins and Pectinases; Elsevier: Amsterdam, 1996; 3-19.
4. Kauss, H.; Swanson, A.L.; Arnold, R.; Odzuck, W. Biochim Biophys Acta 1969, 192, 55-61.
5. Mohnen, D.; Lou Doong R.; Liljebjelke, K.; Fralish, G.; Chan, J. In: Visser, J.; Voragen, A.G.J. (Eds.) Progress in Biotechnology, Vol 14, Pectins and Pectinases; Elsevier: Amsterdam, 1996; 109-126.
6. Zhan, D.; Janssen, P.; Mort, A.J. Carbohydr res 1998, 308, 373-380.
7. Peugnet, I.; Goubet, F.; Bruyant-Vannier, M.P.; Thoiron, B.; Morvan, C.; Schols, H.A.; Voragen, A.G.J. Planta 2001, 213, 435-445.
8. Huisman, M.M.H.; Fransen, C.T.M.; Kamerling, J.P.; Vliegenthart, J.F.G.; Schols, H.A.; Voragen, A.G.J. Biopolymers 2001, 58, 279-294.
9. Pilnik, W.; Voragen, A.G.J. In: Fox, P.F. (Ed.) Food Enzymology, Vol. 1; Elsevier Applied Science: London, 1991; 303-336.
10. Knox, J.P.; Linstead, P.J.; King, J.; Cooper, C.; Roberts, K. Planta 1990, 181, 512-521.
11. Willats, W.G.T.; Gilmartin, P.M.; Dalgaard Mikkelsen, J.; Knox, J.P. Plant J 1999, 18, 57-65.
12. Talmadge, K.W.; Keegstra, K.; Bauer, W.D.; Albersheim, P. Plant Physiol 1973, 51, 158-173.
13. Darvill, A.G.; McNeil, M.; Albersheim, P. Plant Physiol 1978, 62, 418-422.
14. McNeil, M.; Darvill, A.G.; Albersheim, P. Plant Physiol 1980, 66,1128-1134.
15. York, W.S.; Darvill, A.G.; McNeil, M.; Stevenson, T.T.; Albersheim, P. Methods Enzymol 1986, 118, 3-40.
16. Kikuchi, T.; Sugimoto, H. Agr Biol Chem 1976, 40, 87-92.

17. de Vries, J.A.; Rombouts, F.M.; Voragen, A.G.J.; Pilnik, W. Carbohydr Polym 1982, 25-33.
18. Schols, H.A.; Vierhuis, E.; Bakx, E.J.; Voragen, A.G.J. Carbohydr Res 1995,275,343-360.
19. Huisman, M.M.H.; Schols, H.A.; Voragen, A.G.J. Carbohydr Polym 1999, 38, 299-307.
20. Voragen, A.G.J.; Pilnik, W.; Thibault, J.-F.; Axelos, M.A.V.; Renard, C.M.G.C. In: Stephen, A.M. (Ed.) Food Science and Technology, Vol 67, Food Polysaccharides and their Applications; Dekker: New York, 1995; 287-339.
21. Baron-Epel, O.; Gharyal, P.K.; Schindler, M. Planta 1988, 175, 389-395.
22. Whitcombe, A.J.; O'Neill, M.A.; Steffan, W.; Albersheim, P.; Darvill, A.G. Carbohydr Res 1995, 271, 15-29.
23. Ishii, T.; Matsunaga, T. Carbohydr Res 1996, 284, 1-9.
24. O'Neill, M.A.; Warrenfeltz, D.; Kates, K.; Pellerin, P.; Doco, T.; Darvill, A.G.; Albersheim, P. J Biol Chem 1996, 271, 22923-22930.
25. Albersheim, P.; Darvill, A.G.; O'Neill, M.A.; Schols, H.A.; Voragen, A.G.J. In: Visser, J.; Voragen, A.G.J. (Eds.) Progress in Biotechnology, Vol 14, Pectins and Pectinases; Elsevier: Amsterdam, 1996; 47-55.
26. Loomis, W.D.; Durst, R.W. BioFactors 1992, 3, 229-239.
27. Aspinall, G.O.; Cottrell, I.W.; Egan, S.V.; Morrison, I.M.; Whyte, J.N.C. J Chem Soc C 1967, 170, 1071-1080.
28. Schols, H.A.; Bakx, E.J.; Schipper, D.; Voragen, A.G.J. Carbohydr Res 1995, 279, 265-279.
29. Yu, L.; Mort, A.J. In: Visser, J.; Voragen, A.G.J. (Eds.) Progress in Biotechnology 14. Pectins and Pectinases; Elsevier Science: Amsterdam, 1996; pp 79-88.
30. Weightman, R.M.; Renard, C.M.G.C.; Thibault, J.-F. Carbohydr Polym 1994, 24, 139-148.
31. Renard, C.M.G.C.; Weightman, R.M.; Thibault, J.-F. Int J Biol Macromol 1997, 21, 155-162.
32. Beldman, G.; van den Broek, L.A.M.; Schols, H.A.; Searle-van Leeuwen, M.J.F.; van Laere, K.M.J.; Voragen, A.G.J. Biotechnol Lett 1996, 18, 707-712.
33. van der Vlugt-Bergmans, C.J.B.; Meeuwsen P.J.A.; Voragen, A.G.J.; van Ooyen, A.J.J. Accepted for publication in Appl Environ Microbiol.
34. Kroef, van den Broek, Voragen, unpublished results.
35. Lau, J.M.; McNeil, M.; Darvill, A.G.; Albersheim, P. Carbohydr Res 1987, 168, 245-274.
36. Bacic, A.; Harris, P.J.; Stone, B.A. The Biochemistry of Plants, Vol 14, Carbohydrates; Academic Press: London, 1988; 297-369.
37. Schols, H.A.; Voragen, A.G.J.; Colquhoun, I.J. Carbohydr Res 1994, 256, 97-111.
38. McNeil, M.; Darvill, A.G.; Albersheim, P. Plant Physiol 1982, 70, 1586-1591.
39. Thomas, J.R.; Darvill, A.G.; Albersheim, P. Carbohydr Res 1989, 185, 279-305.
40. Huisman, M.M.H.; Brüll, L.P.; Thomas-Oates, J.E.; Haverkamp, J.; Schols, H.A.; Voragen, A.G.J. Carbohydr Res 2001, 330, 103-114.
41. Clarke, A.E.; Anderson, R.L.; Stone, B.A. Phytochemistry 1979, 18, 521-540.
42. Bouveng, H.O.; Lindberg, B. Acta Chem Scand 1958, 12, 1977.
43. Young, R.A.; Sarkanen, K.V. Carbohydr Res 1977, 59, 193-201.
44. Selvendran, R.R. J Cell Sci Suppl 1985, 2, 51-88.
45. Kiyohara, H.; Yamada, H.; Otsuka, Y. Carbohydr Res 1987, 167, 221-237.
46. van Laar, H.; Tamminga, S.; Williams, B.A.; Verstegen, M.W.A.; Engels, F.M. Anim Feed Sci Technol 1999, 79, 179-193.
47. van Laar, H. Soya beans and maize: The effect of chemical and physical structure of cell wall polysaccharides on fermentation kinetics; Thesis Wageningen Agricultural University: Wageningen, 2000.

REGULATION OF PECTINOLYTIC GENE EXPRESSION IN *ASPERGILLUS*

R.P. DE VRIES[1] AND L. PAŘENICOVÁ[2]
[1]*Molecular Genetics of Industrial Microorganisms, Wageningen University, Dreijenlaan 2, 6703 HA Wageningen, The Netherlands*
Current address: Microbiology, Utrecht University, Padualaan 8, 3584 CH Utrecht, The Netherlands, e-mail: r.p.devries@bio.uu.nl
[2] *Dipartimento di Biologia, Universita degli studi di Milano, Via Celoria 26, 20133 Milano, Italy*

Abstract

This chapter discusses the current knowledge of the regulation of pectinolytic genes in *Aspergillus* and presents a summary of recent data obtained on this topic. So far, little is known about regulatory factors involved in pectin degradation. It appears that the pectinolytic regulatory system is more complex than the xylanolytic regulatory system, in which a general activating factor (XlnR) plays a central role. Some pectinolytic genes are expressed in response to the presence of pectin or pectin derived compounds, but constitutively expressed pectinolytic genes have also been reported. Indications for the involvement of general regulatory factors such as carbon catabolite repression and pH regulation have also been reported. Specific regulatory systems have been proposed for the regulation of genes encoding pectinolytic enzymes acting on the side chains of pectin.

1. Introduction

Due to the complex structure of the pectic polysaccharides, biodegradation of these compounds requires a wide range of enzymes. These enzymes can be divided into two groups: those acting on the pectin main chain (endo-and exopolygalacturonases, pectin and pectate lyases, rhamnogalacturonan hydrolases and lyases, pectin methyl and acetyl esterases, and rhamnogalacturonan acetyl esterase) and those acting on side chains of the pectic hairy regions (arabinofuranosidases, endoarabinanase, β-galactosidase, endogalactanases, and feruloyl esterases).

The production of such a large number of enzymes is a significant energetic burden for microorganisms. Therefore, many microorganisms have developed complex regulatory systems for the production of these enzymes. In this chapter the regulatory factors affecting the production of pectinolytic enzymes by species of the filamentous fungus *Aspergillus* will be discussed. These fungi are widely studied as a source of pectinolytic enzymes and serve as a model system for phytopathogenic fungi such as *Botrytis cinerea*.

The genus *Aspergillus* can be divided into several subgroups, some of which consist of pathogenic fungi (*e.g. A. fumigatus*, *A. flavus* and *A. parasiticus*). Most important for industrial applications are members of the group of black aspergilli. This group has been

F. Voragen et al. (eds.), Advances in Pectin and Pectinase Research, 169-181.
© 2003 *Kluwer Academic Publishers. Printed in the Netherlands.*

classified in detail over the last 10 years, which resulted in the clear distinction of eight groups of black aspergilli (*A. niger*, *A. tubingensis*, *A. foetidus* var., *A. carbonarius*, *A. japonicus*, *A. aculeatus*, *A. heteromorphus*, and *A. ellipticus*) (67) and to the classification of two new species represented by the isolates *A. brasiliensis* IMI381727 and *A. aculeatus* CBS114.80 (68). Products of several of these species have obtained a GRAS (Generally Regarded As Safe) status, which allows them to be used in food and feed applications. The black aspergilli have good fermentation capabilities and secrete high levels of protein making them interesting organisms for industrial applications.

The expression of genes encoding polysaccharide degrading enzymes is usually induced by a monomeric or dimeric degradation product of the polymeric substrate. In aspergilli the regulatory systems involved in (hemi-)cellulose degradation are among the best studied. In *A. niger* it has been shown that the xylanolytic system is induced in the presence of D-xylose (84) and the regulation of the expression occurs at the transcriptional level via the activator XlnR (86). Recently, a model was suggested for the role of XlnR in the regulation of (hemi-)cellulose degradation by *A. niger* (Fig. 1) (23). During growth of *A. niger* in the presence of arabinoxylan XlnR is activated by monomeric xylose which is already present in the substrate. Alternatively, xylose can be released by endoxylanase B and β-xylosidase, both of which are present at low constitutive levels. The expression of (hemi-) cellulolytic genes is then activated by XlnR (29, 41, 85).

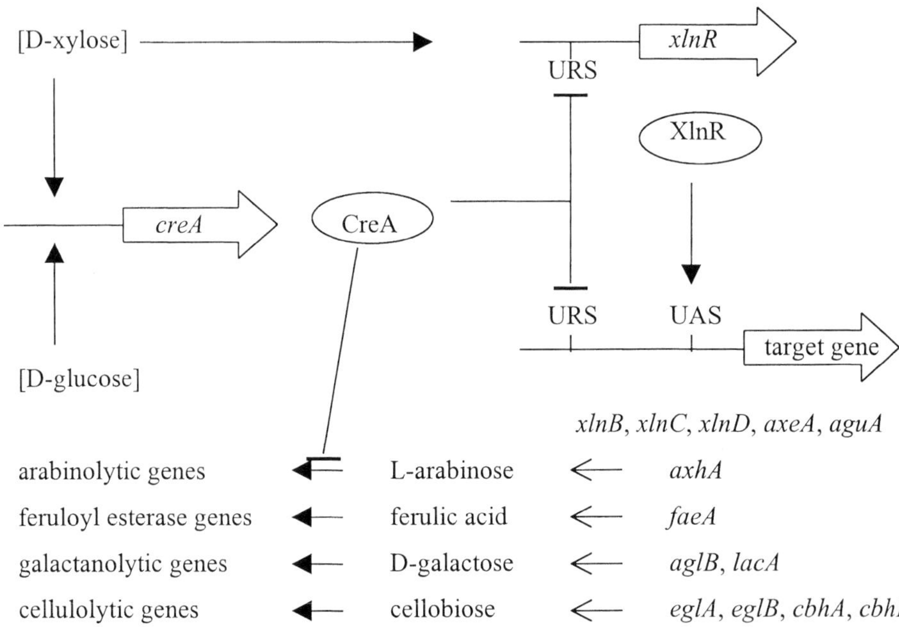

FIG. 1. Model for the role of XlnR and CreA in the regulation of the genes encoding (hemi-) cellulose degrading enzymes by *Aspergillus niger*. UAS, upstream activating sequence; URS, upstream repressing sequence; ⟶ activation; ⟶| repression; ⟶ released by corresponding enzyme. Reprinted from de Vries *et al.* (23) with permission from the publisher.

High concentrations of xylose cause CreA mediated repression, which results in reduced expression levels of these genes (31). The presence of glucose prevents the expression of (hemi-) cellulolytic genes to an even greater extent (21, 28). XlnR mediated expression of (hemi-)cellulolytic genes results in the production of enzymes that release not only xylose, but also arabinose, cellobiose, ferulic acid, and galactose. These compounds induce the expression of sub-sets of genes encoding polysaccharide degrading enzymes. This chapter reviews the current knowledge on the regulation of pectinolytic genes by *Aspergillus*, with special emphasis on *A. niger*. Comparison will be made between regulation of xylanolytic genes and pectinolytic genes in *A. niger*.

2. Production of pectinolytic enzymes by *Aspergillus*.

Papers describing the characteristics of purified pectinases also provide insights into the conditions under which these enzymes are produced by *Aspergillus*. The production of pectin main chain cleaving enzymes (polygalacturonases, pectin lyases, rhamnogalacturonan hydrolases and lyases) has been detected on sugar beet pectin (14), apple pectin (45, 79), polygalacturonic acid (75), a combination of rhamnose and galacturonic acid (80), and soybean flour (52). This suggests that production of these enzymes by *Aspergillus* is linked to the presence of pectin or pectin derived compounds. However some pectinolytic enzymes have also been reported to be produced constitutively (63, 66). These may be essential to liberate monomeric sugars from pectin that induce the production of the other pectinolytic enzymes.
The enzymes acting on the side chains of the pectic hairy regions (arabinofuranosidases, endoarabinanase, β-galactosidase, endogalactanases, and feruloyl esterases) have also been purified from cultures grown on a variety of carbon sources. These are not all related to pectin, indicating that some of the enzymes also play a role in the degradation of other polysaccharides. Production of these enzymes by *Aspergillus* has been observed on arabinoxylan and wheat bran (β-galactosidase (44, 61), arabinofuranosidase (48)), sugar beet pulp and polygalacturonic acid (β-galactosidase (62), endogalactanase (51), arabinofuranosidase and endoarabinanase (37, 83)), lactose (β-galactosidase (72)), soybean and locust bean gum (endogalactanase (2, 19) and L-arabinose and L-arabitol (arabinofuranosidase and endoarabinanase (71, 82)).
Although these data give some indications about the regulation of pectinase production by *Aspergillus*, they can also be misleading. Many of the compounds mentioned above are complex structures and contain different monomeric building blocks. To analyse which of these monomeric compounds trigger the expression of pectinolytic genes and the subsequent production of pectinases, studies have to be performed at the molecular level.

3. Expression of pectinolytic genes by *Aspergillus*.

The expression of pectinases in *A. niger* has been studied mainly with the purpose of optimising growth conditions to increase the yield of enzymes. These studies usually compare the production of pectinases in solid-state and submerged fermentation (SsF and SmF, respectively) under different physiological conditions, i.e. varying the composition of culture media, pH, temperature and days of cultivation. Generally, in SsF higher production of pectinases was observed on pectin between 1-2 days after inoculation and the synthesis of enzymes was less sensitive to carbon repression compared to SmF (59, 77). A low pH (pH 3-4) favoured PG production on complex substrates like citrus pectin (39, 40) and wheat bran (17), and variation in the pectinolytic enzyme composition has also been observed. Pectin, polygalacturonic and galacturonic acid induced synthesis of pectin methyl esterase (PME) and endopolygalacturonase (PG) in *A. niger* by increasing transcription, whereas glucose was found to repress the synthesis of both enzymes by a carbon repression mechanism (59).

In *A. niger* N400 a low level of expression of PGI and PGII was detected on pectin by western blot analysis and it was shown that multicopy transformants of the PGI-, PGII- and PGC-encoding genes, under the control of their own promoters, produced higher amounts of the corresponding proteins (10, 11, 13). *A. niger* and *A. tubingensis pga*II gene expression was detected in the presence of 1 % pectin and 1 % sugar beet pulp and repressed on 2 % glucose. Furthermore, *A. tubingensis pga*II multicopy transformants of *A. niger* showed the same induction pattern as in *A. tubingensis* suggesting a common regulatory system operating in both species (12). Similar patterns of expression, i.e. pectin induced and sucrose repressed, were observed for the pectin lyases of *A. niger* N400 (55, 56).

A promoter deletion study of the *A. niger pga*II gene demonstrated that the gene is subject to positive control, which is not limiting for its overexpression, and identified a 223 bp promoter fragment responsible for the high expression of PGII on pectin (14). A comparison of this promoter DNA sequence with yeast regulatory DNA sequences revealed a region of high identity with the upstream activation site UAS2 of the *Saccharomyces cerevisiae CYC1* gene, which is under control of the HAP2/3/4 system. A similar sequence was identified in the *A. niger pga*I and *pga*C genes from which the *A. niger pga*-consensus, 5' TYATTGGTGGA 3', was derived (11). The *A. niger* pectinase gene promoters were also analysed for the presence of a hexanucleotide sequence (11). The sequence 5' CCCTGA 3' occurs at least once in all of these genes. However, as yet there is no further experimental evidence to prove the function of both DNA-promoter sequences.

The pectinase induction has also been studied in other *Aspergillus* species. In *A. aculeatus*, the rhamnogalacturonan hydrolase encoding gene (*rgh*A) was found to be expressed on apple pectin and a combination of galacturonic acid and rhamnose (80). *A. parasiticus* and *A. flavus* both possess a *pga* gene that is expressed on sucrose and pectin (16, 87). In addition, *A. flavus* also possesses a *pga* gene that is specifically induced on pectin. In *A. tubingensis* expression of the exopolygalacturonase-encoding gene was detected in the presence of galacturonic acid in the medium and was absent when glucose was added (50).

Expression of the genes encoding arabinanases has been observed on monomeric carbon sources, such as L-arabinose and L-arabitol (38, 42, 43, 71, 82). The expression of two arabinofuranosidase encoding genes (*abf*A and *abf*B) and one endoarabinanase encoding gene (*abn*A) has been studied in *A. niger* (38). Transformants containing additional copies of one of the genes showed reduced expression levels of the other two genes compared to a wild type strain. This indicated that these genes are most likely under the co-ordinated control of a specific, positively acting transcription factor. The reduced expression would then be caused by titration of the transcription factor by the additional copies of one of the genes. Indications that L-arabitol, and not L-arabinose, is the true inducing compound of this regulatory system were obtained from an *A. nidulans* mutant defective in L-arabitol dehydrogenase activity (24). This mutant accumulated L-arabitol when grown on media containing glycerol and L-arabitol. Under these conditions an increased arabinofuranosidase, endoarabinanase, and L-arabinose reductase activity was observed compared to a wild-type strain. These data also suggest co-ordinated induction of extracellular arabinose releasing enzymes and enzymes involved in L-arabinose catabolism. This has been confirmed in *A. niger* where induction of arabinofuranosidases and endoarabinanases occurred simultaneously with the induction of intracellular enzyme activities involved in L-arabinose catabolism (L-arabinose reductase and L-arabitol dehydrogenase) (82).

4. Carbon catabolite repression.

The carbon catabolite repressor protein CreA (34, 74) is the major factor responsible for carbon repression in *Aspergillus*. CreA is a zinc finger protein which binds to specific sites in the promoters (SYGGRG) of a wide range of target genes (53). In the presence of glucose, fructose, or other easily metabolisable substrates, CreA inhibits or decreases the expression of the target genes. Several CreA mutants have been isolated from *A. nidulans* and *A. niger* with (partially) de-repressed phenotypes (6, 73, 76). Using these mutants, CreA mediated repression of gene expression has been detected for several extracellular proteases in *A. niger* (47) and for the proline and alcohol gene clusters in *A. nidulans* (20, 65). Recently, the mechanism of CreA mediated repression has been investigated in more detail (78). It was demonstrated that the *cre*A gene is strongly expressed when repressing monomeric carbon sources are added to the culture. CreA possesses a strong capacity of autoregulation and diminishes the expression of its corresponding gene (*creA*) rapidly after an induction peak caused by sugar monomers. This phenomenon has been demonstrated to be dependent on the formation of glucose-6-phosphate (78). Although a significantly higher *cre*A expression was observed under carbon de-repressing conditions, this did not result in the formation of the CreA-DNA complex. Formation of this complex was dependent on *de novo* protein synthesis (78). Whether the change from inactive to active CreA and *vice versa* is caused by covalent modification of CreA or by protein degradation is not clear at this point.

The influence of CreA on the plant cell wall degrading enzyme systems from *Aspergillus* has been studied for a large number of genes. CreA mediated repression in *Aspergillus* has been detected for genes encoding arabinanases and L-arabinose catabolic enzymes (73), several endoxylanases (21, 36, 69, 70, 73), β-xylosidase (54), arabinoxylan arabinofuranohydrolase (43) and feruloyl esterase (30). Several of the pectinolytic genes from *Aspergillus* are repressed in the presence of glucose (12, 49, 50, 60, 77), indicating involvement of CreA in the regulation of pectin degradation. Gel mobility shift analysis of a fragment of the promoter of the *A. nidulans* pectate lyase encoding gene (*pel*A) using a fusion protein containing the N-terminal part of CreA demonstrated binding of the fusion protein to this fragment (46) and supplied additional evidence for CreA in the regulation of (some) pectinolytic genes.

CreA mediated repression is not restricted to glucose and fructose alone. Other monomeric carbon sources also result in (weaker) CreA mediated repression of gene expression. Xylose is commonly regarded as the monomeric inducer of xylanolytic gene expression in *Aspergillus* (see 3.1). Recently, the effect of different xylose concentrations on xylanolytic gene expression was studied (31) demonstrating that high concentrations of xylose result in CreA mediated repression of the expression of xylanolytic genes. At lower xylose concentrations higher expression levels were observed. A similar phenomenon was observed for the regulation of cellulase biosynthesis in *A. terreus* (1). At high concentrations of glucose, xylose, and cellobiose a decrease in cellulase production was observed. Addition of cyclic AMP (cAMP) relieved catabolite expression of cellulase production in glycerol repressed *A. nidulans*, indicating a role for this compound in carbon catabolite repression (7). An effect of increased induction of pectin lyase (PL) of *Penicillium griseoroseum* in the presence of cAMP was also reported (8, 9), which might indicate that this secondary messenger is also involved in the regulation of pectinase expression in fungi.

The ferulic acid induced expression of the feruloyl esterase encoding genes (*faeA* and *faeB*) from *A. niger* was studied in combination with a number of different monomeric carbon sources (R.P. de Vries, unpublished results). For nearly all combinations (including galacturonic acid) higher *faeA* and *faeB* expression was detected in a CreA derepressed mutant than in a wild-type strain, indicating that CreA mediated repression occurred in the presence of all these carbon sources.

5. pH regulation.

The major factor involved in pH dependent expression in *Aspergillus* is the pH regulatory protein PacC. At alkaline pH PacC activates alkaline expressed genes and represses acid expressed genes, therefore PacC has a dual function (81). The gene (*pacC*) encoding this protein has been cloned from *A. nidulans* (81), and was shown to complement *A. nidulans* pH regulatory mutants, previously isolated as mutants defective in phophatase expression (3, 5, 33). An *A. niger pacC* homologue has also been cloned (58). PacC is activated under alkaline conditions by proteolytic modification of the C-terminal region, which enables it to bind to specific sites (GCCARG) in the promoter of target genes (64). Six genes (*pal*A, *pal*B, *pal*C, *pal*F, *pal*H and *pal*I) are believed to be involved in the signal transduction pathway leading to the activation of PacC (4, 15, 81). A more detailed description of pH regulation of gene expression in fungi can be found in a recent review (32).

pH regulation of genes encoding cell-wall degrading enzymes has not been studied in detail in *Aspergillus*. However indications for pH dependent expression of xylanolytic and pectinolytic genes have been obtained. The production of different polygalacturonases was studied in *A. kawachii* using culture media buffered to different pH values (52). From culture medium at pH 2 two polygalacturonases were purified, whereas only one polygalacturonase could be purified from culture medium at pH 5. The N-terminal amino acid sequence of these enzymes were different demonstrating that the pH of the culture medium influences which polygalacturonase is produced by *A. kawachii*. Using PacC mutants it was shown that in *A. nidulans* the expression of the major arabinofuranosidase encoding gene (*abf*B) (42) and two endoxylanase encoding genes (*xln*A and *xln*B) (57) are regulated by PacC. These two endoxylanase encoding genes have opposite expression patterns, *xln*A being expressed under alkaline conditions and *xln*B under acidic conditions.

6. Recent data on the expression of pectinolytic genes from *Aspergillus niger*.

6.1. EXPRESSION PROFILING.

The data available so far on the expression of pectinolytic genes from *A. niger* does not give a clear picture of the regulatory system(s) involved. Unlike the xylanolytic genes, pectinolytic genes are not all expressed in response to a single monomeric inducer. Recently a detailed analysis of the expression of 28 pectinolytic genes from *A. niger* was performed using two *A. niger* strains (a wild type and a CreA derepressed mutant) and 16 different conditions (25). The conditions included transfers of the glucose pregrown mycelia to the media containing either the monomeric building blocks of pectin or polygalacturonic acid or sugar beet pectin, with samples being harvested at three time points. This resulted in the identification of groups of genes with similar expression profiles (Table 1), suggesting that they are co-regulated.

TABLE 1. Grouping of pectinolytic genes according to their expression profile.

Expression profile	Gene	Enzyme
I	*pgaI, pgaII, pgaC, pgaE*	endopolygalacturonase
	pelB, pelC	pectin lyase
	plyA	pectate lyase
	rhgA	rhamnogalacturonan hydrolase
	rgaeA	rhamnogalacturonan acetyl esterase
II	*pelA, pelD*	pectin lyase
III	*pgaA, pgaB, pgaD*	endopolygalacturonase
IV	*pgaX*	exopolygalacturonase
V	*rglA*	rhamnogalacturonan lyase
VI	*rhgB*	rhamnogalacturonan hydrolase
VII	*pmeA*	pectin methyl esterase
VIII	*abfA, abfB*	arabinofuranosidase
	abnA	endoarabinanase
	lacA	β-galactosidase
IX	*faeA, faeB*	feruloyl esterase

Profile I is characterised by expression in the presence of galacturonic acid, polygalacturonic acid and sugar beet pectin. The expression increases in time and is significantly effected by CreA. In profile II expression is also observed in the presence of galacturonic acid and sugar beet pectin. However, expression is highest at early time points and decreases with time. Profile III displays constitutively expression. In profile IV expression is again detected in the presence of galacturonic acid, polygalacturonic acid and sugar beet pectin. No CreA effect is detectable and the expression levels are nearly constant in time. In profile V expression is only detected 24 h after transfer in the presence of polygalacturonic acid or sugar beet pectin, while profile VI shows expression on the same carbon sources 8 h after transfer. Profile VII is very specific in that expression was only detected 2 h after transfer in the presence of polygalacturonic acid or a combination of arabinose or xylose with galacturonic acid. Profile VIII is specific for the hydrolases degrading the side chains of the hairy regions of pectin. Expression is observed in the presence of galacturonic acid and sugar beet pectin, but higher levels were detected in the presence of L-arabinose. Profile IX is specific for the feruloyl esterases in that expression is specifically detected in the presence of ferulic acid. For *faeA* expression was also observed in the presence of xylose, which was previously demonstrated to be due to regulation by XlnR (30).

These data demonstrate that the presence of galacturonic acid or polysaccharides containing galacturonic acid indeed induce the expression of all genes encoding enzymes active on the pectin main chain. However, other factors (e.g. CreA) have a modulating effect on the expression of these genes. As a result of this, the pectinolytic genes are not all expressed under the same conditions and at the same time. The grouping of the pectinolytic genes is an important step in a strategic approach to identify the regulatory factors involved in pectin degradation. A preliminary hypothesis can be proposed based on these data. The early expression of the gene encoding the exopolygalacturonase (*pgaX*) suggests a key role for this enzyme in liberating the actual inducer, most likely galacturonic acid or a metabolite derived from it. This is supported by results that showed that *A. niger* is not able to import digalacturonate (L. Pařenicová, personal communication), which indicates that galacturono-oligosaccharides can not be the inducer of the pectinolytic genes. This compound is then able to induce the expression of the other pectinolytic genes at the time points determined by the other regulatory factors involved.

6.2. INDUCTION BY L-ARABINOSE/L-ARABITOL.

As mentioned in the previous section, genes encoding two arabinofuranosidases (*abfA* and *abfB*), an endoarabinanase (*abnA*) and a β-galactosidase (*lacA*) from *A. niger* are expressed in the presence of arabinose. A recent study resulted in the isolation of two *A. niger* mutants with reduced arabinofuranosidase and endoarabinanase activity (22). More detailed analysis revealed that these mutants were in fact regulatory mutants in which the expression of *abfA*, *abfB*, *abnA* and *lacA* was strongly reduced. Further analysis of these mutants revealed that the expression of the D-xylulose kinase gene (*xkiA*), an enzyme of the arabinose catabolic pathway, was negatively affected and that the activity of two other enzymes of this pathway, L-arabinose reductase and L-arabitol dehydrogenase, were reduced. This indicates that there is a common regulatory system in *A. niger* for the genes encoding pectin side chain cleaving enzymes as well as enzymes from the L-arabinose catabolic pathway (22).

6.3 INDUCTION BY FERULIC ACID.

The induction of a feruloyl esterase encoding gene (*faeA*) from *A. niger* by ferulic acid, has recently been studied (30, 35). This gene is expressed during growth on xylan and xylose, which is mediated by XlnR (see 3.1.). The addition of ferulic acid to media containing xylan or xylose increased the expression of *faeA* whereas other xylanolytic genes were unaffected, indicating a second regulatory system for the induction of *faeA*. Although the two systems positively influence each other with respect to the expression of *faeA*, ferulic acid induced expression is not dependent on XlnR (30). In an XlnR negative mutant, the level of expression of *faeA* was similar on ferulic acid and a combination of ferulic acid and xylose, while no expression of other xylanolytic genes was detected under these conditions. Recently, a second feruloyl esterase gene has been cloned from *A. niger* (*faeB*) and its regulation studied (27). This gene has a low constitutive expression level, but is also specifically induced by ferulic acid. During the expression profiling study (25), expression of both *faeA* and *faeB* was detected when ferulic acid was present in the medium. However, no expression of these genes was detected in the presence of arabinose, indicating that they are not under the control of the same regulator as the other genes encoding pectin side chain cleaving enzymes.

Concluding remarks.

This chapter presents an overview of the current knowledge of the regulatory systems involved in pectin degradation by *Aspergillus*. The pectinolytic regulatory system appears to be more complex than the xylanolytic regulatory system. Expression profiling allowed a number of the pectinolytic genes to be grouped, which is a first step towards our understanding of the regulation of pectinases in fungi. However, these data also demonstrate significant differences in the expression pattern between individual genes. These data should be compared to the synergistic relations that have been reported for a number of the pectinolytic enzymes.

Pectin methyl esterase from *A. aculeatus* strongly enhanced the degradation and depolymerization of pectin by polygalacturonases (18). This could explain why the pectin methyl esterase encoding gene (*pmeA*) is expressed early during growth. Recently synergy in the degradation of hairy regions from sugar beet pectin was studied using six accessory enzymes and a main chain-cleaving enzyme (26). Synergistic effects between β-1,4-endogalactanase, β-galactosidase, endoarabinanase, and arabinofuranosidase from *A. niger* were detected. This correlates with the co-regulation of the genes encoding the later three enzymes and suggests that the endogalactanase encoding gene could also be under the same regulatory control.

It is clear that we have just begun to understand some of the factors involved in the regulation of pectinolytic genes in *Aspergillus*. Although recent developments, such as the sequencing of the *A. niger* genome, will facilitate the identification of additional pectinolytic genes and possible regulatory factors, to generate and to analyse *Aspergillus* mutants affected in the regulation of pectinolytic genes will still be a major benefit to the understanding of this system.

References

1. Ali, S., and A. Sayed. 1992. Regulation of cellulase biosynthesis in *Aspergillus terreus*. World J. Microbiol. Biotechnol. 8:73-75.
2. Araujo, A., and O. P. Ward. 1990. Extracellular mannanases and galactanases from selected fungi. J. Ind. Microbiol. 6:171-178.
3. Arst, H. N., Jr., C. R. Bailey, and H. A. Penfold. 1980. A possible role for acid phosphatase in g-amino-n-butyrate uptake in *Aspergillus nidulans*. Arch. Microbiol. 125:153-158.
4. Arst, H. N., Jr., E. Bignell, and J. Tilburn. 1994. Two new genes involved in signalling ambient pH in *Aspergillus nidulans*. Mol. Gen. Genet. 245:787-790.
5. Arst, H. N., Jr., and D. J. Cove. 1970. Molybdate metabolism in *Aspergillus nidulans*. II. Mutations affecting phosphatase activity or galactose utilization. Mol. Gen. Genet. 108:146-153.
6. Arst, H. N., Jr., D. Tollervey, C. E. A. Dowzer, and J. M. Kelly. 1990. An inversion truncating the *creA* gene of *Aspergillus nidulans* results in carbon catabolite derepression. Mol. Microbiol. 4:851-854.
7. Bagga, P. S., D. K. Sandhu, and S. Sharma. 1991. Effect of exogenous cyclic AMP on catabolite repression of cellulase formation in *Aspergillus nidulans*. Acta Biotechnol. 11:395-402.
8. Baracat-Pereira, M. C., J. L. C. Coelho, R. C. Minussi, V. M. Chaves-ALves, R. L. Brandao, and D. O. Silva. 1999. Cyclic AMP and low molecular weight effector(s) present in yeast extract are involved in pectin lyase production by *Penicillium griseoroseum* cultures on sucrose. Appl. Biochem. Biotechnol. 76:129-141.
9. Baracat-Pereira, M. C., R. C. Minussi, J. L. C. Coelho, and D. O. Silva. 1997. Tea extract as an inexpensive inducer of pectin lyase in *Penicillium griseoroseum* cultured on sucrose. J. Industr. Microbiol. Biotechnol. 18:308-311.
10. Bussink, H. J., K. B. Brouwer, L. H. de Graaff, H. C. Kester, and J. Visser. 1991. Identification and characterization of a second polygalacturonase gene of *Aspergillus niger*. Curr. Genet. 20:301-307.
11. Bussink, H. J. D., F. P. Buxton, B. A. Fraaye, L. H. de Graaff, and J. Visser. 1992. The polygalacturonases of *Aspergillus niger* are encoded by a family of diverged genes. Eur. J. Biochem. 208:83-90.
12. Bussink, H. J. D., F. P. Buxton, and J. Visser. 1991. Expression and sequence comparison of the *Aspergillus niger* and *Aspergillus tubingensis* genes encoding polygalacturonase II. Curr. Genet. 19:467-474.
13. Bussink, H. J. D., H. C. M. Kester, and J. Visser. 1990. Molecular cloning, nucleotide sequence and expression of the gene encoding prepro-polygalacturonase II of *Aspergillus niger*. FEBS Lett. 273:127-130.
14. Bussink, H. J. D., J. P. T. W. van den Hombergh, P. R. L. A. van den IJssel, and J. Visser. 1992. Characterization of polygalacturonase-overproducing *Aspergillus niger* transformants. Appl. Microbiol. Biotechnol. 37:324-329.
15. Caddick, M. X., A. G. Brownlee, and H. N. J. Arst. 1986. Regulation of gene expression by pH of the growth medium in *Aspergillus nidulans*. Mol. Gen. Genet. 203:346-353.
16. Cary, J. W., R. Brown, T. E. Cleveland, M. Whitehead, and R. A. Dean. 1995. Cloning and characterization of a novel polygalacturonase-encoding gene from *Aspergillus parasiticus*. Gene. 153:129-133.
17. Cavalito, S. F., J. A. Arcas, and R. A. Hours. 1996. Pectinase production profile of *Aspergillus foetidus* in solid state cultures at different acidities. Biotechnol. Lett. 18:251-256.
18. Christgau, S., L. V. Kofod, T. Halkier, L. N. Anderson, M. Hockauf, K. Dorreich, H. Dalboge, and S. Kauppinen. 1996. Pectin methyl esterase from *Aspergillus aculeatus*: expression cloning in yeast and characterization of the recombinant enzyme. Biochem. J. 319:705-712.
19. Christgau, S., T. Sandal, L. V. Kofod, and H. Dalboge. 1995. Expression cloning, purification and characterization of a b-1,4-galactanase from *Aspergillus aculeatus*. Curr. Genet. 27:135-141.
20. Cubero, B., and C. Scazzocchio. 1994. Two different, adjacent and divergent zinc finger binding sites are necessary for CreA-mediated carbon catabolite repression in the proline gene cluster of *Aspergillus nidulans*. EMBO J. 13:407-415.
21. de Graaff, L. H., H. C. van den Broeck, A. J. J. van Ooijen, and J. Visser. 1994. Regulation of the xylanase-encoding *xlnA* gene of *Aspergillus tubingensis*. Mol. Microbiol. 12:479-490.
22. de Groot, M. J. L., P. J. I. van de Vondervoort, R. P. de Vries, P. A. vanKuyk, G. J. G. Ruijter, and J. Visser. Specific regulation of *Aspergillus niger* arabinan metabolism. Microbiol. In preparation.
23. de Vries, R. P., J. A. E. Benen, L. H. de Graaff, and J. Visser. 2001. Plant cell wall degrading enzymes produced by *Aspergillus.*, p. In press. *In* H. D. Osiewacz (ed.), Industrial Applications, vol. X.
24. de Vries, R. P., M. J. A. Flipphi, C. F. B. Witteveen, and J. Visser. 1994. Characterisation of an *Aspergillus nidulans* L-arabitol dehydrogenase mutant. FEMS Microbiol. Lett. 123:83-90.
25. de Vries, R. P., J. Jansen, G. Aguilar, L. Pařenicová , J. A. E. Benen, and J. Visser. Expression profiling of pectinolytic genes from *Aspergillus niger*. Submitted.

26. de Vries, R. P., H. C. M. Kester, C. H. Poulsen, J. A. E. Benen, and J. Visser. 2000. Synergy between accessory enzymes from *Aspergillus* in the degradation of plant cell wall polysaccharides. Carbohydr. Res. 327:401-410.
27. de Vries, R. P., H. C. M. Kester, P. A. vanKuyk, and J. Visser. The *Aspergillus niger faeB* gene encodes a second feruloyl esterase involved in pectin and xylan degradation, and is specifically induced on aromatic compounds. Biochem. J. Submitted.
28. de Vries, R. P., C. H. Poulsen, S. Madrid, and J. Visser. 1998. *aguA*, the gene encoding an extracellular a-glucuronidase from *Aspergillus tubingensis*, is specifically induced on xylose and not on glucuronic acid. J. Bacteriol. 180:243-249.
29. de Vries, R. P., H. C. van den Broeck, E. Dekkers, P. Manzanares, L. H. de Graaff, and J. Visser. 1999. Differential expression of three a-galactosidase genes and a single b-galactosidase gene from *Apergillus niger*. Appl. Environ. Microbiol. 65:2453-2460.
30. de Vries, R. P., and J. Visser. 1999. Regulation of the feruloyl esterase (*faeA*) gene from *Aspergillus niger*. Appl. Environ. Microbiol. 65(12):5500-5503.
31. de Vries, R. P., J. Visser, and L. H. de Graaff. 1999. CreA modulates the XlnR induced expression on xylose of *Aspergillus niger* genes involved in xylan degradation. Res. Microbiol. 150:281-285.
32. Denison, S. H. 2000. pH regulation of gene expression in fungi. Fung. Genet. Biol. 29:61-71.
33. Dorn, G. 1965. Phosphatase mutants in *Aspergillus nidulans*. Science. 150:1183-1184.
34. Dowzer, C. E. A., and J. M. Kelly. 1991. Analysis of the *cre*A gene, a regulator of carbon catabolite repression in *Aspergillus nidulans*. Mol. Cell. Biol. 11:5701-5709.
35. Faulds, C. B., R. P. de Vries, P. A. Kroon, J. Visser, and G. Williamson. 1997. Influence of ferulic acid on the production of feruloyl esterases by *Aspergillus niger*. FEMS Microbiol. Lett. 157:239-244.
36. Fernandez-Espinar, M., F. Pinaga, L. de Graaff, J. Visser, D. Ramon, and S. Valles. 1994. Purification, characterization, and regulation of the synthesis of an *Aspergillus nidulans* acidic xylanase. Appl. Microbiol. Biotechnol. 42:555-562.
37. Flipphi, M. J. A., M. van Heuvel, P. van der Veen, J. Visser, and L. H. de Graaff. 1993. Cloning and characterisation of the *abfB* gene coding for the major a-L-arabinofuranosidase (AbfB) of *Aspergillus niger*. Curr. Genet. 24:525-532.
38. Flipphi, M. J. A., J. Visser, P. van der Veen, and L. H. de Graaff. 1994. Arabinase gene expression in *Aspergillus niger*: indications for co-ordinated gene expression. Microbiol. 140:2673-2682.
39. Galiotou-Panayotou, M., M. Kapantai, and O. Kalantzi. 1997. Growth conditions of *Aspergillus* sp. ATHUM-3482, for polygalacturonase production. Appl. Microbiol. Biotechnol. 47:425-429.
40. Galiotou-Panayotou, M., P. Rodis, and M. Kapantai. 1993. Enhanced polygalacturonase production by *Aspergillus niger* NRRL-362 grown on supplemented citrus pectin. Lett. Appl. Microbiol. 17:145-148.
41. Gielkens, M. M. C., E. Dekkers, J. Visser, and L. H. de Graaff. 1999. Two cellobiohydrolase-encoding genes from Aspergillus niger require D-xylose and the xylanolytic transcriptional activator XlnR for their expression. Appl. Environ. Microbiol. 65:4340-4345.
42. Gielkens, M. M. C., L. Gonzales-Candelas, P. Sanchez-Torres, P. J. I. van de Vondervoort, L. H. de Graaf, and J. Visser. 1999. The *abfB* gene encoding the major a-L-arabinofuranosidase of *Aspergillus nidulans*: nucleotide sequence, regulation and construction of a disrupted strain. Microbiol. 145:735-741.
43. Gielkens, M. M. C., J. Visser, and L. H. de Graaff. 1997. Arabinoxylan degradation by fungi: Characterisation of the arabinoxylan arabinofuranohydrolase encoding genes from *Aspergillus niger* and *Aspergillus tubingensis*. Curr. Genet. 31:22-29.
44. Gonzalez, R. R., and P. Monsan. 1991. Purification and some characteristics of b-galactosidase from *Aspergillus fonsecaeus*. Enzyme Microb. Technol. 13:349-352.
45. Harmsen, J. A. M., M. A. Kusters-van Someren, and J. Visser. 1990. Cloning and expression of a second *Aspergillus niger* pectin lyase gene (*pelA*): indications of a pectin lyase gene family in *A. niger*. Curr. Genet. 18:161-166.
46. Ho, M.-C., M. P. Whitehead, T. E. Cleveland, and R. A. Dean. 1995. Sequence analysis of the *Aspergillus nidulans* pectate lyase *pel*A gene and evidence for binding of promoter regions to CREA, a regulator of carbon catabolite repression. Curr. Genet. 27:142-149.
47. Jarai, G., and F. Buxton. 1994. Nitrogen, carbon, and pH regulation of extracellular acidic proteases of *Aspergillus niger*. Curr. Genet. 26:238-244.
48. Kaneko, S., T. Shimasaki, and I. Kusakabe. 1993. Purification and some properties of intracellular a-L-arabinofuranosidase from *Aspergillus niger* 5-16. Biosci. Biotech. Biochem. 57:1161-1165.
49. Kauppinen, S., S. Christgau, L. V. Kofod, T. Halkier, K. Dorreich, and H. Dalboge. 1995. Molecular cloning and characterization of a rhamnogalacturonan acetylesterase from *Aspergillus aculeatus*. Synergism between rhamnogalacturonan degrading enzymes. J. Biol. Chem. 270:27172-27178.

50. Kester, H. C. M., M. A. Kusters-van Someren, Y. Muller, and J. Visser. 1996. Primary structure and characterization of an exopolygalacturonase from *Aspergillus tubingensis*. Eur. J. Biochem. 240:738-746.
51. Kimura, I., N. Yoshioka, and S. Tajima. 1998. Purification and characterization of an endo-1,4-b-D-galactanase from *Aspergillus sojae*. J. Ferment. Bioeng. 85:48-52.
52. Kojima, Y., T. Sakamoto, M. Kishida, T. Sakai, and H. Kawasaki. 1999. Acidic condition-inducible polygalacturonase of *Aspergillus kawachii*. J. Mol. Catalysis B: Enzymatic. 6:351-357.
53. Kulmburg, P., M. Mathieu, C. Dowzer, J. Kelly, and B. Felenbok. 1993. Specific binding sites in the *alcR* and *alcA* promoters of the ethanol regulon for the CreA repressor mediating carbon catabolite repression in *Aspergillus nidulans*. Mol. Microbiol. 7:847-857.
54. Kumar, S., and D. Ramon. 1996. Purification and regulation of the synthesis of a b-xylosidase from *Aspergillus nidulans*. FEMS Microb. Lett. 135:287-293.
55. Kusters-van Someren, M., M. Flipphi, L. H. de Graaff, H. van den Broeck, H. Kester, A. Hinnen, and J. Visser. 1992. Characterisation of the *Aspergillus niger pelB* gene: structure and regulation of expression. Mol. Gen. Genet. 234:113-120.
56. Kusters-van Someren, M. A., J. A. M. Harmsen, H. C. M. Kester, and J. Visser. 1991. The structure of the *Aspergillus niger pelA* gene and its expression in *Aspergillus niger* and *Aspergillus nidulans*. Curr. Genet. 20:293-299.
57. MacCabe, A. P., M. Orejas, J. A. Perez-Gonzalez, and D. Ramon. 1998. Opposite patterns of expression of two *Aspergillus nidulans* xylanase genes with respect to ambient pH. J. Bact. 180:1331-1333.
58. MacCabe, A. P., J. P. T. W. van den Hombergh, J. Tilburn, H. N. J. Arst, and J. Visser. 1996. Identification, cloning and analysis of the *Aspergillus niger* gene *pac*C, a wide domain regulatory gene responsive to ambient pH. Mol. Gen. Genet. 250:367-374.
59. Maldanado, M. C., and A. M. de Saad. 1998. Production of pectinesterase and polygalacturonase by *Aspergillus niger* in submerged and solid state systems. J. Industr. Microbiol. Biotechnol. 20:34-38.
60. Maldonado, M. C., A. M. Strasser de Saad, and D. Callieri. 1989. Catabolite repression of the synthesis of inducible polygalacturonase and pectinesterase by *Aspergillus* sp. Curr. Microbiol. 18:303-306.
61. Manzanares, P., L. H. de Graaff, and J. Visser. 1998. Characterization of galactosidases from *Aspergillus niger*: purification of a novel a-galactosidase activity. Enzyme Microb. Technol. 22:383-390.
62. McKay, A. M. 1991. Extracellular b-galactosidase production during growth of filamentous fungi on polygalacturonic acid. Lett. Appl. Microbiol. 12:75-77.
63. Mikhailova, R. V., L. I. Sapunova, and A. G. Lobanok. 1995. Three polygalacturonases constitutively synthesized by *Aspergillus alliaceus*. World J. Microbiol. Biotechnol. 11:330-332.
64. Orejas, M., E. A. Espeso, J. Tilburn, S. Sarkar, H. N. J. Arst, and M. A. Penalva. 1995. Activation of the *Aspergillus* PacC transcription factor in response to alkaline ambient pH requires proteolysis of the carboxy-terminal moiety. Genes & Development. 9:1622-1632.
65. Panozzo, C., E. Cornillot, and B. Felenbok. 1998. The CreA repressor is the sole DNA-binding protein responsible for carbon catabolite repression of the *alc*A gene in *Aspergillus nidulans* via its binding to a couple of specific sites. J. Biol. Chem. 273:6367-6372.
66. Pařenicová, L., J. A. E. Benen, H. C. M. Kester, and J. Visser. 2000. *pga*A and *pga*B encode two constitutively expressed endopolygalacturonases of *Aspergillus niger*. Biochem. J. 345:637-644.
67. Pařenicová, L., J. A. E. Benen, R. A. Samson, and J. Visser. 1997. Evaluation of RFLP analysis of the classification of selected black *Aspergilli*. Mycol. Res. 101:810-814.
68. Pařenicová, L., P. Skouboe, J. Frisvad, R. A. Samson, L. Rossen, M. ten Hoor-Suykerbuyk, and J. Visser. 2001. Combined molecular and biochemical approach identifies *Aspergillus japonicus* and *Aspergillus aculeatus* as two species. Appl. Environ. Microbiol. 67:521-527.
69. Perez-Gonzalez, J. A., N. N. M. E. van Peij, A. Bezoen, A. P. MacCabe, D. Ramon, and L. H. de Graaff. 1998. Molecular cloning and transcriptional regulation of the *Aspergillus nidulans xlnD* gene encoding b-xylosidase. Appl. Environ. Microbiol. 64:1412-1419.
70. Pinaga, F., M. T. Fernandez-Espinar, S. Valles, and D. Ramon. 1994. Xylanase production in *Aspergillus nidulans*: Induction and carbon catabolite repression. FEMS Microbiol. Lett. 115:319-324.
71. Ramon, D., P. van der Veen, and J. Visser. 1993. Arabinan degrading enzymes from *Aspergillus nidulans*: induction and purification. FEMS Microbiology Letters. 113:15-22.
72. Reczey, K., H. Stalbrand, B. Hahn-Hagerdal, and F. Tjerneld. 1992. Mycelia-associated b-galactosidase activity in microbial pellets of *Aspergillus* and *Penicillium* strains. Appl. Microbiol. Biotechnol. 38:393-397.
73. Ruijter, G. J. G., S. I. Vanhanen, M. M. C. Gielkins, P. J. I. van de Vondervoort, and J. Visser. 1997. Isolation of *Aspergillus niger creA* mutants: Effects on expression of arabinases and L-arabinose catabolic enzymes. Microbiol. 143:2991-2998.
74. Ruijter, G. J. G., and J. Visser. 1997. Carbon repression in *Aspergilli*. FEMS Microbiol. Lett. 151:103-114.

75. Ruttkowski, E., N. Q. Khanh, F.-J. Wientjes, and M. Gottschalk. 1991. Characterisation of a polygalacturonase gene of *Aspergillus niger* RH5344. Molec. Microbiol. 5:1353-1361.
76. Shroff, R. A., R. A. Lockington, and J. M. Kelly. 1996. Analysis of mutations in the *cre*A gene involved in carbon catabolite repression in *Aspergillus nidulans*. Can. J. Microbiol. 42:950-959.
77. Solis-Pereira, S., E. Favela-Torres, G. Viniegra-Gonzalez, and M. Gutierrez-Rojas. 1993. Effects of different carbon sources on the synthesis of pectinase by *Aspergillus niger* in submerged and solid state fermentations. Appl. Microbiol. Biotechnol. 39:36-41.
78. Strauss, J., H. K. Horvath, B. M. Abdallah, J. Kindermann, R. L. Mach, and C. P. Kubicek. 1999. The function of CreA, the carbon catabolite repressor of *Aspergillus nidulans*, is regulated at the transcriptional and post-transcriptional level. Mol. Microbiol. 32:169-178.
79. Suykerbuyk, M. E. G., H. C. M. Kester, P. J. Schaap, H. Stam, W. Musters, and J. Visser. 1997. Cloning and characterization of two rhamnogalacturonan hydrolase genes from *Aspergillus niger*. Appl. Environ. Microbiol. 63:2507-2515.
80. Suykerbuyk, M. E. G., P. J. I. van de Vondervoort, P. J. Schaap, and J. Visser. 1996. Identification of regulatory mutants of *Aspergillus aculeatus* affected in rhamnogalacturonan hydrolase expression. Curr. Genet. 30:439-446.
81. Tilburn, J., S. Sarkar, D. A. Widdick, E. A. Espeso, M. Orejas, J. Mungroo, M. A. Penalva, and H. A. Arst Jr. 1995. The *Aspergillus* PacC zinc finger transcription factor mediates regulation of both acidic and alkaline expressed genes by ambient pH. EMBO J. 14:779-790.
82. van der Veen, P., M. J. A. Flipphi, A. G. J. Voragen, and J. Visser. 1993. Induction of extracellular arabinases on monomeric substrates in *Aspergillus niger*. Arch. Microbiol. 159:66-71.
83. van der Veen, P., M. J. A. Flipphi, A. G. J. Voragen, and J. Visser. 1991. Induction, purification and characterisation of arabinases produced by *Aspergillus niger*. Arch. Microbiol. 157:23-28.
84. van Peij, N. N. M. E., J. Brinkmann, M. Vrsanska, J. Visser, and L. H. de Graaff. 1997. b-xylosidase activity, encoded by *xlnD*, is essential for complete hydrolysis of xylan by *Aspergillus niger* but not for induction of the xylanolytic enzyme spectrum. Eur. J. Biochem. 245:164-173.
85. van Peij, N. N. M. E., M. M. C. Gielkens, R. P. de Vries, J. Visser, and L. H. de Graaff. 1998. The transcriptional activator *XlnR* regulates both xylanolytic and endoglucanase expression in *Aspergillus niger*. Appl. Environ. Microbiol. 64:3615-3619.
86. van Peij, N. N. M. E., J. Visser, and L. H. de Graaff. 1998. Isolation and analysis of *xlnR*, encoding a transcriptional activator coordinating xylanolytic expression in *Aspergillus niger*. Mol. Microbiol. 27:131-142.
87. Whitehead, M. P., M. T. Shieh, T. E. Cleveland, J. W. Cary, and R. A. Dean. 1995. Isolation and characterization of polygalacturonase genes (*pec*A and *pec*B) from *Aspergillus flavus*. Appl. Environ. Microbiol. 61:3316-3322.

TRANSGENESIS WITH AN ANTISENSE CONSTRUCT AND A PROMOTER-REPORTER FUSION FOR THE STUDY OF *Lupme3*, A FLAX PECTIN METHYLESTERASE GENE

J. LACOUX[1], D. ROGER[1], JM. DOMON[1], I. DUVAL[1], C. BUREL[2], D. KLEIN-EUDE[2], AP. BALANGE[2], C. MORVAN[2], E. LAINE[1]

[1] *Laboratoire de Biotechnologies et Physiologie Végétales, Faculté des Sciences, Université de Picardie Jules Verne; 33, rue Saint-Leu, 80039 Amiens cedex, France.*

[2] *Equipe Parois et Polymères Pariétaux, ESA CNRS 6037, 76821 Mt Saint Aignan cedex, France.*

E-mail address : eric.laine@sc.u-picardie.fr

Abstract

The *Lupme3* flax (*Linum usitatissimum*) gene encodes a putative pectin methylesterase (PME). A cDNA was generated by RACE-PCR using two consensus sequences to generate primers. PMEs are encoded by numerous genes (more than 60 ORFs in *Arabidopsis* contain PME consensus sequences) and little information is available on their expression regulation, and the role of the different isoforms in the plant, except for those that are active in fruits.

A construct using a partial sequence of this gene inserted in antisense orientation downstream of the CaMV 35S promoter was inserted in the flax genome *via Agrobacterium tumefaciens*. Transgenic lines were analysed for gene expression (northern blots) and PME activity *in vitro*. PME activity in calli was decreased at basic and neutral experimental pH (6.5, 7.5, 8.5) but increased at pH 5.5. This slight increase at pH 5.5 could be due to compensation by other isoforms, a similar (although opposite) phenomenon has been encountered by other workers who performed sense transgenesis with PME in potato (Pilling *et al.*, 2000). For all transgenic lines but one, PME activity at high pH (pH 8.0) was reduced in calli. On the contrary, activity in mature shoots of plants grown in greenhouse was unchanged as well as the overall growth and aspect of the plants. Similar experiments were conducted with tobacco in which the transgenic plants growth was delayed as compared to the standard plants.

Another type of experiment was performed using a "*Lupme3* promoter-reporter gus" fusion construct, in order to analyse the expression of *Lupme3* gene *in planta*: this construct was introduced in an heterologous system i.e tobacco, and in the homologous system i.e. flax. Expression in transgenic tobacco was observed in small leaves veins and in the root central cylinder as well as in calli, but not in mature shoots. In flax also, expression was detected in calli but not in mature shoots.

These results correlate well with those of the PME activity in transgenic antisense flax plants and calli and indicate that this PME isoform is active in young vascular tissues and in undifferentiated cells but is not produced in mature stems.

F. Voragen et al. (eds.), Advances in Pectin and Pectinase Research, 183-199.
© 2003 *Kluwer Academic Publishers. Printed in the Netherlands.*

1. Introduction

1.1 PME ACTION

Pectins, major polymers present in plant cell walls, are thought to be synthesized in a highly methylesterified state in the Golgi apparatus (Zhang and Staehelin, 1992; Doong *et al.*, 1995). Following their incorporation into the extracellular matrix, they might be enzymatically deesterified *in muro* by pectin methylesterases (PMEs, EC 3.1.1.11), ubiquitous enzymes, found in various plant tissues, organs and species. PME activity is mainly associated with proteins ionically-bound to cell walls (Gaffé *et al.*, 1989), although some soluble apoplastic (Bordenave and Goldberg, 1994) or cellular forms (Vannier, 1994) have also been detected. The action patterns of plant PMEs can be influenced by local cell wall pH, the existing balance of methyl and free carboxyl groups on homogalacturonan substrates and metal ion concentration (Catoire *et al.*, 1998; Nari *et al.*, 1986; Denès *et al.*, 2000).

The generation of free carboxyl groups is responsible for changes in the apoplastic pH and in the interchain chelation of calcium that affects gelling ability. The proportion of methylesterified pectins changes during cell elongation and is supposed to have profound effects on cell wall strength and extensibility (McCann *et al.*, 1994).

1.2 PME ISOFORMS

A number of PME isoforms can be revealed and classified by isofocusing electrophoresis into very alkaline, alkaline and neutral or slightly acid forms. Detailed studies of these multiple isoforms have been undertaken on model plants and organs such as poplar bark by Guglielmino *et al.* (1997), mung bean hypocotyls by Bordenave and Goldberg (1994) and flax seedlings by Alexandre *et al.* (1997).

In flax, PMEs are distributed into 3 groups differing at least by their pI (neutral: n, pI 7 and 7.5; moderately basic: b1, pI 8.5; and very basic: b2, pI>9.5). The relative amount of each group differs in cell cultures (all groups), calli (2 groups: neutral and b2) or plants (2 groups, b1 and b2 mainly) (Gaffé *et al.*, 1992; Mareck *et al.*, 1995). Another difference is in the optimum pH for activity: neutral and b1 forms had broad optimum pH from 5 to 8. The b2 isoform, extracted from calli exhibited high optimum pH (around 8.5) while that extracted from hypocotyls was characterized by an acidic optimum pH (around 5.5). Spatial distribution of isoforms has been reported with the basic forms (b1) being located in the inner tissues of flax seedlings and the very basic (b2) form only occurring in the epidermis or outer tissues (Alexandre *et al.*, 1997). Gugliemino *et al.* (1997) also reported a temporal (seasonal) change in PME isoforms activities in poplar bark.

1.3 PMEs ROLE

Pme genes occur in multigene families and encode isoforms with differing action patterns with respect to the removal of methyl-esters. PMEs have been suggested to modulate cell wall loosening and extensibility (Goldberg *et al.*,1996), as well as cell to cell cohesion (Liners *et al.*, 1994). Ricard and Noat (1986) have underlined the relationships between the degree of esterification of the galacturonic units and both the plastic potential of the cell wall and the cell to cell adhesion. However, the specific functions of the PME populations in the context of cell expansion and other processes are not well understood (Bordenave *et al.*, 1996; Micheli *et al.*, 1998).

The role of PME in elongating organs has been hypothesized by a number of authors. Alexandre *et al.* (1997) suggested that the basic b1 isoform of flax PME, that is mostly active in the basal part of hypocotyls, could be involved in the acidification of pectins in cell junctions of phloem and xylem during the active period of cambial differentiation. A similar observation was reported by Gugliemino *et al.* (1997), who detected basic isoforms in the cambium of poplar.

Pme genes transcription occurs also during development of pollen grains (Wakeley *et al.*, 1998; Futamura *et al.*, 2000), or seed germination (Ren and Kermode, 2000) where it could be related to weekening of the seed coats that allow root radicle emergence. The involvement of PMEs in the diffusion of plant viruses was also shown by immunodetection (Dorokhov *et al.*, 1999; Chen *et al.*, 2000). Wen *et al.* (1999), using transgenesis (antisense strategy) with *pme* gene, were able to demonstrate a role for PME in the root cap border cells separation.

1.4 PME GENES

A number of pectin methylesterase genes seem to exist in the plant genome. A growing number of *pme* gene sequences was published in the past decade, and in 1998, Micheli *et al.* reported the presence of 12 genes encoding PMEs in *Arabidopsis*. Now, with the completion of the arabidopsis genome sequencing, the analysis of the whole genome allowed to identify not less than 79 ORF putatively coding for pectin esterases (PE) because they contain their consensus sequences (The Arabidopsis Genome Initiative, 2000), out of which, a vast majority are probably *pme* genes (Richard, 2000). This very large size of the gene family for a single enzyme type is apparently the rule for most of the enzymes whose activity is related to cell wall modification: dozens of genes also do exist for polygalacturonases or cellulases (*table 1*).

Table 1: number of genes putatively coding for cell wall enzymes in *Arabidopsis thaliana*. Enzymes involved in cell wall components modification are very numerous (at least 420 in *Arabidopsis*) (The Arabidopsis Genome Initiative, 2000).

Enzyme type	Gene Number
Cellulose synthase	12
ß-1,3-glucanase	39
Endoxyloglucane transglycosylase	20
Cellulase	50
Peroxidase	69
Polygalacturonase	52
Pectate lyase	20
Pectin esterase	79

Genomic sequences and cDNAs have been also reported in other species, such as *Populus* [Micheli], *Petunia* [Mu *et al.*] and *Lycopersicon* [Hall *et al.*; Gaffé *et al.*], *Citrus* [Nairn *et al.*], *Linum* [Dauchel] (table 4).

In most of the *pme* genes cloned so far, the mature protein is encoded by the C terminal part of the ORF, a cleavage occuring downstream of a $R^{R}/_{K}$ $L^{L}/_{M}$ consensus sequence. The large N terminal sequence could play a role during the trafficking through the Golgi apparatus. Such a large number of ORF containing the consensus sequences characterizing PMEs has been related to a very variable mature protein coding sequence, but in fact, very strong homologies do exist within the *pme* cloned, irrespectively of their organ or tissue of expression (*table 4*). As similarities in coding sequences for the mature protein do exist, such a large number of PME encoding gene could be due to the requirement for very various expression pattern, which could be linked to the variability found in the fourty different cellular types that exist within every plant, as emphasized by The Arabidopsis Genome Initiative (2000).

A wide variation effectively exists in the N terminal coding region of sequenced *pme* genes, but little is known about the variation occuring in the promoter region, as at the time being, only a limited number of promoter regions of PE genes have been isolated and analysed: Kim *et al.* (1997) studied a *Brassica napus* pectinesterase gene promoter and Roger *et al.* (2001) studied the *Lupme3* gene expression following promoter-reporter fusion transgenesis into tobacco, i.e. an heterologous system.

Functional genomic study of *pme* is also hindered by the fact that very few mutants affected for pectins have been isolated so far (*table 2*). One can hypothetize that this is perharps due to this very large number of genes, and thus to the possible compensation of the lack of one functional enzyme by the overexpression of another enzyme with similar activity, as it has been showm to occur in antisense transgenesis with glucanase (Beffa *et al.*, 1993), such a phenomenon would presumably be able to reduce or suppress the phenotypic impact of mutations into *pme* genes.

Table 2: Mutants with altered pectins.

•sku5: phenotype: modified root geotropism; gene altered: pme encoding (Rutherford et Masson, 1996; Sedbrock *et al.* 1998)
•qrt (quartet): phenotype: no tetrade splitting, due to the persistence of pectin embedding; gene altered: unknown; (Rhee *et al.* 1998)
•kor (korrigan temperature sensitive): phenotype: reduced hypocotyle size, -30% cellulose but overproduction of pectins; gene altered: glucanase (Nicol *et al.* 1998; Sato *et al.* 1999)

1.5 TRANSGENESIS, AN USEFUL TOOL FOR CELL WALL METABOLISM STUDIES, AND POTENTIALY FOR CELL WALL MODIFICATION

Transgenesis could be a good tool to assign a role for PMEs: with either sense (Pilling *et al.*, 2000) or antisense PME constructs (Tieman *et al.*, 1992; Hall *et al.*, 1993; Wen *et al.*, 1999) the intervention of PMEs in respectively shoot elongation, in fruit softening during maturation or in root cap border cell separation was demonstrated (*table 3*). Transgenesis with genes others than those coding for pectin metabolism related enzymes has also yielded plants with altered pectin contents (*table 3*), thus demonstrating the possibility of compensation made by the plant beetween the major cell wall components.

Table 3: Reported effects of transgenesis on pectins.

•Sense *pme* (potato): no impact on DME, ion exchange capacity modification: Phenotype: larger tubers, quicker stem growth (Pilling *et al.*, 2000).
•Antisense *pme* (green bean): Phenotype: root cap cell separation inhibited. (Wen *et al.*, 1999).
•Antisense *pme* (tomato): Phenotype: ion exchange capacity modification, altered fruit integrity during senescence (no effect on maturation), DW/FW increased (Tieman et al., 1992, 1994; Mishra and Handa, 1998).
•Silencing of Cellulose synthase (*cbd*) (tobacco): less cellulose, "compensation" by an increase in homogalacturonan (with a reduced degree of esterification (Burton *et al.*, 2000)).
•Sense *sam* (S-adenosyl-methionine synthase) (flax): increased DME of pectin (Lamblin *et al.*, 2001).

Flax is a convenient model plant for genetic studies on cell wall metabolism due to its small genome, and its amenability to *Agrobacterium* mediated transformation. An effect on the DME of pectins was observed when an heterologous *sam* gene was overexpressed in flax (*sam* encodes S-adenosylmethionine synthetase, an enzyme involved in the synthesis of SAM, the methyl donor for the pectin methylation operated by pectin methyltransferases in Golgi apparatus prior to their exportation into the cell wall) (Lamblin *et al.*, 2001).

Moreover, antisense strategy using *pme* gene sequence could be a tool for retting improvement. The retting process of flax on the field is dependent of wheather, and need improvement as one out of seven harvests suffers from excessive retting that lead to poor fiber quality. Now, one parameter of retting is the susceptibility of pectin to degradation by soil microorganisms pectolytic enzymes; and this susceptibility depends to a great extend on the methylation degree of the cell wall pectins, itself determined by the PME activity.

1.6 LUPME3, A GENE ENCODING A PUTATIVE FLAX PME

In order to elucidate both the role and specificity of expression of a PME isoform, we have performed work using transgenesis as a tool for investigating its expression regulation and localization, and to link genes and isoform only characterized so far by biochemical means.

The present study was conducted using the *Lupme3* sequence (AF 188895; Dauchel), this cDNA was isolated by RACE PCR performed on RNA extracted from flax seedling hypocotyls, and is coding for a putative 34 kDa PME3 isoform with a very basic pI 9.7 (data deduced from its sequence). *Lupme3* shares a strong homology with a number of cloned pme (*table 4*). This sequence was subsequently used for generating both transformed flax (cv. Barbara) and tobacco plants with *Lupme3* antisense construct driven by a constitutive promoter.

To further elucidate the role of PME3 in flax, we investigated whether regulation did exist at the transcriptional level for the *Lupme3* gene. For this purpose, a 5' flanking region of the *Lupme3* gene was isolated by PCR walking (promoter sequence accession number AF 308812), and cloned. Then, a construct with a *gus* reporter gene driven by this potential regulatory upstream sequence was introduced into tobacco and flax via *Agrobacterium tumefaciens*.

Table 4: Cloned genes that share homology with *Lupme3*

Organism	Name	Accession #	Expressed in	Identity % (amino acids in mature part)	Authors
Solanum tuberosum	Pest2	AF152172	Epidermis	79	Pilling *et al.*
Citrus sinensis	PECS-c1	U82976	Fruit	79	Nairn *et al.*
Populus tremula x	Putative pme	AJ277547	Cambium	78	Micheli
Citrus sinensis	PECS1.1	U82973	Fruit	79	Nairn *et al.*
Nicotiana tabacum	pme	AJ401158	Bind to TMV movement protein	78	Dorokhov
Lycopersicon esculentum	pme u1	U49330	Leaf	78	Gaffé *et al.*
Arabidopsis thaliana	ATPME2 precursor	U25649	Genomic	72	Richard
Vigna radiata	pe precursor	X94443	Hypocotyle	58	Breton
Pisum sativum	pe	AF081457	Root cap	56	Zhu
Nicotiana plumbaginifolia	pme	Z71754	Somatic embryo	52	Seger
Salix gilgiana	pme	AB029461	Male flower	52	Futamura; Shinohara
Zea mays	ZmC5	Y13285	Pollen	52	Hussey
Brassica napus	Bp19	X56195	Pollen	44	Fabijanski

2. Material And Methods

2.1 SEQUENCES

2.1.1 Isolation of Lupme3 cDNA by RACE-PCR

RACE-PCR cloning was performed as described by Frohman *et al.* (1988) and Loh (1991). Briefly, 5'end amplification was carried out from a G-tailed cDNA template using a poly(dC) anchor primer, and a degenerated reverse gene primer deduced from a conserved PME sequence (DFIFGN). 3'end amplification was performed on the first-strand cDNA using an oligo (dT) adapter primer and a specific forward primer deduced from the sequenced PME cDNA. In order to increase the specificity, nested PCR was performed using internal primers.

2.1.2 Isolation of a promoter sequence by PCR walking on genomic DNA

Flax DNA was isolated and digested with *Pvu*II restriction enzyme. Adaptator oligonucleotide duplex were ligated to each end of the generated fragments in order to obtain a library that was then used as DNA template for PCR walking. For this purpose a nested strategy was used. Two PCR reactions were carried out using two sets of primers designed after reference to the specific 5' end coding region of *Lupme3* cDNA and to the adaptator sequence. The resulting 2.0 kb PCR fragment was cloned.

2.2 VECTOR PLASMIDS

2.2.1 pBIPMEF

The partial cDNA insert was excised and ligated into compatible restriction sites inducing an antisense orientation in the binary pBI121 plasmid (from which the *gus* gene has been previously deleted, Clontech). This binary vector (pBIPMEF) (*figure 1 A*) carrying a part of the cDNA (*Lupme3*), in antisense orientation and under the control of the CaMV 35S promoter and the *nptII* selectable gene (neomycin phosphotransferase conferring kanamycin resistance) was used to transform flax cells.

2.2.2 pPME3.gf

The PME3.gf fragment (a 1.5 kb sequence (-1542 to –11)) upstream from the *Lupme3* gene was inserted into the polylinker region of the binary vector pBI101 upstream from a promoterless *gus* gene cassette in order to create a transcriptional fusion with the *gus* reporter gene. The resulting chimeric construct was designated pPME3.gf (*figure 1 B*), and the construction confirmed by PCR analysis using a 5' primer chosen within the *nptII* selectable marker gene and a 3' primer within the *gus* reporter gene. The promoter: *gus* junctions were sequenced on both strands. It was mobilized by triparental mating into the disarmed *Agrobacterium tumefaciens* strain GV3101.

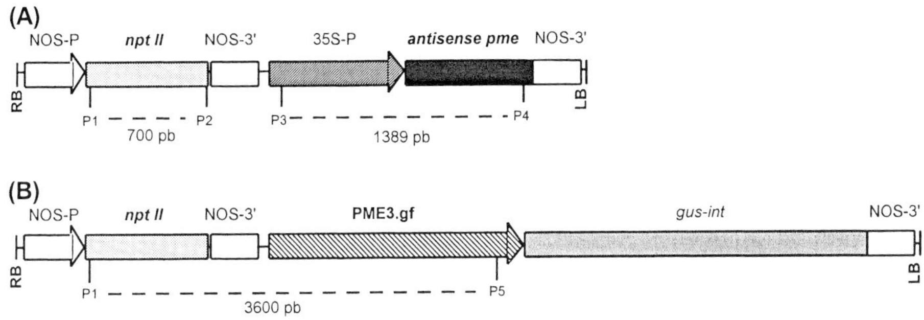

Fig. 1: Schematic structures of plant transformation vectors (A) binary vector pBI-PMEF with the *nptII* selectable gene (neomycin phosphotransferase conferring kanamycin resistance) and a partial cDNA of a flax *pme* gene in antisense orientation. (B) binary vector pPME3.gf with the gus reporter gene under the control of the PME3 promoter sequence. NOS-P, NOS-3': promoter and 3' end of the nopaline synthase gene; 35S-P: promoter of the CaMV 35S RNA gene. RB and LB: right and left borders ot the T-DNA.

2.3 TRANSGENESIS OF FLAX

Flax (cv. Barbara) transgenic cell lines were obtained following the protocol described by Mlynarova *et al.* (1994). Briefly: 6 days old seedling hypocotyls were excised and cocultivated for 2 days with *A. tumefaciens*. Green kanamycine-resistant buds were then isolated and grown on selective medium: Murashige and Skoog (1962) macronutrients and micronutrients, Morel and Wetmore (1951) vitamins, 15 $g.l^{-1}$ sucrose, 3.5 mM MES, 100 $mg.l^{-1}$ kanamycin, 0.09 μM benzyladenine, 0.01 μM naphtalene acetic acid, 100 $mg.l^{-1}$ kanamycin, 250 $mg.l^{-1}$ Timentin and 8 $g.l^{-1}$ Bacto agar (Difco), pH 5.7). Then rooted on the same medium (but hormone free). A single vigorous elongating shoot was taken per kanamycin-resistant callus and multiplied by auxilary budding to ensure the multiplication of cloned plantlets. A « recallusing » test was performed to select putatively transgenic shoots. For this purpose, leaves of elongated shoot were transferred onto medium supplemented with 200 $mg.l^{-1}$ kanamycin to test their ability to form callus; this callus was then subcultured every two weeks to produce the callus material.

Flax plants rooted *in vitro* were transferred and acclimated in greenhouse and grown in pots. Weaning was facilited by covering the plants with a plastic bottle until growth had resumed.

2.4 TRANSGENESIS OF TOBACCO

Transformation of tobacco (*Nicotiana tabacum*, line PBD6, kindly provided by Seita, France) was performed as described by Horsch *et al.* (1985). Then, plants were transferred *ex vitro* in pots and grown in a greenhouse.

2.5 MOLECULAR ANALYSIS OF TRANSGENIC PLANTS

The presence of the transgene in the regenerated plants and calli was verified by isolating DNA and using PCR to detect the presence of the constructs after verification of the absence of remaining bacteria using YEB medium. DNA was isolated from material frozen in liquid nitrogen.

PCR experiments were performed on genomic DNA using P1 and P2 primers designed for the amplification of the *nptII* gene. To confirm the presence of the *antisense pme* sequence and its orientation, P3 and P4 primers were used *(figure 1 A)*. The presence of the promoter sequence was confirmed using P1 and P5 (*figure 1 B*). Products were separated by agarose gel electrophoresis, stained with ethidium bromide, and visualised by UV transillumination. The absence of *Agrobacterium* was assessed by subculturing a piece of callus on YEB *Agrobacterium* medium culture.

For Southern blot: genomic DNA (20 μg) was digested with *Eco*RI or *Bam*HI. The fragments were fractionated on a 0.8 % agarose gel and transferred to Hybond-N membranes (Amersham). Hybridizations were performed at 42 °C with a 0.7 kb digoxigenin labelled *nptII* probe. Membranes were detected as notified by the instruction manual (Dig High Prime DNA Labeling and Detection Starter kit II, Roche). Hybridization was detected by exposure to X-Omat films (Kodak).

In order to check both transgene and endogenous *Lupme3* gene expression in the transgenic stems, an hybridization (northern type) was carried out, using double strand probe (a 755 bp purified fragment of the *Lupme3* coding sequence labelled with [α-^{32}P] dCTP), able to detect both sense and antisense RNAs.

2.6 PME ACTIVITY

Protein extracts were prepared as previously reported (Gaffé *et al.*, 1992) from 6 transgenic lines (CB1, CB3, CB6, CB7, CB9 and CB33), and control (untransformed control). Material was immersed for 15 min in 0.5 M mannitol, then treated overnight with 0.1% Triton X100, and extensively washed with 1 mM $CaCl_2$. Pectin methylesterases and others proteins ionically-bound to the cell walls were extracted by treatment with an high salt concentration (1 M NaCl).

Pectin methylesterase activity was then titrimetrically measured by following the increase in free carboxyl groups (Alexandre *et al.*, 1997). The carboxyl groups liberated by PME from *Citrus* pectin (Sigma) were measured using 10 mM NaOH under nitrogen. The measures were performed at pH 8.0 (the pH being maintained at 8.0) using a titrator (Mehtrom Stat 718). PME activity was measured in cell lines on day 10 after subculture (exponential growth phase). For each plant and cell line the PME activity value was the mean of 5 independent measurements (5 extractions). Cell wall protein concentrations in the extracts were determined according to the micromethod of Bradford (1976) using a Biorad kit, with bovine serum albumine in 1M NaCl as standard.

2.7 REPORTER GENE ACTIVITY ASSAY: HISTOCHEMICAL ASSAYS OF β-GLUCURONIDASE

Histochemical assays were performed using the protocol of Jefferson *et al.* (1987) modified by Kosugi *et al.* (1990) to minimize any endogenous activity. Green tissues were cleared by immersion in 95% ethanol to visualize the blue staining.

3. Results

3.1 PRODUCTION OF FLAX AND TOBACCO TRANSGENIC MATERIAL

Calli and plants, actively growing on selective medium containing kanamycin, were obtained from both flax and tobacco. In order to study the impact of the antisense transgenesis, six flax clones (referred to as CB1, CB3, CB6, CB7, CB9 and CB33) and two tobacco clones (referred to as TA1, and TA7) were used.

The presence of the transgenes, *pme antisense* chimeric gene and *gus* reporter gene under the control of the promoter sequence, were confirmed by PCR and/or Southern.

3.2 ANALYSIS OF THE ORF AND PROMOTER SEQUENCE

The complete cDNA sequence *Lupme3* (Genbank, AF188895), was 1808 bp long and contained a putative open reading frame (ORF) of 1665 bp. It would encode a polypeptide composed of 555 aminoacids, but a predicted cleavage site might be located between Q_{236} and T_{239}, giving a polypeptide around 34 kDa with a pI of 9.7 as mature protein.

Sequencing results confirmed that this sequence was upstream from the *Lupme3* gene and sequence analyses were in agreement with promoter features. A TATA box present 30 bases upstream from the transcription start. Database-assisted promoter analysis indicated the presence of potential regulatory boxes involved in many light-regulated plant promoters. Besides, some regulatory sequences playing a role in targeting expression to pollen were also identified.

3.3 EFFECT OF THE ANTISENSE TRANSGENESIS

3.3.1 Gene expression and PME activity in flax stems

A signal corresponding to the antisense transcripts in the different transgenic clones was detected. However, in both the transgenic clones and the control, an hybridization signal corresponding to the endogenous *Lupme3* gene expression was observed (data not shown).

This PME activity was not decreased in the transgenic stems as compared to the non transformed control (*figure 2 A*).

3.3.2 Gene expression and PME activity in flax calli

For this undifferentiated material, northern blots gave different results: two signals were detected: one corresponding to the antisense transcripts was observed in all the transgenic clones but was absent in the control, and one corresponding to the expression of the endogenous gene was observed (data not shown).

Five out of the six transgenic clones exhibited a reduction in the level of PME at pH 8 activity as compared to the control (*figure 2 B*). One clone (CB9) presents an higher PME activity than that of the control. This strong activity is correlated to more abundant sense RNA, associated to a weak signal corresponding to the antisense transcripts (data not shown).

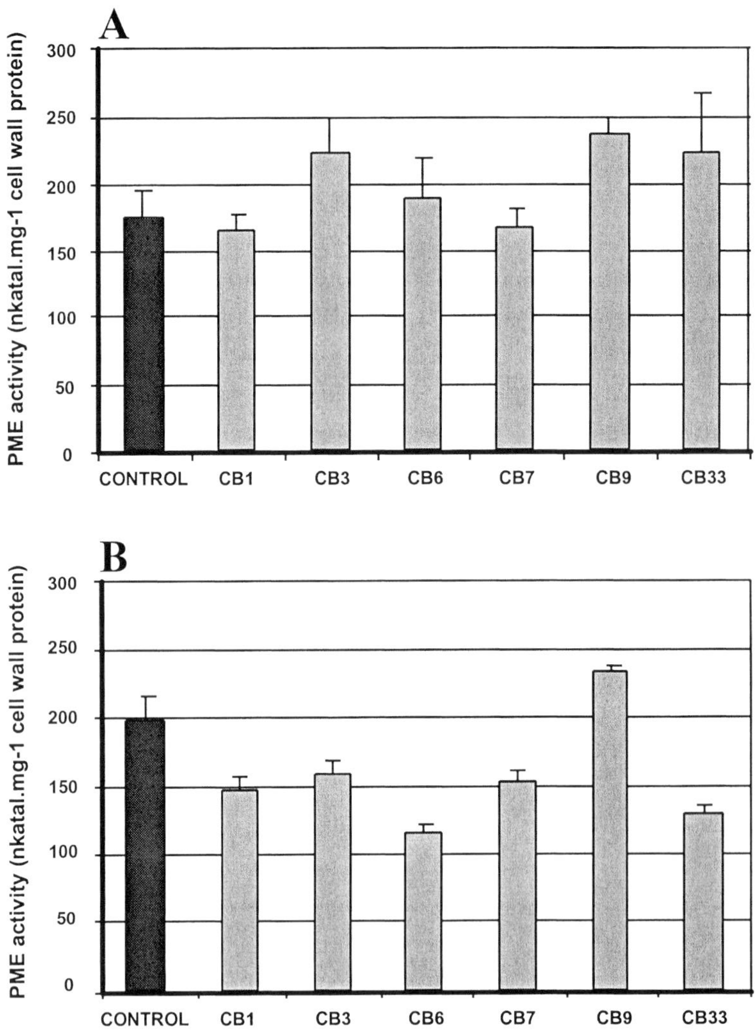

Fig. 2: **A**: PME activity of NaCl protein extracts from cell walls of control and transgenic stems. **B**: PME activity of NaCl protein extracts from cell walls of control and transgenic calli.

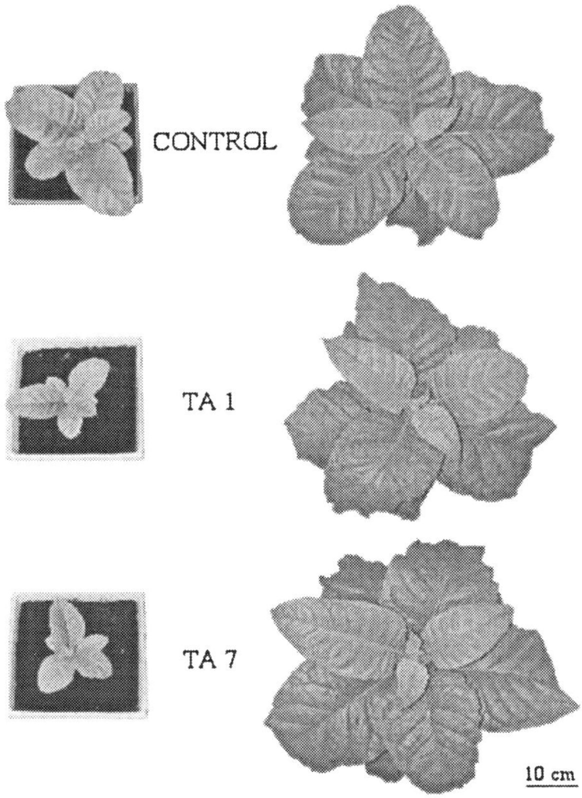

Left: 17 days after weaning: transgenic plants are less developed as the control
Right: 54 days after wearing: control and transgenic plants show a similar development

Fig. 3.

3.3.3 Transgenic tobacco phenotype

Seventeen days after the transfer of *in vitro* grown plants in greenhouse, differences in the growth rates of the leaves of the transgenic tobacco (TA1, TA7 and TA13) as compared to the control plant were observed (*figure 3*). But then, fifty four days later, the growth rates of transgenic plants increased and became larger than those of the control, and as a result, both ended up similar in size.

But this growth delay was not observed with flax.

Denes, J.M., Baron, A., Renard, C.M., Pean, C. and Drilleau, J.F. (2000) Different action patterns for apple pectin methylesterase at pH 7.0 and 4.5, *Carbohydr Res.* **327(4)** 385-393.

Doong, L.R., Liljebjelke, K, Fralish, G., Kumar, A. and Mohnen D. (1995) Cell-free synthesis of pectin, *Plant Physiol.* **109**, 141-152.

Dorokhov, Y.L., Makimen, K., Frolova, O.Y., Merits, A., Saarinen, J., Kalkkinen, N. Atabekov, J.G. and Saarma, M. (1999) A novel function for a ubiquitous plant enzyme pectin methylesterase: the host-cell receptor for the tobacco mosaic virus movement protein, *FEBS Letters* **461**, 223-228.

Frohman, M.A., Dush, M.K. and Martin, G.R. (1988) Rapid production of full-length cDNAs from rare transcripts: amplification using a single gene-specific oligonuclotide primer, Proc. Natl. Acad. Sci. *U.S.A.* ***85**, 8998-9002.*

Futamura, N., Mori, H., Kouchi, H. and Shinohara, K. (2000) Male flower-specific expression of genes for polygalacturonase, pectin methylesterase and β -1,3-glucanase in a dioecious willow (*Salix gilgiana* Seemen) , *Plant Cell Physiol.* **41(1)**, 16-26.

Gaffé, J., Morvan, C., Jauneau, A. and Demarty M. (1989) Mise en évidence d'activités polygalacturonases et pectines méthylestérases dans les parois d'hypocotyles et de cals de lin, *C.R. Acad. Sci.*, Série 3, **308**, 107-113.

Gaffé, J., Morvan, C., Jauneau, A. and Demarty M. (1992) Partial purification of flax cell wall pectin methylesterase, *Phytochemistry* **31**, 761-765.

Gaffé, J., Tiznado, M.E. and Handa, A.K. (1997) Characterization and functional expression of a ubiquitously expressed tomato pectin methylesterase, *Plant Physiol* **114**, 1547-1556.

Goldberg, R., Morvan, C., Jauneau, A. and Jarvis, M.C. (1996) Methylesterification, de-esterification and gelation of pectins in the primary cell wall, In: Visser J, Voragen AGJ (eds) *Pectins and Pectinases*, Elsevier, Amsterdam, 151-171.

Guglielmino, N., Liberman, M., Catesson, A.M., Mareck, A., Prat, R., Mutaftschiev, S. and Goldberg, R. (1997) Pectin methylesterases from poplar cambium and inner bark: localization, properties and seasonal changes, *Planta* **202**, 70-75.

Hall, L.N., Bird, C.R., Picton, S., Tucker, G.A., Seymour, G.B. and Grierson D. (1994) Molecular characterization of cDNA clones representing pectin esterase isoenzymes from tomato, Plant Mol. Biol. *25, 313-318.*

Hall, L.N., Tucker, G.A., and Smith, C.J.S. (1993) Antisens inhibition of pectin esterase gene expression in transgenic tomatoe, Plant J. ***3**, 121-129.*

Horsch, R.B., Fry, J.E., Hoffmann, N.L., Eichholtz, D., Rogers, S.G., and Fraley, R.T. (1985) A simple and general method for transferring genes into plants, *Science* **277**, 1229-1231.

Jefferson, R.A., Kavanagh, T.A. and Bevan, M.W. (1987) GUS fusions: β-glucuronidase as a sensitive and versatile gene fusion marker in higher plants, *EMBO J.* **6**, 3901-3907.

Kim, H.U., Park, B.S., Jin, Y.M. and Chung, T.Y. (1997) Promoter sequences of two homologous pectin esterase genes from chinese cabbage (*Brassica napus* L. ssp. *pekinensis*) and pollen specific expression of the *GUS* gene driven by a promoter in tobacco plants, *Mol. Cells*, **7**, 21-27.

Kosugi, S., Ohashi, Y., Nakajima, K. and Arai, Y., (1990) An improved assay for β-glucuronidase in transformed cells: Methanol almost completely suppresses a putative endogenous -glucuronidase activity, *Plant Sci.* **70**, 133-140.

Lamblin, F., Saladin, G., Dehorter, B., Cronier, D., Grenier, E., Lacoux, J., Bruyant, P., Lainé, E., Chabbert, B., Girault, F., Monties, B., Morvan, C., David, H. and David, A. (2001) Overexpression of a heterologous *sam* gene encoding S-adenosylmethionine synthetase in flax (*Linum usitatissimum*) cells: consequences on methylation of lignin precursors and pectins, *Physiol. Plant.* **112(2),** 223-232.

Liners, F., Gaspar, T. and Van Cutsem, P. (1994) Acetyl- and methyl-esterification of pectins of friable and compact sugar-beet calli: consequences for intercellular adhesion, *Planta* **192**, 545-556.

Loh, E. (1991) Anchored PCR: amplification with single-sided specificity methods: *a companion to Methods in Enzymology* **2(2)** 11-19.

Mareck, A., Gaffé, J., Morvan, O., Alexandre,C. and Morvan C. (1995) Characterization of isoforms of pectin methylsterase using polyclonal antibodies, Plant Cell Physiol. ***36(3)** 409-417.*

McCann, M.C., Shi, J., Roberts, K. and Carpita, N.C. (1994) Changes in pectin structure and localization during the growth of unadapted and NaCl-adapted tobacco cells, *Plant J.* **5**,773-785.

Micheli, F., Hollinger, C., Goldberg, R. and Richard, L. (1998) Characterisation of the pectin methylesterase-like gene AtPME3: a new member of a family comprising at least 12 genes in Arabidopsis thaliana, *Gene* **220**, 13-20.

Mishra, K.K. and Handa A.K. (1998) Post-transcriptional silencing of pectin methylesterase gene in transgenic tomato fruits results from impaired pre-mRNA processing *The Plant J.* **14**, 583-592.

Mlynarova, L., Bauer, M., Nap, J. P. and Pretova, A. (1994) High efficiency *Agrobacterium*-mediated gene transfer to flax, *Plant Cell Rep.* **13**, 282-285.

Morel, G. and Wetmore, R.M. (1951) Fern callus tissue culture, *Am. J. Bot.* **38**, 141-143.

Murashige, T and Skoog, F (1962) A revised medium for rapid growth and bioassays with tobacco tissue cultures, Physiol. Plant. *15, 473-497.*

Nari, J., Noat, G., Diamantidis, G., Woudstra, M. and Ricard, J. (1986) Electrostatic effects and the dynamics of enzyme reactions at the surface of plant cell. 3. Interplay between limited cell-wall autolysis, pectin methylesterase activity and electrostatic effects in soybean cell walls, *Eur. J. Biochem.* **155(1)** 199-202.

Nicol, F. and Höfte, H. (1998) Plant cell expansion: scaling the wall, *Curr. Opin. Plant Biol.* 12-17.

Nicol, F., His, I., Jauneau, A., Vernhettes S., Canut, H. and Höfte H. (1998) A plasma membrane-bound putative endo-1,4-β-D-glucanase is required for normal wall assembly and cell elongation in *Arabidopsis*, *EMBO J.* **17** 5563-5576.

Pilling, J., Willmitzer, L. and Fisahn, J., (2000) Expression of a *Petunia inflata* pectin methyl esterase in *Solanum tuberosum* L. enhances stem elongation and modifies cation distribution, *Planta* **210**, 391-399.

Ren, C. and Kermode, A.R., An increase in pectin methyl esterase activity accompanies dormancy breakage and germination of yellow cedar seeds, *Plant Physiol.* **124(1)**, 231-242.

Rhee, S. and Somerville C. (1988) Tetrad pollen formation in *quartet* mutants of *Arabidopsis thaliana* is associated with persistence of pectic polysaccharides of the pollen mother cell wall, *Plant J.* **15** 79-88.

Ricard, J. and Noat, G., (1986) Electrostatic effects and dynamics of enzyme reactions at the surface of plant cells. A theory of the ionic control of a complex multi-enzyme system, *Eur. J. Biochem.* **155**, 183-190.

Richard, L. (2000) Effet de la contrainte hydrique sur la régulation de l'expression des gènes codant les pectines méthylestérases au niveau de suspensions cellulaires d'*Arabidopsis*, *Journées du Réseau Français des Parois*, Versailles, France.

Roger, D., Lacoux, J., Lamblin, F., Gaillet, D., Dauchel, H., Klein, D., Balangé, AP., David, A. and Lainé, E. (2001) Isolation of a flax pectin methylesterase promoter and its expression in transgenic tobacco, *Plant Science.* **160**, 713-721.

Rutherford, R. and Masson, P. (1996) *Arabidopsis thaliana sku* mutant seedlings show exaggerated surface-dependent alteration in root growth vector, *Plant Physiol.* **111** 987-998.

Sato, S., Kato, T., Kakegawa, K., Ishii, T., Liu, Y.G., Takabe, K., Kuga, S., Sato, S., Nakamura, Y., Tabata, S. and Shibata, D. (1999) Cell wall analysis of temperature sensitive mutant, *Annual Meeting of Japanese Society for Plant Physiologists*, abstract.

Sedbrook, J., Hung, K., Carroll, K. and Masson, P. (1998) *sku5*, a mutation in a pectin esterase-like gene, confers an exaggerated right slanting phenotype on agar surfaces, 9th *International Conference on Arabidopsis Research*, University of Wisconsin-Madison, USA, abstract p109.

The Arabidopsis *genome initiative (2000) Analysis of the genome sequence of the flowering plant* Arabidopsis thaliana, Nature ***408**, 796-815.*

Tieman, D.M., Harriman, R.W., Ramamohan, G. and Handa, A.K. (1992) An antisense pectin methylesterase gene alters pectin chemistry and soluble solids in tomato fruit, *Plant Cell* **4**, 667-679.

Vannier, M.P. (1994) Mise en évidence des pectines méthyltransférases chez le lin. Caractérisation des pectines méthyltransférases, *PhD thesis*, Université de Rouen, France.

Wakeley, P.R., Rogers, H.J., Rozycka, M., Greenland, A.J. and Hussey, P.J. (1998) A maize pectin methylesterase-like gene, ZmC5, specifically expressed in pollen, *Plant Mol. Biol.* **37**, 187-192.

Wen, F., Zhu, Y. and Hawes, MC. (1999) Effect of pectin methylesterase gene expression on pea root development, *Plant Cell* **11**, 1129-1140.

Zhang, G.F. and Staehelin, L.A. (1992) Functional compartimentation of the Golgi apparatus of plant cells. Immunocytochemical analysis of high-pressure frozen and freeze-substituted sycamore maple suspension culture cells, *Plant Physiol.* **99**, 1070-1083.

REGULATION AND ROLE OF PECTINASES IN PHYTOPATHOGENIC FUNGI

C. HERBERT, G. BOUDART, C. BOREL, C. JACQUET,
M.-T. ESQUERRE-TUGAYE and B. DUMAS
UMR 5546 CNRS-Université Paul Sabatier, Pôle de Biotechnologie Végétale, Chemin de Borde Rouge, BP17 Auzeville, F-31326, Castanet-Tolosan, France
e-mail: dumas@smcv.ups-tlse.fr

Abstract

Fungal plant pathogens secrete a large array of cell wall hydrolases. Among them, pectinases play a dual role during the interaction between the fungus and the host plant. Pectin degradation contributes to fungal virulence but also induces defence gene expression in the host plant. To evaluate the role of pectinases in these different aspects of the interaction between plants and filamentous fungi, strategies involving gene disruption, reporter genes and site-directed mutagenesis are being used. In this chapter, we will review recent studies aiming to analyse the role of pectinases in the outcome of different plant-fungus interactions.

1. Introduction

The plant cell wall is a potential barrier to the penetration and spread of fungal pathogens in plant tissues and most of plant parasites produce enzymes that can degrade cell wall polymers. Pectinases, including pectin and pectate lyases, polygalacturonases, pectin-methyl esterases and rhamnogalacturonases, are the first cell wall-degrading enzymes produced during infections and are thought to play a major role in pathogenesis. Pectinases are encoded by gene families whose size appears to vary with the type of infection. For instance, necrotrophic pathogens such as *Botrytis cinerea* and *Sclerotinia sclerotiorum* which attack a large number of plant species produced a large set of endopolygalacturonase (endoPG) isoforms (Fraissinet-Tachet *et al.*, 1995; Wubben *et al.*, 1999) whereas biotrophic and hemibiotrophic pathogens with a restricted host range such as *Colletotrichum lindemuthianum* and *Claviceps purpurea* contain only two different endoPG genes (Centis *et al.*, 1996; Centis *et al.*, 1997; Tenberge *et al.*, 1996). On the other hand, pectinases appear to induce very efficiently plant defense reactions probably through the release of pectic fragments which could act as endogenous elicitors. In this article, we will review recent progress made in the understanding of the regulation of pectinase production in phytopathogenic fungi, and the role of these enzymes during the interaction with the host plant.

F. Voragen et al. (eds.), Advances in Pectin and Pectinase Research, 201-208.
© 2003 *Kluwer Academic Publishers. Printed in the Netherlands.*

2. Role of pectinases in pathogenesis

Gene disruption experiments of pectinase genes were performed in several phytopathogenic fungi (Table 1). Most of the mutant strains retain their pathogenicity (Gao *et al.*, 1996; Garcia-Maceira *et al.*, 2001; Scott-Craig *et al.*, 1998; Scott-Craig *et al.*, 1990) but a clear involvement of pectinase genes has been established in macerating fungi such as *Alternaria citri* (Isshiki *et al.*, 2001) and *B. cinerea* (ten Have *et al.*, 1998). From these different studies it can be concluded that the importance of pectinolytic enzymes in pathogenicity will vary with the type of disease and the host plant. However, it must be pointed out that residual pectinolytic activities were detected in the disrupted strains suggesting that functional pectinase genes were still present.

Actually, gene redundancy is a major obstacle to determine the importance of pectinolytic activity in pathogenesis. To circumvent this problem, all the genes coding a given enzymatic activity could be disrupted. Recently, two pectate lyase genes were inactivated in the pea pathogen *Nectria hematococca* (Rogers *et al.*, 2000) leading to a strong reduction of virulence. However, this strategy is feasible only in the case of a small number of genes. An other approach is to inactivate genes involved in the transcriptional induction of pectinases genes. This strategy was successful in the case of the maize pathogen *Cochliobolus carbonum*. The gene *ccSFN1*, an ortholog of the yeast regulator *SFN1*, was shown to be essential for the expression of a large set of cell-wall degrading enzymes and for pathogenesis (Tonukari *et al.*, 2000). However, *ccSFN1* is a general transcriptional regulator for glucose-repressed genes, and mutations in *ccSFN1* can affect expression of unknown genes specifically involved in pathogenesis. Thus, isolation of regulators specific for pectin-induced genes is necessary to investigate the role of pectinases in pathogenesis.

Finally, expression of a new pectinase in a pathogenic fungus can be sufficient to promote pathogenicity. This was shown in the case of *Colletotrichum magna*, a pathogen of cucurbits, which was transformed with a pectate lyase gene from the avocado pathogen *Colletotrichum gloesporioides*. Transformed isolates became more aggressive on watermelon and avocado (Yakoby et al., 2000). However, expression of an endoPG gene from *Fusarium oxysporum* f.sp. *lycopersici* in deficient strains of *F. oxysporum* f.sp. *melonis* did not induce any change in pathogenicity (Di Pietro and Roncero, 1998).

TABLE 1. Disrupted pectinase genes in phytopathogenic fungi and effect on pathogenicity

Fungus	Disrupted Gene (s)	Enzymatic actvity	Host plant	Effect on pathogenicity	Ref.
Alternaria citri	**Acpg1**	***endoPG***	Citrus	Yes	Isshiki *et al.*, 2001
Alternaria alternata	**Aapg1**	***endoPG***	Citrus	No	Isshiki *et al.*, 2001
Aspergillus flavus	*P2c*	endoPG	Cotton	Yes	Shieh *et al.*, 1997
Botrytis cinerea	Bcpg1	*endoPG*	Tomato	Yes	Ten Have *et al.*, 1998
Cryphonectria parasitica	EnPG1	*endoPG*	Chestnut	No	Gao *et al.*, 1996
Cochliobolus carbonum	PGN1	*endoPG*	Maize	No	Scott-Craig *et al.*, 1990
	PGX1	*exoPG*		No	Scott-Craig *et al.*, 1998
	PGN1/PGX1			No	Scott-Craig *et al.*, 1998
Fusarium oxysporum	Pgx4	*exoPG*	Tomato	No	Garcia-Maceira *et al.*, 2000
	Pg5	*endoPG*		No	Garcia-Maceira *et al.*, 2000
Glomerella cingulata	PnlA	*Pectin lyase*		No	Bowen *et al.*, 1995
Nectria hematococca	PelA	*Pectate lyase*	Pea	No	Rogers *et al.*, 2000
	PelD	*Pectate lyase*		No	Rogers *et al.*, 2000
	PelA/PelD			Yes	Rogers *et al.*, 2000

3. Regulation of endoPG production

3.1. TRANSCRIPTIONAL REGULATION OF ENDOPG GENES

Since endoPGs are strong elicitors of plant defense responses, a tight control of endoPG production during plant infection should play an important role in the outcome of the interaction. Pectinases are not only produced during necrotrophic growth of pathogens. Biochemical and molecular studies have shown that pectinolytic enzymes are secreted during the first stage of infection as well. Pectinolytic activities have been detected during the penetration and development of the biotrophic fungus *Uromyces viciae-fabae* (Mendgen *et al.*, 1996) and are present in the extracellular matrix secreted by germlings of *B. cinerea* (Doss, 1999). Recently, we have used the gene coding the green fluorescent protein from jellyfish as reporter gene to study the expression of *CLPG2*, one of the endoPG gene of the bean pathogen *C. lindemuthianum* (Dumas *et al.*, 1999). A promoter fragment of *CLPG2* was fused to the GFP gene and the construct was introduced into the *C. lindemuthianum* genome. Detection of GFP by fluorescent microscopy revealed that *CLPG2* is expressed during appressorium formation *in vitro* on glass slides and during bean infection. Taken together, these results suggest that pectin degradation occurs during the penetration of the fungal cell into the plant tissues. As it was shown for *C. lindemuthianum*, pectin or molecules derived from the degradation of the polymer are probably not involved in the induction of pectinases production during the very first stages of pathogenesis since endoPG gene induction was observed during the differentiation of infection structures *in vitro*, without any plant molecules in the medium.

In the case of facultative parasites, endoPG production can be induced in axenic culture. Generally, endoPG genes are repressed by glucose and induced by pectin or molecules derived from the pectic polymer in a similar way as for the saprohytic fungi such as *Aspergillus* spp. However, a complex pattern of expression has been observed for necrotrophic fungi such as *B. cinerea*. Each endoPG gene showed a different pattern of expression (Wubben *et al.*, 1999; Wubben *et al.*, 2000). For instance, *Bcpg1* is constitutively expressed in glucose and polygalacturonic acid medium whereas other endoPG genes showed an inducible expression in polygalacturonic acid and are subjected to carbon catabolite repression. The biological significance of the complexity of pectinase genes expression is not known. However, disruption of the constitutively expressed gene *Bcpg1* decreases significantly the virulence of *B. cinerea* on tomato (ten Have *et al.*, 1998) and disruption of P2c, a constitutively expressed endoPG gene from *A. flavus* caused a significant reduction in aggressiveness. This suggest that constitutive endoPGs may be involved in the primary release of pectic fragments which in turn induced expression of other endoPG isoforms. Other inducing molecules include simple sugars such as arabinose and rhamnose. In *C. lindemuthianum*, a strong induction of one endoPG gene, *CLPG1*, has been observed in medium containing rhamnose and arabinose (Hugouvieux *et al.*, 1997).

From these studies, it appears that endoPG genes are controlled by different stimuli, comprising pectin and molecules derived from pectin, environmental conditions such as pH and developmental processes such as appressorial differentiation. It remains to identify the protein factors responsible for the induction of endoPG genes during these different situations. Recently, a detailed analysis of the *CLPG2* promoter from *C. lindemuthianum* has been performed. This analysis led to the identification of a cis-acting sequence which is necessary and sufficient for the induction of the reporter gene coding the GFP in pectin medium, and during appressorial differentiation both *in vitro* and during pathogenesis (Herbert *et al.*, manuscript in preparation). This result suggests that pectinase genes expression could be under the control of a limited number of transcriptional regulators whose activity is induced by multiple stimuli. Future work will be aimed to identify these factors by using the cis-acting sequence we have isolated as a probe.

3.2 POST-TRANSLATIONAL REGULATION OF PECTINASES PRODUCTION

Besides being induced at the level of protein synthesis through transcriptional regulation of the corresponding genes, production of pectinases in the extracellular medium can also be controlled at the level of protein maturation, glycosylation and vesicular transport. Unfortunately, little is known about these processes in filamentous fungi. In eukaryotic cells, a highly conserved exocytic pathway mediates secretion of extracellular proteins. Secreted proteins are correctly folded in the endoplasmic reticulum, transported through the Golgi apparatus where they are glycosylated and targeted to the plasma membrane. The regulation of vesicular transport has been extensively studied in mammalian cells and in the budding yeast *Saccharomyces cerevisiae*. Specificity of the transport is controlled in part by a set of proteins, the Rab-proteins. Within the super-family of Ras, Rab proteins constitute a family of small GTPases required for vesicle docking and fusion. Rab-proteins, in their active form, are attached to the external face of vesicules at their C-terminus by isoprenoid lipids. Through protein-protein interactions, their function is to tether the vesicle to the site where the fusion is going to occur (Chavrier and Goud, 1999). In filamentous fungi, secretion of extracellular proteins probably occurs at the expanding tip where the newly formed cell wall is synthesised (Wessels, 1994).

In *C. lindemuthianum*, endoPG has been detected intracellularly during growth in glucose medium whereas no pectinolytic activity was present in the medium. Transfer of the mycelium grown in glucose medium to pectin medium rapidly induced expression of endoPG genes and production of extracellular endoPG. We have hypothesised that in pectin medium, an induction of the secretory pathway could be involved in the rapid secretion of endoPG. To test this hypothesis, we have isolated a gene coding for a Rab/GTPase, *CLPT1*, from *C. lindemuthianum* (Dumas *et al.*, 2001). CLPT1 is a functional homolog of the yeast SEC4p. SEC4p is specifically involved in the vesicular transport of post-Golgi vesicles to the plasma membrane (Salminen and Novick, 1987). In glucose medium, expression of *CLPT1* is very low. However, a strong expression of *CLPT1* was detected in pectin medium supporting the idea that pectin can induce protein secretion through an activation of the vesicular transport. Disruption of *CLPT1* is now under way to evaluate more precisely the regulatory role of this gene in the secretion of extracellular enzymes.

4. Elicitor activity of endoPG

It is well established that defensive plant responses can be induced by products released during cell-wall degradation leading to the concept of endogenous elicitors (Hahn, 1996). Among these, pectic fragments generated by enzymatic digestion of cell walls by pure endoPG have been shown to be very potent elicitors of plant defence (Boudart *et al.*, 1995). Moreover, the presence of an inhibitor of fungal endoPG located in the cell wall, PGIP (Polygalacturonase Inhibiting Protein), is able to favour the accumulation, *in vitro*, of oligogalacturonates having an elicitor activity (Cervone *et al.*, 1989). While pectic elicitors are generally viewed as non specific elicitors (Heath, 2000), several lines of evidence have shown that in the *C. lindemuthianum- Phaseolus vulgaris* interaction, induction of defense by endoPG is more intense in the cultivar resistant to the fungus compared to suceptible cultivars (Lafitte *et al.*, 1993). In response to endoPG, β-1,3-glucanase is induced early in resistant seedlings, but only in a delayed manner in susceptibles ones. Thus, endoPG mimics the effect of the fungus itself on defence induction. Since the differential effect of endoPG might result either from the protein itself or from the nature of the fragments it releases from the cell walls, elicitor activity of pectic fragments generated through the action of endoPG on susceptible or resistant bean cell walls was analysed (Boudart *et al.*, 1998). The fragments solubilized from the resistant plant induced an increase of PR proteins (β-1,3-glucanases and chitinases) when challenged on resistant or susceptible bean seedlings. On the other hand, pectic fragments released from susceptible bean cell walls only weakly elicited defence reactions on both cultivars. Sugar composition revealed that neutral sugars were more abundant in the fragments released from the resistant plant than from the susceptible one and more importantly, a strong difference in the esterification degree of pectic fragments was noticed.

The situation is probably more complex *in planta*, since suppressors of defence responses can also be released from the enzymatic digestion of pectin. This effect has been studied in wheat (Moerschbacher *et al.*, 1999) and bean (Boudart *et al.*, 1995). Up to now, no evidence that pectic molecules with elicitor activity are produced *in planta* have been obtained. Recently, a gene coding for the major endoPG of *C. lindemuthianum* was cloned into a plant-expression vector and the protein was transiently expressed into tobacco cells through *Agrobacterium tumefaciens* infiltration (Boudart *et al.* unpublished results). A strong plant response was detected only when the protein was located in the cell wall and mutations which abolished enzymatic activity of CLPG1, abolished the eliciting activity. These results suggest that, at least, degradation of pectin is necessary to induce the plant defence responses.

5. Conclusion

Phytopathogenic fungi produce a large set of pectinolytic enzymes whose production is controlled at the level of gene expression and probably at a post-translational level. Unfortunatly, gene disruption studies did not allow to draw a general role for pectinases in pathogenicity probably because most of the enzymatic activities are encoded by redundant genes. However, the use of disrupted mutants have pointed out that fungal pathogens have adapted the production of pectinolytic enzymes to the type of infection (necrotrophic or biotrophic) and to the chemical nature of the cell wall present in the infected tissue.

Studying pathogenicity and defence has shown that the cell wall and cell-wall degrading enzyme can also play an important role in the molecular dialogue between plant and pathogens. Pectin and pectic fragments are able to induce on one hand pectinase genes expression and on the other hand plant defence responses. A better understanding of signalling pathways leading to these responses should help to unravel the role of this molecular dialogue in the outcome of plant-fungus interactions.

6. References

Boudart, G., Déchamp-guillaume, G., Lafitte, C., Ricart, G., Barthe, J. P., Mazau, D., and Esquerré-Tugayé, M.-T. (1995) Elicitors and suppressors of hydroxyproline-rich glycoprotein accumulation are solubilized from plant cell walls by endopolygalacturonase, *Eur. J. Biochem.* **232**, 449-457.

Boudart, G., Lafitte, C., Barthe, J. P., Frasez, D., and Esquerré-Tugayé, M. T. (1998) Differential elicitation of defense responses by pectic fragments in bean seedlings, *Planta* **206**, 86-94.

Bowen, J. K., Templeton, M. D., Sharrock, K. R., Crowhurst, R. N., and Rikkerink, E. H. A. (1995) Gene activation in the plant pathogen *Glomerella cingulata*: three strategies for the disruption of the pectin lyase gene *pnlA*, *Mol. Gen. Genet.* **246**, 196-205.

Centis, S., Dumas, B., Fournier, J., Marolda, M., and Esquerré-Tugayé, M.-T. (1996) Isolation and sequence analysis of *Clpg1*, a gene coding for an endopolygalacturonase of the phytopathogenic fungus *Colletotrichum lindemuthianum*, *Gene* **170**, 125-129.

Centis, S., Guillas, I., Séjalon, N., Esquerré-Tugayé, M.-T., and Dumas, B. (1997) Endopolygalacturonase genes from *Colletotrichum lindemuthianum*: cloning of *CLPG2* and comparison of its expression to that of *CLPG1* during saprophytic and parasitic growth of the fungus, *Mol Plant-Microbe Interact* **10**, 769-775.

Cervone, F., Hahn, M. G., De Lorenzo, G., Darvill, A., and Albersheim, P. (1989) Host-pathogen interactions XXXIII. A plant protein converts a fungal pathogenesis factor into an elicitor of plant defense responses, *Plant Physiol.* **90**, 542-548.

Chavrier, P., and Goud, B. (1999) The role of ARF and Rab GTPases in membrane transport., *Cur. Opin. Cell Biol.* **11**, 466-475.

Di Pietro, A., and Roncero, I. G. (1998) Cloning, expression and role in pathogenicity of *pg1* encoding the major extracellular endopolygalacturonase of the vascular wilt pathogen *Fusarium oxysporum*, *Mol. Plant-Microbe Interact.* **11**, 91-98.

Doss, R. P. (1999) Composition and enzymatic activity of the extracellular matrix secreted by germlings of *Botrytis cinerea*, *Appl. Environ. Microbiol.* **65**, 404-408.

Dumas, B., Centis, S., Sarrazin, N., and Esquerré-Tugayé, M. T. (1999) Use of green fluorescent protein to detect expression of an endopolygalacturonase gene of *Colletotrichum lindemuthianum* during bean infection, *Appl. Environ. Microbiol.* **65**, 1769-1771.

Dumas, B., Borel, C., Herbert, C., Maury, J., Jacquet, C., Balsse, R., and Esquerré-Tugayé, M. T. (2001) Molecular characterization of *CLPT1*, a Sec4-like Rab/GTPase of the phytopathogenic fungus *Colletotrichum lindemuthianum* which is regulated by the carbon source, *Gene* **272**, 219-225.

Fraissinet-tachet, L., Reymond-cotton, P., and Fèvre, M. (1995) Characterization of a multigene family encoding an endopolygalacturonase in *Sclerotinia sclerotiorum*, *Curr. Genet.* **29**, 96-99.

Gao, S., Choi, G. H., Shain, L., and Nuss, D. L. (1996) Cloning and targeted disruption of *enpg-1*, encoding the major *in vitro* extracellular endopolygalacturonase of the chestnut blight fungus, *Cryphonectria parasitica*, *Appl. Environ. Microbiol.* **62**, 1984-1990.

Garcia-Maceira, F. I., Di Pietro, A., and Roncero, M. I. (2000) Cloning and disruption of *pgx4* encoding an *in planta* expressed exopolygalacturonase from *Fusarium oxysporum*, *Mol. Plant-Microbe Interact.* **13**, 359-365.

Garcia-Maceira, F. I., Di Pietro, A., Huertas-Gonzalez, M. D., Ruiz-Roldan, M. C., and Roncero, M. I. (2001) Molecular characterization of an endopolygalacturonase from *Fusarium oxysporum* expressed during early stages of infection, *Appl. Environ. Microbiol.* **67**, 2191-2196.

Hahn, M. (1996) Microbial elicitors and their receptors in plants, *Ann. Rev. Phytopathol.* **34**, 387-412.

Heath, M. C. (2000) Nonhost resistance and nonspecific plant defenses, *Curr. Opin. Plant Biol.* **3**, 315-319.

Hugouvieux, V., Centis, S., Lafitte, C., and Esquerré-Tugayé, M.-T. (1997) Induction by α-L-arabinose and α-L-rhamnose of endopolygalacturonase gene expression in *Colletotrichum lindemuthianum*, *Appl. Environ. Microbiol.* **63**, 2287-2292.

Isshiki, A., Akimitsu, K., Yamamoto, M., and Yamamoto, H. (2001) Endopolygalacturonase is essential for citrus black rot caused by *Alternaria citri* but not brown spot caused by *Alternaria alternata*, *Mol. Plant-Microbe Interact.* **14**, 749-757.

Lafitte, C., Barthe, J. P., Gansel, X., Déchamp-guillaume, G., Faucher, C., Mazau, D., and Esquerré-Tugayé, M.-T. (1993) Differential induction by endopolygalacturonase of β-1,3 glucanases in *Phaseolus vulgaris* isolines susceptible and resistant to *Colletotrichum lindemuthianum* race β, *Mol. Plant-Microbe Interact.* **6**, 628-634.

Mendgen, K., Hahn, M., and Deising, H. (1996) Morphogenesis and mechanisms of penetration by plant pathogenic fungi, *Ann. Rev. Phytopathol.* **34**, 367-386.

Moerschbacher, B. M., Mierau, M., Graessner, B., Noll, U., and Mort, A. J. (1999) Small oligomers of galacturonic acid are endogenous suppresors of disease resistance reactions in wheat leaves, *J. Exp. Bot.* **50**, 605-612.

Rogers, L. M., Kim, Y. K., Guo, W., Gonzalez-Candelas, L., Li, D., and Kolattukudy, P. E. (2000) Requirement for either a host- or pectin-induced pectate lyase for infection of *Pisum sativum* by *Nectria hematococca*, *Proc. Natl. Acad. Sci. U S A* **97**, 9813-9818.

Salminen, A., and Novick, P. J. (1987) A ras-like protein is required for a post-Golgi event in yeast secretion, *Cell* **49**, 527-538.

Scott-Craig, J. S., Cheng, Y. Q., Cervone, F., De Lorenzo, G., Pitkin, J. W., and Walton, J. D. (1998) Targeted mutants of *Cochliobolus carbonum* lacking the two major extracellular polygalacturonases, *Appl. Environ. Microbiol.* **64**, 1497-1503.

Scott-Craig, J. S., Panaccione, D. G., Cervone, F., and Walton, J. D. (1990) Endopolygalacturonase is not required for pathogenicity of *Cochliobolus carbonum* on maize, *Plant Cell* **2**, 1191-1200.

Shieh, M. T., Brown, R. L., Whitehead, M. P., Cary, J. W., Cotty, P. J., Cleveland, T. E., and Dean, R. A. (1997) Molecular genetic evidence for the involvement of a specific polygalacturonase, P2c, in the invasion and spread of *Aspergillus flavus* in cotton bolls, *Appl Environ Microbiol* **63**, 3548-3552.

ten Have, A., Mulder, W., Visser, J., and Van Kan, J. A. (1998) The endopolygalacturonase gene Bcpg1 is required for full virulence of *Botrytis cinerea*, *Mol. Plant-Microbe Interact.* **11**, 1009-1016.

Tenberge, K. B., Homann, V., Oeser, B., and Tudzynski, P. (1996) Structure and expression of two polygalacturonase genes of *Claviceps purpurea* oriented in tandem and cytological evidence for pectinolytic enzyme activity during infection of rye, *Phytopathology* **86**, 1084-1097.

Tonukari, N. J., Scott-Craig, J. S., and Walton, J. D. (2000) The *Cochliobolus carbonum SNF*1 gene is required for cell wall-degrading enzyme expression and virulence on maize, *Plant Cell* **12**, 237-248.

Wessels, J. G. H. (1994) Developmental regulation of fungal cell wall formation, *Annu. Rev. Phytopathol.* **32**, 413-437.

Wubben, J. P., Mulder, W., ten Have, A., van Kan, J. A., and Visser, J. (1999) Cloning and partial characterization of endopolygalacturonase genes from *Botrytis cinerea*, *Appl. Environ. Microbiol.* **65**, 1596-1602.

Wubben, J. P., ten Have, A., van Kan, J. A., and Visser, J. (2000) Regulation of endopolygalacturonase gene expression in *Botrytis cinerea* by galacturonic acid, ambient pH and carbon catabolite repression, *Curr. Genet.* **37**, 152-157.

Yakoby, N., Freeman, S., Dinoor, A., Keen, N. T., and Prusky, D. (2000) Expression of pectate lyase from *Colletotrichum gloeosporioides* in *C. magna* promotes pathogenicity, *Mol. Plant Microbe Interact.* **13**, 887-891.

SYNTHESIS OF PECTIN FRAGMENTS BY MODULAR DESIGN PRINCIPLE

C. VOGEL, B. NOLTING, S. KRAMER, W. STEFFAN, A.-J. OTT
University of Rostock, Department of Chemistry, Division of Organic Chemistry, Albert-Einstein-Str. 3a, 18059 Rostock, Germany,
christian.vogel@chemie.uni-rostock.de

Abstract

The article describes the methods of step- and blockwise synthesis of homo- and rhamnogalacturonan fragments. Thus, glycosylation of the D-galacturonic acid ester derivative **17**, which was prepared directly from D-galacturonic acid, with the thioglycoside **5**, derived from the same sugar, provides α(1→4')-linked dimer **18** in 64% yield. Selective deprotection of the 4'-*O*-position of **18** furnished the glycosyl acceptor **19**, which was coupled again with the donor **5** to yield the α(1''→4')-linked trimer **20** in 48% yield.

For a stepwise synthesis of rhamnogalacturonan I fragments, the rhamnoglycosyl 1-acetate **25** was coupled with the galacturonate glycosyl acceptor **17** to gave the desired disaccharide methyl (4-*O*-benzoyl-3-*O*-benzyl-2-*O*-acetyl-α-L-rhamnopyranosyl)-(1→4)-(allyl 2,3-di-*O*-benzyl-β-D-galactopyranosid)uronate **(33)** in 88% yield. Selective deprotection of the 2'-*O*-postion of **33** led to the acceptor **34**. The coupling of the galacturonate donor **5** with **34** gave rise to the α(1''→2')-linked trimer **35**. For the block synthesis, methyl (allyl 4-*O*-acetyl-2,3-di-*O*-benzyl-β-D-galactopyranosid)uronate (**36**) and the disaccharide **33** are key intermediates that can be used either as glycosyl acceptors or glycosyl donors after a minimum of manipulations. Deallylation and introduction of the trichloroacetimidate function at the anomeric center gave the glycosyl donors **38** and **44**, respectively. On the other hand, deacetylation of **36** and **33** provided the acceptors **17** and **43**, respectively. Generally, on condition that no neighboring group effect influenced the reaction at the anomeric center of the α-trichloroacetimidate galacturonate glycosyl donors **38**, **41**, and **44**, α-glycosidic linkages were nearly exclusively formed and offer for the first time the possibility to use D-galacturonate oligomers in a block synthesis.

1. Stepwise Buildup of Homogalacturonan Fragments

Pursuing our program directed at the synthesis of pectin fragments, which have a very complex structure and several biological functions and activities [1], we looked for an option to use oligomeric galacturonates in glycosidic reactions to form higher oligomers of defined structure. The repeating units in pectin fragments are often α(1→2) or α(1→4) linked between D-galacturonic acid itself and L-rhamnose residues. Until now, a synthetic method to prepare stepwise a D-GalA-α(1→4)-D-GalA disaccharide or

F. Voragen et al. (eds.), Advances in Pectin and Pectinase Research, 209-220.
© 2003 *Kluwer Academic Publishers. Printed in the Netherlands.*

a D-GalA-α(1→2)-L-Rha-α(1→4)-D-GalA trisaccharide is application of D-galactopyranosyluronic thioglycosides, in which the neighboring group in the 2-*O*-position does not exercise an influence on the anomeric center of the glycosyl donor [2-5].

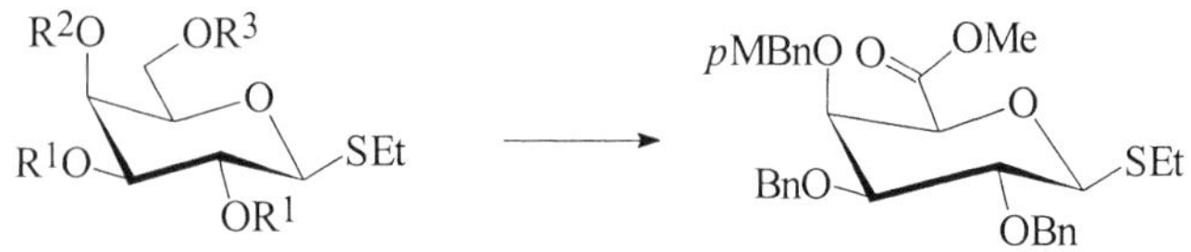

1: R^1= R^2= R^3= H

2: R^1= H R^2, R^3= >CHC_6H_4-*p*-OMe

3: R^1= Bn R^2, R^3= >CHC_6H_4-*p*-OMe

4: R^1= Bn R^2= *p*MBn R^3 = H

5

Scheme 1

Presently, we not found a suitable approach to prepare a thioglycoside of a galacturonic acid ester starting directly from galacturonic acid [6]. Furthermore, for a stepwise build-up of galacturonic acid oligomers we need the option of selective deprotection of the 4-*O*-position of a glycosyl donor. The *p*-methoxybenzyl group accomplishes the dual role either as temporary protecting group or as reactivity increasing group. Treatment of **1** with *p*-methoxybenzaldehyde dimethylacetal and *p*-toluenesulfonic acid at 40 °C under vacuum (40 mbar) provided the *p*-methoxybenzylidene acetal **2** in excellent yield (88%, Scheme 1) [7]. Benzylation of **2** in the presence of sodium hydride yielded the dibenzyl derivative **3** (82%). Reductive opening of the 4,6-*p*-methoxybenzylidene ring in **3** was performed by treatment with sodium cyanoborohydride and carefully purified trimethylchlorosilane to give exclusively the 4-*p*-methoxybenzyl ether **4** (92%) [4,8,9]. The critical point in this approach is the oxidation of the C-6 primary hydroxyl group in the presence of a thioglycoside functionality. In our experience, a stepwise procedure achieves the best results [10]. Thus, the Pfitzner-Moffat oxidation of **4** led to the corresponding aldehyde which was, after separation of precipitated *N*,*N*'-dicyclohexylurea, further oxidized in the presence of methanol and pyridinium dichromate. After this two-step procedure, the isolated yield of the desired methyl(thiogalactosid)uronate **5** amounted to 39% [4].

For the preparation of galacturonic acid derivatives suitable as glycosyl acceptors, it is possible to proceed directly from commercially available D-galacturonic acid. This seems to be advantageous in comparison to an approach involving D-galactose-derived intermediates, because the crucial oxidation step can be avoided. As reported previously [11,12], crystalline D-galacturonic acid (**6**) was transformed into its tetra-*O*-acetyl derivative **8**. In order to ensure the highest reactivity from the acceptors, the partially benzylated derivatives of galacturonic acid were produced by the following optimized approach (Scheme 2). The Helferich glycosylation with galactopyranosyluronate bromide **9** led to the allyl glycosides **10** in 98% yield [4]. Cleavage of the acetyl groups of **10** was achieved with methanolic 0.28 M hydrochloric acid in 90% yield. The products **11** were treated with acetone/2,2-dimethoxypropane/*p*-toluenesulfonic acid to give **12** in nearly 95% yield [13]. The benzylation of **12** with benzyl 2,2,2-trichloroacetimidate and a catalytic amount of trifluoromethanesulfonic acid [14] provided the fully protected derivatives **13** in 76% yield. The isopropylidene groups of **13** were removed within one hour with 90% trifluoroacetic acid in acetic acid at room temperature [15]. Crystallization and chromatographic purification of the mother liquor resulted in **14** in 78% yield [4].

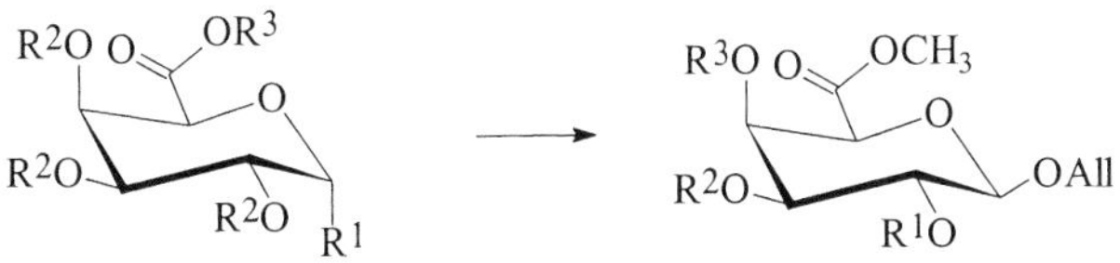

6: R^1 = OH R^2 = R^3 H (crystalline)
7: R^1 = OAc R^2 = Ac R^3 = H
8: R^1 = OAc R^2 = Ac R^3 = CH_3
9: R^1 = Br R^2 = Ac R^3 = CH_3

10: R^1 = R^2 = R^3 = Ac
11: R^1 = R^2 = R^3 = H
12: R^1 = H R^2,R^3 = >C(CH_3)
13: R^1 = Bn R^2,R^3 = >C(CH_3)
14: R^1 = Bn R^2 = R^3 = H
15: R^1 = H R^2 = R^3 = Ac
16: R^1 = Bn R^2 = R^3 = Ac
17: R^1 = R^2 = Bn R^3 = H

Scheme 2

An alternative route to **14** started from compound **10** using an enzyme-catalyzed regioselective deacetylation step [16]. We could optimize the workup procedure at the end of the enzymatic process by using pure ethanol as organic solvent during the reaction. So, the 2-*O*-unprotected derivative **15** was obtained on preparative scale in 95% yield. After acid-catalyzed benzylation (**16**, 62%) and deacetylation (90%) **14** was obtained with an overall yield of 53% based on **10** [4].

Generally it is to be noted, that in all reactions with esters of galacturonic acid in basic medium there is a high tendency for β-elimination. Therefore, most protection and deprotection operations must be performed under acidic conditions. One of the exceptions was the regioselective benzylation of **14** via 3,4-*O*-butylstannyl intermediates [15]. The desired glycosyl acceptor **17** was obtained in 58% yield. However, this reaction was connected with a partial formation of the corresponding benzyl galacturonate (25%) which can also be used as suitable acceptor in glycosylation reactions [4].

Next, we used the *p*-methoxybenzyl derivative **5** as glycosyl donor and compound **17** as acceptor (Scheme 3). In order to manage the lability of the *p*-methoxybenzyl function, we checked different promotors. The best result was obtained with an iodonium di-*sym*-collidine perchlorate. Now, the glycosylation of **17** with a slight excess of **5** provided the disaccharide **18** in 64% isolated yield based on **17**. Traces of a β-linked disaccharide could not be detected [4].

For a stepwise buildup of higher homogalacturonan fragments, the *p*-methoxybenzyl group in **18** was removed with 2,3-dichloro-5,6-dicyano-1,4-benzoquinone yielding the compound **19** in 88%, whereas cerium ammonium nitrate in this case gave only poor results [4]. Finally, the coupling of **19** with **5** was performed in the presence of iodonium di-*sym*-collidine perchlorate in such a way that after three and six hours additional amounts of glycosyl donor and promotor were added in order to ultimately reach a ratio between **5** and **19** of 2:1 (Scheme 3). After standard workup and purification by HPLC, the trisaccharide **20** was isolated in 48% yield related to **19** [4].

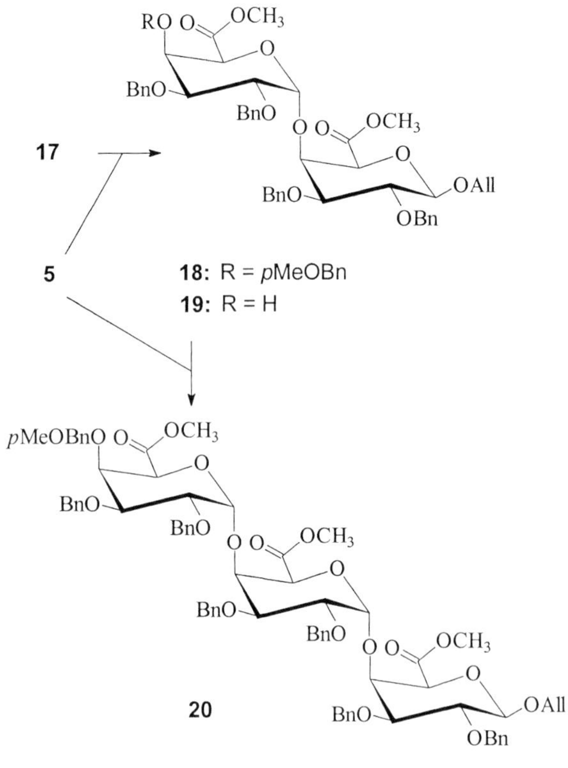

Scheme 3

In summary, a synthetic route to the α(1→4)-linked D-galacturonic acid dimer **18** (64%) and the trimer **20** (48%) was described in the past, providing the option for a stepwise buildup of homogalacturonans. It seems noteworthy that the monosaccharide glycosyl acceptors **17** was obtained from free D-galacturonic acid in a few efficient steps (average yield in each step was 89%, most intermediates were crystalline). Unfortunately, a supply of the thioglycoside from galacturonic acid, which served as glycosyl donor, still requires the crucial C-6 hydroxyl group oxidation step of suitable galactose precursors.

2. Stepwise Buildup of Rhamnogalacturonan Fragments

For a stepwise buildup of rhamnogalacturonan I fragments, the formation of a L-Rha-α(1→4)-D-GalA disaccharide was systematically examined with several suitable rhamnopyranosyl donors (**25**, **27-30**, Scheme 4), synthesis of which started from the orthoester derivative **21** [17]. The 4-*O*-position of **21** was protected with a benzoyl group (**22**).

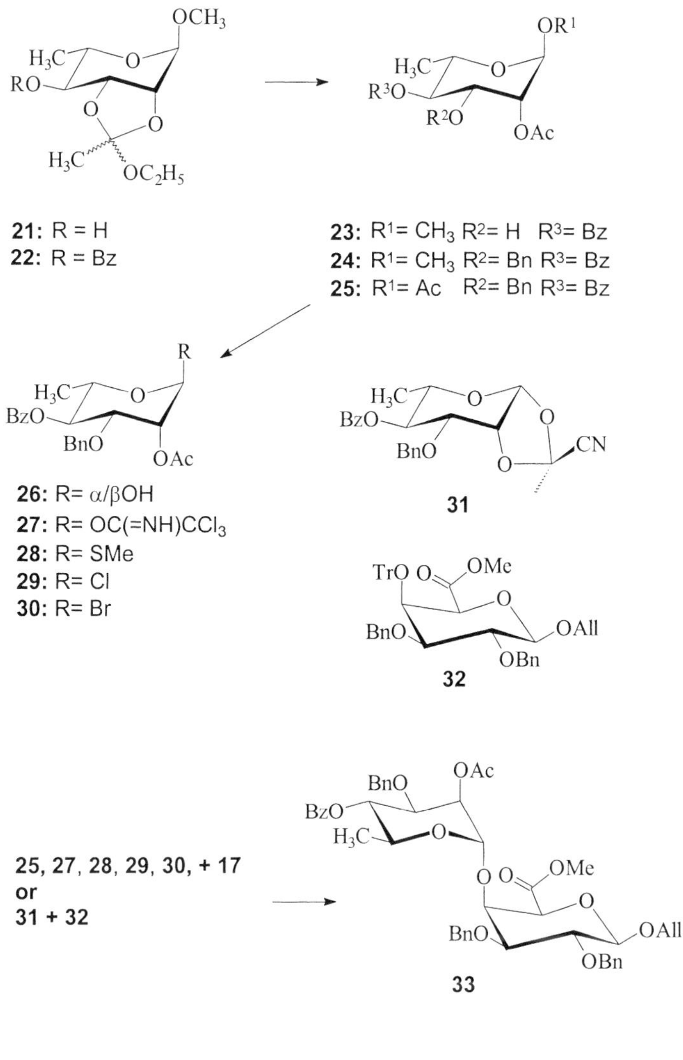

Scheme 4

Regioselective opening of the orthoester structure in **22** led to the 2-*O*-acetyl protecting group of compounds **23** which were involved in neighboring participation in the later glycosylation, and favored formation of the desired 1,2-*trans* linked products [17].

Benzylation of **23** with benzyl trichloroacetimidate and a catalytic amount of trifluoromethanesulfonic acid provided **24** in 63% yield. The introduction of a benzyl ether at the 3-*O*-position functions both as a permanent protecting group and for activation of the later glycosyl donors. Acetolysis of **24** gave nearly exclusively the α-acetates **25** in 84% yields [17].

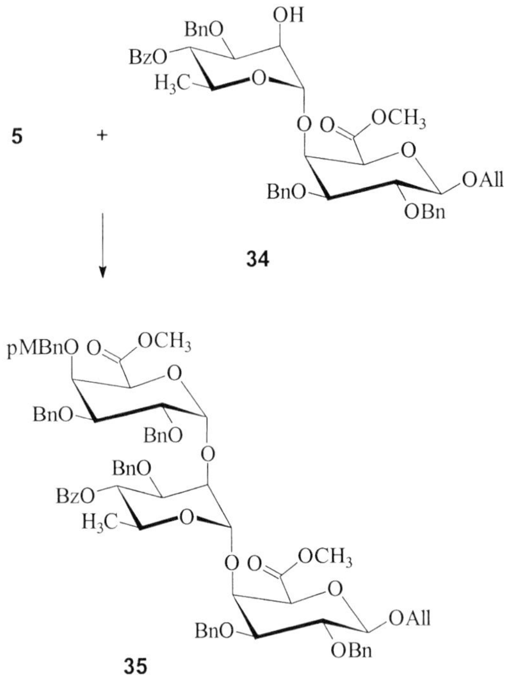

Scheme 5

For the preparation of the trichloroacetimidate **27,** the acetyl group at the anomeric center was removed with hydrazine acetate and the resulting compound **26** treated with trichloroacetonitrile and DBU to provide the compound **27** in 92% yield [17]. The reaction of **25** with methylthiomethylsilane provided the corresponding thioglycoside **28** in nearly quantitative yield [17]. Finally, the rhamnosyl chloride **29** and the bromide **30** could be obtained in comparably good yields by treatment of **25** with dichloromethyl methyl ether and oxalyl bromide, respectively [17]. In all cases, ^{13}C-^{1}H coupling constant $J_{C\text{-}1,H\text{-}1}$ with values higher than 169 Hz indicated the 1,2-*trans* configuration of the substituents at the rhamnose derivatives **27-30**.

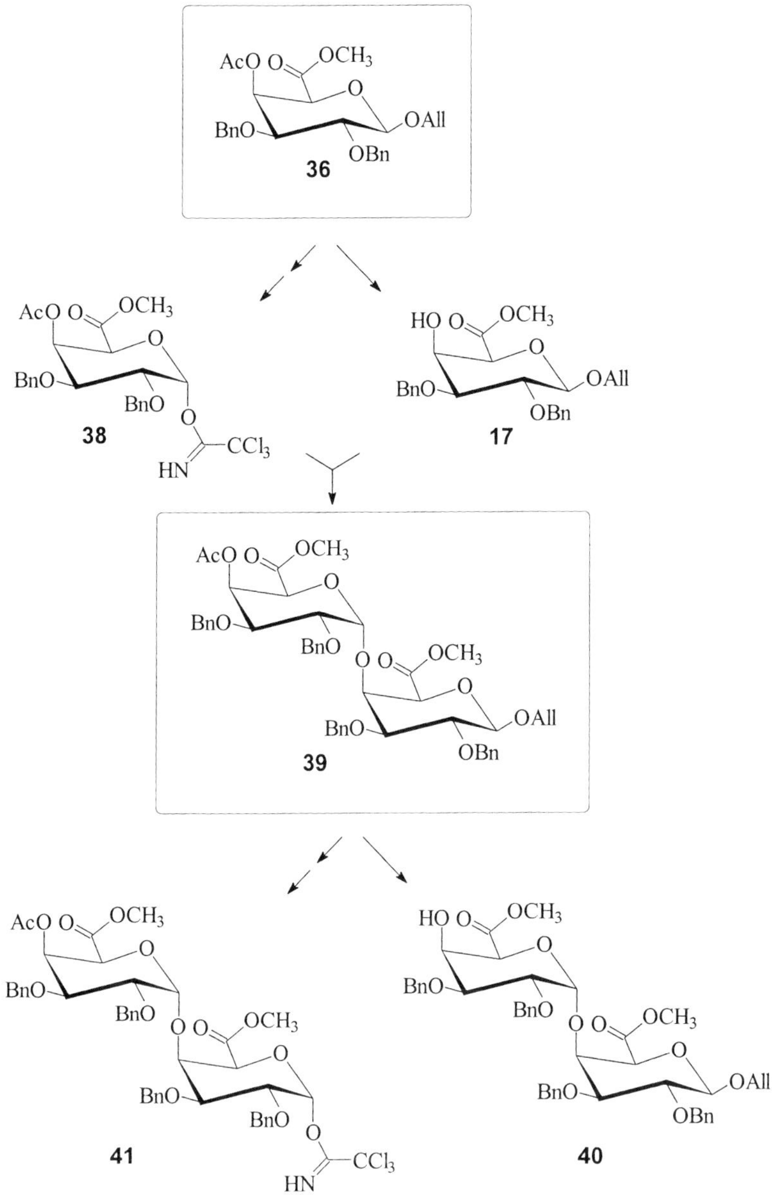

Scheme 6

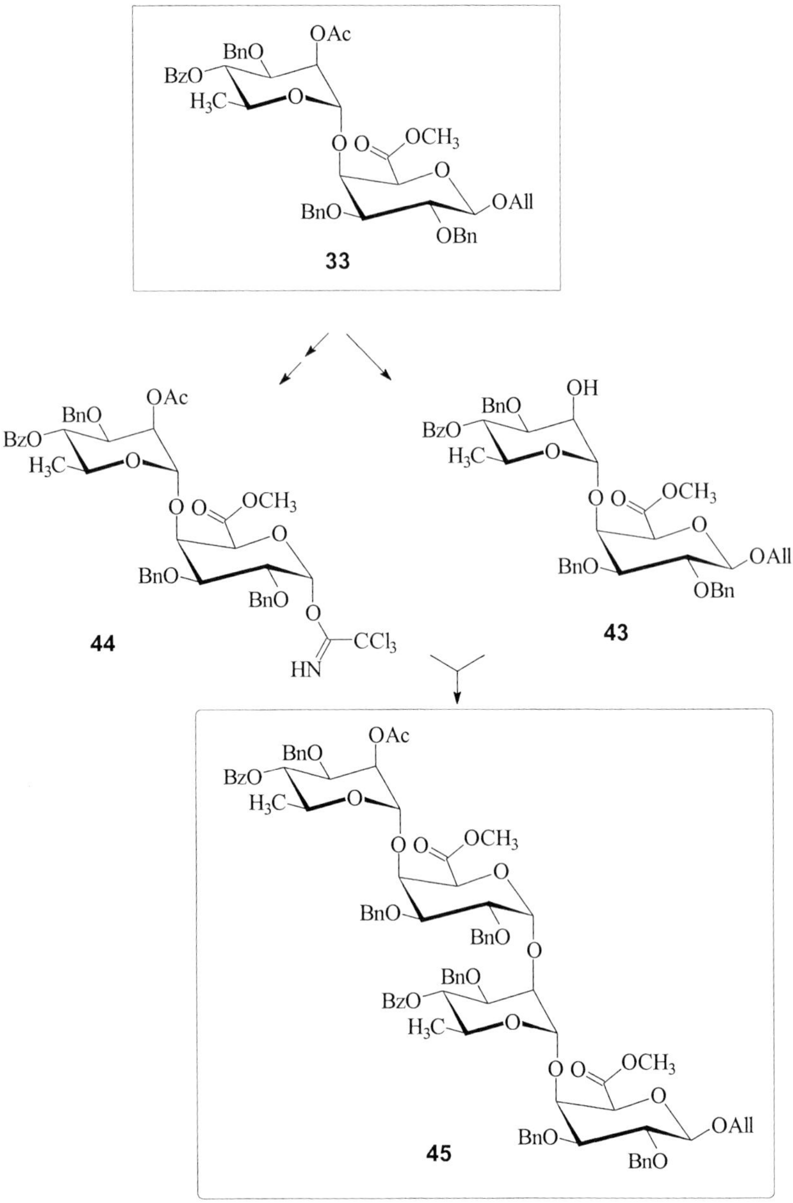

Scheme 7

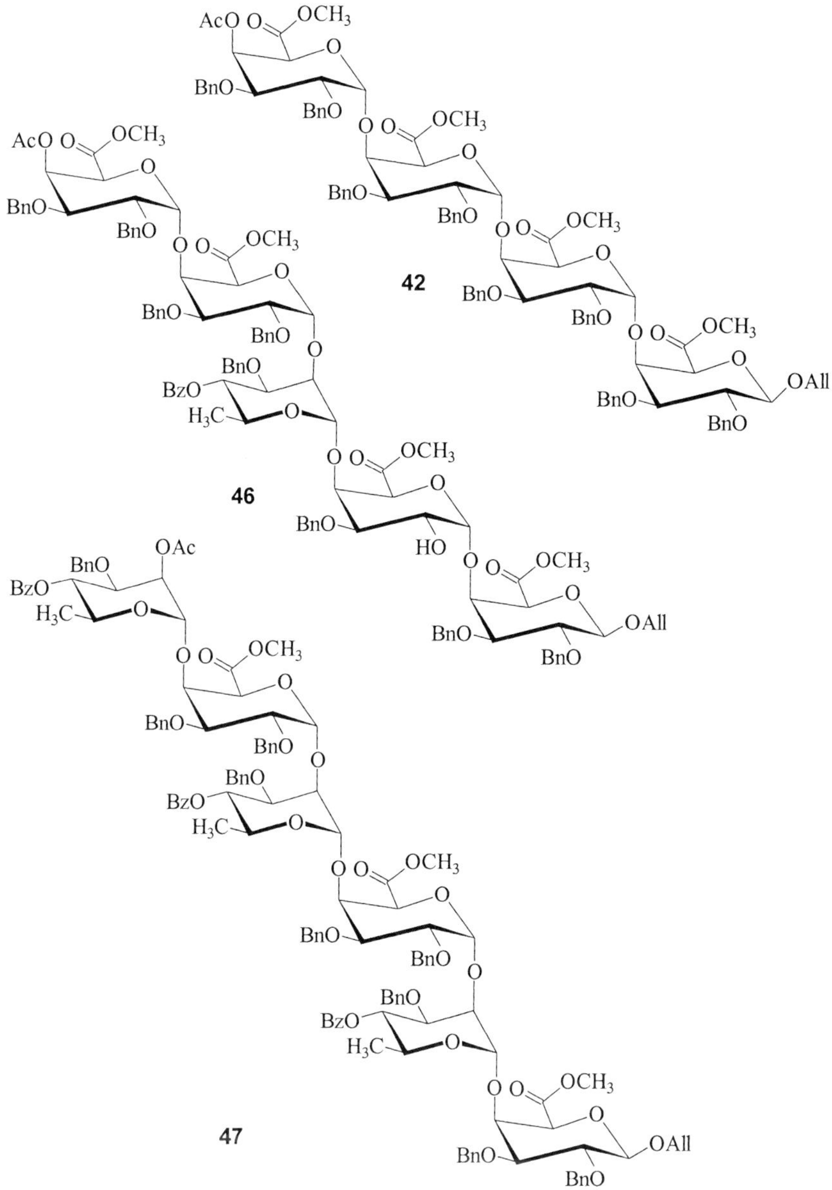

Scheme 8

For the comparative study, the rhamnoglycosyl donors **25**, **27-30** were coupled with the galacturonate glycosyl acceptor **17** in a ratio of 1:1. For every glycosyl donor several promoters were tested, and the best result of each reaction type was compared. Thus, the highest yields of the α(1→4)-coupled disaccharide **33** were achieved with the glycosyl chloride **29** (73%) and the glycosyl bromide **30** (76%) in the presence of silver trifluoromethanesulfonate [17].

Comparable results afforded the coupling of the 1-acetate **25** with the aid of trimethylsilyl trifluoromethanesulfonate (66%), in the course of which the utilization of a 1.3:1 molar ratio of donor **25** and acceptor **17** increased effectively the yield of disaccharide **33** to 88% [17]. In contrast, the glycosylation with the rhamnosyl trichloroacetimidate **27** in the presence of boron trifluoride diethyl etherate gave only comparable results (50% yield) [17] due to the trityl-cyanoethylidene condensation of the 4-*O*-tritylated acceptor **32** [5] with the donor **31** [5]. Suprisingly low was the yield of disaccharide **33** when the thioglycoside **28** was used as glycosyl donor. Promotion with *N*-iodosuccinimide/trifluoromethanesulfonic acid as well as with dimethyl(methyl-thio)sulfonium triflate gave approximately 30% yields only, whereas iodonium di-*sym*-collidine perchlorate and iodonium di-*sym*-collidine triflate were completely inactive [17].

For a stepwise buildup of a higher rhamnogalacturonan I fragment, the acetyl group at 2-*O*'-position of **33** was selectively removed by methanolic hydrogen chloride resulting quantitatively in the glycosyl acceptor **34**. The coupling of the D-galactopyranosyluronic thioglycoside **5** as glycosyl donor with acceptor **34** was performed in the presence of freshly prepared iodonium di-*sym*-collidine perchlorate. After standard workup and purification by MPLC, the trisaccharide **35** was isolated in 48% yield. Again, the glycosylation was strictly stereoselective and provided exclusively the α(1''→2')-linked oligomer [5].

3. Blockwise Buildup of Homogalacturonan Fragments by a Modular Design Principle

Since there is no suitable preparative procedure to obtain a thioglycoside starting from a D-galacturonate precursor, the corresponding D-galacturonate trichloroacetimidates were synthesized. The glycosylation of the galacturonate acceptor **17** with donor **38** in a ratio of 1:1 promoted by trimethylsilyl trifluoromethanesulfonate furnished the corresponding disaccharides **39** in a yield of 59% and no β-coupled disaccharide **39**β was observed [17]. Subsequent experiments have shown that the α- or β-configuration of the trichloroacetimidate group at the anomeric center of the donor **38** exerted no influence on the outcome of stereoselectivity of the glycosylations here investigated. Therefore, in the following experiments we used the α/β mixtures of galacturonate trichloroacetimidates [17].

Now, the preparation of galacturonates suitable as donors in α-glycosylation reactions can be carried out directly from commercially available D-galacturonic acid (D-GalA→**36**→**38**) avoiding the crucial oxidation step in comparison to an approach involving D-galactose-derived intermediates.

Next, we made sure that disaccharide **39** could serve as a module of a glycosyl acceptor and donor. Thus, the deacetylation of **39** with 0.28 M methanolic hydrochloric acid provided the deprotected 4'-*O*-position of the disaccharide **40** in a quantitative yield. The disaccharide **39** was then transformed *via* deallylation into the α-trichloroacetimidate **41** and coupled with the glycosyl acceptor **40** under exactly the same conditions described above. Fortunately, the tetrasaccharide **42** (Scheme 8) was obtained after chromatographic purification in 60% isolated yield. Only traces of a tetrasaccharide formed as a result of a β-glycosylation were noticed in the NMR spectra of the reaction mixture [18].

4. Blockwise Buildup of Rhamnogalacturonan Fragments by a Modular Design Principle

Encouraged by the good results of the glycosylations in the homogalacturonan series, the disaccharide **33** was used as a module for the synthesis of higher rhamnogalacturonan fragments. Selective deacetylation of the 2'-*O*-position provided the glycosyl acceptor **43** (Scheme 7). On the other hand, introduction of the trichloroacetimidate function at the anomeric center gave the disaccharide glycosyl donor **44**, which could be applied in a glycosylation reaction with **43** to form the tetrasaccharide **45** in 60% yield [18].

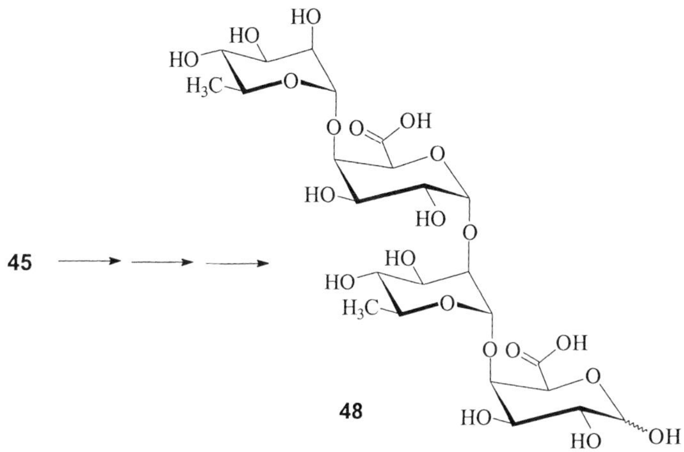

Scheme 9

Using the module **33**, **36**, **39**, **45** the pentasacharide **46** and the hexasaccharide **47** were prepared in 61 and 59% yield, respectively (Scheme 8) [18].
In a model reaction, the allyl and benzyl protective groups of **45** were removed by palladium chloride catalyzed deallylation and hydrogenolysis over Pd-C to result in the partially deprotected tetrasaccharide (Scheme 9). Finally, the cleavage of the ester linkages was achieved in methanol/water in the presense of lithium hydroxide at room temperature to provide the fully deprotected oligosaccharide **48** [18].
Thus, the trichloroacetimidate glycosylation procedure offers for the first time the possibility to use D-galacturonate oligomers in a block synthesis in the course of which α-glycosidic linkages were realized. Furthermore, the protective group pattern of the galacturonates and rhamnogalacturonates allowed to combine these compounds by a modular design principle.

5. References and Notes

[1] Darvill, A., Augur, C., Bergmann, C., Carlson, R.W., Cheong, J.-J., Eberhard, S., Hahn, M.G., Ló, V.-M., Marfá, V., Meyer, B., Mohnen, D., O'Neill, M.A., Spiro, M.D., van Halbeek, H., York, W.S., and Albersheim, P. (1992) Oligosaccharins-oligosaccharides that regulate growth, development and defense responses in plants, *Glycobiology* **2**, 181-198.

[2] Magaud, D., Grandjean, C., Doutheau, A., Anker, D., Shevchik, V., Cotte-Pattat, N., and Robert-Baudouy, J. (1997) An efficient and highly stereoselective α(1→4) glycosylation between two D-galacturonic acid ester derivatives, *Tetrahedron Lett.* **38**, 241-244.

[3] Magaud, D., Grandjean, C., Doutheau, A., Anker, D., Shevchik, V., Cotte-Pattat, N., and Robert-Baudouy, J. (1998) Synthesis of the two monomethyl esters of the disaccharide 4-*O*-α-D-galacturonosyl-D-galacturonic acid and of precursors for the preparation of higher oligomers methyl uronated in definite sequences. *Carbohydr. Res.* **314**, 189-199.

[4] Kramer, S., Nolting, B., Ott, A.-J., and Vogel, C. (2000) Synthesis of homogalacturonan fragments, *J. Carbohydr. Chem.* **19**, 891-921.

[5] Nolting, B., Boye, H., Vogel, C. (2000) Synthesis of rhamnogalacturonan I fragments, *J. Carbohydr. Chem.* **19**, 923-938.

[6] Kramer, S. (1997) α-Glycosidierungen mit D-Galacturonsäurederivaten, *Ph.D. Dissertation*, University of Rostock.

[7] Nilsson, S., Lönn, H., and Norberg, T. (1991), Synthesis of two tumor-associated oligosaccharides: di- and trifucosylated para-lacto-*N*-hexose, *J. Carbohydr. Chem.* **10**, 1023-1048.

[8] Nilsson, M., Svahn, C.-M., and Westman, J. (1993) Synthesis of the methyl glycosides of a tri- and a tetra-saccharide related to heparin and heparan sulphate, *Carbohydr. Res.* **246**, 161-172.

[9] Johansson, R. and Samuelsson, B. (1984) Regioselective reductive ring-opening of 4-methoxybenzylidene acetals of hexopyranosides. Access to a novel protecting group strategy, *J. Chem. Soc., Perkin Trans. I*, 2371-2374.

[10] Garegg, P.J., Olsson, L., and Oscarson, S. (1995) Synthesis of methyl(ethyl 2-*O*-acetyl-3,4-di-*O*-benzyl-1-thio-β-D-glucopyranosid)uronates and evaluation of their use as reactive β-selective glucuronic acid donors, *J. Org. Chem.* **60**, 2200-2204.

[11] Tajima, K. (1985) Degradation reaction of α-D-galactopyranuronic acid in acetic anhydride-base system, *Chem. Lett.*, 49-52.

[12] Vogel, C., Jeschke, U., Kramer, S., and Ott, A.-J. (1997) Synthesis of *N*-(D-galacturonoyl) amino acids and dipeptides, *Liebigs Ann./Recueil*, 737-743.

[13] Steffan, W., Vogel, C. and Kristen, H. (1990) D-Galactose-derived D-galacturonic acid derivatives suitable as glycosyl acceptors, *Carbohydr. Res.* **204**, 109-120.

[14] Iversen, T. and Bundle, D.R. (1981) Benzyl trichloroacetimidate, a versatile reagent for acid-catalysed benzylation of hydroxy-groups, *J. Chem. Soc., Chem. Commun.*, 1240-1241.

[15] Vogel, C., Steffan, W., Ott, A.-J., Betaneli, V.I. (1992) D-Galacturonic acid derivatives as acceptors and donors in glycosylation reactions, *Carbohydr. Res.* **237**, 115-129.

[16] Vogel, C., Kramer, S., and Ott, A.-J. (1997) Regioselective enzymatic acetyl ester hydrolysis of D-galacturonic acid derivatives and related reactions, *Liebigs Ann./Recueil*, 1425-1428.

[17] Nolting, B.; Boye, H.; Vogel, C. (2001) Block synthesis with galacturonate trichloroacetimidates. *J. Carbohydr. Chem.* **20**, 565-610.

[18] Nolting, B. (2001) Synthese von Rhamnogalacturonan-Fragmenten, *Ph.D. Dissertation*, University of Rostock.

MECHANISTIC LESSONS FROM STRUCTURAL STUDIES OF THE PECTATE LYASES

STEVEN R. HERRON AND FRANCES JURNAK
Department of Physiology and Biophysics
College of Medicine
University of California
Irvine, CA 92697
e-mail: jurnak@uci.edu

Pectate lyases are depolymerizing enzymes that degrade plant cell walls, causing tissue maceration and death. The laboratory has been engaged in structural studies of several members of the pectate lyase superfamily. Most effort has been focused on *Erwinia chrysanthemi* pectate lyase C (PelC). There is now an accumulation of data that supports a detailed description of the enzymatic mechanism. The data include structural studies of native PelC, PelC-Ca^{2+} complexes, and an inactive PelC mutant complexed with an oligogalacturonic acid fragment. The structural studies have been supported by amino acid sequence alignments, site-specific mutagenesis experiments, kinetic studies at various pHs, and deuterium labeling studies. Together, the data implicate an invariant arginine as the amino acid that initiates proton abstraction in the β-elimination cleavage of pectate. The data suggest several potential sources of the proton donated to the scissile glycosidic bond to complete the reaction, but do not provide sufficient evidence for any particular one. An overview of the research and its implications for other members of the pectate lyase superfamily are presented.

1. Introduction

Pectate lyases are enzymes that catalyze the exolytic or random endolytic cleavage of the pectate component of the plant cell wall [1-3]. The pectate component is a homogalacturonan, consisting of α-1,4-linked galacturonic acid (Gal*p*A) units. Most of the scientific studies on pectate lyases have been carried out on enzymes that are secreted by plant pathogenic organisms, such as those from the *Erwinia* spp. family. These pectate lyases have been shown to be causally involved in soft-rot diseases, causing economic damage to a wide variety of plant crops. The pathogenic organisms typically enter the plant through wounds produced by harvesting or by insects and subsequently secrete large amounts of pectolytic enzymes into the intercellular spaces. The pectolytic enzymes degrade the middle lamella portion of the plant cell wall in leaves, stem, fruit, tuber, or roots, causing tissue damage, cell rupture and complete maceration within three to five days. *Erwinia* bacterial strains, deficient in one or more pectate lyase genes, have decreased virulence [4-6]. Mutations affecting the extracellular secretion of the pectic enzymes also reduce virulence [7]. The introduction of pectate lyase genes into *Escherichia coli* enables *E. coli* to macerate plant tissue and display soft rot-like symptoms [8] as do highly purified preparations of

F. Voragen et al. (eds.), Advances in Pectin and Pectinase Research, 221-233
© 2003 *Kluwer Academic Publishers. Printed in the Netherlands.*

pectate lyases [9]. In many pathogenic organisms, the pectate lyases are coded by multiple, independently regulated genes. The isozymes share sequence similarities, ranging from 20% to 90%, but generally fall into two subclasses, depending upon the number of disulfide bonds and the length of the final cleavage product [10, 11]. Pectate lyases in plants were first detected in style and pollen proteins by sequence similarities with the virulence forms [12-14]. Pectate lyase genes have now been identified in a wide range of plants, including corn, tomatoes, strawberries, bananas, and *Arabidopsis thaliana* [15]. In *Arabidopsis* alone, four pectate lyase genes have been conclusively identified and another 21 have been proposed from gene annotation. Two of the former pectate lyase genes are associated with a symbiotic relationship between *Arabidopsis thaliana* and the obligate fungal pathogen *Erysiphe cichoracearum*. Inactivation of either of these two pectate lyases in *Arabidopsis* disrupts the symbiotic relationship, causing stunted growth and cupped leaves in plants [16]. Presumably, the plant pectate lyases are involved in a normal developmental process, such as modifying the pectate component of the plant cell wall as the plant grows. However, this assumption remains to be tested or proven.

The pectate lyase superfamily includes pectate lyases, plant pollen/style proteins, and pectin lyases [10, 11]. The latter proteins catalyze the cleavage of pectin, a neutral form of pectate in which the uronic acid moiety of Gal*p*A has been esterified with a methyl group. Three-dimensional structures of five members of the pectate lyase superfamily have been determined. These include *Erwinia chrysanthemi* pectate lyase C (PelC) [17], *Bacillus subtilis* pectate lyase (*B. subtilis* Pel) [18], *E. chrysanthemi* pectate lyase E (PelE) [19], *Aspergillus niger* pectin lyase A (PLA) [20], and *A. niger* pectin lyase B (PLB) [21]. All of the enzymes fold into a parallel β helix, a topology in which parallel β strands are wound into a large right-handed coil. Differences among the structures occur in the number, size, and conformation of the loops that protrude from the core parallel β helix. These loops are believed to form the substrate-binding region, with the differences reflecting altered preferences for saccharide units distal to the catalytic site [22]. A detailed comparison of the PelC and PelE structures has permitted an accurate sequence alignment of the entire superfamily [10, 11]. The multiple sequence alignment has revealed the presence of ten invariant amino acids in the superfamily. The invariant residues cluster into two distinct regions on the structures. Site-specific mutagenesis has demonstrated that the region containing three invariant amino acids, Asp-131, Arg-218, and Pro-220 in PelC nomenclature, is the pectolytic site [23]. The function of the second region, clustered around the invariant WiDH sequence, Trp-142, Asp-144, and His-145 in PelC nomenclature, has not been identified. The location of the active site was further confirmed by the structure of a mutant PelC-substrate complex called R218K-$(Ca^{2+})_4$-pentaGal*p*A [24]. By correlating the structural results with end product analyses, the location of the scissile bond has been identified. The structural results have suggested functional roles for all conserved and invariant amino acids in the region. Moreover, the structure has offered the first view of a pectate fragment at atomic resolution, revealing surprising details. In contrast to the popular "egg-box" model of pectate [25, 26], the oligosaccharide fragment is composed of a mixture of 2_1 and 3_1 helical conformations [24]. Four Ca^{2+} ions link the uronic acid groups to the protein, but the arrangement of the Ca^{2+} ions is distinctly different than that found in the "egg-box" model.

Pectate lyases catalyze the cleavage of α-1,4-polyglacturonic acid (PGA) by a β-elimination mechanism. The cleavage results in an unsaturated bond between C-4 and C-5 of the saccharide unit at the non-reducing end of the final product. The formation of a double bond in the product and the subsequent increase in the absorbance at 232 nm forms the basis of a simple spectroscopic assay for quantitating pectate lyase activity [27]. Ca^{2+} is required for *in vitro* cleavage of the polygalacturonic substrate and presumably is required for *in vivo* cleavage of the pectate component of the plant cell wall. The pH optimum of the reaction is high, occurring between 8.5 and 9.5 for all isozymes [28]. It is a long-standing conundrum as to why pectate lyases are more active at a basic pH when, in fact, the pH of the plant cell wall is acidic. Most pectate lyases are only marginally active at 5.75, the approximate pH of the cell wall [29]. However, it has been observed that the pH increases as the plant cell wall is lysed, thereby increasing the activity of the pectate lyases. Although this explanation may satisfactorily rationalize the alkaline behavior of pectate lyases from pathogenic organisms, it does not explain why plant pectate lyases routinely function at a pH of 5.75 during normal plant processes. An alternate explanation may lie in the observations that some pectate lyases function as hydrolases, cleaving PGA by the addition of water to the glycosidic oxygen [30, 31]. Hydrolase activity is routinely missed by the standard spectrophotometric assay, but can be detected by HPLC analyses of cleavage products. In fact, saturated oligosaccharide cleavage products of pectate lyase reactions have been detected, but only under conditions of acidic pH. Thus, some pectate lyases may function as both hydrolases and lyases, depending upon the pH and other factors in the local environment. As described in this presentation, structural and additional biochemical studies of PelC are revealing the atomic details of the lyase and hydrolase mechanisms applicable to all pectate lyases.

2. Details of the PelC Lyase Mechanism

The β-elimination reaction is a common mechanism found in rapid enzyme-catalyzed cleavage of polymeric substrates, in which an α proton is abstracted from a carbon atom adjacent to a carbonyl or carboxylic acid group [32]. Although the mechanism has been best characterized in epimerases and enolases, Gacesa has recognized that the same mechanism is shared by all polysaccharide lyases [33]. In essence, the glycosidic bond is cleaved by a β-elimination reaction that removes a proton, yielding an end product that contains an unsaturated bond. As applied to the pectate lyase family, the β-elimination reaction consists of the three steps shown in Figure 1. In the first step, the negative charge of the uronic acid moiety is neutralized. The PelC R218K-$(Ca^{2+})_4$-pentaGal*p*A complex clearly demonstrates that neutralization of the uronic acid group adjacent to the scissile bond occurs through interactions with Lys-190 and two Ca^{2+}

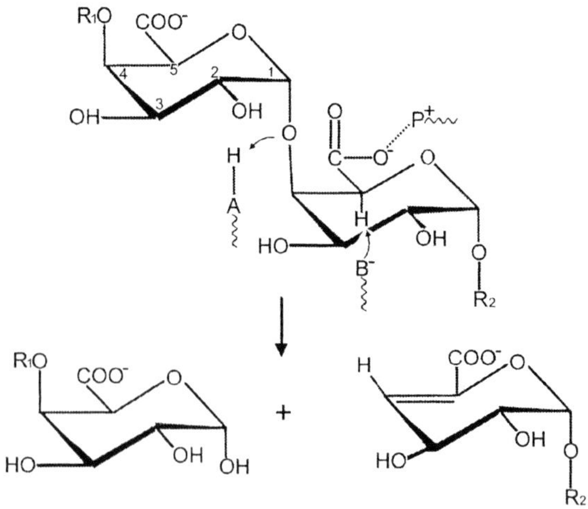

Figure 1. A schematic diagram of the β-elimination reaction mechanism of the cleavage of α-1, 4-polygalaturonic acid. Three chemical groups are required for the catalytic mechanism: P^+, which neutralizes the uronic acid group; B^-, a general base that abstracts the α-proton from C-5; and HA, a general acid that donates a proton to the glycosidic oxygen.

ions. Lysine, analogous to Lys-190 in PelC, is an invariant amino acid in the pectate lyase family, but not the pectin lyase family [10, 11]. Indeed, the invariance of a lysine at this position would serve no role in pectin lyases, because pectin does not carry a negative charge on the uronate moiety. The neutralization of the uronic acid moiety in pectate lyases also reduces the pK_a of C-5, from approximately 32 to a pK_a that is more compatible with the physiological pH. The reduction in the pK_a of C-5 is necessary for the second step, that of proton abstraction. In the R218K-$(Ca^{2+})_4$-pentaGal*p*A structure, no group capable of abstracting a proton is found near C-5. However, if the structure of the wildtype PelC is superimposed upon the R218K-$(Ca^{2+})_4$-pentaGal*p*A structure, as shown in Figure 2, Arg-218 is in a perfect position to abstract a proton from C-5 adjacent to the scissile bond. Proton abstraction is required across the superfamily; thus it is not surprising to find that an arginine, analogous to Arg-218, is invariant [10].

A proton abstraction role by an arginine is unusual and has not previously been observed in another native enzyme. However, proton abstraction by an arginine has been documented for two mutant enzymes: mandelate racemase [34] and the acyl-acyl carrier protein thioesterase [35]. Because the unusual role for an arginine is the consequence of a modeling study, further evidence has been sought to confirm this hypothesis. Kinetic studies at several substrate concentrations have been carried out and the pH profile of lyase activity determined. The pH optimum for PelC occurs at pH 9.0. The pH profile is bell-shaped, suggesting that two ionization events, essential for catalysis, occur within two pH units. As noted previously, the structural studies suggest that the two ionizable groups in PelC are Lys-190 and Arg-218, but with an unexpected twist. If Arg-218 carries out the proton abstraction, the arginine must be in the neutral

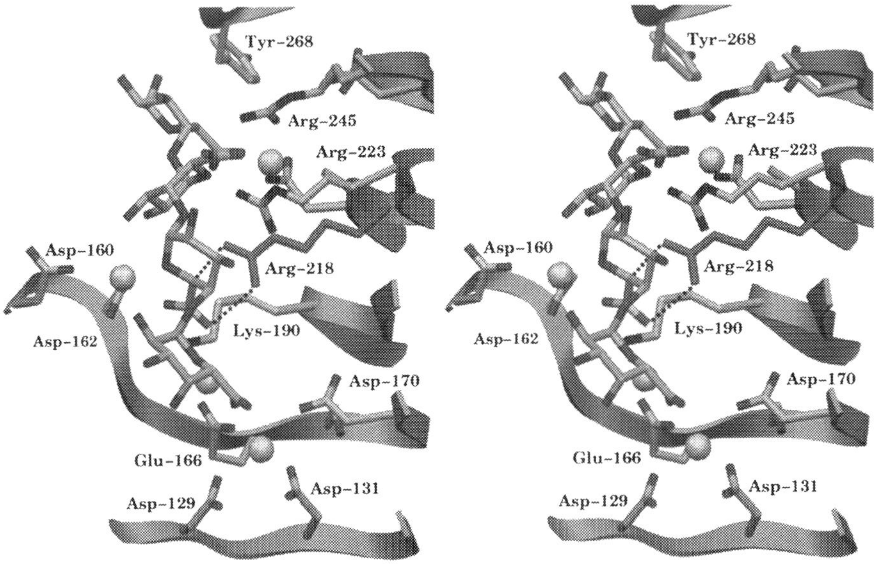

Figure 2. A stereo view of the R218K-$(Ca^{2+})_4$-pentaGal*p*A substrate superimposed upon the structure of wildtype PelC. The only difference in the protein is that the mutant Lys-218 of the actual protein structure has been replaced with Arg-218 by modeling techniques. One of the guanidinium nitrogens of Arg-218 lies within 2.6 Å of C-5 of Gal*p*A-3 and the other nitrogen, within 2.7 Å of O-6B of Gal*p*A-3.

state. Similarly, Lys-190 can only serve in the neutralization step if the amino acid is protonated. Such a situation can only occur if the pK_a of Arg-218 is lower than that of Lys-190. As shown in Figure 3, the pK_a of Arg-218 has been determined from the pH activity profile to be 8.29 and the pK_a of Lys-190 to be 9.65. To further support these assignments, the pK_a of both amino acids have been estimated using macroscopic electrostatic calculations encoded in the program MEAD [36] and applied to the PelC-$(Ca^{2+})_4$-tetraGal*p*A model. The computed pK_a values are 8.65 for Arg-218 and 9.46 for Lys-190. Both values agree reasonably well with the experimentally determined values.

Further substantiation of a proton abstraction role for an arginine comes from a comparison of PelC with mandelate racemase (MR). MR is a well-characterized enzyme that catalyzes the interconversion of the R and S chiral forms of mandelate via an enolic intermediate [34]. Although the three-dimensional structures of the enzymes and substrates are very different, both MR and PelC catalyze the abstraction of an α proton from a carbon atom adjacent to carboxylic acid. Thus, the initial reaction steps, leading to an enolic intermediate, are similar. To determine if there is a structural basis for this mechanistic similarity, the active site region of the wildtype PelC-$(Ca^{2+})_4$-tetraGal*p*A model has been superimposed upon the structure of a MR-Mg^{2+}-α-phenylglycidate complex shown in Figure 4. The similarity is striking. The Mg^{2+} ion in

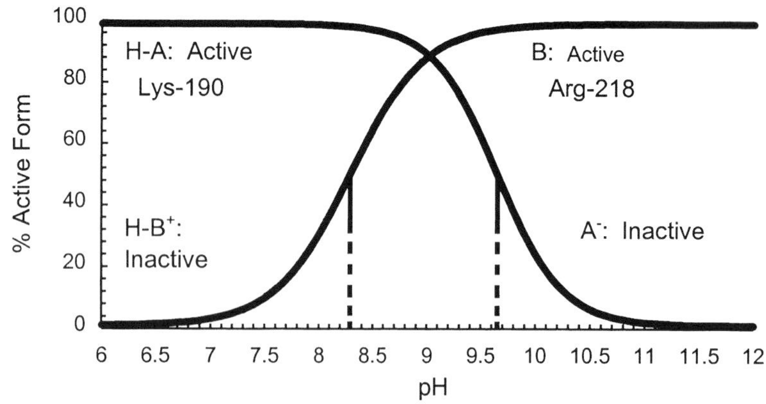

Figure 3. PelC pectolytic activity as a function of pH. The pK_a of the proton donor (H-A), Lys-190, and the proton acceptor (B), Arg-218, have been experimentally determined. The ionization of each proton is represented by a titration curve centered on the respective pK_a. The superposition of the two titration curves results in a plot that represents the theoretical pH profile of PelC activity. The theoretical profile is in good agreement with the experimentally determined profile.

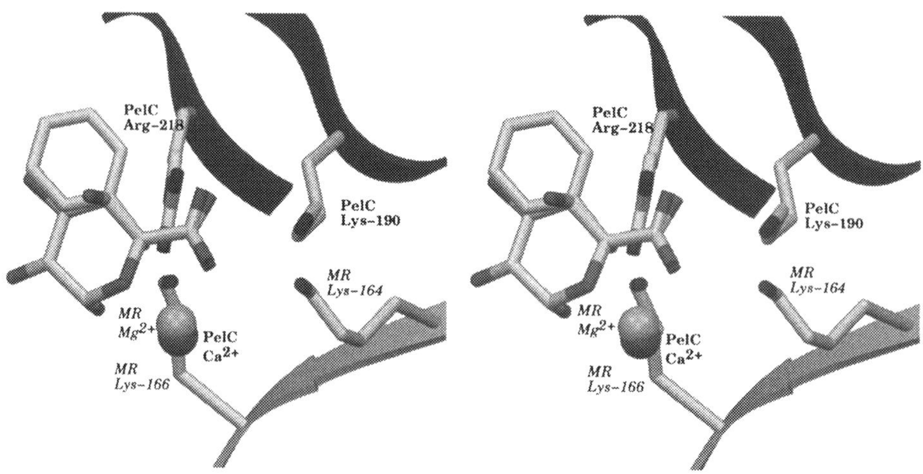

Figure 4. A stereo view of the superposition of the carbon acid groups of the substrates from the wildtype PelC-$(Ca^{2+})_4$-tetraGal*p*A model and the mandelate racemase-substrate complex. The PelC backbone is shown as a black ribbon and the PelC labels, as bold script. The MR backbone is shown as a gray ribbon and the MR labels as italicized script. The Ca^{2+} of PelC, shown as a black sphere, superimposes upon the Mg^{2+} of MR, shown as a gray sphere. The positively charged groups that neutralize the carbon acid, Lys-190 in PelC and Lys-164 in MR, are spatially aligned. The groups that initiate proton abstraction, Arg-218 in PelC and Lys-166 in MR, are in a similar orientation.

MR superimposes upon the Ca^{2+} ion in PelC. The neutralizing residue, Lys-164 in MR,is spatially equivalent to Lys-190 in PelC. Similarly, the proton abstractor, Lys-166 in MR, is analogous to Arg-218 in PelC. The similarity in the active site regions of the two proteins is an excellent example of convergent evolution, which applies not to the entire reaction, but only to the first few steps of the reaction. Further evidence of similarity comes from an MR mutant in which the proton abstractor, Lys-166 has been changed to an arginine [34]. Although the activity of MR is reduced, the K166R retains activity, but shifts its pH profile to match that of PelC, as shown in Figure 5.

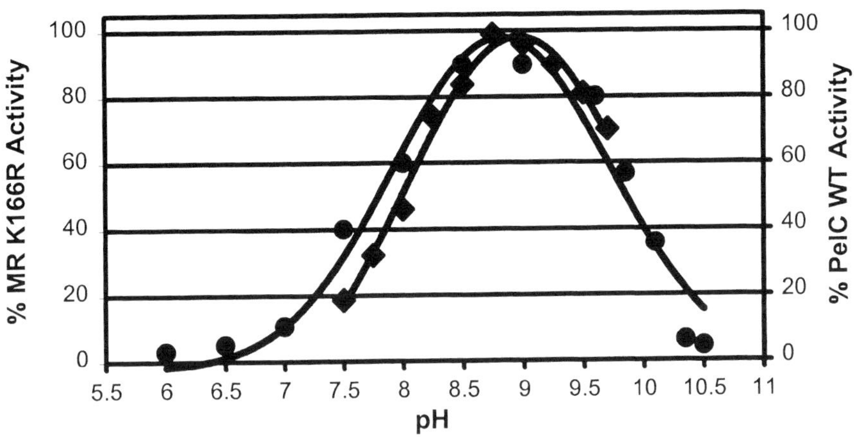

Figure 5. Comparison of the pH profiles of the MR-K166R mutant and wildtype PelC. The K166R-MR pH data (•) and the wildtype-PelC pH data (♦) were fit to a sigmoidal curve using regression analysis.

The third step in the β-elimination mechanism is the transfer of a proton to the glycosidic oxygen. This part of the reaction remains an enigma. In the R218K-$(Ca^{2+})_4$-pentaGal*p*A complex, no amino acids or solvent molecules are close enough to transfer a proton to the scissile glycosidic oxygen. The closest groups, each at a distance of 2.7 Å, are two hydroxyl groups on adjacent Gal*p*A units, as shown in Figure 6. One of the hydroxyl groups, OH-3 on Gal*p*A-3, forms a hydrogen bond with Arg-223. This amino acid is invariant only within the pectate lyase family, but not within the pectin lyase family. Thus it is unlikely that a nonconserved amino acid would have a critical role in a reaction step shared by all members of the superfamily. The other hydroxyl group, OH-2 on Gal*p*A-4, interacts with three water molecules, through which proton exchange is possible. The first hydroxyl group does not interact with a solvent molecule but the second hydroxyl group does. To distinguish between the two possibilities, a D_2O experiment has been carried out, with the results graphed as a Lineweaver-Burke reciprocal plot in Figure 7. There is a two-fold decrease in V_{max} and a 5.6-fold change in the slope in D_2O.

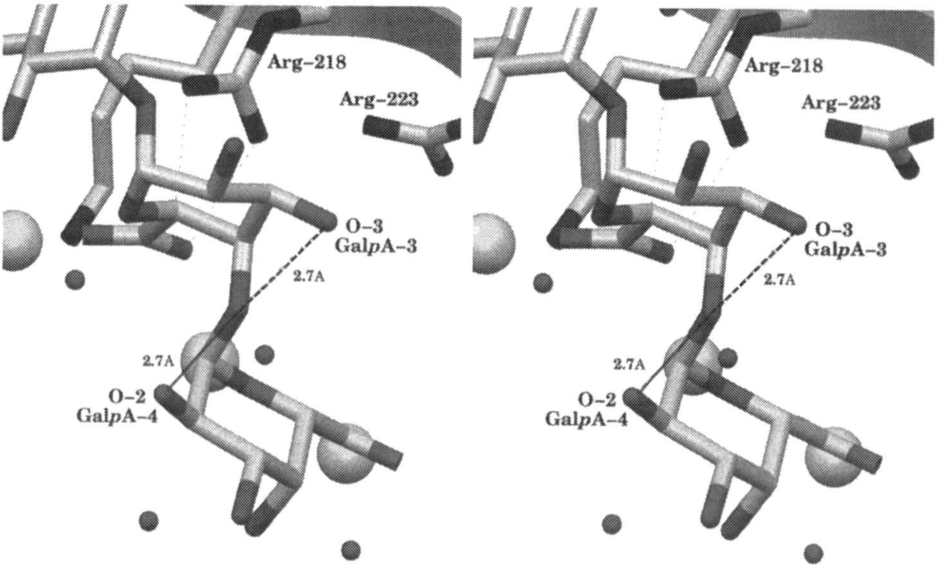

Figure 6. Stereo view of the region around the glycosidic scissile bond in the PelC-(Ca^{2+})$_4$-tetraGalpA model. The closest neighbors to the glycosidic oxygen are two hydroxyl groups, O-3 on GalpA-3 and O-2 on GalpA-4, both at a distance of 2.7 Å. The Ca^{2+} ions are shown as large spheres and the waters, as small, black, solid spheres.

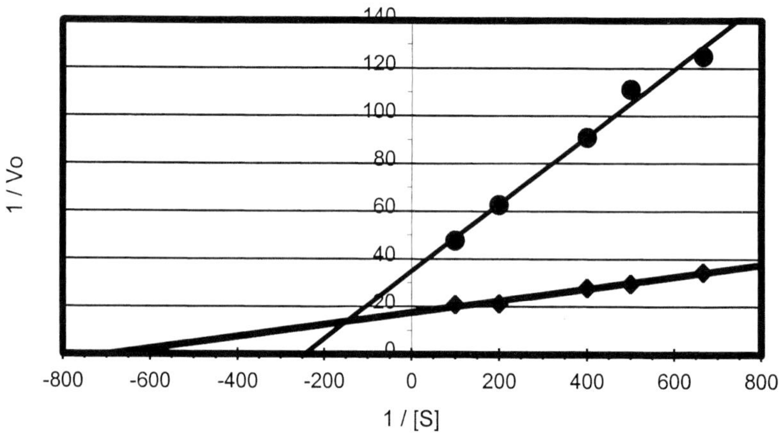

Figure 7. The effect of D_2O on the activity of PelC shown on a Lineweaver-Burke reciprocal plot. The D_2O data (•) and the H_2O data (♦) were fit to a linear line using regression analysis. The PelC concentration was 0.01 μ/ml and the five PGA substrate concentrations in gm/l were 0.15%, 0.20%, 0.25%, 0.50%, and 1.0%. The reaction was buffered at pH 9.5 in bis-tris propane and contained 0.4 mM $CaCl_2$.

The magnitude of the D_2O effect is similar to other enzymes [37] in which a water molecule is involved in a step that precedes the first irreversible step of the catalytic reaction, that is, glycosidic bond cleavage in the pectate lyases. Based on the D_2O experiment, OH-2 of Gal*p*A-4 is more likely to be the group that transfers the proton to the scissile glycosidic oxygen. Further experiments are underway to determine if a proton transfer from a water molecule to this hydroxyl group is activated by one of the Ca^{2+} ions.

3. Putative PelC Hydrolase Mechanism

Many of the enzymes that cleave the pectate or cellulose components of the plant cell wall belong to a superfamily of enzymes known as glycosyl hydrolases [38, 39]. These enzymes function in normal plant processes but can also function as the primary virulence factors secreted from pathogenic organisms. Glycosyl hydrolases incorporate a water molecule via a general acid catalysis during cleavage of the glycosidic bond between two saccharide units as shown in Figure 8. Generally, hydrolases have acidic pH optima, using only aspartic and glutamic acid groups during catalysis to function as either the essential proton donor or the nucleophile or base. The mechanism is classified as either retaining or inverting, depending upon the stereochemical outcome

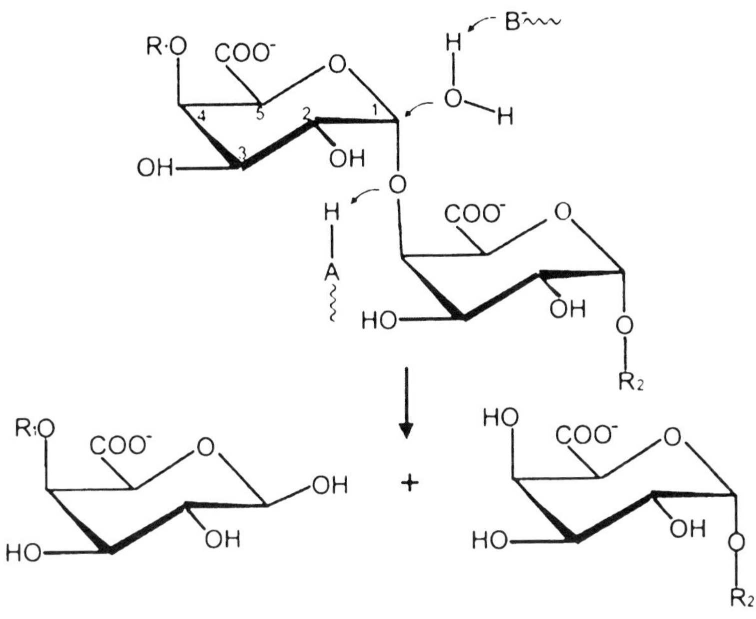

Figure 8. A schematic diagram of the inverting hydrolase reaction mechanism of the cleavage of α-1, 4-polygalaturonic acid by PelC. Two chemical groups are required for the catalytic mechanism: B^-, a general base that activates a water molecule to make a nucleophilic attack on C-1; and HA, a general acid that donates a proton to the glycosidic oxygen.

of the anomeric configuration about C-1 adjacent to the scissile glycosidic oxygen [39]. Possibly because these enzymes function best at an acidic pH near that of the plant cell wall, the glycosyl hydrolases appear to be more prevalent than the polysaccharide lyases in the plant kingdom. In light of the predominance of glycosyl hydrolases, the observations of Hotchkiss and colleagues [31] that some pectate lyases also have hydrolase activity should come as no surprise. Yet, fundamental questions remain as to how and why dual functions could be built into one enzyme.

If PelC also functions as a hydrolase, where is the active site and which amino acids are involved in catalysis? To address the question, PelC was compared to *A. niger* endopolygalacturonase II (PG-II) [40]. Structural studies of PG-II demonstrate that the enzyme folds into a parallel β-helix with the active site region in a cleft that is analogous to the lyase active site of PelC. Multiple sequence alignments of the superfamily have identified a cluster of invariant and conserved amino acids in this cleft. Site-specific mutagenesis studies have identified three aspartic acids as the probable catalytic residues. Asp-180 and Asp-202 are believed to activate a water molecule, which makes a nucleophilic attack upon C-1 adjacent to the scissile glycosidic bond. Asp-201 serves the role of the proton donor to complete the reaction [40]. The hydrolase mechanism is inverting, despite the observation that the spatial arrangement of the aspartic acids are closer to those found in retaining mechanisms. The structure of Ca^{2+}-free PelC was superimposed upon the PG-II structure, with the results shown in Figure 9. Most significantly, the conserved and invariant amino acids involved in the lyase active site of PelC align with the invariant amino acids involved in the hydrolase activity of PG-II.

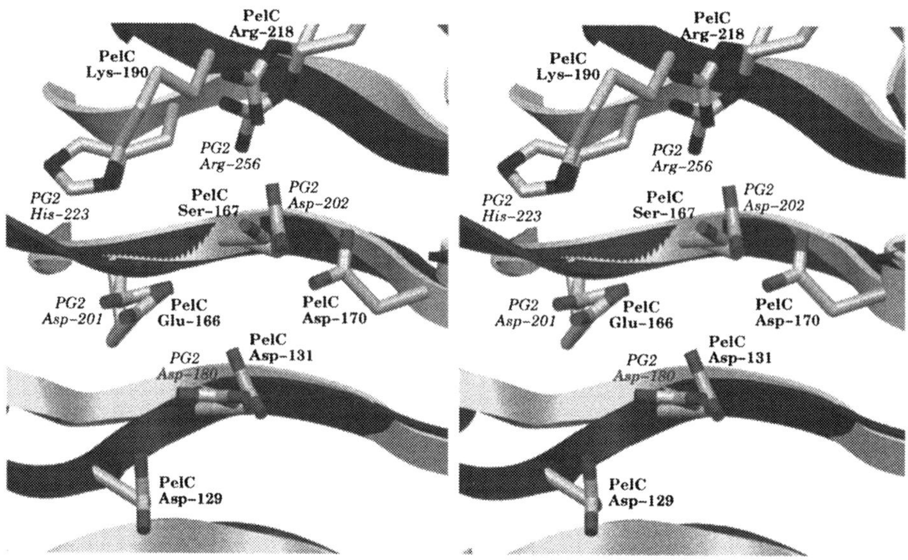

Figure 9. A stereo view of PelC superimposed upon the hydrolase active site of PG-II. The PelC backbone is shown as a black ribbon and the PelC labels, as bold script. The PG-II backbone is shown as a gray ribbon and the PG-II labels as italicized script.

Asp-131, Glu-166, and Ser-167 in PelC superimpose upon Asp-180, Asp-201, and Asp-202, all aspartic acids implicated in PG-II catalysis. Furthermore, Asp-170 in PelC superimposes upon Ala-205 in PG-II, a residue that is an aspartic acid in many other polyglacturonases. PelC amino acids, which participate in lyase activity and in interactions with the substrate, Lys-190, Arg-218, and Arg-223, also superimpose upon invariant PG-II residues, His-223, Arg-256, and Lys-258, respectively. The latter residues in PG-II have been implicated in substrate interactions [41]. Other positively charged amino acids in the substrate-binding cleft of PelC also superimpose upon similar groups in PG-II, although many are not necessarily conserved. For example, Arg-134 and Lys-172 in PelC superimpose upon Lys-127 and Arg-233 in PG-II respectively. Thus, the putative PGA binding site in PG-II is very similar to the PGA substrate site identified in PelC. These results suggest that the amino acids involved in PelC lyase activity, are likely to be the same set used in PelC hydrolase activity, albeit in an altered functional mode.

There is one significant difference between PelC and PG-II. Wildtype PelC binds a single Ca^{2+} ion that coordinates the amino acids analogous to the catalytic residues in PG-II. In PelC nomenclature, the primary Ca^{2+} site coordinates to Asp-129, Asp-131, Glu-166, and Asp-170. This particular Ca^{2+} site has been the focus of intensive investigations in our laboratory. When Ca^{2+} is removed from PelC, a water molecule assumes its position and two amino acids undergo a conformational change. Moreover, the Ca^{2+} site has a pH-dependent dissociation constant. The K_d is 0.17 mM at pH 9.5, but drops to 9.5 mM at pH 4.5. Thus, at the pH of the plant cell wall, this primary Ca^{2+} is only partially occupied. Consequently, Asp-131, Glu-166, and Asp-170 would be available for hydrolase activity in that population of PelC molecules that are Ca^{2+}-free. Further studies are underway to determine if the occupancy of the primary Ca^{2+} site is a regulatory switch that determines whether PelC acts as a hydrolase or a lyase.

4. Summary

Structural studies of PelC have led to many surprises, starting with the novel parallel β helix topology of the protein. Before the first structure, very little had been known about the enzymatic mechanism. Comparative structural analyses of PelC and PelE have led to the identification of invariant amino acids in the superfamily and later to the confirmation of the pectolytic site by site-specific mutagenesis [23]. One of the PelC mutants, R218K, proved to be particularly invaluable because its pectolytic activity is extremely low [23]. With the R218K mutant, it has been possible to trap a substrate in the crystals, leading to the three-dimensional structure of a mutant PelC-substrate complex [24]. Not only do the structural results reveal the only atomic view of a pectate fragment, but the results also lead to the assignment of functional roles of all the invariant and conserved amino acids in the region. Particularly surprising are the additional Ca^{2+} sites, linking the pectate fragment to the protein and the identification of Arg-218 as the amino acid most likely to be responsible for a proton abstraction step in the reaction. The complex structure suggests other details of the β-elimination mechanism, details that have now been confirmed by kinetic analyses or are under investigation with targeted experiments. The PelC structures have also explained the recent observations that some pectate lyases have two types of enzymatic activities, hydrolase as well as lyase activity. Again structural superposition studies, aligning PelC to PG-II, have suggested that PelC is capable of using the same active site residues, but in an altered fashion, to behave as a hydrolase or a lyase. The structural studies are also suggesting that the occupancy of the primary Ca^{2+} site is a regulatory switch controlling which enzymatic activity of PelC is used in the cell wall. One area of research, in which the structural studies will not likely have a role, is that of determining the function of pectate lyases in normal plant processes.

5. Acknowledgements

The authors gratefully acknowledge many helpful discussions with Dr. Ernie Rose and the generous research support by the United States Department of Agriculture (Grant # 98-02966). F. J. is a member of the Program in Macromolecular Structure and the Chao Family Comprehensive Cancer Center at the University of California, Irvine.

6. References

1. Barras, F., Van Gijsegem, F., and Chatterjee, A.K. (1994) Extracellular enzymes and pathogenesis of soft rot *Erwinia*. *Annu. Rev. Phytopathol.* **32**, 201-234.
2. Kotoujansky, A. (1987) Molecular genetics of soft-rot *Erwinias*. *Annu. Rev. Phytopathol.* **25**, 405-430.
3. Collmer, A. and Keen, N.T. (1986) The role of pectin enzymes in plant pathogenesis. *Annu. Rev. Phytopathol.* **24**, 383-409.
4. Ried, J.L. and Collmer, A. (1988) Construction and characterization of an *Erwinia chrysanthemi* mutant with directed deletions in all of the pectate lyase structural genes. *Molec. Plant-Microbe Inter.* **1**, 32-38.
5. Boccara, M., Diolez, A., Rouve, M., and Kotoujansky, A. (1988) The role of individual pectate lyases of *Erwinia* strain 3937 in pathogenicity on saintpaulia plants. *Physiol. Molec. Plant Pathol.* **33**, 95-104.
6. Boccara, M. and Chatain, V. (1989) Regulation and role in pathogenicity of *Erwinia chrysanthemi* 3937 pectin methylesterase. *J. Bacteriol.* **171**, 4085-4087.
7. Thurn, K.K. and Chatterjee, A. K. (1985) Single-site chromosomal Tn5 insertions affect the export of pectolytic and cellulolytic enzymes in *Erwinia chrysanthemi* EC16. *Appl. Environ. Microbiol.* **50**, 894-898.
8. Keen, N.T. and Tamaki, S.J. (1986) Structure of two pectate lyase genes from *Erwinia chrysanthemi* EC16 and their high-level expression in *Escherichia coli*. *J. Bacteriol.* **168**, 595-606.
9. Tamaki, S.J., Gold, S., Robeson, M., Manulis, S., and Keen, N.T. (1988) Structure and organization of the *pel* genes from *Erwinia chrysanthemi* EC16. *J. Bacteriol.* **170**, 3468-3478.
10. Henrissat, B., Heffron, S.E., Yoder, M.D., Lietzke, S.E., and Jurnak, F. (1995) Functional implications of structure-based sequence alignment of proteins in the extracellular pectate lyase superfamily. *Plant Physiol.* **107**, 963-976.
11. Heffron, S., Henrissat, B., Yoder, M.D., Lietzke, S., and Jurnak, F. (1995) Structure-based multiple alignment of extracellular pectate lyase sequences. *Molec.Plant-Microbe Inter.* **8**, 331-334.
12. Wing, R.A., Yamaguchi, J., Larabell, S. K., Ursin, V. M., and McCormick, S. (1989) Molecular and genetic characterization of two pollen-expressed genes that have sequence similarity to pectate lyases of the plant pathogen *Erwinia*. *Plant Mol. Biol.* **14**, 17-28.
13. Budelier, K.A., Smith, A. G., and Gasser, C. S. (1990) Regulation of a stylar transmitting tissue-specific gene in wild type and transgenic tomato and tobacco. *Mol. Gen. Genet.* **224**, 183-192.
14. Rafnar, T., Griffith, I.J., Juo, M.C., Bond, J.F., Rogers, B.L., and Klapper, D.G. (1991) Cloning of *Amb aI* (antigen E), the major family of short ragweed pollen. *J. Biol. Chem.* **266**, 1229-1236.
15. Coutinho, P.M. and Henrissat, B. (2001) Carbohydrate-active enzymes server at URL: http://afmb.cnrs-mrs.fr/~pedro/CAZY/db.html, July 18.
16. Vogel, J., and Somerville, S. (2001) Powdery mildew-resistant *Arabidopsis* mutants. 12th International Conference on *Arabidopsis* Research, Poster Abstract 195, Madison Wisconsin, June 23-27.
17. Yoder, M.D., Keen, N.T., and Jurnak, F. (1993) New domain motif: the structure of pectate lyase C, a secreted plant virulence factor. *Science* **260**, 1503-1507.
18. Pickersgill, R., Smith, D., Worboys, K., and Jenkins, J. (1998) Crystal structure of polygalacturonase from *Erwinia carotovora* ssp. *carotovora*. *J. Biol. Chem.* **273**, 24660-24664.19. Lietzke, S.E., Keen, N.T., Yoder, M.D., and Jurnak, F. (1994) The three-dimensional structure of pectate lyase E, a plant virulence factor from *Erwinia chrysanthemi*. *Plant Physiol.* **106**, 849-862.

20. Mayans, O., Scott, M., Connerton, I., Gravesen, T., Benen, J., Visser, J., Pickersgill, R., and Jenkins, J. (1997) Two crystal structures of pectin lyase A from *Aspergillus* reveal a pH driven conformational change and striking divergence in the substrate-binding clefts of pectin and pectate lyases. *Structure* **5**, 677-689.
21. Vitali, J., Schick, B., Kester, H.C.M., Visser, J., and Jurnak, F (1998) The three-dimensional structure of *Aspergillus niger* pectin lyase B at 1.7-angstrom resolution. *Plant Physiol.* **116**, 1-13.
22. Herron, S.R., Benen, J.A.E., Scavetta, R.D., Visser, J., and Jurnak, F. (2000) Structure and function of pectic enzymes: virulence factors of plant pathogens. *Proc. Natl. Acad. Sci.* **97**, 8762-8769.
23. Kita, N., Boyd, C.M., Garrett, M.R., Jurnak, F., and Keen, N.T. (1996) Differential effect of site-directed mutations in PelC on pectate lyase activity, plant tissue maceration, and elicitor activity. *J. Biol. Chem.* **271**, 26529-26535.
24. Scavetta, R.D., Herron, S.R., Hotchkiss, A.T., Kita, N., Keen, N.T., Benen, J.A.E., Kester, H.C.M., Visser, J., and Jurnak, F. (1999) Structure of a plant cell wall fragment complexed to pectate lyase C. *Plant Cell* **11**, 1081-1092.
25. Walkinshaw, M.D., and Arnott, S. (1981) Conformations and interactions of pectins. II. Models of junction zones in pectinic acids and Ca^{2+} pectate gels. *J. Mol. Biol.* **153**, 1075-1085.
26. Grant, G.T., Morris, E.R., Rees, D.A., Smith, P.J.C., and Thom, D. (1973) Biological interactions between polysaccharides and divalent cations. *FEBS Lett.* **32**, 195-198.
27. Keen, N.T., Dahlbeck, D. Staskawicz, B., and Belser, W. (1984) Molecular cloning of pectate lyase genes from *Erwinia chrysanthemi* and their expression in *Escherichia coli. J. Bacteriol.* **159**, 825-831.
28. Tardy, F., Nasser, W., Robert-Baudouy, J., and Hugouvieux-Cotte-Pattat, N. (1997) Comparative analysis of the five major *Erwinia chrysanthemi* pectate lyases: enzyme characteristics and potential inhibitors. *J. Bacteriol.* **179**, 2503-2511.
29. De Lorenzo, G., Cervone, F., Hahn, M.G., Darvill, A., and Albersheim, P. (1991) Bacterial endopectate lyase: evidence that plant cell wall pH prevents tissue maceration and increases the half-life of elicitor-active oligogalacturonides. *Physiol. Molec. Plant Pathol.* **39**, 335-344.
30. Hotchkiss, A.T., and Hicks, K.B. (1993) Analysis of pectate lyase-generated oligogalacturonic acids by high-performance anion-exchange chromatography with pulsed amperometric detection. *Carbohydr. Res.* **247**, 1-7.
31. Hotchkiss, A.T., Revear, L.G., and Hicks, K.B. (1996) Substrate depolymerization pattern of *Pseudomonas viridiflava* SF-312 pectate lyase. *Physiol. Mol. Plant Pathol.* **48**, 1-9.
32. Gerlt, J.A. and Gassman, P.G. (1992) Understanding enzyme-catalyzed proton abstraction from carbon acids: importance of late transition states in concerted mechanisms. *J. Am. Chem. Soc.* **115**, 11552-11568.
33. Gacesa, P. (1987) Alginate-modifying enzymes – a proposed unified mechanism of action for the lyases and epimerases. *FEBS Lett.* **212**, 199-202.
34. Kallarakal, A.T., Mitra, B., Kozarich, J.W., Gerlt, J.A., Clifton, J.G., Petsko, G.A., and Kenyon, G.L (1995) Mechanism of the reaction catalyzed by mandelate racemase - structure and mechanistic properties of the K166R mutant. *Biochem.* **34**, 2788-2797.
35. Yuan, L., Nelson, B.A., and Caryl, G. (1995) The catalytic cysteine and histidine in the plant acyl-acyl carrier protein thioesterases. *J. Biol. Chem.* **271**, 3417-3419.
36. You, T.J. and Bashford, D. (1995) Conformation and hydrogen ion titration of proteins: a continuum electrostatic model with conformational flexibility. *Biophys. J.* **69**, 1721-1733.
37. Pocker, Y. and Bjorkquist, D.W. (1997) Comparative studies of bovine carbonic anhydrase in H_20 and D_20. Stopped-flow studies of the kinetics of interconversion of CO_2 and HCO_3^-. *Biochem.* **16**, 5698-5707.
38. Withers, S.G. (2001) Mechanisms of glycosyl transferases and hydrolases. *Carbohydr. Polymers* **44**, 325-337
39. Davies, G. and Henrissat, B. (1995) Structures and mechanisms of glycosyl hydrolases. *Structure* **3**, 853 859.
40. Van Santen, Y., Benen, J.A., Schröter, K.H., Kalk, K.H., Armand, S., Visser, J., and Dijkstra, B.W. (1999) 1.68-Å crystal structure of endopolygalacturonase II from *Aspergillus niger* and identification of active site residues by site-directed mutagenesis. *J. Biol. Chem.* **274**, 30474-30480.
41. Pagès, S., Heijne, W.H., Kester, H.C., Visser, J., and Benen, J.A. (2000) Subsite mapping of Aspergillus niger endopolygalacturonase II by site-directed mutagenesis. *J. Biol. Chem.* **275**, 29348-29353.

MODE OF ACTION ANALYSIS AND STRUCTURE-FUNCTION RELATIONSHIPS OF *ASPERGILLUS NIGER* PECTINOLYTIC ENZYMES

J.A.E. Benen[1] G.J.W.M. van Alebeek[2] A.G.J. Voragen[2] and J. Visser[3]

Dept. Agrotechnology and Food Sciences, Wageningen University, [1] Laboratory of Microbiology, section Molecular Genetics of Industrial Microorganisms, Dreijenlaan 2, 6703 HA Wageningen, The Netherlands, and [2]Laboratory of Food Chemistry, Bomenweg 2, 6703 HD Wageningen, The Netherlands. [3]Present address: P.O. Box 396, 6700 AJ Wageningen, The Netherlands.

Abstract

As a result of the cloning and overexpression of individual *Aspergillus niger* pectinase genes the corresponding enzymes are now readily available for detailed characterisation using defined substrates, including derivatised oligogalacturonides. In this chapter we give an overview of the results of these studies for a pectate and pectine lyase and a polygalacturonase.

Detailed kinetic studies on pectate lyase have revealed that the substrate binds to the enzyme as a Ca^{2+}-substrate complex, thus explaining the absolute requirement of pectate lyases for Ca^{2+}-ions.

Using derivatised oligogalacturonides it was shown that pectin lyase A cannot cleave substrates that are: 1) fully de-esterified, 2) fully ethyl-esterified, 3) fully amidated, and 4) fully methylamidated. Furthermore, partial de-esterification of fully methyl esterified oligogalacturonides, one methyl group on average, resulted in a strong reduction of the catalytic efficiency and a change in the preferred binding mode. From these studies it can be concluded that pectin lyase A is highly specific for pectin with a high degree of methylesterification.

Endopolygalacturonases (endoPGs) were thouroughly characterised kinetically which allowed to calculate subsite affinities. Site directed mutagenesis of endoPGII in combination with the 3D-structure allowed to pin-point amino acids involved in catalysis as well as amino acids that are important for binding of the substrate. At subsite –5 the presence of an arginine appeared critical for displaying multiple attack attack on a single chain (processive) behavior.

1. Introduction

The chapters on pectin structure in this book demonstrate that pectin is the most complex carbohydrate in nature. A microorganism like the saprophytic filamentous fungus *Aspergillus niger*, is required to have a broad arsenal of pectinolytic enzymes at its disposal to take full advantage of this carbon source. Indeed, presently twenty-one

F. Voragen et al. (eds.), Advances in Pectin and Pectinase Research, 235-256.
© 2003 *Kluwer Academic Publishers. Printed in the Netherlands.*

genes encoding such pectinases have been cloned and individually overexpressed. These genes comprise a family of seven endopolygalacturonase encoding genes [1], a family of six pectin lyase encoding genes [2], two rhamnogalacturonase genes [3], a rhamnogalacturonan acetylesterase [4], a rhamnogalacturonan lyase gene, a pectate lyase gene [5], a pectin methylesterase gene and an exopolygalacturonase gene. Several of these genes were cloned by reverse genetics or by antibody screening of cDNA expression libraries whereas others were cloned by screening phage libraries under heterologous conditions using probes based on known genes from *A. niger* or other organisms. The recent cloning of a xylogalacturonan hydrolase gene from *A. tubingensis* [6], which proved to be present in *A. niger* as well when probed for, demonstrates that most likely only a fraction of the total pectinolytic enzyme spectrum of *A. niger* is known. In addition to the pectinolytic enzymes that act on the pectin main-chain, there are numerous enzymes that act on side-chains attached to the rhamnose residues of the rhamnogalacturonan I. These enzymes comprise arabinofuranosidases, endoarabinanases, galactosidases, galactanases and feruloylesterase.

To obtain a complete inventory of enzymes involved in pectin degradation and insight into their interplay will require a full-scale functional genomics and proteomics analysis of the *A. niger* pectinase (gene) spectrum addressing regulation at the transcriptional level as well as enzyme levels and enzyme specificity. Undoubtedly, such an analysis will be initiated. We have been studying *A. niger* pectinase gene expression and pectinase specificity, with emphasis on endopolygalacturonases, for several years and will present our findings on mode of action analysis and structure function relations here. Elsewhere in this volume studies on the regulation of *A. niger* pectinase gene expression will be presented (see De Vries and Parenicová on page 249).

2. Homogalacturonan Degrading Enzymes

It has been generally accepted by now that pectin consists of homogalacturonan and rhamnogalacturonan regions. Recently, the challenging idea has been put forward that homo- and rhamnogalacturonan do not alternate, as has long been the dogma, but that homogalacturonan is in fact a side chain of rhamnogalacturonan (see Voragen and co-workers elsewhere in this volume). Despite the potential change in view of the general concept of the overall pectin structure, it will not affect the presence of homogalacturonan as discrete regions. Homogalacturonan consists of α-(1,4) linked galacturonic acid residues that can be esterified with methanol at the uronate functionality (methyl esterification at O6) or esterified with acetate at O2 and/or O3 (acetyl esterification). The degree of methyl esterification (DM) can be as low as 0 % in polygalacturonic acid to as high as 95 % in high methoxyl (HM) commercial pectins. Generally, homogalacturonan is obtained from plants at 50-70 % DM. The degree of acetyl esterification (DA) is generally much lower and depends heavily on the source of the pectin. In citrus pectin DA is generally very low (< 2%) [7], whereas in sugar beet pectin DA up to 35% has been reported [8].

The distribution of methyl- and acetyl esters over the homogalacturonan chain can be fully random to completely blockwise and any form of distribution in between. It can easily be envisaged that the type and distribution of esterification can greatly influence the activity of enzymes acting on the homogalacturonan part. The best known examples of de-polymerizing enzymes acting on either HM-pectin or polygalacturonic acid are pectin lyases (HM pectin) or polygalacturonases (PGs) and pectate lyases, respectively. To date no enzyme is known that is absolutely specific for either high or low DM pectin. Generally the activity of the enzyme either decreases (polygalacturonases or pectate lyases) or increases (pectin lyases) gradually with increasing DM depending on the type of enzyme. In addition, the distribution of the methyl groups is very important. In blockwise de-esterified pectin both polygalacturonase as well as pectin lyase have a higher activity than on random de-esterified pectin with the same DM. This is exemplified in Table 1.

Table 1. Degradation of pectin with different degrees of methyl esterification by endopolygalacturonase II and pectin lyase A. Lemon pectin with degree of esterification of 81% (E81) was de-esterified with fungal PME (F43, F58, random distribution) or plant PME (P41, P60, block distribution). pga: polygalacturonic acid. ND: not determined. Activity is based on initial rates using 2.5 mg/ml substrate at pH 4.2 (PGII) or 3.0 mg/ml substrate at pH 6.0 (PLA).

Enzyme	Substrate					
	pga	P41	F43	P60	F58	E81
	Relative activity (%)					
PGII	100	84	48	41	24	7
PLA	ND	77	19	108	58	100

Although the data suggest that the tolerance for deviation from the preferred substrate (high or low DM) is quite high, it should be kept in mind that polymer substrates were used and initial reaction rates were determined. By mass spectrometry (MS) techniques Körner and co-workers [9] analyzed the tolerance for absence or presence of methylesters at particular subsites of PGII by sequencing partially methyl esterified oligogalacturonides oligo(Gal*p*A) obtained after prolonged hydrolysis of the polymer substrates. This study revealed that some subsites of PGII can accommodate a methyl esterified galacturonic acid (Gal*p*A) quite well. To definitely establish the actual tolerance for methyl esterification of Gal*p*A at a particular subsite has to await the synthesis of a set of partially methylesterified oligoGal*p*A with known chain length and known position of esterification.

Both acetyl- and methyl esterification greatly influence the physical properties of the pectin molecule. These decorations of the pectin are probably very important to the plants that synthesize such pectins. However, to a microorganism with a limited number of depolymerases these decorations form obstacles for effective degradation. To reduce the DM and DA many microorganisms, including *A. niger*, have esterases at their disposal. From *A. niger* both a pectin methyl esterase (PME) and a rhamnogalacturonan acetyl esterase (RGAE) have been cloned and overexpressed. In addition, a pectin acetyl esterase (PAE), specific for homogalacturonan was purified [10]. The cloning of this gene is currently in progress (van Alebeek and Benen, unpublished). It is not known whether PME also de-esterifies rhamnogalacturonan or whether a specific RGME exists.

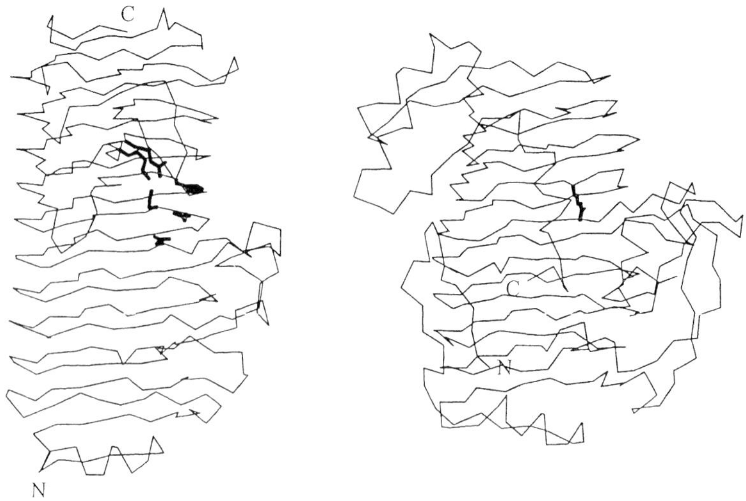

Figure 1. Cα-traces of endopolygalacturonase II (left panel) and pectin lyase A. Shown in bold are the catalytic site residues.

The *A. niger* PME is optimally active at pH 5.0. This implies that only the *A. niger* polygalacturonases, with similar pH optima, will benefit from PME action. To ensure pectin degradation over a broad pH range *A. niger* has a complement of enzymes at its disposal with optimal activities between pH 3.8 to 8.5. This includes a pectate lyase (PLYA), pectin lyases (PLs) and PGs. Although the pectic lyases and the PGs catalyze completely different reactions, β-elimination versus hydrolysis, and their sequence identity is marginal (less than 18 %), there is a striking similarity in structural topology: a right handed parallel β-helix, first described for pectate lyase C from *Erwinia chrysanthemi*. As an example of this topology the Cα-trace of PGII is presented in Fig. 1. Not only the pectic lyases and hydrolases (PGs and rhamnogalacturonases) show this topology, also for PME from *E. chrysanthemi* this type of structure was reported [11]. Elsewhere in this volume Jurnak and coworkers and Pickersgill and coworkers discuss more details on these structures. The major differences, apart from amino acid replacements, are found in the number of rungs of the β-helix and the length of the loops extending from the helix core. Very recently, the structure of a family 10 polysaccharide lyase (pectate lyase) has been solved that shows a topology is completely different from the other pectic lyases (notably the very large family 1) (GJ Davies, and GW Black, Pers. Comm.). An overview of the classification of polysaccharide degrading enzymes, including esterases, compiled by Coutinho and Henrissat [12] can be found at the following URL: http://afmb.cnrs-mrs.fr/~pedro/CAZY/db.html.

The 3D structures reveal a cleft that extends from the N- to the C-terminus and this complies with the endolytic mode of action of these enzymes. The length of the cleft, related to the size of the (methyl esterified) Gal*p*A units, indicates that multiple subsites are involved in the binding of the substrate. Each of the subsites binds a Gal*p*A unit with a particular affinity and the sum of binding energy of the subsites involved determines the mode of action on oligoGal*p*A when not all subsites are covered. Analysis of the products obtained after cleavage of a set of (methyl esterified) oligoGal*p*A of defined length and the corresponding reaction rates gives insight into the affinity of individual subsites and understanding of the specificity of the enzyme. These analyses were carried out for the *A. niger* PLYA, PLA and PGs. Specific features of these enzymes will be discussed in the following sections.

3. Mode of Action of *A. niger* Pectate Lyase A

The overexpressed *A. niger* plyA gene resulted in the production of an N-glycosylated form and an N-glycon free form [13] of PLYA. These enzymes showed identical specific activities of 40.5 and 41.1 U/mg, respectively under standard conditions [5]. For further analysis the N-glycon free enzyme was chosen to avoid any interference of the N-glycosylation. During pH optimum determination, using 0.25 % (mass/vol) polygalacturonic acid as a substrate in the presence of 1 mM $CaCl_2$, two distinct pH optima were found, one at pH 7.5 in Tris/HCl buffer and one at pH 8.0 in 2-amino methyl propanol-buffer. Since the activity decrease at the basic side of the optima coincided with the pK_a of the respective buffers, an effect of the charge of the buffer on the activity of the enzyme was suspected. Supplementation of the buffers with various concentrations of NaCl revealed that the pH optimum was indeed related to the ionic strength and charge of the buffer. In the presence of 100 mM NaCl the pH optimum was at pH 8.45 in both buffers [5]

Before analyzing the mode of action of PLYA the effect of NaCl on the cleavage rate of $(GalpA)_6$ was studied. It turned out that $V_{max\ app}$ (appr 300 U/mg) was independent of the NaCl concentration but that the K_m increased from 0.29 mM to 1.16 mM when [NaCl] increased from 0 mM to 125 mM. The fact that only K_m increased and $V_{max\ app}$ remained constant demonstrates that the K_d for $(GalpA)_6$ is affected. Thus, whereas NaCl resulted in an increased activity on polymer substrate, a decrease was observed on hexagalacturonic acid. The difference was accounted for by the swamping effect on polygalacturonic acid by electrolytes. This swamping effect results in the liberation of loosely bound Ca^{2+} ions and the reduction of inter-strand dimer formation of polygalacturonic acid [14] and thus increases the actual concentration of polymer substrate and free Ca^{2+} ions. Such a swamping effect is only to be expected when the chain length exceeds 25 Gal*p*A units [15], hence no stimulatory effect is to be expected for oligoGal*p*A.

The fact that pectate lyases require Ca^{2+} ions for catalysis has long been known. Not until the structure of the *Bacillus subtilis* pectate lyase was solved [16], was there formal prove of Ca^{2+} being bound to the enzyme. This Ca^{2+} ion is very likely involved in the catalytic mechanism (see Jurnak elsewhere in this volume). However, the high affinity of the *B. subtilis* pectate lyase for Ca^{2+} (Kd = 2.3 μM, Pickersgill Pers. Comm.)

can not explain why reaction rates on polymer substrate increase up to [$CaCl_2$] is 1 mM and on oligomer substrates up to 10 mM $CaCl_2$ (see below).

Table 2. Mode of action of pectate lyase A.
Bond cleavage frequencies are given in percentages. The reducing end is printed in bold face type script. 'n' signifies the degree of polymerization. 'G' signifies Gal*p*A.

n	**$CaCl_2$**	**Rate**											
	mM	**$U.mg^{-1}$**											
4							G	-	G	-	G	-	**G**
	0.1	0.04						100					
	1.0	0.8						100					
	10.0	9.1						100					
5					G	-	G	-	G	-	G	-	**G**
	0.1	1.2				2.4		97.6					
	1.0	16.8				8.2		91.8					
	10.0	106.0				17.6		82.3					
6			G	-	G	-	G	-	G	-	G	-	**G**
	0.1	6.2		10.8		60.3		29.1					
	1.0	95.4		36.2		41.6		22.3					
	10.0	332.1		59.0		25.6		15.4					

A likely role for Ca^{2+} ions that also explains the [$CaCl_2$] dependence, is the direct involvement in substrate binding. This was studied in two ways: by analyzing the mode of action as a function of [$CaCl_2$] and by studying the kinetic behavior as a function of [$CaCl_2$]. In Table 2 the bond cleavage frequencies (BCFs) and reaction rates of PLYA are listed using various oligoGal*p*A of defined degree of polymerization (DP). The enzyme does not cleave dimers and trimers and therefore these are not included in the Table. The tetramer is only cleaved at the third glycosidic linkage from the reducing end and this is independent of [$CaCl_2$] although a strong rate increase was recorded with increasing [$CaCl_2$]. However, for both the pentamer and the hexamer the BCFs changed in favor of cleavage of the substrate at the ultimate glycosidic linkage counting from the reducing end when [$CaCl_2$] increased. Here again, a strong rate increase was observed. The shift in the mode of action on penta- and hexamer as a function of [$CaCl_2$] can only result from a direct role of Ca^{2+} ions in the binding of the substrate at (a) particular subsite(s).

A kinetic analysis using tetra and hexamers revealed strong sigmoidal behavior when $CaCl_2$ was the varied substrate (see Fig. 2, only tetramer data shown). For tetramer cleavage only one binding mode was observed. The sigmoidal kinetics strongly indicate that there are at least two enzyme substrate-Ca^{2+}-complexes, in the same binding mode, but with different rates of turnover. For the hexamer there are multiple possibilities of enzyme substrate-Ca^{2+} complexes as there are three productive binding modes which each can have different affinities for Ca^{2+} and different turnover rates.

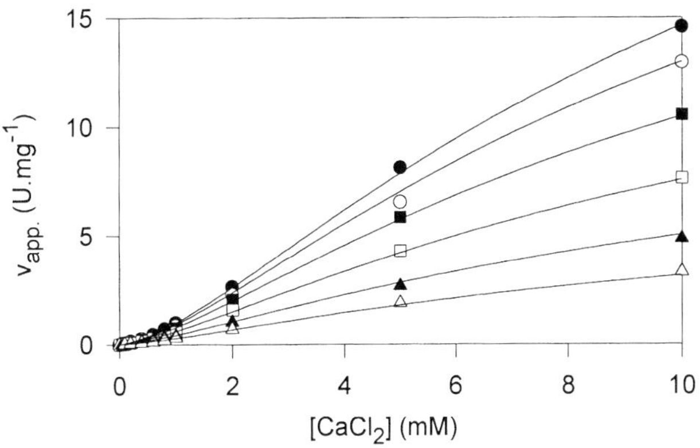

Figure 2. A. niger pectate lyase A kinetics. $CaCl_2$ as the varied substrate at fixed [(Gal*p*A)$_4$]: △, 0.075 mM; ▲, 0.15 mM; □, 0.25 mM; ■, 0.4 mM; ○, 0.8 mM; ●, 1.3 mM.

These studies thus demonstrate that Ca^{2+} ions are directly involved in the binding of the substrate. In the recently solved structure of an enzyme substrate-Ca^{2+} complex of pectate lyase C from *E. chrysanthemi*, complexed with pentamer, the four Gal*p*A units visible all appeared to make contact with the enzyme via a bridging Ca^{2+} ion between the uronate function of the substrate units and the enzyme [17]. The Ca^{2+} ions in this structure are in different positions than found in inter-strand contacts and the substrate appeared to have adopted a partial 3_1- 2_1 helix that is different from the normal 2_1 helix found in solution. Combining these data with the kinetic and mode of action studies for *A. niger* PLYA, it can be concluded that the substrate enters the enzyme as a substrate-Ca^{2+} complex and that the rate limiting step is the distortion of the helix upon binding of the substrate with concomitant rearrangement of the Ca^{2+} ions.

4. *A. niger* Pectin Lyase A

In contrast to pectate lyase A, pectin lyase A (PLA) has a preference for HM pectin and is independent of Ca^{2+} ions. However, the activity of PLA on pectins is highly dependent on the distribution of the methyl-esters over the homogalacturonan backbone (see Table 1). To study the mode of action of PLA pectins are to heterogeneous, therefore oligoGal*p*A of defined DP were purified and derivatized under controlled conditions.

4.1. PREPARATION OF MODEL SUBSTRATES OF DEFINED DP

Saturated oligoGal*p*A were produced through the digestion of polygalacturonic acid by endo-polygalacturonase, whereas unsaturated oligoGal*p*A were produced by pectin lyase digestion of high DM pectin [18].

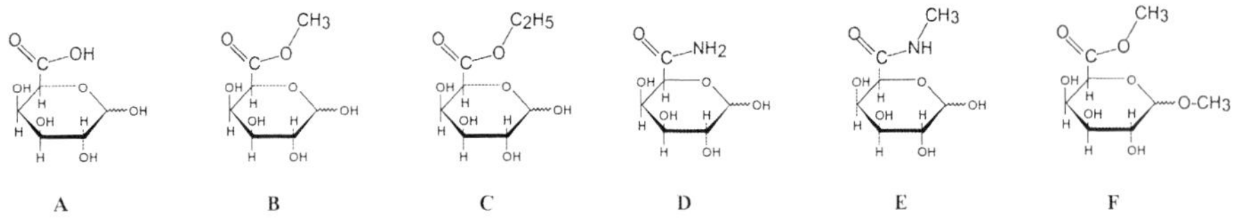

Figure 3. Structures of (un)derivatized galacturonic acid: Gal*p*A (A), 6-O-methyl-Gal*p*A (B), 6-O-ethyl-Gal*p*A (C), amide-Gal*p*A (D), 6-N-methyl-Gal*p*A (E), and 1-O-methyl, 6-O-methyl-Gal*p*A (F).

Large-scale separation was performed on a source-Q column (1.2 L), followed by a precipitation to recover the purified oligomers. The purity was checked on HPAEC at pH 12 and the molecular weight was determined by MALDI-TOF MS as described [18].

The oligoGal*p*A were derivatized at the C_6-position and/or the C_1-position (Fig. 3). Methyl- and ethyl-esterification was performed in acid methanol (0.02N H_2SO_4) at 4 °C. After 14 days of incubation fully esterified oligoGal*p*A were recovered and stored at -20 °C. The purity and molecular weight were checked by MALDI-TOF MS [18]. Amidated oligoGal*p*A were synthesized by amidation of fully methyl-esterified oligoGal*p*A. Using ammonia up to 70% of the methyl-esters could be amidated, whereas amidation using methylamine resulted in conversions of 95-97% [19]. The amidated oligoGal*p*A, amide-Gal*p*A and 6-N-methyl Gal*p*A, resemble Gal*p*A and 6-O-methyl-Gal*p*A, respectively, and may be able to mimic these compounds in their digestibility. Glycosidation of the C_1-position (reducing end) was carried out in acid methanol (0.1N H_2SO_4) at ambient temperature. At these higher acid concentrations both methyl-esterification and methyl-glycosidation occurred. All reactions were monitored using MALDI-TOF MS. Furthermore, the purity (degree of amidation, degree of esterification) was determined from the mass spectra.

4.2. MODE OF ACTION OF *A. niger* PECTIN LYASE A

The mode of action of pectin lyase A (PLA) was studied using the model oligoGal*p*A generated. Since PLA is known to have a preference for pectin with a high degree of methyl-esterification, the mode of action of PLA was first investigated using fully methyl-esterified oligoGal*p*A. The reactions were monitored for each of the substrates using HPAEC at pH12. PAD detection revealed all products, whereas UV detection only shows the unsaturated products as is demonstrated in Figure 4A for a PLA digestion of (6-O-CH_3-Gal*p*A)$_6$.

This figure shows that Gal*p*A$_6$-6CH_3 is degraded into saturated (Gal*p*A)$_2$, (Gal*p*A)$_3$, and (Gal*p*A)$_4$, with concomitant formation of unsaturated (Gal*p*A)$_2$, (Gal*p*A)$_3$, and (Gal*p*A)$_4$. After quantification of the products, progression profiles were constructed (Figure 4B). From the progression profiles initial reaction rates were determined based on the production of both the saturated products and the unsaturated products. It appeared that the reaction rates increased with increasing DP up to a DP of 8, which suggests that at least 8 subsites are present on the enzyme. Digestion of the methyl-esterified oligoGal*p*A (DP 4-10) always resulted in unsaturated $(GalpA)_3$ as the major product.

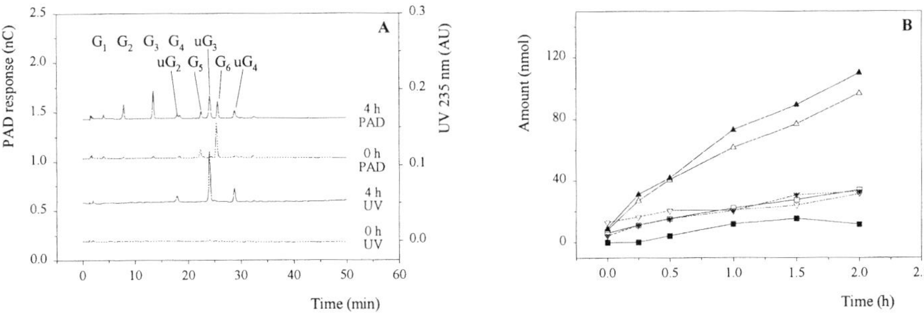

Figure 4. HPAEC elution patterns (A) of a PLA digest of $(6\text{-O-CH}_3\text{-Gal}p\text{A})_6$ after 0 and 4h of incubation detected by either PAD or UV (235 nm) and corresponding progression profile (B). Saturated (open symbols) and unsaturated (closed symbols) products are indicated as follows: $(GalpA)_2$ (,), (Gal *p*A)$_3$ (,), and $(GalpA)_4$ (∇,).

Subsequently, the influence of the carboxyl substituent (C_6) was investigated. In Table 3 the results are shown for $(6\text{-O-CH}_3\text{-Gal}p\text{A})_6$, which is representative for all compounds tested with a different DP. When no methyl-esters were present PLA activity was not detected anymore. When the methyl-ester was replaced by an ethyl-ester, the PLA activity dropped severely, indicating that the length of the ester is restricted to a methyl-group. Replacement of the O-linked methyl-ester by an N-linked hydroxyl or methyl group, also resulted in complete loss of PLA-activity. So PLA is very specific for methyl-esters and does not tolerate any other substituent at the carboxyl function.

PLA is not active on oligoGal*p*A, but it is active on partially methyl esterified pectin suggesting several subsites can accommodate an non-esterified Gal*p*A. To gain more insight into which subsite(s) this involves, the mode of action of PLA behavior on partially methyl-esterified oligoGal*p*A was analyzed. Fully-methyl-esterified oligoGal*p*A were quantitatively de-esterified using a dilute sodium hydroxide solution. Using this controlled de-esterification procedure $(6\text{-O-CH}_3\text{-Gal}p\text{A})_5$ and $(6\text{-O-CH}_3\text{-Gal}p\text{A})_6$ were de-esterified to a DM of 80 and 86%, respectively, corresponding to on

Table 3. The effect of the C_1- and C_6-substituent on the PLA activity.

Substrate	PLA activity (mU/mg)
$(\text{Gal}p\text{A})_6$	0
$(6\text{-O-CH}_3\text{-Gal}p\text{A})_6$	538
$(6\text{-O-C}_2\text{H}_6\text{-Gal}p\text{A})_6$	3
$(6\text{-NH}_2\text{-Gal}p\text{A})_6$	0
$(6\text{-N-CH}_3\text{-Gal}p\text{A})_6$	0
$1\text{-O-CH}_3\text{-}(6\text{-O-CH}_3\text{-Gal}p\text{A})_6$	107

average 1 free carboxyl group per molecule for both the pentamer and the hexamer. The location of de-esterification can not be controlled, thus in theory all isomers can be present. However, during esterification it appeared that the reducing and non-reducing end are preferentially esterified [20], therefore it is assumed that the chemical de-esterification is in reverse order compared to the chemical esterification and therefore only the isomers with one carboxyl-group of the inner Gal*p*A-residues are most likely present. Incubation of PLA with these substrates had two effects. First, the reaction rate dropped severely as shown for the hexamer (Table 4). Secondly, the bond cleavage frequency (BCF), which represents the different binding modes of the substrate, is changed. The preferred binding mode resulting in the formation of unsaturated $(GalpA)_3$ decreases and more unsaturated $(GalpA)_2$ and $(GalpA)_4$ is formed now. At the end of incubation, when almost all substrate (approximately 80% of the total initial amount present) is converted, this decrease in unsaturated $(GalpA)_3$ is even more pronounced. This demonstrates that the tolerance for only one free carboxyl group is very low.

Table 4. Effect of DM on BCF and reaction rate on $(6\text{-O-CH}_3\text{-Gal}p\text{A})_6$ and $1\text{-O-CH}_3\text{-}(6\text{-O-CH}_3\text{-Gal}p\text{A})_6$. The reducing end is indicated with boldface typescript (**G**).

DP	DM (%)	Rate (mU/mg											
6	97	538	G	-	G	-	G	-	G	-	G	-	**G**
				5		21		65		9			
6	86	43	G	-	G	-	G	-	G	-	G	-	**G**
						29		59		12			
6	97	107	G	-	G	-	G	-	G	-	G	-	**G-O-CH$_3$**
						28		72					
6	73	5	G	-	G	-	G	-	G	-	G	-	**G-O-CH$_3$**
						35		47		18			

In addition to the influence of the C_6 substituent, methyl-glycosidation of the reducing end (C_1) was investigated as well. Methyl-glycosidated oligoGal*p*A (DP 5-8) were subjected to PLA digestion. The activity increased with increasing DP, although the activity was much lower compared to the non-glycosidated oligoGal*p*A (Table 3). Furthermore, the only products found were unsaturated $1\text{-O-CH}_3\text{-}(\text{Gal}p\text{A})_3$ and $1\text{-O-CH}_3\text{-}(\text{Gal}p\text{A})_4$. After partial de-esterification the BCF changed and unsaturated $1\text{-O-CH}_3\text{-}(\text{Gal}p\text{A})_2$ was found as well, but larger unsaturated products were never found. The consequence of the de-esterification was a large drop in the activity as was shown for the non-glycosidated equivalent (Table 4). HPAEC (pH12) analysis also revealed the presence of two forms of the methyl-glycosidated substrates and products that have tentatively been identified as the α and β anomers. It turned out that only one of the anomers was degraded to either α or β anomer of the unsaturated 1-O-CH_3-(Gal*p*A)$_3$. Although both anomers of the unsaturated 1-O-CH_3-(Gal*p*A)$_4$ were formed, also here preference for one of the anomers was found. Currently, a detailed analysis of this phenomenon is in progress.

4.3. STRUCTURE FUNCTION RELATIONSHIPS IN *A. niger* PECTIN LYASES

At this moment there are no PL structures available complexed with either a substrate or an inhibitor. Only, pectate lyase C from *E. chrysanthemi* has been complexed with a pentamer in which 4 out 5 Gal*p*A residues could be identified [17]. By superimposing the structures of the substrate and that of PLB, Herron and co-workers were able to model a (6-O-CH_3-Gal*p*A)$_4$ into the active site cleft of PLB [21]. Since PLA shows a very high sequence and structural identity [22, 23], this model can also be applied to PLA. The PLB model revealed several amino acids which could be involved in catalysis and / or binding of the substrate. One of the residues indicated is the Arg-236, which is invariant in the pectate lyase superfamily, and which is most likely the 'catalytic base' (See Jurnak elsewhere in this volume). Site directed mutagenesis of Arg-236 demonstrated complete loss of activity, which confirmed the hypothesis. In PLA the R236A mutant all activity was lost as well (not shown). This residue is located between subsites three and four counting from the reducing end. Upon catalysis this would result in the formation of unsaturated (Gal*p*A)$_3$ and saturated Gal*p*A, which is in agreement with the mode of action of PLA. Based upon this model, amino acids have been mutated in order to identify the different subsites in the active site. The first analyses on pectin look promising. However, insight into the subsite architecture can only be obtained from detailed mode of action analyses with defined oligoGal*p*A. These experiments are currently in progress and may confirm the correctness of the modeled substrate.

5. Mode of Action of *A. niger* Polygalacturonases

After removal of methyl and acetyl esters from homogalacturonan by the respective esterases, a relatively simple molecule, polygalacturonic acid, remains. Despite the simplicity of this molecule, *A. niger* possesses at least 7 endoPGs. By studying gene expression and by analyzing the biochemical properties of the individual enzymes the role of each enzyme in the pectin degradation was investigated [24 – 27].

The basic characterization was started by determination of the pH optimum and the kinetic parameters K_m and $V_{max\ app}$. These are presented in Table 5. For all enzymes, except PGB, the activity region is between pH 3 and 5 with an optimum near pH 4. However, for PGB the activity region is between pH 4 and 6 with an optimum at pH 5.0. Unlike the pH optima, which are quite similar, the $V_{max\ app}$ values differ greatly. Since all seven enzymes have a complete catalytic machinery (see section structure-function relationships in *A. niger* PGs), the diversity in $V_{max\ app}$, notably for PGC, PGD and PGE must be attributed to preference of another substrate than polygalacturonic acid. By determination of the specific activity on partially methyl esterified pectins it was established that PGA and PGB prefer partially methyl esterified substrates (DM between 22-45 %) whereas the remaining enzymes prefer the non-methyl esterified substrate; the activity gradually decreased when DM increased.

Table 5. pH optimum and kinetics of *A. niger* endopolygalacturonases

enzyme	pH[a]	K_m [b] (mg/ml)	V_{max} [b] (U/mg)
PG I	4.2	< 0.15	800
PG II	4.2	< 0.15	4000
PG A	4.0	< 0.15	1200
PG B	5.0	0.9	900
PG C	4.1	< 0.15	25
PG D	4.2	0.2	96
PG E	3.8	2.5	80

To understand the different properties of the enzymes an in depth analysis of the mode of action was performed [24 – 27]. As an initial inventory the product progression profiles (PPPs) during polygalacturonic acid hydrolysis were determined. The PPPs can be grouped into two categories. The one being represented by PGI holds for PGA, PGC and PGD and the other, being represented by PGII, holds for PGB and PGE. Those two typical PPPs are presented in Figs 5A and 5B, respectively. The PPP of PGII is the most easily to understand, as it is typical for random endo acting enzymes. When a polymer substrate is attacked randomly, a statistical distribution of products is expected that is characterized by large products in the early stages of the reaction. Subsequently, these larger products are converted into smaller products until the smallest substrate efficiently hydrolyzable is consumed. The PGII PPP is indeed reflecting these events. The trimer formed is a poor substrate and will only be degraded at very long incubation times.

Inspection of the PPP of PGI (Fig. 5A) learns that the monomer accumulation sets off already at the beginning of the reaction. Furthermore, there is hardly any accumulation of oligoGal*p*A of DP > 5. From this it can be concluded that PGI (and PGA, PGC and PGD) are not fully random acting enzymes. To disclose the origin of this type of product progression a detailed mode of action analysis was initiated which included all seven PGs [24-27].

Unlike pectate and pectin lyases, which generate an unsaturated bond that serves as a tag to identify the direction of binding of the substrate, PGs form products that can not be linked to either originating from the reducing or non-reducing end. To this end a set of borohydride reduced oligoGal*p*A was prepared and used in the mode of action analysis as well. The reduced end served as a tag to one of the two products generated from a particular reduced oligogalacturonide and this allowed establishing the direction of binding [24]. These data are not shown.

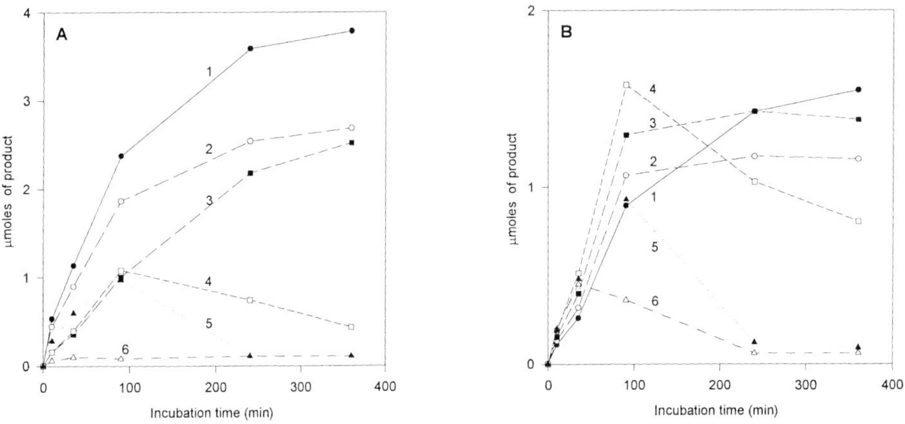

Figure 5. Product progression during polygalacturonic acid hydrolysis by PGI (Panel A) and PGII (Panel B). The numbers to the curves signify the degree of polymerization.

As examples of the mode of action analyses the data for PGI and PGII are presented in Table 8 (end of chapter). Except for PGD, which hydrolyzed the dimer [27], the smallest substrate hydrolyzable by the *A. niger* PGs is the trimer. Hydrolysis of the trimer always takes place at the first glycosidic linkage from the reducing end and proceeds very slowly. Also the tetramer is exclusively or with high preference hydrolyzed at the first glycosidic bond. Compared to the trimer, this proceeds almost two orders of magnitude faster and demonstrates that subsite –3 has a very high affinity. For pentamer and hexamer hydrolysis it was observed that the BCFs become specific for each enzyme [24-27]. For PGI the BCFs for hexamer could not be determined as the product pairs deviated from stoichiometry. The data shown are taken from the reduced heptamer (see [25] for details). This deviation may be the result of secondary attack or multiple attack/processivity. Secondary attack means that both products are released from the enzyme and the larger product serves again as a substrate. Processivity is characterized by the release of only one product, usually the smaller, whereas the other is retained at the enzyme and subsequently shifts over the active site for another catalytic event.

To distinguish between secondary attack and processivity Robyt and French [28] recommended the use of a ratio plot. Such a plot for PGI acting on hexamer substrate is shown in Fig. 6. In case of secondary attack the lines representing monomer and pentamer production coincide early in the reaction (close to the origin) whereas in case of processivity they run separately direct from the start of the reaction. From this plot it

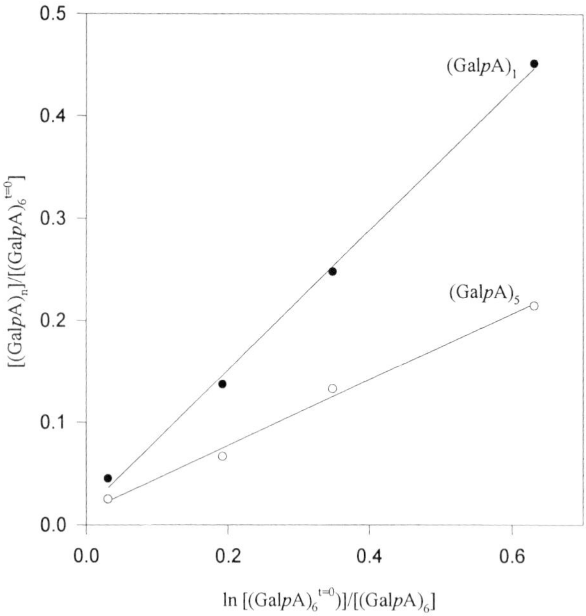

Figure 6. Ratio plot according to Robyt and French [28] for hexamer hydrolysis by PGI.

was concluded that PGI is a processive enzyme. An identical conclusion was reached for PGA, PGC and PGD albeit that PGA and PGC are processive from DP 7 onward and PGD from DP 4 onward [25-27]. Based on this knowledge the PPP for PGI can easily be understood. The strong formation of monomers and the lack of accumulation of oligoGal*p*A of DP >5 is a result of the retaining of larger oligomers at the enzyme which are degraded in a processive way that yields monomers.

The underlying principle of processivity, retaining the product at the enzyme, has as a consequence that during steady state turnover part of the enzyme population is tight up in an inactive/unproductive EP complex. In the mode of action analysis this is manifested as a decreased rate of turnover by PGI of the pentamer compared to the tetramer. Normally an increase in turnover is associated with every increase of DP until all subsites on the enzyme are filled. When the hexamer is hydrolyzed in the preferred binding mode, from subsite –5 to +1, the pentamer product remains bound to subsites –5 to –1. Hence subsite –5 to –1 have high affinity for a pentamer. When pentamer is used as a substrate, not only will there be productive complexes, covering subsites –4 to +1 and –3 to +2, but also the unproductive complex which binds from –5 to –1. Thus, pentamer substrate is a competitive inhibitor of PGI.

Based on the BCF analyses for PGI and PGII, a provisionary subsite map was calculated that is shown in Fig. 7. The major difference between these enzymes is the relatively high affinity at subsite –5 of PGI. This high affinity is at the origin of the processive behavior. This not only applies to PGI but also to PGA and PGC as will be presented in the section on structure function relationships in PGs.

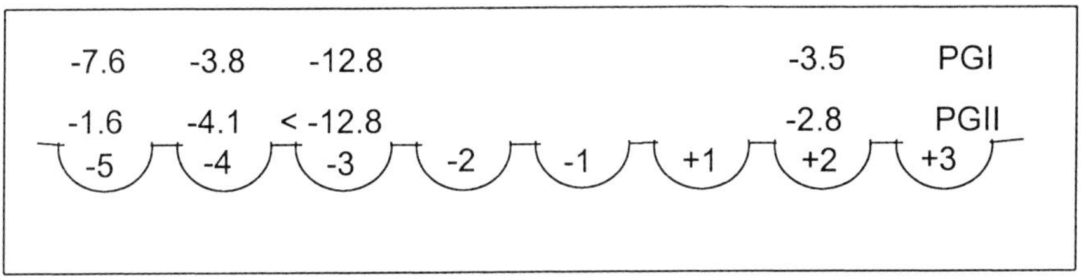

Figure 7. Provisional subsite maps of PGI and PGII. The number of the subsite is indicated in the 'pocket'. The binding energies are in kJ/mole and are indicated above the corresponding subsites.

For PGD processivity became apparent when DP > 4 [27]. Furthermore, the mode of action analysis suggested that this enzyme is composed of only four functional subsites (-3 to +1) and in this respect differs from the other PGs which are composed of at least 7 subsites (-5 to +2). An additional difference between PGD and the other PGs is the fact that it can hydrolyze dimers. Based on these data Parenicova and coworkers suggested that PGD is in fact an oligogalacturonan hydrolase rather than a PG [27].

To suggest roles for the other PGs also requires taking regulation studies into consideration. It turned out that PGA and PGB are constitutively expressed [26] whereas the other PGs are specifically induced during growth on pectic substrates. In view of the fact that PGA and PGB both prefer partially methyl esterified substrate, combined with there complementary pH optima, suggests that these are scouting enzymes that make the fungus sense the presence of pectin by generating low molecular weight inducers. Next, the full array of PGs is expressed in which PGII, with the highest specific activity is the major force generating oligoGal*p*A which are quickly converted to monomers by the processive PGI. The 'low activity' PGC and PGE most likely attack parts of the homogalacturonan which have not yet been identified.

6. Structure Function Relationships in *A. niger* Endopolygalacturonases

To further the insight into the mechanism of the PGs and to assess determinants of specificity a site directed mutagenesis program was started. As a model enzyme PGII was chosen, as the 3D-structure of this particular enzyme would become available during the course of our studies [29].

6.1. IDENTIFICATION OF THE CATALYTIC MACHINERY

At the time this study started the PGs and rhamnogalacturonases (RHGs) were already grouped into family 28 of glycosyl hydrolases and a plethora of sequences was available in the databases. An elaborate sequence alignment revealed the strict conservation of only a limited number of amino acids among PGs and an even smaller number when the RHGs were included. For catalysis only charged residues were considered relevant as generally only Asp and Glu residues are involved in catalysis in glycosyl hydrolases, although for polygalacturonases a His was implicated in catalysis

[30]. Furthermore, since the substrate is a poly-uronic acid, positively charged residues were considered to be involved in binding of the substrate close to the active site. Thus, residues chosen for mutagenesis were (numbering according PGII): D180, D201, D202, H223, R256 and K258 [31].

Table 6. Kinetic parameters of endopolygalacturonase II catalytic mutants
Kinetic parameters were determined in 50 mM NaAc, pH 4.2 at 30 °C.

Enzyme	Kinetic parameters	
	V_{max} U mg^{-1}	K_m mg ml^{-1}
Wild type	2000	<0.15
D180A	0.17	0.15
D180E	0.24	0.3
D180N	1.4	1.5
D201E	0.04	0.3
D201N	0.19	0.3
D202E	12.7	0.7
D202N	0.3	1.5
D180E/D201E	0.04	<0.15
H223A	10.0	0.15
H223C	21.5	0.80
H223Q	0.36	1.1
H223S	1.7	1.5
R256Q	278	1.7
K258N	16.2	2.8

The complete list of mutant enzymes prepared and studied by Armand and coworkers is listed in Table 6. This table shows that all mutants are affected in catalysis. The worst performers however, were those mutagenized at D180 or D201. For D202 and H223, depending on the side chain engineered, still reasonable activities were obtained. Also enzyme R256Q showed a good activity. Not surprisingly, those three amino acids are not conserved throughout family 28. In the RHGs D202 is replaced by Glu, and R256 by Met or Leu. For His223 there is no clear counterpart. Based on the dogma that in a family the catalytic residues are conserved, those three residues can never be part of the catalytic machinery. However, when D202 was replaced by Glu, the RHG counterpart, still a reasonable activity was achieved, suggesting that this residue is involved in catalysis. Also the strong impact of mutagenesis of H223 indicates a prominent role.
Based on detailed mode of action analysis and careful interpretation of the results, Armand *et al.* [31] suggested the following roles for the individual amino acids of the catalytic machinery. R256 and K258 are involved in the binding of the substrate at subsites +1 and −1 respectively. D180, assisted by D202, activates H_2O that acts as a nucleophile. D201 serves as the acid that protonates the leaving group. H223 most likely shares a proton with D201 and thus helps maintain the proper ionization state of D201.

Having identified the critical residues and following inspection of the 3D structure, Armand *et al.* [31] and van Santen *et al.* [29] concluded that the arrangement of catalytic residues is completely different from the 'normal' arrangement found in inverting enzymes. Instead of a 9 to 9.5 Å distance between acid and base, in PGs this distance is in the order of 4 to 5 Å. This implies that the mechanism of PGs is different from inverting enzymes in that the nucleophilic attack and the protonation occur from the same side instead of opposite sides from the glycosidic bond.

6.2. SUBSITE MAPPING BY SITE-DIRECTED MUTAGENESIS

The catalytic machinery being identical in all PGs little if any specificity determination is to be expected with respect mode of action or sensitivity toward methyl esterification from these amino acids. However, the elaborate mode of action analysis of all seven *A. niger* PGs had revealed definite differences in BCFs and PPPs. Based on sequence alignments in conjunction with the mode of action studies, site-directed mutagenesis studies were carried out by Pagès and coworkers to assess the position with respect to subsite and role of selected amino acids [32]. In Fig. 8 the actual position in the 3D structure of the amino acids mutagenized is depicted. In Table 7 the kinetic parameters of the mutant enzymes are listed.

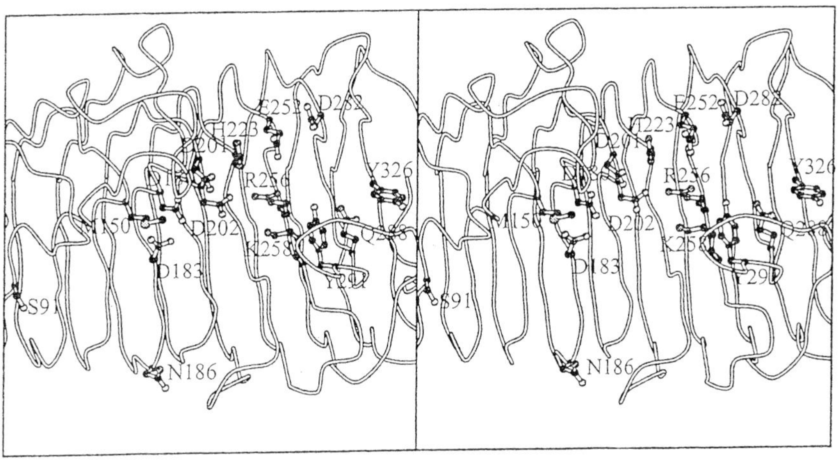

Figure 8. Stereo picture of endopolygalacturonase II depicting amino acid residues mutagenized.

Inspection of Table 7 shows that all mutant enzymes are kinetically competent except for Y291L and Y291F. Also enzyme D183N is quite strongly affected. For the mutants of Y291 a significant effect was more or less expected, as this residue is one of the strictly conserved amino acids in all PGs. From its nature and position in the structure, it can be inferred that Y291 stacks with the sugar ring of Gal*p*A at subsite +1. Mutant Y291F not only shows a stacking effect of the ring to be important but an additional strong interaction with the phenolic function of Y291 as well. The carboxylate oxygens of D183 are on average only 3.6 Å away from D180. In the previous section it was shown that D180 is critical for catalysis. The close intimacy of D180 and D183 strongly indicates the sharing of a proton. The least role expected for D183 is the maintenance of the proper ionization state of D180 to play its H_2O activating role [32].

Table 7. Kinetic parameters of endopolygalacturonase II subsite mutants
Kinetic parameters were determined in 50 mM NaAc, pH 4.2 at 30 °C.

Enzyme	$V_{max. app}$ $U.mg^{-1}$	$K_{m. app}$ $mg.ml^{-1}$
Wild type	4000	0.15
M150Q	1700	0.4
D183N	500	0.5
N186E	3000	0.5
E252A	3000	0.6
D282K	2500	0.3
Q288E	2600	1
Y291L	35	1.9
Y291F	250	1.8
Y326L	4200	0.2

For the other mutant enzymes the effect on overall catalysis was less dramatic although some profound changes in PPPs and mode of action were recorded (see Table 7). Taken all data into account the amino acids mutagenized were assigned to individual subsites (See Fig 9) [32]. Interestingly, when E252 was mutagenized into A252, the enzyme became more tolerant for methyl esterified substrate. Most probably because of the change in polarity in combination with providing enough space to accommodate the methyl esterified Gal*p*A at subsite +2. In contrast, enzyme N186E showed a decreased tolerance for methyl esterified substrate. A good explanation for this can as yet not be provided and has to await the elucidation of the 3D-structure of an enzyme-substrate complex.

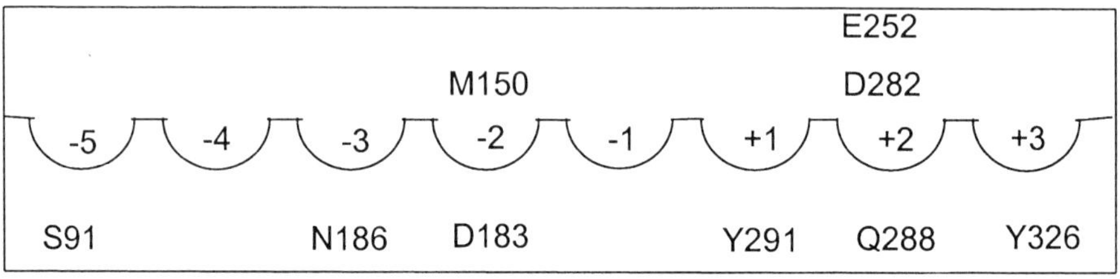

Figure 9. Schematic overview of subsites of PGII and amino acids assigned based on site directed mutagenesis.

6.3. UNDERSTANDING PROCESSIVE BEHAVIOR

In section 5 we described that processive behavior of PGI probably originates from the high affinity at subsite –5. Although no subsite maps for the other processive enzymes,

PGA and PGC, were calculated, the BCF analyses indicated a similar origin for processive behavior (PGD is an exception as it only has 4 subsites). The high sequence identity among *A. niger* PGs and the availability of a 3D structure, prompted Pagès *et al.* [33], to explore both the structure and the primary sequence alignment for amino acids that may stand at the heart of the observed behavior. Assuming the substrate to bind linearly in the substrate binding cleft, in combination with the fact that each Gal*p*A residue will span approximately 4.4 Å, and knowing the position of subsites –1 and +1, the location of subsite –5 could easily be identified in the structure of PGII [33]. Amino acid residues constituting subsite –5 were subsequently identified and compared to the other five PGs in a multiple sequence alignment. This comparison revealed the presence of a Ser in the non-processive enzymes (see Fig. 8) whereas an Arg was present at the equivalent position in the processive enzymes. Arginine being positively charged, the presence of this residue to interact with the carboxylate of the substrate, could explain the high affinity at subsite –5 in the processive enzymes. To test this hypothesis mutants PGII-S91R and PGI-R95S were prepared and characterized with respect to PPP and mode of action [33].

Comparison of the PPPs (not shown) learned that mutant PGII-S91R now belonged to the PGI (processive) type whereas mutant PGI-R95S belonged to the PGII (non-processive) type. Also the mode of action analyses (Table 8) demonstrated that the mutations involved subsite –5 and that the presence of Arg at position 91 in PGII or Ser at position 95 in PGI cause dramatic changes in BCFs on hexamer substrate. Furthermore, in PGII-S91R severe substrate inhibition by pentamer was observed (to a lesser extend even for tetramer that apparently also binds unproductively) whereas this inhibition was relieved in PGI-R95S. These data demonstrate that the absence or presence of one single residue, Arg, at subsite –5 is of crucial importance to determine processive or non-processive behavior. The fact that PGI is processive from DP 6 onward and PGA, PGC and PGII-S91R only from DP 7 onward can be explained by the difference in total affinity of subsites –5 to –1 and –4 to +1. In PGI the total affinity of –5 to –1 is almost the same as for –4 to +1. Hence the pentamer product retained at subsites –5 to –1 can easily shift to –4 to +1. However, in PGA, PGC and PGII-S91R the total affinity from –4 to +1 is lower than from –5 to –1. Thus, a retained pentamer will not shift easily. Only when a hexamer is retained at –6 to –1 (resulting from heptamer hydrolysis), shifting to –5 to +1 again involves the high affinity subsite –5 (subsite –6 is most likely imaginary or has only very low affinity).

Table 8. BCFs for wild type PGI and PGII and mutated forms PGI-R95S and PGII-S91R

Enzyme		OligoGal*p*A					
						G -	G -
PGI	1.6						100
PGI-R95S	2.3						100
PGII	0.4						100
PGII-S91R	0.3						100
					G -	G -	G -
PGI	150					3	97
PGI-R95S	114					3	97
PGII	155						100
PGII-S91R	63						100
				G -	G -	G -	G -
PGI	96					48	52
PGI-R95S	315					57	43
PGII	663					32	68
PGII-S91R	35					10	90
			G -	G -	G -	G -	G -
PGI	244				3	15	82
PGI-R95S	179				17	60	23
PGII	552				8	57	35
PGII-S91R	513				2	15	83
		G -	G -	G -	G -	G -	G -
PGI	324						
PGI-R95S	285						
PGII	1200						
PGII-S91R	510						

Acknowledgments

This work was financially supported by the European Community grants no's. AIR2-CT-941345, ERBBIO4CT960685, and QLK3-CT-1999-00089. We thank Dr Gertie van Pouderooyen for preparing figures 1 and 8. Reprint permission for figure 5 by the Eur. J. Biochem. is acknowledged.

7. References

1. Bussink, H.J., Buxton, F.P., Fraaye, B.A., de Graaff, L.H., and Visser, J. (1991) The polygalacturonases of *Aspergillus niger* are encoded by a family of diverged genes, *Eur. J. Biochem.* **208**, 83-90.
2. Harmsen, J.A., Kusters-van Someren, M.A., and Visser, J. (1990) Cloning and expression of a second *Aspergillus niger* pectin lyase gene (*pelA*): indications of a pectin lyase gene family in *A. niger*, *Curr. Genet.* **18**, 161-166.
3. Suykerbuyk, M.E.G., Kester, H.C.M, Schaap, P.J., Stam, H., Musters, W., and Visser, J. (1997) Cloning and characterization of two rhamnogalacturonan hydrolase genes from *Aspergillus niger*, *Appl. Environ. Microbiol.* **63**, 2507-2515.
4. De Vries, R.P., Kester, H.C.M., Poulsen, C.H., Benen, J.A.E., and Visser, J. (2000) Synergy between enzymes from *Aspergillus* in the degradation of plant cell wall polysaccharides, *Carbohydrate Res.* **327**, 401-410.
5. Benen, J.A.E., Parenicová, L., Kester, H.C.M., and Visser, J. (2000) Characterization of pectate lyase A from *Aspergillus niger*, *Biochemistry*, **39**, 15563-15569.
6. Van der Vlugt-Bergmans, C.J.B., Meeuwsen, P.J.A., Voragen, A.G.J., and Van Ooyen., A.J.J. (2000). Endo-xylogalacturonan hydrolase, a novel pectinolytic enzyme, *Appl. Environ. Microbiol.* **66**, 36-41.
7. Kravtchenko, T.P., Voragen, A.G.J., and Pilnik, W. (1992) Analytical comparison of three industrial pectin preparations, *Carb. Polymers* **18**, 17-25.
8. Voragen, A.G.J., Pilnik, W., Thibault, J-F., Axelos, M.A.V., and Renard, C.M.C.G. (1995) Pectins. *In* A.M. Stephen (ed.) Polysaccharides and their applications, Marcel Dekker Inc., New York, Basle, Hong Kong, pp. 287-339.
9. Körner, R., Limberg, G., Christensen, T.M.I.E., Mikkelsen, J.D., and Roepstorff, P. (1999) Sequencing of partially methyl-esterified oligogalacturonates by tandem mass spectrometry and its use to determine pectinase specificities, *Anal. Chem.* **71**, 1421-1427.
10. Searle-van Leeuwen, M.J.F., Vincken, J.-P,. Schipper, D,. Voragen, A. G. J., and Beldman, G. (1996). Acetyl esterases of *Aspergillus niger*: purification and mode of action on pectins. *In* J. Visser, and A. G. J. Voragen (eds.) Progress in Biotechnology 14: Pectins and Pectinases, Elsevier Science, Amsterdam, pp. 793-798.
11. Jenkins, J. Mayans, O, Smith, D. Worboys, K., and Pickersgill, R.W. (2001) Three-dimensional structure of *Erwinia chrysanthemi* pectin methylesterase reveals a novel esterase active site, *J. Mol. Biol.* **305**, 951-960.
12. Coutinho, P.M. and Henrissat, B. (1999) Carbohydrate-active enzymes: an integrated database approach. H.J. Gilbert, G.J. Davies, B. Henrissat and B. Svensson, (eds). Recent advances in Carbohydrate Bionengineering, The royal Society of Chemistry, Cambridge, UK, pp 3 – 12.
13. Colangelo, J., Licon, V., Benen, J., Visser, J., Bergmann, C., and Orlando, R. (1999) Characterization of the N-linked glycosylation site of recombinant pectate lyase, *Rapid Comm. Mass Spec.* **13**, 2382-2387.
14. Morris, E.R., Powell, D.A., Gidley, M.J., and Rees, D.A. (1982) Conformations and interactions of pectins. I. Polymorphism between gel and solid states of calcium polygalacturonate, *J. Mol. Biol.* **155**, 507-16.
15. Kohn, R. (1975) Ion binding on polyuronates- Alginate and Pectin, *Pure Appl. Chem.* **42**, 371-397.
16. Pickersgill, R., J Jenkins, J., Harris, G., Nasser, W. and Robert-Baudouy, J. (1994) The structure of *Bacillus subtilis* pectate lyase in complex with calcium. *Nature Struct. Biol.* **1**, 717-723.
17. Scavetta, R.D., Herron, S.R., Hotchkiss, A.T., Kita, N., Keen, N.T., Benen, J.A.E., Kester, H.C.M., Visser, J. and Jurnak, F. (1999) Structure of plant cell wall fragment complexed to PelC, *Plant Cell* **11**, 1081-1092
18. Van Alebeek, G.J.W.M., Zabotina, O., Schols, H.A., Beldman, G., and Voragen, A.G.J. (2000) Esterification and glycosidation of oligogalacturonides: examination of the reaction products using MALDI-TOF MS and HPAEC, *Carb. Polymers* **43**, 39-46.
19. Van Alebeek, G.J.W.M., Schols, H.A., and Voragen, A.G.J. (2001) Amidation of methyl-esterified oligogalacturonides: examination of the reaction products using MALDI-TOF MS, *Carb. Polymers*, in press.
20. Van Alebeek, G.J.W.M., Zabotina, O., Schols, H.A., Beldman, G., and Voragen, A.G.J. (2001) Structural analysis of (methyl-esterified) oligogalacturonides using post-source decay matrix-assisted laser desorption/ionization time-of-flight mass spectrometry, *J Mass Spectrom* **35**, 831-840.
21. Herron, S.R. Benen, J.A.E., Scavetta, R.D., Visser, J., and Jurnak, F. (2000) Structure and function of pectic enzymes: Virulence factors of plant pathogens, *Proc. Natl. Acad. Sci. USA* **97**, 8762-8769.
22. Mayans, O., Connerton, I., Gravesen, T., Benen, J.A.E., Visser, J., Pickersgill, R., and Jenkins, J. (1997) Crystal structure of pectin lyase A from *Aspergillus niger, Structure* **5**, 677-689.
23. Vitali, J., Schick, B., Kester, H.C.M., Visser, J., and Jurnak, F. (1998) The three-dimensional structure of *Aspergillus niger* pectin lyase B at 1.7-Å resolution, *Plant Physiol.* **116**, 69-80.

24. Parenicová, L., Benen, J.A.E., Kester, H.C.M., and Visser, J. (1998) *pga*E Encodes a fourth member of the endopolygalacturonase gene family from *Aspergillus niger*, *Eur. J. Biochem.* **251**, 72-80.
25. Benen, J.A.E., Kester, H.C.M. and Visser, J. (1999) Kinetic characterization of *Aspergillus niger* N400 endopolygalacturonases I, II and C, *Eur. J. Biochem.* **259**, 577-585.
26. Parenicová, L., Benen, J.A.E., Kester, H.C.M., and Visser, J. (2000) *pga*A and *pga*B encode constitutively expressed members of the *Aspergillus niger* endopolygalacturonase gene family, *Biochem. J.* **345**, 637-644.
27. Parenicová, L., Kester, H.C.M., Benen, J.A.E., and Visser, J. (2000) *pga*D encodes a new type of endopolygalacturonase from *Aspergillus niger*, *FEBS* **467**, 333-336
28. Robyt, J.F. and French, D. (1967) Multiple attack hypothesis of α-amylase action: action of porcine pancreatic, human salivary, and *Aspergillus orizae* α-amylases, *Arch. Biochem. Biophys.* **122**, 8-16.
29. Santen van, Y., Benen, J.A.E., Schroter, K.-H., Kalk, K.H., Armand, S., Visser, J., and Dijkstra, B.W. (1999) 1.86A crystal structure of endopolygalacturonase II from *Aspergillus niger* and identification of active site residues by site-directed mutagenesis, *J. Biol. Chem.* **274**, 30474-30480.
30. Rexová-Benková, L. and Marcková, M. (1978) Active groups of an extracellular endo-D-galacturonase of *Aspergillus niger* derived from pH effect and kinetic data, *Biochem. Biophys. Acta* **523**, 162-169.
31. Armand, S., Wagemaker, M.J.M., Kester, H.C.M., Sanchez-Torres, P., van Santen, Y., Dijkstra, B.W., Visser, J., and Benen, J.A.E. (2000) The active-site topology of *Aspergillus niger* endopolygalacturonase II as studied by site-directed mutagenesis, *J. Biol. Chem.* **275**, 691-696.
32. Pagès, S., Heijne, W.H.M., Kester H.C.M., Visser, J., and Benen, J.A.E. (2000) Subsite mapping of *Aspergillus niger* endopolygalacturonase II by site-directed mutagenesis, *J. Biol. Chem.* 275, 29348-29353.
33. Pagès, S., Kester, H.C.M., Visser, J., and Benen, J.A.E. (2001) Changing a single amino acid residue switches processive and non-processive behavior of *Aspergillus niger* endopolygalacturonase I and II, *J. Biol. Chem.* 276, 33652-33656.

ENDO-XYLOGALACTURONAN HYDROLASE

A novel enzyme for fruit processing

M.A. HERWEIJER[1], J.-P. VINCKEN[2], P.J.A. MEEUWSEN[2], C.J.B. VAN DER VLUGT-BERGMANS[2], G. BELDMAN[2], A.J.J. VAN OOYEN[1,2], A.G.J. VORAGEN[2]

[1] *DSM Food Specialties, Research and Development, P.O. Box 1, 2600 MA Delft, marga.herweijer@dsm.com*

[2] *Wageningen University, Agrotechnology and Food Sciences, Laboratory of Food Chemistry, P.O. Box 8129, 6700 EV Wageningen*

Abstract

Commercial pectinases used in apple juice manufacturing contain a mixture of pectin-degrading enzyme activities. Nevertheless, fragments of branched pectic molecules (or pectic hairy regions) are resistant to degradation, and can cause membrane fouling in the final ultrafiltration step of concentrated apple juice. The pectic hairy regions contain both rhamno- and xylogalacturonan. Whereas rhamnogalacturonan-degrading enzymes were already identified, no enzyme was known for a fast degradation of xylogalacturonan. Apparently, these enzymes are very minor components of technical pectinases. Since *Aspergillus* strains are able to completely degrade pectin, including xylogalacturonan, we decided to use expression cloning to obtain xylogalacturonan-degrading enzymes.

A cDNA library of *Aspergillus tubingensis* was constructed in the yeast *Kluyveromyces lactis*, using a carbon source rich in xylogalacturonan. The library was screened using a hairy regions preparation from apple, and xylogalacturonan prepared from gum tragacanth as substrates. A novel endo-xylogalacturonase was found, XGH. The enzyme specifically degrades xylose-substituted galacturonic acid backbones. In lab scale filtration experiments, XGH was able to decrease membrane fouling caused by hairy regions from apple.

1. The role of enzymes in fruit juice processing

Clear apple juice is one of the most popular fruit juices, and worldwide apple is the most important raw material for the production of clear juice. In 2000 around 10 Mio tonnes of the apples were processed into concentrate, which is about 20% of the world production of apples. Juice concentrates are shipped all over the world, from apple-growing countries to juice-consuming countries. Cell wall polysaccharides play an important role in fruit juice manufacturing. For instance, fruit contains pectin, which due to its viscous nature and gel-forming properties can cause problems during processing. Application of enzyme preparations in fruit processing is already a quite old practice. In particular, pectinase complexes have been used to facilitate processing.

F. Voragen et al. (eds.), Advances in Pectin and Pectinase Research, 257-266.
© 2003 *Kluwer Academic Publishers. Printed in the Netherlands.*

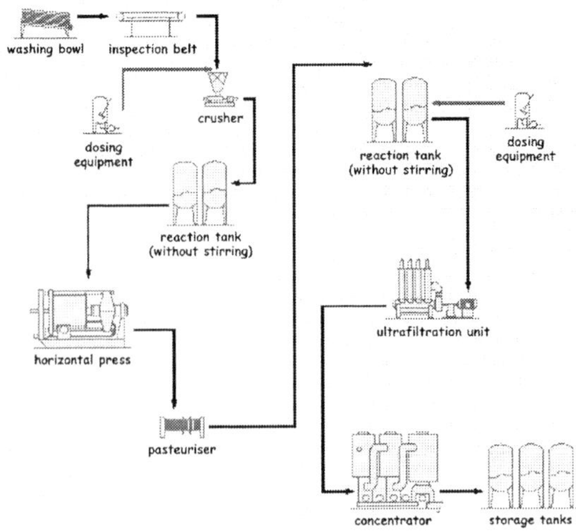

Figure 1. Schematic overview of the production of apple juice concentrate.

Often, these technical pectinase preparations are derived from the food grade fungus *Aspergillus niger*. These preparations contain a mixture of enzymes aimed at the optimal degradation of pectin.

There is a great variety in fruits, and in processing methods (reviewed in Grassin and Fauquembergue, 1999). Of these, the production of apple juice concentrate is probably the most important one. This process (Fig. 1) starts by grinding the apples. After addition of the enzyme preparation, containing a complex of pectin-degrading enzymes, the pulp is incubated for 30-60 min at room temperature. This step is called maceration. Maceration is followed by pressing and flash pasteurization of the raw juice. After cooling to 50 °C, pectinase is added again; the purpose of this is to degrade the pectin in the raw juice, and to promote the flocculation of cell wall particles, necessary for clarification of the juice. In the beginning of the season amylase is also included in the technical preparations in order to degrade the starch present in unripe apples. If the starch is not removed aggregation of starch polymers (retrogradation) may take place in the concentrate, causing undesirable hazes. After approximately 2 hours of depectinization and clarification the juice is ready for ultrafiltration followed by concentration.

Pectinases have several roles in juice production. (i) During the maceration step these enzymes degrade the pectins in the apple cell wall. As a result, the cell walls loose their strength, which enhances the efflux of the juice from the cell. In addition, the solubilized pectin in crushed fruits can form a highly viscous mass, which is difficult to press. Pectinases play an important role in reducing the viscosity of this mass. These two effects together greatly improve the pressability of the juice. (ii)

During the depectinization step pectin is degraded, resulting in clarification of the juice as described above. (iii) Last but not least, complete removal of pectin is needed to facilitate concentration of the juice, since gellification of residual pectin prevents efficient concentration.

With pectinases the manufacturers of apple juice concentrates are able to reach much higher yields, and they can obtain a better juice quality. Table 1 shows an example of the effects of enzyme addition on the juice yields. With fresh apples the effect of enzymes on yield is only a few percent. However on stored apples, which are used during the major part of the year, the effect is much larger, up to 10-12%. This difference is due to the changes that occur in the cell wall structure during storage.

Table 1. Typical yields in the production of concentrated apple juice.

	fresh apples	stored apples
no enzyme	82 – 84 %	72 - 75 %
with enzyme	85 – 87 %	83 - 85 %

2. Product improvement

Technical pectinases are subject to continuous improvements in order to keep up with new developments in the juice-manufacturing industry. For example, not so long ago ultrafiltration has been introduced as an extra step in juice processing to achieve an improved clarification of the depectinized juice. A problem inherent to the use of ultrafiltration is fouling of membranes. It has been recognized that a major cause of membrane fouling is the accumulation of highly branched parts of pectin, the so-called hairy regions (Schols *et al.*, 1990a). Briefly, the hairy regions are composed of different polysaccharides, which are covalently linked. Two of the constituent polysaccharides

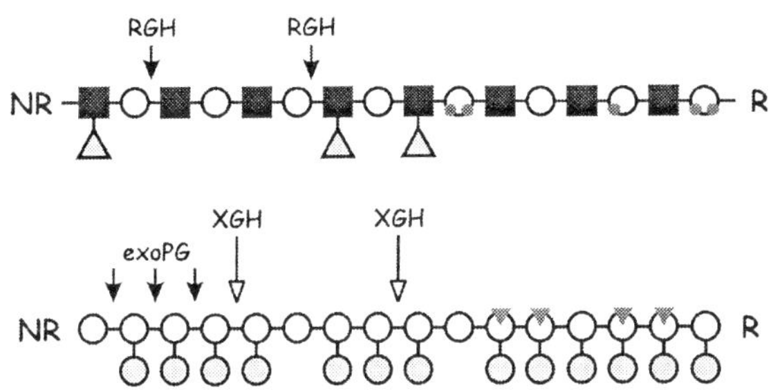

Figure 2. Schematic illustration of RG- and XGA-degradation by their respective enzymes. RGH is hindered by the presence of acetyl groups, and cleaves the non-acetylated regions of RG. Both exoPG and XGH are hindered by the presence of methoxyl groups. Open circles represent α-D-Gal*p*A residues, closed boxes α-L-Rha*p* residues, shaded triangles β-D-Gal*p*, shaded circles β-D-Xyl*p* residues, small shaded circles acetyl groups, and small closed triangles methoxyl groups. "NR" indicates the non-reducing end, "R" the reducing end.

are rhamnogalacturonan (RG) and xylogalacturonan (XGA), the structure of which is schematically shown in Fig. 2. More details on the structure of these pectic hairy regions can be found elsewhere in this volume (see also Schols and Voragen, 1996).

For a long time, the main target of the technical pectinases was homogalacturonan or the so-called smooth region of apple pectin. This part of the pectin is responsible for the gelling properties of the molecule, which are undesirable in juice processing as we have discussed above. Therefore, the enzyme cocktails contain endopectin lyase (endoPL), endopolygalacturonase (endoPG) and pectin methylesterase (PME). These enzymes do not act on the hairy regions, since they are not able to degrade the prevailing backbones in the hairy regions, RG and XGA. Two types of endo-enzymes that are able to degrade the RG backbone have been identified, *i.e.* rhamnogalacturonan hydrolase (RGH) (Schols *et al.*, 1990b), and rhamnogalacturonan lyase (RGL) (Mutter *et al.*, 1996). The mode of action of RGH is illustrated in Fig. 2. RGH cleaves the RG backbone between a galacturonic acid (GalA) and a rhamnose (Rha) residue (thus, at the reducing side of the galacturonosyl residue). Contrary to this, RGL cleaves the RG backbone between Rha and GalA (not shown). Both RGH and RGL help in reducing the fouling problem, particularly when they are used in combination with rhamnogalacturonan acetylesterase (Searle-van Leeuwen *et al.*, 1992). However, application of these three enzymes does not solve the fouling problem, mainly because of the presence of high-molecular weight XGA. Sofar, no endo-enzyme with the ability to degrade XGA had been identified. Although the backbone of XGA consists only of galacturonosyl residues, its substitution with xylose (Xyl) residues prohibits the action of all known endoPGs. One enzyme, an exo-polygalacturonase (exoPG) was shown to be able to remove GalA or the disaccharide GalA-Xyl in an exo-fashion from XGA (Beldman *et al.*, 1996). However, the degradation of XGA proceeds slowly in this way, and therefore an endo-acting enzyme would be preferred. Since many black Aspergilli

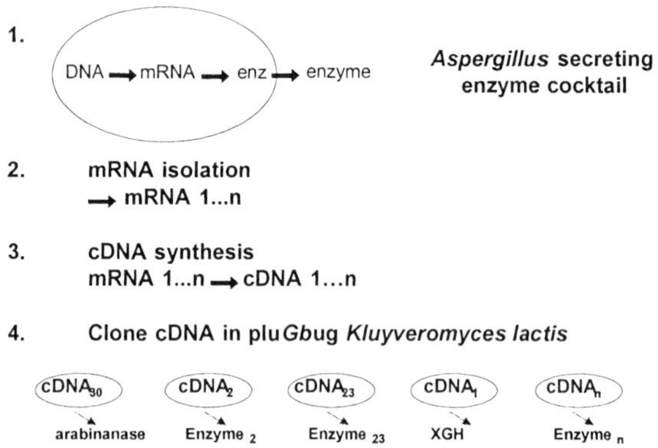

Figure 3. Schematic overview of expression cloning.

are able to degrade plant materials completely, we hypothesized that endo-xylogalacturonases (XGHs) are part of the cell wall degrading toolbox of these fungi. Our objective was to find such an activity to solve the fouling problem.

3. Screening for xylogalacturonan-degrading enzymes from *Aspergillus* using expression cloning

A well-known method to clone fungal hydrolytic enzymes is expression cloning. How to find a xylogalacturonan-degrading enzyme? A good starting point to find an enzyme that degrades a particular polysaccharide is to grow a fungus, in our case *Aspergillus tubingensis*, on the substrate to be degraded. Defatted soybean meal was selected as a carbon source, since the carbohydrate fraction of soy contains pectin which is relatively rich in XGA-containing pectic polymers (Huisman *et al.*, 2001). It was anticipated that growth of the fungus on soybean meal induced the expression of XGA-degrading enzymes.

To clone these putative enzymes, an expression library was constructed as described in van der Vlugt-Bergmans and Van Ooyen (1999). The procedure is schematically shown in Fig. 3. First, mRNA from the soy-carbohydrate-grown *A. tubingensis* culture was isolated and converted into cDNA. Each mRNA corresponds to one cDNA. The cDNA was cloned into a high-productivity host strain, the yeast *K. lactis*, resulting in a library of *K. lactis* transformants, each overexpressing a different *A. tubingensis* enzyme.

In order to increase the chance of finding the desired enzyme, we prepared a substrate that was enriched in XGA. For this purpose, the commercially available gum tragacanth was used as a starting point. Gum tragacanth essentially consists of two polysaccharides, XGA and an arabinogalactan. Further, it contains small amounts of fucose residues, probably attached to the XGA (see Table 2). The gum tragacanth was first saponified; this procedure removed methyl groups from the GalA residues, yielding a polysaccharide referred to as sGT. Subsequently, sGT was treated with dilute acid (0.1 M trifluoric acetic acid at 121 °C for 1 hour), taking advantage of the fact that the arabinosyl and fucosyl linkages are more labile under these conditions than the xylosyl linkages. Table 2 shows that both arabinosyl and fucosyl are effectively removed, yielding the polysaccharide referred to as sXGA. For comparison, the sugar composition of XGA as present in apple pectic hairy regions (XGA-MHR) is also indicated in Table 2. The GalA/Xyl ratio shows that our XGA-analogue has a lower degree of substitution. In contrast to XGA-MHR, sXGA could be easily prepared in large quantities, which was clearly an advantage for screening. Saponified modified hairy regions (sMHR), prepared according to Schols *et al.* (1990a), were used as a second substrate for screening of the library.

Table 2. Comparison of the sugar compositions (mol%) of various xylogalacturonan preparations: saponified gum tragacanth (sGT), saponified and acid-hydrolyzed gum tragacanth (sXGA), and a xylogalacturonan derived from the pectic hairy regions (XGA-MHR).

	Rha	Fuc	Ara	Xyl	Gal	Glc	GalA	GalA/Xyl
sGT	2	5	26	17	7	7	36	2.1
sXGA	3	0	1	25	9	6	56	2.2
XGA-MHR	4	0	8	38	4	5	41	1.1

In order to screen the cDNA library successfully, a sensitive assay for the detection of enzyme activity is required. A universal screening assay was developed, based on the detection of increasing amounts of reducing end groups with bicinchoninic acid (BCA) (Meeuwsen *et al.*, 2000). With this assay not only XGA-degrading activity can be detected, but any kind of polysaccharide-degrading activity. Thus, in the case of sMHR as a substrate, also enzymes degrading other parts of the MHR (such as arabinan and galactan) can be found. The individual *K. lactis* transformants of the library were transferred to 96-well plates, and grown for two days at 30 °C in a chemically defined medium. This medium contained mannitol, a non-reducing sugar, as a carbon source to minimize background staining in the BCA assay. After this, the plates were centrifuged,

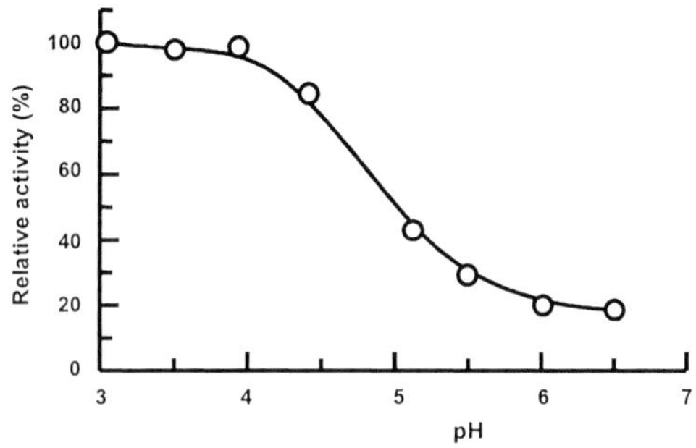

Figure 4. Dependence of the XGH activity on the pH of the assay medium. XGH activity was determined by measuring the release of reducing sugars from sXGA in time.

and the supernatants were screened for sXGA- and sMHR-degrading activity using the BCA assay. To reduce the protein background the supernatant had to be diluted appropriately. Since the sXGA and sMHR polymers also contained a small amount of reducing end groups a correction for the substrate background was made.

Screening of 3500 individual transformants on sXGA and sMHR resulted in 3 and 38 positive transformants, respectively. From all positive transformants, supernatants were tested for polymer-degrading activity by size exclusion chromatography on a panel of substrates: sXGA, arabinan, soy galactan, polygalacturonic acid, carboxymethyl-cellulose and oat spelts xylan. The 3 positive transformants detected on sXGA clearly showed endo-xylogalacturonase activity, as well as 1 out of the 38 positive transformants detected on sMHR. The transformants that were active on sXGA showed little activity on polygalacturonic acid. This clearly confirmed the discovery of a novel enzyme activity, an endo-xylogalacturonan hydrolase (XGH). Sequencing of the cDNA inserts of all XGH-positive transformants showed that the cDNA inserts were identical. Thus, one gene encoding XGH was identified. From the other 37 transformants that scored positive on sMHR in the BCA assay, 13 transformants showed endo-arabinanase activity, and 24 colonies did not show activity on the substrates tested.

4. Characterization of XGH – gene and protein

The gene encoding the novel enzyme was called xghA (van der Vlugt-Bergmans *et al.*, 2000). It contains an open reading frame of 1,218-bp, encoding a 406-residue polypeptide of which the first 18 amino acids probably form a signal sequence. Two potential N-glycosylation sites are present in the sequence. The calculated molecular mass of the protein is 40,112 Da, and the calculated isoelectric point 4.1. The enzyme is remarkably stable at low pH, with an optimum between pH 3 and 4 (Fig. 4). This makes the enzyme very suitable for low pH applications like apple juice processing.

The protein sequence is 31-39% homologous to microbial and plant endoPGs, and 30-32% to the *Aspergillus* RGHs (van der Vlugt-Bergmans *et al.*, 2000). These sequence comparisons suggest that XGH is more closely related to endoPGs than to RGHs. The XGH had highest homology to exoPG of *Aspergillus tubingensis* (44%).

Interestingly, exoPG from *Aspergillus aculeatus* was shown to remove GalA-Xyl disaccharides from saponified gum tragacanth (Beldman *et al.*, 1996).

5. Degradation of pectic polymers by XGH

Degradation of sXGA by XGH was followed by high-performance size-exclusion chromatography (HPSEC). Figure 5A shows that XGH can degrade the polymer completely into oligosaccharides. The molecular masses of these oligosaccharides are revealed by MALDI-TOF mass spectroscopy (Fig. 6). Most of the oligosaccharides are composed of galacturonosyl and xylosyl residues, ranging from the disaccharide GaX to the octasaccharide Ga_6X_2. Based on the sugar composition of sXGA in Table 2, also galactosyl-containing oligosaccharides may be expected. Between Ga_2X_2 and Ga_3X in Fig. 6 a peak with a molecular mass of 687 is present. This oligosaccharide may correspond to Ga_2X with a galactosyl residue attached to the xylosyl residue. The release of GalA monosaccharides was evidenced with high-performance anion-exchange chromatography (HPAEC, data not shown). This, together with the absence of oligogalacturonides in the digest (see Fig. 6, which is confirmed by HPAEC) and the poor activity of XGH towards polygalacturonic acid, suggests that XGH shows processivity, *i.e.* XGH cleaves more than one glycosidic bond upon each encounter with the substrate. Currently, the mode of action of XGH is investigated in more detail.

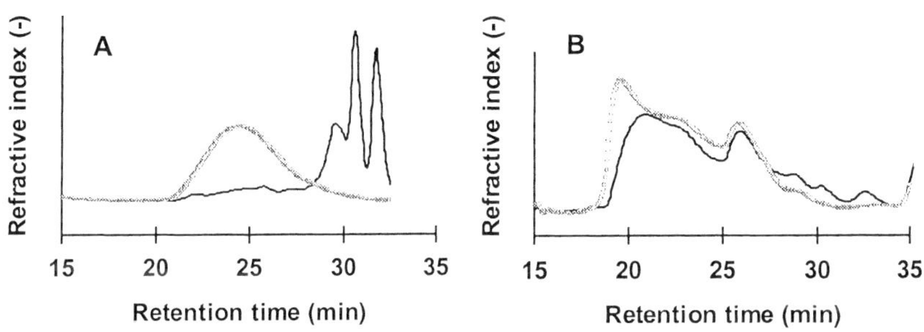

Figure 5. Size-exclusion chromatography of enzyme-treated sXGA (A) and sMHR (B). The substrates were incubated for 18 hours at 30 °C in the presence of XGH (solid black line) or water (gray line).

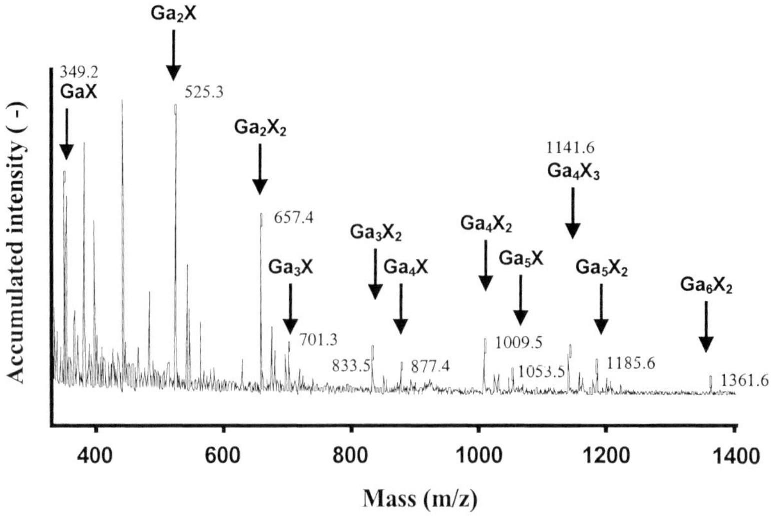

Figure 6. MALDI-TOF mass spectrum of sXGA incubated for 18 hrs at 30 °C. Ga_3X_2 indicates an oligosaccharide composed of 3 hexuronic acid residues (GalA) and 2 pentose residues (Xyl). The molecular mass of the sodium-cationized form of the various oligosaccharides is indicated in the spectrum.

Degradation of sMHR by XGH was also monitored by HPSEC. A small but consistent decrease in the molecular weight of the largest MHR population was observed (Fig. 5B). This small effect suggests a so-called external position of XGA in MHR, *i.e.* it is unlikely that XGA interconnects polysaccharides with a RG-backbone (Schols and Voragen, 1996). Both XGH and exoPG will prove useful tools in the further elucidation of the fine structure of MHR.

6. Lab-scale application trials

The importance of XGH for application in apple juice manufacturing was addressed in small-scale laboratory trials. Ultrafiltration filter fouling was mimicked in an Amicon filter device equiped with an ultrafiltration membrane with a cut-off of 30 kDa, operated at a pressure of 2 bars. A solution of 0.5% (w/w) sMHR was incubated with various MHR-degrading enzymes (RGH, XGH, and endo-arabinanase) or combinations thereof. The treated sMHR solutions were subjected to ultrafiltration, and the flux through the membrane was recorded in time (Fig. 7). It is apparent that XGH can retard the filter fouling. Improvement of the flux is more significant when XGH is used in combination with other enzymes (endoarabinanase, RGH). These results suggest that XGH may provide an important contribution to solving the fouling problem. Application of XGH in pilot-scale trials is currently in progress.

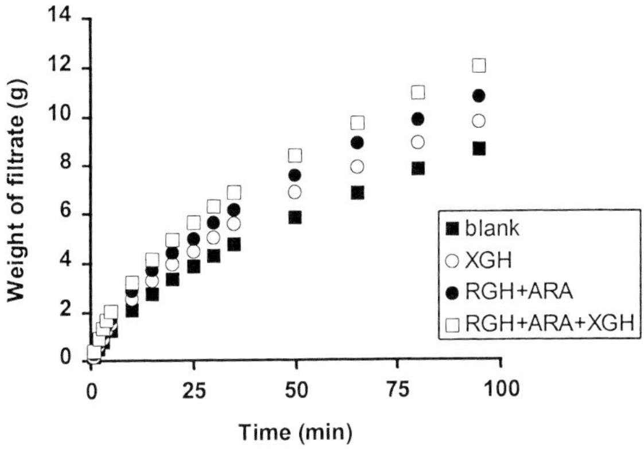

Figure 7. Permeate flux through an ultrafiltration membrane (cut-off value 30 kDa) of a 0.5% (w/w) sMHR solution incubated for 17 hrs at 30 °C with various enzymes as a function of time.

7. Conclusion

This work has demonstrated that screening of expression libraries can be a powerful tool in the development of new enzyme preparations for fruit juice processing. Prerequisites for successful enzyme-mining are (i) a tailor-made cDNA library, (ii) the availability of a "clever" substrate (enriched in the structures causing the problems), and (iii) a sensitive assay to screen the library. The advantage of this approach is that it leads directly to the gene(s) of the desired enzyme(s), which makes over-production of the enzyme(s) relatively easy. In our case, an endo-xylogalacturonan hydrolase was identified, the presence of which has not previously been demonstrated. The enzyme efficiently degrades the enzyme-resistant xylogalacturonan regions of ramified pectins, and provides an important contribution to solving the problem of membrane-fouling in apple juice manufacturing.

References

Beldman, G., van den Broek, L.A.M., Schols, H.A., Searle-van Leeuwen, M.J.F., van Laere, K.M.J., and Voragen, A.G.J. (1996) An exogalacturonase from *Aspergillus aculeatus* able to degrade xylogalacturonan, *Biotechnol. Lett.* **18**, 707-712

Grassin, C.G., and Fauquembergue, P. (1999) Enzymes, Fruit juice processing, in M.C. Flickinger and S.W. Drew (eds.) *Encyclopedia of Bioprocess Technology: Fermentation, Biocatalysis and Bioseparation*, John Wiley & Sons, pp. 1030-1061

Huisman, M.H.M., Fransen, C.T.M., Kamerling, J.P., Vliegenthart, J.G.F., Schols, H.A., and Voragen, A.G.J. (2001) The CDTA-soluble pectic substances from soybean meal are composed of rhamnogalacturonan and xylogalacturonan but not homogalacturonan, *Biopolymers* **58**, 279-294

Meeuwsen, P.J.A., Vincken, J.-P., Beldman, G., and Voragen A.G.J. (2000) A universal assay for screening expression libraries for carbohydrases, *J. Biosci. Bioeng.* **89**, 107-109

Mutter, M., Colquhoun, I.J., Schols, H.A., Beldman, G., and Voragen, A.G.J. (1996) Rhamnogalacturonase B from *Aspergillus aculeatus* is a rhamnogalacturonan α-L-rhamnopyranosyl-(1,4)-α-D-galactopyranosyluronide lyase, *Plant. Physiol.* **110**, 73-77

Schols, H.A., Posthumus, M.A., and Voragen, A.G.J. (1990a) Structural features of hairy regions of pectins isolated from apple juice produced by the liquefaction process, *Carbohydr. Res.* **206**, 117-129

Schols, H.A., Geraeds, C.J.M., Searle-van Leeuwen, M.F., Kormelink, F.J.M., and Voragen, A.G.J. (1990b) Rhamnogalacturonase: a novel enzyme that degrades the hairy regions of pectins, *Carbohydr. Res.* **206**, 105-115

Schols, H.A., and Voragen, A.G.J. (1996) Complex pectins: structure elucidation using enzymes, in J. Visser and A.G.J. Voragen (eds.) *Progress in Biotechnology 14: Pectins and Pectinases*, Elsevier, pp. 3-19

Searle-van Leeuwen, M.J.F., van den Broek, L.A.M., Schols, H.A., Beldman, G., and Voragen, A.G.J. (1992) Rhamnogalacturonan acetylesterase: a novel enzyme from *Aspergillus aculeatus*, specific for the deacetylation of hairy (ramified) regions of pectins, *Appl. Microbiol. Biotechnol.* **38**, 347-349

Van der Vlugt-Bergmans, C.J.B., and van Ooyen, A.J.J. (1999) Expression cloning in *Kluyveromyces lactis*, *Biotechnol. Techniques* **13**, 87-92

Van der Vlugt-Bergmans, C.J.B., Meeuwsen, P.J.A., Voragen, A.G.J., and van Ooyen, A.J.J. (2000) Endo-xylogalacturonan hydrolase, a novel pectinolytic enzyme, *Appl. Environ. Microbiol.* **66**, 36-41

THE STRUCTURES AND ACTIVE SITES OF PECTINASES

R.W. PICKERSGILL[1] AND J.A. JENKINS[2]
[1] *Biological Sciences, Queen Mary, University of London, London E1 4NS, UK*
[2] *Institute of Food Research, Norwich Research Park, Norwich NR4 7UA, UK*
Email: r.w.pickersgill@qmul.ac.uk

Abstract

We describe the structures of four pectinases that bind and modify the polysaccharide pectin. These pectinases belong to three distinct sequence families but share right-handed parallel β-helix architecture. They possess constellations of aspartates and an arginine that catalyse the lyase, hydrolase, and esterase activities.

1. Introduction

Since the first pectate lyase structure by Marilyn Yoder, Noel Keen, and Francis Jurnak was published in June 1993, there has been an explosion of structures and structural information (Table 1). Two observations are remarkable and will be discussed in this paper: firstly the repeated occurrence of the right-handed parallel β-helix domain architecture in these enzymes and secondly the unexpected active sites these enzymes possess.

TABLE 1. Crystal structures of pectinases.

Enzyme	Origin	PDB code	Reference
Lyases			
Pectate lyase	*Erwinia chrysanthemi*	2PEC	Yoder *et al.* 1993a,b
Pectate lyase	*Erwinia chrysanthemi*	1PCL	Lietke *et al.* 1994
Pectate lyase / Ca^{2+}	*Bacillus subtilis*	1BN8	Pickersgill *et al.* 1994
Pectate lyase / substrate	*Erwinia chrysanthemi*		Scarvetta *et al.* 1999
Pectate lyase	*Bacillus sp.*	1EE6	Akita *et al.* 2000
Pectin lyase A	*Aspergillus niger*	1IDJ /K	Mayans *et al.* 1997
Pectin lyase B	*Aspergillus niger*	1QCX	Vitali *et al.* 1998
Glycoside hydrolases			
Rhamnogalacturonase	*Aspergillus aculeatus*	1RMG	Petersen *et al.* 1997
Polygalacturonase	*Erwinia carotavora*	1BHE	Pickersgill *et al.* 1998
Polygalacturonase	*Aspergillus niger*	1CZF	van Santen *et al.* 1999
Esterases			
RGAE	*Aspergillus aculeatus*	1DEX /O	Mølgaard *et al.* 2000
Pectin methylesterase	*Erwinia chrysanthemi*	1QJV	Jenkins *et al.* 2001

F. Voragen et al. (eds.), Advances in Pectin and Pectinase Research, 267-275.
© 2003 *Kluwer Academic Publishers. Printed in the Netherlands.*

With the exception of rhamnogalacturonan acetylesterase (RGAE), whose structure is a variation on the classical α/β hydrolase fold, all the structures listed in Table 1 are parallel β-helix proteins. Where coordinates are available from the protein databank the PDB identifier is given (http://www.rcsb.org/pdb/index.html). Where a code is not given we understand the coordinates are available from the authors and where a forward slash is used a second coordinate set is available, for instance 1IDJ/K implies two structures 1IDJ and 1IDK. The lyases are all family 1 with the exception of 1EE6 which is a family 3 lyase. The glycoside hydrolases are family 28. The esterases are family 1 (RGAE) and 8 (pectin methylesterase) carbohydrate esterases (http://afmb.cnrs-mrs.fr/~pedro/CAZY/).
Our contributions to this activity include the first pectin lyase, polygalacturonase, and pectin methylesterase structures in addition to the structure of pectate lyase in complex with calcium (Table 1; Figure 1).

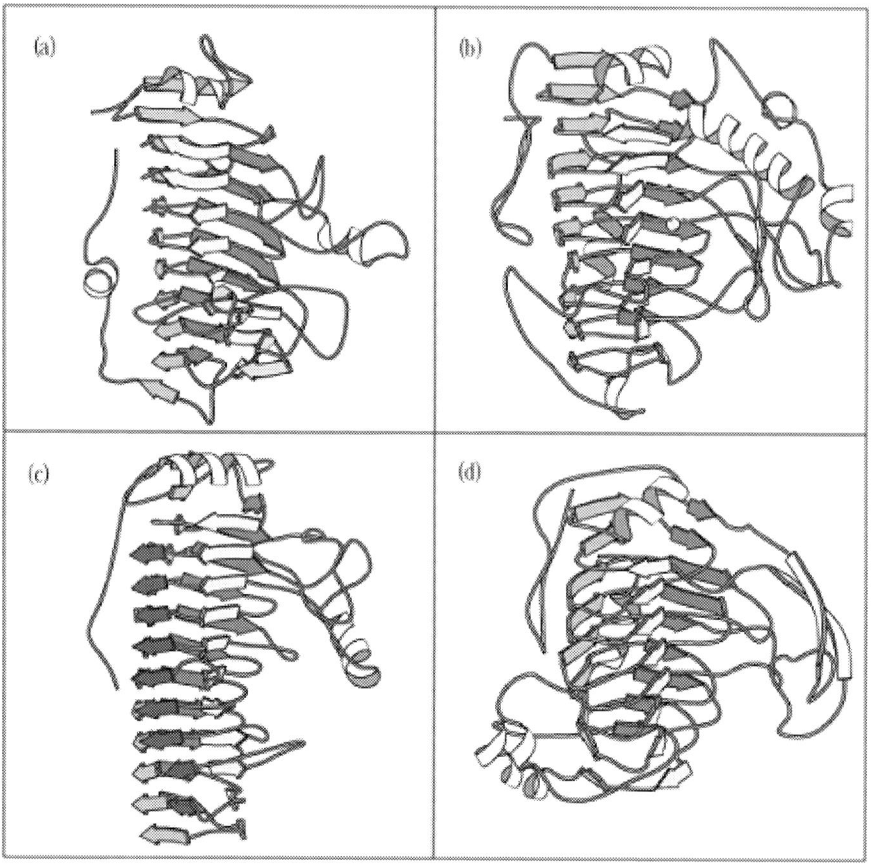

Figure 1. Structures of pectinases solved by the authors and their collaborators: (a) *Erwinia chrysanthemi* pectin methylesterase; (b) *Bacillus subtilis* pectate lyase; (c) *Erwinia carotavora* polygalacturonase; (d) *Aspergillus niger* pectin lyase A. The PDB codes and references to these structures are given in Table 1. Figures 1, 2, 3b, 3c, 4, and 6d were prepared using MOLSCRIPT (Kraulis 1991).

Unlike the lyases and pectin methylesterase that have 8 complete coils forming the parallel β-helix, polygalacturonase and rhamnogalacturonase have 10 coils, and a fourth β-sheet named PB1a as it occurs between PB1 and PB2 (the sheet nomenclature is given in Figure 2b).

2. Uniquely the right-handed parallel β-helix has no sequence fingerprint

The parallel β-helix domain has been discussed in several reviews (Yoder *et al.* 1993; Jurnak *et al.* 1994; Yoder & Jurnak 1995; Jenkins *et al.* 1998). Here we focus on one remarkable feature. Superhelical proteins from the simple coiled coil proteins and collagen through to the more complex leucine rich repeat structures such as ribonuclease inhibitor protein all have their own characteristic underlying sequence repeat. While the parallel β-helix structures are all characterised by striking like-on-like stacking of amino acid residues both within and on the exterior of the β-helix there is no single repeated pattern across the pectinases. The first stacks seen in the pectate lyases were aromatic stacks and the asparagine ladder so called because the asparagines hydrogen bond to one another within the interior of the parallel β-helix domain (Figure 2a). The pectin methylesterase structure reveals strikingly that neither the aromatic stack on PB3 nor the asparagine ladder is essential for formation of the domain architecture (Figure 2b) and there is no sequence fingerprint for the right-handed parallel β-helix proteins. This unique feature sets the right-handed parallel β-helix architecture apart from all other superhelical architectures established to date.

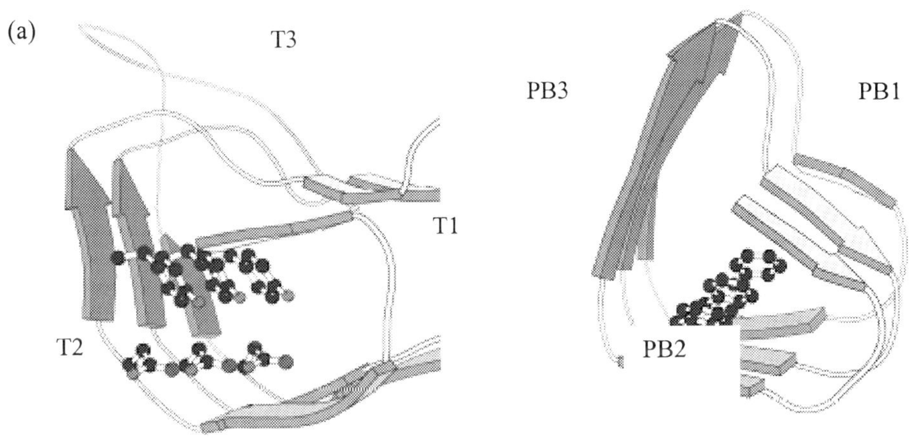

Figure 2. Like-on-like side chain stacks in pectate lyase and pectin methylesterase. (a) The pectate lyase aromatic stack on PB3 and asparagine ladder at the T2 turn. (b) Pectin methylesterase has no internal asparagine ladder and its aromatic stack is on PB2. The turn nomenclature is given in (a) and sheet nomenclature in (b).

3. The Active Sites of the Lyases

The family 1 pectin and pectate lyases, as classified by Henrissat and Continho (http://afmb.cnrs-mrs.fr/~pedro/CAZY/), have at least 15% sequence identity. Pectate lyases cleave the α-1,4 glycosidic bond of pectate (Figure 3a) by β-elimination, calcium is essential, and these enzymes have pH optima in the range 8.5-9.5. Pectin lyases cleave a substrate methylated at C6 again by β-elimination but do not require calcium and have pH optima in the range 6.0-8.5.

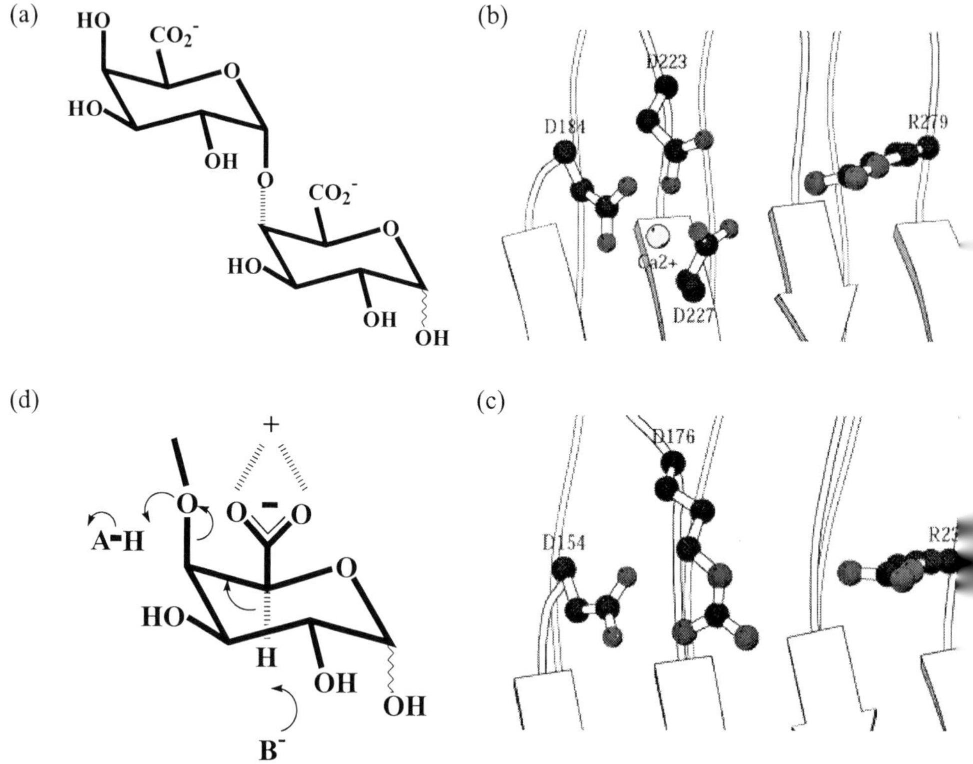

Figure 3. (a) Pectate lyase cleaves the dashed bond in polygalacturonic acid. (b) The active site of *Bacillus subtilis* pectate lyase showing conserved carboxylates involved in calcium binding and conserved arginine 279. (c) The pectin lyase A active site. (d) The proposed mechanism for pectate lyase. Chemical structures were drawn using ChemDraw (http://www.camsoft.com).

The pectate lyases have a conserved cluster of carboxylates that form a calcium-binding site and a conserved arginine at their active centre (Figure 3b). Only one aspartate and the arginine are conserved across the pectin and pectate lyases (Figure 3c). A proposed mechanism for pectate lyase would have a proton acceptor and proton donor as key elements, as well as a positive charge to reduce the pKa of the C5 proton (Figure 3d). Abstraction of the C5 proton results in β-elimination and formation of an unsaturated C4-C5 bond. Which residue could be the proton acceptor? Francis Jurnak has convincingly demonstrated that there is no appreciable structural rearrangement on formation of the complex, that is the aspartate is not released from its role in calcium-binding so the only remaining candidate is the conserved arginine. The conclusion that an arginine abstracts the C5 proton and the obvious lack of a candidate for proton donation to the glycosidic oxygen is remarkable. Perhaps for the elimination reaction to occur efficiently maintaining the correct conformation for cleavage is more important than the potency of the proton acceptor and donor. That the arginine is the proton abstractor is even more surprising when it is remembered that pectin lyase has a pH optimum around 6.0-8.5.

The mechanism of pectate lyase is described by Francis Jurnak elsewhere in this volume and has been recently reviewed by her and colleagues (Herron *et al.* 2000).

4. The Polygalacturonase Active Site

The active site of polygalacturonase, a family 28 glycoside hydrolase, which catalyses the cleavage of the α-1,4 bond of polygalacturonic acid by hydrolysis with pH optima around 5.5 is shown in Figure 4.

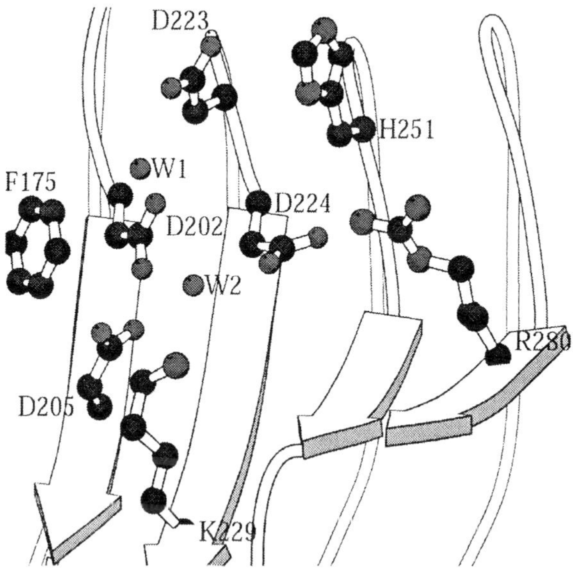

Figure 4. The active site of polygalacturonase.

Aspartates 202 and 223 are invariant and approximately 5Å apart (Figure 4). Residue 224 is either aspartate, as here, or glutamate in the polygalacturonases. Residue 205 is either aspartate, as here, or histidine. His 251 is important for the activity of polygalacturonase but not for that of RGAE. Hydrolysis of the glycosidic bond can occur via one of two possible mechanisms retention or inversion (Koshland 1953) and polygalacturonase is an inverting enzyme (Biely *et al.* 1996). The bond cleaved is indicated in Figure 5 along with a brief description of inverting and retaining mechanisms. The separation of the active site aspartates in polygalacturonase does not fit that expected for an inverting enzyme so we conclude that for an α-linked carbohydrate the nucleophile and acid/base can be on the same side of the cleaved bond as opposed to on opposite sides as previously seen in hydrolases cleaving β-linkages. This is illustrated in figure 5d where the nucleophile and acid/base are close but the mechanism is inverting. The positions of the invariant aspartates 202 and 223, and arginine 280 (Figure 4) are remarkably similar to those seen for aspartates 184 and 223, and arginine 279 in pectate lyase (Figure 3b).

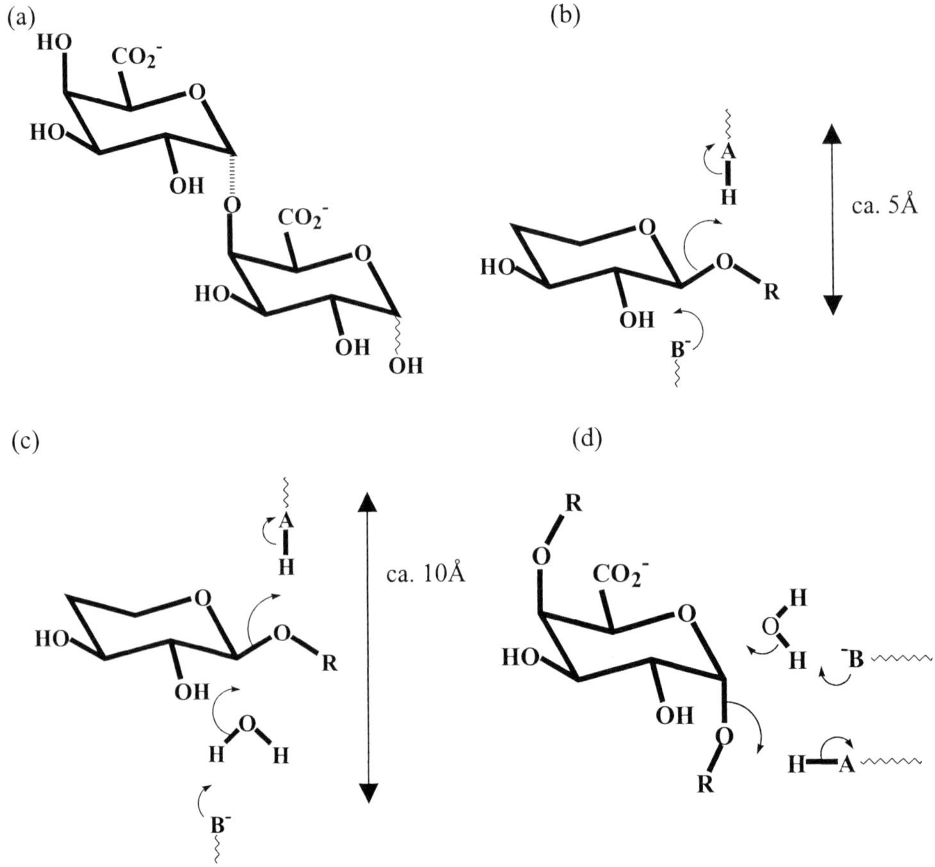

Figure 5. (a) The bond cleaved by polygalacturonase is drawn dashed. (b) The textbook geometry of hydrolysis by retention has the acid/base and nucleophile separated by about 5Å. (c) The textbook geometry of hydrolysis by inversion has the acid/base and nucleophile separated by about 10Å. (d) Polygalacturonase inverts but with close aspartates and this panel shows how we propose this occurs.

5. The active site of pectin methylesterase

Most structures of neutral lipases and esterases have been found to adopt the common α/β hydrolase fold and contain a catalytic Ser-His-Asp catalytic triad (Ollis *et al.* 1992; Heikinheimo *et al.* 1999). Some variation occurs in both the overall protein fold and in the location of the catalytic triad, and in some enzymes the role of the aspartate is replaced by a main-chain carbonyl oxygen (Wei *et al.* 1995). We were therefore surprised to discover that pectin methylesterase folded into a right-handed parallel β-helix protein and has two aspartates at its active site.

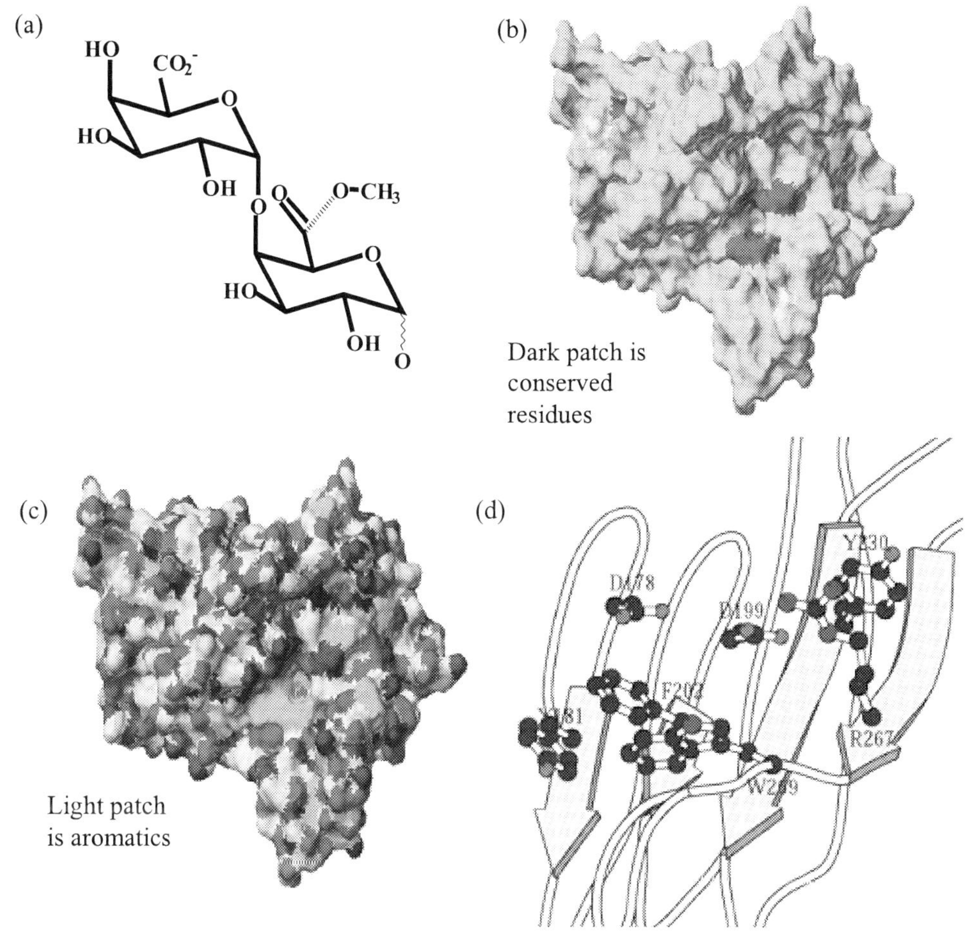

Figure 6. The active site of pectin methylesterase. (a) Cleaved ester linkage drawn as dashed line. (b) Conservation maps to a cleft in the surface of pectin methylesterase (c) The cleft is also rich in aromatics expected to bind the hydrophobic methylated pectin substrate. (d) The conserved aspartates 178 and 199, and conserved arginine 267 proposed to be the catalytic residues.

The activity of pectin methylesterase profoundly affects the properties of the pectins by converting the C6 methylester group to a carboxylate (Figure 6a). The active site of pectin methylesterase was located on the basis of sequence conservation across 26 pectin methylesterases.
The conserved residues cluster on the surface of PB1 at the same location as seen in the other parallel β-helix pectinases (Figure 6b). This identification of the active site is further supported by the clustering of aromatics involved in binding the hydrophobic methylated pectin molecule (Figure 6c).
We propose that the conserved residues Asp 178, Asp 199, and Arg 267 are the catalytic residues in the esterase activity of pectin methylesterase (Figure 6d). Tove Christensen and her colleagues in Jørn Mikkelsen's Laboratory (Danisco Cultor, Copenhagen) have established *Erwinia chrysanthemi* pectin methylesterase as a block-acting enzyme (results presented as a poster at the Pectins and Pectinases 2001 meeting). Previously it was thought all bacterial enzymes were random acting. The block action may be related to the deep substrate-binding cleft of this pectin methylesterase. The structure of a plant pectin methylesterase with a shallower substrate-binding cleft has recently been solved (results presented as a poster at the by Pectins and Pectinases 2001 meeting by Kenth Johansson and colleagues from Hans Eklund's Laboratory in Uppsala).

6. Concluding comment

The common architecture, similar location of active sites, and possibility that one activity could evolve from another by relatively few amino-acid residue substitutions makes the case for these enzymes being a superfamily (superfamily meaning probable common evolutionary origin) compelling. It is unusual to see so many different methods of substrate recognition and mechanisms in one superfamily. Pectate lyase interacts with substrate via calciums while pectin lyase and pectin methylesterase use aromatics. These pectinase structures have revealed novel, and surprising, active sites including: a proton abstracting arginine in pectate lyase, two close aspartates in the inverting endo-polygalacturonase, and the active site of pectin methylesterase shows that catalysis involves two aspartates rather than a serine.

7. Acknowledgements

We thank the BBSRC and EU for support and Nicole Hugouvieux-Cotte-Pattat, Vladimir Shevchik, Jacques Benen, and Jaap Visser for valuable collaborations. We also acknowledge valuable collaborations with Jørn Mikkelsenn, Tove Christensen (Danisco-Cultor), and Jørgen Søderburg (CP Kelco) and thank Francis Jurnak for stimulating discussions.

8. References

Akita M., Suzuki A., Kobayashi T., Ito S. & Yamane T. Crystallization and preliminary X-ray analysis of high-alkaline pectate lyase (2000) *Acta Cryst.* **D56**, 749-750.

Biely, P., Benen, J., Heinrichova, K., Kester, H.C.M., & Jisser, J. (1996) Inversion of configuration during hydrolysis of alpha-1,4-galacturonidic linkage by three *Aspergillus* polygalacturonases. *FEBS Lett.* **382**, 249-255.

Heikinheimo, P., Goldman, A., Jeffries, C. & Ollis, D. (1999) Of barn owls and bankers: a lush variety of α/β hydrolases. *Structure* **7**, R141-R146.

Herron, S.R., Benen, J.A.E., Scavetta, R.D., Visser, J. & Jurnak, F. (2000) Structure and function of pectic enzymes: Virulence factors of plant pathogens. *Proc. Natl. Acad. Sci.* **97**, 8762-8769.

Jenkins, J., Mayans, O. & Pickersgill, R. (1998) The structure and evolution of parallel β-helix proteins. *J. Struct. Biol.* **122**, 236-246.

Jenkins, J., Mayans, O., Smith, D., Worboys, K. & Pickersgill, R.W. (2001) Three-dimensional structure of *Erwinia chrysanthemi* pectin methylesterase reveals a novel esterase active site. *J. Mol. Biol.* **305**, 951-960.

Jurnak, F., Yoder, M.D., Pickersgill, R. & Jenkins, J. (1994) Parallel β-domains: a new fold in protein structures. *Current Opinion in Structural Biology*, **4**, 802-806.

Koshland, Jr., D.E. (1953) Stereochemistry and the mechanism of enzymatic reactions. *Biol. Rev.* **28**, 416-436.

Kraulis, P.J. (1991) MOLSCRIPT; a program to produce both detailed and schematic plots of proteins. *J. Appl. Cryst.* **24**, 946-950.

Lietzke, S.E., Scavetta, R.D., Yoder, M.D. & Jurnak, F. (1996) The refined three-dimensional structure of pectate lyase E from *Erwinia chrysanthemi* at 2.2Å resolution. *Plant Physiol.* **111**, 73-92.

Mayans, O., Scott, M., Connerton, I., Gravesen, T., Benen, J., Visser, J., Pickersgill, R. & Jenkins, J. (1997) Two crystal structures of pectin lyase A from *Aspergillus* reveal a pH driven conformational change and striking divergence in the substrate-binding clefts of pectin and pectate lyases. *Structure* **5**, 677-689.

Mølgaard, A. Kauppinen, S. & Larsen, S. (2000) Rhamnogalacturonan acetylesterase elucidates the structure and function of a new family of hydrolases. *Structure* **8**, 373-383.

Ollis, D., Cheah, E., Cygler, M., Dijkstra, B., Frolow, F., Franken, S., Harel, M., Remington, S., Silman, I., Schrag, J., Sussman, J., Verschueren, K. & Goldman, A. (1992) The α/β hydrolase fold. *Prot. Eng.* **5** 197-211.

Petersen, T. N., Kauppinen, S. & Larsen, S. (1997) The crystal structure of rhamnogalacturonase A from *Aspergillus aculeatus*: a right-handed parallel β-helix. *Structure* **5**, 533-544.

Pickersgill, R., Jenkins, J., Harris, G., Nasser, W. & Robert-Baudouy, J. (1994) The structure of *Bacillus subtilis* pectate lyase in complex with calcium. *Nat. Struct. Biol.* **1**, 717-723.

Pickersgill, R., Smith, D., Worboys, K. & Jenkins, J. (1998) Crystal structure of Polygalacturonase from *Erwinia carotovora* ssp. *carotovora*. *J. Biol. Chem.* **273**, 24660-24664.

Scavetta, R.D., Herron, S.R., Hotchkiss, A.T., Kita, N., Keen, N.T., Benen, J.A.E., Kester, H.C.M., Visser, J. & Jurnak, F. (1999) Structure of a plant cell wall fragment complexed to pectate lyase C. *Plant Cell*, **11**, 1081-1092.

van Santen, Y., Benen, J.A.E., Schroter, K.H., Kalk, K.H., Armand, S., Visser, J. & Dijkstra, B.W. (1999) 1.68 Å crystal structure of endopolygalacturonase II from Aspergillus niger and identification of active site residues by site directed mutagenesis. *J. Biol. Chem.* **274**, 30474-30480.

Vitali, J., Schick, B., Kester, H.C.M., Visser, J. & Jurnak, F. (1998) The three-dimensional structure of *Aspergillus niger* pectin lyase B at 1.7Å resolution *Plant Physiol.* **116**, 69-80.

Wei, Y., Schottel, J., Derewenda, U., Swenson, L., Patkar, S. & Derewenda, Z. (1995) A novel variant of the catalytic triad in the *Streptomyces scabies* esterase. *Nat. Struct. Biol.*, **2**, 218-223.

Yoder, M.D., Keen, N.T. & Jurnak, F. (1993a) New domain motif - the structure of pectate lyase C, a secreted plant virulence factor. *Science* **260**, 1503-1507.

Yoder, M.D., Lietzke, S.E. & Jurnak, F. (1993b) Unusual structural features in the parallel β-helix pectate lyases. *Structure* **4**, 241-251.

Yoder, M.D. & Jurnak, F. (1995). The parallel β helix and other coiled folds. *FASEB J.* **9**, 335-342.

RECENT OBSERVATIONS ON THE SPECIFICITY AND STRUCTURAL CONFORMATION OF THE POLYGALACTURONASE -POLYGALACTURONASE INHIBITING PROTEIN SYSTEM

C.W. BERGMANN, L. STANTON, D. KING, R.P. CLAY*, G. KEMP[+]
R. ORLANDO, A. DARVILL and P. ALBERSHEIM
*Complex Carbohydrate Research Center, University of Georgia, Athens GA, 30602, *Dept. of Natural Science, Gainesville College, Gainesville GA, 30503, [+] Dept. of Botany and Genetics, University of the Orange Free State, Bloemfontein, South Africa*

Corresponding author: Carl Bergmann; cberg@ccrc.uga.edu

Abstract

The interaction between fungal polygalacturonases (PGs) and plant cell wall proteins inhibitory to these pectin-degrading enzymes has been well established. By convention, these inhibitory proteins have been referred to as polygalacturonase-inhibiting proteins or PGIPs, with the typical PG/PGIP interaction being characterized by high-affinity, reversibility, and a one:one stoichiometry. Different combinations of various PG and PGIP isoforms and/or glycoforms yield differential or combination-specific degrees of enzyme inhibition, with some PG/PGIP pairings producing enzyme inhibition of as much as 99%. Recent studies in our laboratory have indicated that in some instances, given certain combinations of PGIP isoforms with certain PGs, the rate of catalysis is actually increased beyond that characteristic of the enzyme alone. Clearly, this activation or enhancement of catalytic rate constitutes a deviation from the conventional inhibitory effect attributed to this class of plant cell wall protein, suggesting the need for a re-evaluation of the conventional moniker applied to PGIPs; suggestions include Polygalacturonase-Binding Protein and Polygalacturonase-Modulating Protein. Additional revelations concerning interactions of PG, polygalacturonic acid, and PGIP arose from structural conformation studies involving amide exchange-MS, pepsin cleavage, and UV fluorescence analysis when applied to PG alone, PG/oligogalacturonide complex and PG/oligogalacturonide/PGIP complex. Indications are that the substrate oligomer contacts the surface of the PG along the cleft of the β-barrel configuration predicted for PGs by X-ray structure and site-specific mutation studies. Evidence also supports a conformational change in the PG when bound to substrate, to the effect that the hydrogen bonding in the β-pleated sheet region on the side of the PG opposite to the substrate binding site is destabilized when the enzyme is bound to the substrate. When PGIP binds the PG/substrate complex, it appears to do so on the side opposite the binding site, resulting in apparent inhibition of the conformational change induced upon binding of substrate. These results, together with previously reported data, begin to point to a location for inhibitor

F. Voragen et al. (eds.), Advances in Pectin and Pectinase Research, 277-291.
© 2003 *Kluwer Academic Publishers. Printed in the Netherlands.*

interaction as well as to possible mechanisms for the non-competitive inhibition often attributed to this system.

1. Introduction

During their evolution, plants have responded to the persistent assault by potentially pathogenic bacteria and fungi by the development of a complex set of biochemical defense mechanisms aimed at either excluding pathogens or limiting the extent of pathogen ingress to a permissible level. In one model of the early events of plant-fungus interactions, plant and fungal cells sense the presence of one another by detecting small fragments of cell wall polysaccharides (known as oligosaccharide elicitors) that have been enzymatically released from the respective host and pathogen cell walls [1,2]. The pectic polymers of plant cell walls are an example of cell wall polysaccharides subject to release as elicitor fragments.

Pectins form a major component of the primary cell walls of dicots and non-graminaceous monocots, being found in high concentrations in the middle lamella [3]. They are also present to a lesser degree in graminaceous monocots, where they are also concentrated in the middle lamella [4]. Pectic polysaccharides include homogalacturonan (HG) or polygalacturonic acid (PGA) that may be methyl-esterified to varying degrees, and rhamnogalacturonans I and II (RG-I and RG-II) [5,6]. This heterogeneity of pectic polymers is reflected in the diversity of pectin degrading enzymes (PDEs), or pectinases, that are secreted by invasive microbial pathogens.

Endopolygalacturonases (EPG) and pectate lyases provide much of the pectin-degrading activity of microbes. EPGs are among the first degradative enzymes to be secreted upon fungal infection of plants [7,8], and have been shown to exist in two disinct catagories relative to the degrees of polymerization (DPs) and quantities of oligogalacturonides released through enzymatic action [9]. These various EPGs hydrolyze (principally) non-esterified regions of wall-bound homogalacturonans, thus solubilizing RG-I and RG-II, which further serves to expose non-pectic cell wall polymers to the action of other exo- and endoglycanases, including the cellulases and hemicellulases [10]. The outcome of *in vitro* enzymatic hydrolysis of polymeric homogalacturonans by EPGs is the formation of a range of oligogalacturonides (OGAs) of varying sizes, a sub-population of which (DPs of 9-16) has been demonstrated to elicit various plant defense responses in bioassays. Such observations suggest that the action of microbial EPGs *in planta* is to generate a transient population of biologically active OGAs with DPs of 9-16, which are eventually further fragmented by EPG-mediated hydrolysis to yield inactive mono-, di-, and trigalacturonides. Thus, according to the oligosaccharin model referred to above, while EPG action is necessary to generate a pool of bioactive OGA signaling molecules, unrestricted EPG activity would ultimately result in a weakening of the elicitation response due to an eventual decline in the population of OGA elicitors.

During pathogenesis, there is the potential for interaction between fungal endopolygalacturonases (EPGs) and plant cell wall-derived inhibitors of EPGs known as polygalacturonase inhibiting proteins (PGIPs) [11,12]. PGIPs are soluble, leucine rich repeat (LRR) glycoproteins, found in the extracellular matrix of dicotyledonous and some non-graminaceous monocotyledonous plants [13]. Evidence to date indicates that PGIPs form high-affinity complexes with EPGs in a reversible, stoichiometrically defined manner. The rate of hydrolysis of PGA by an EPG/PGIP complex is, depending upon the organismal sources of the EPG and PGIP, between one and two orders of magnitude slower than that effected by the uncomplexed EPG [9]. This observation suggests that the inhibition of EPGs by PGIPs may not only slow down the gross catalytic rate of EPGs on the solubilization and fragmentation of pectic wall polymers, but in doing so, may also extend the lifetime of biologically active oligogalacturonides (i.e. fragments of homogalacturonan) released from the cell wall by the action of the EPGs. Since oligogalacturonides have been implicated as elicitors of the defense response of plants, extending their lifetime likely affords the plant more time to mount a successful defense response than would be the case in the absence of these PGIPs [12].

Endopolygalacturonases from a single strain of fungus appear to exist in a variety of isoforms; multiple isoforms have been demonstrated in *Colletotrichum lindemuthianum*, *Cryphonectria parasitica*, *Cochliobolus carbonum*, *Sclerotinia sclerotorium*, *Aspergillus niger*, *Botrytis cinerea*, *Fusarium oxysporum* [9,14,15], and several other fungi. Additionally, these EPG isoforms may each be expressed in multiple glycoforms, and may vary in their modes of action as well as in their abilities to interact with, and be inhibited by, PGIPs [9,16]. This variability suggests that the catalytic mode of action of a particular fungal EPG, coupled with its particular inhibitability by PGIPs, may be critical factors in determining the virulence of a potential fungal pathogen.

Among the numerous isoforms and glycoforms of pectin degrading hydrolytic enzymes, evidence suggests that some variants are inducable according to physiologic/nutritional factors, while others appear to be constituitively produced. *Aspergillus niger* encodes a family of seven different EPGs, six of which have been characterized [17-22]. Among these are two isoforms, termed PG A and PG B, which are constitutively expressed [18]. A recent study indicates the majority of *Botrytis cinerea* EPGs to be constitutive [23]. These observations also suggest different functionalities of these molecules, within the sphere of host-pathogen interactions.

PGIP genes have been cloned from several species [24,25] and sequence analysis indicates that they encode proteins with a leucine rich repeat (LRR) structural motif. LRRs are characteristic of protein-protein interactions [26] and have been identified as conserved features in many disease resistance genes [25], suggesting that PGIPs may be involved in pathogen recognition. Although PGIPs from a range of species show considerable homology, individual isoforms appear to exhibit a degree of specificity with respect to the EPGs that they inhibit *in vitro* [27-29]. Not only are there differences in degree of inhibition for various EPG-PGIP pairings, but the type of inhibition may change as well, indicating that the location of the interaction may vary, and is, at least in some cases, not at the active site [16]. It is not known whether specificity of inhibition *in vivo* is a contributing factor in determining host cultivar/pathogen race specificity, as experiments have only addressed *in vitro* systems. An excellent example of the exquisite specificity of PGIPs for PGs is presented in Figure 1, which shows the ability of a PGIP purified from wheat to be able to inhibit, of all fungal PGs tested, only the PG isolated from the gramineous pathogen *Cochliobolus sativus*.

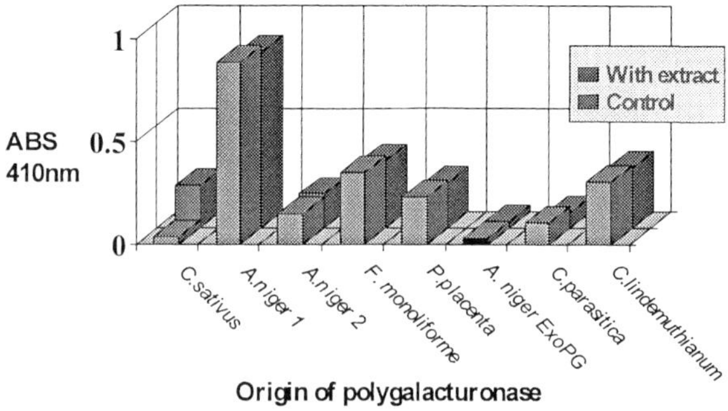

Figure1. Specificity of crude PGIP from wheat.

While it is apparent that the potential interactions of EPGs and PGIPs from different species are quite varied, it has also been recently demonstrated that two PGIPs from *P. vulgaris* (PGIP-1 and PGIP-2) are encoded by separate genes that are over 97% homologous, yet have different specificities of inhibition. The reported finding of a third PGIP gene (PGIP-3) with only 40% homology to PGIPs-1 and -2 [30] suggests that variations in PGIP activity within a source species may be as divergent as those between different species.

Structural information for EPGs and PGIPs that might contribute to our understanding of these variations in interactions are limited. The crystal structure for a bacterial EPG [31] and *Aspergillus niger* PG II [32] have been determined, but no crystal structure exists for a PGIP, and no complete model of the EPG-PGIP complex has been proposed. A proposed model for the structure of a bean PGIP, based on its membership in the plant-specific LRR class of proteins [33,34] has been published, and site-specific mutation experiments indicate a role for several specific amino acids within the bean PGIP during an EPG-PGIP interaction [35]. A recent evolutionary analysis identified 9 amino acids in both EPGs and PGIPs which are more likely to evolve in response to natural selection, and which therefore may play a role in the specificity of EPG-PGIP interactions [16].

As part of an ongoing study of EPGs and PGIPs, we have recently obtained evidence that PGIPs may in some cases act as activators of *A. niger* PG A and PG B. Micro-environmental parameters of the apoplast, such as variation in pH, appear to have the potential to alter the nature of the PGIP/EPG interaction from inhibition to activation. These observations have served to suggest a more complex regulatory mechanism in plant pathogenesis than was previously contemplated for EPG/PGIP interactions. The studies in the present work also allow us to begin to describe the structure of the *A. niger* EPG/*P. vulgaris* PGIP complex, as well as providing insight into the mechanism of inhibition. From this starting point, other EPG/PGIP complexes with different attributes may be studied, as a means toward understanding the variations in EPG/PGIP interactions and the possible role of this protein-protein interaction in pathogenicity.

2. Materials and Methods

2.1. EPGs

EPGs from *Aspergillus niger* were all kind gifts from the laboratory of Jaap Visser, Wageningen University. PG I, PG II, PG A and PG B, were all produced as previously reported [36,37]. The D201E mutant form of PG-II was prepared as published [38].

2.2. PGIPs

Native *P. vulgaris* cv. Pinto PGIPs were separated into two separate pools, corresponding in activity to PGIP-1 and -2, by affinity chromatography on a *F. moniliforme* EPG affinity column, as reported in Desiderio et.al [29]. Native PGIP-1 was further purified on an *A. niger* ANI affinity column. All native bean PGIPs, including a 33KD putative bean PGIP, were purified as a final step by ion exchange on a Hi-Trap S cartridge [39]. PGIP from pear and bean were purified according to Stotz et. al. [28], using Hi-Trap S in place of the mono-S column.

2.3. REDUCING SUGAR ASSAY

Generation of reducing sugar during PG-catalyzed hydrolysis of PGA was determined by the PAHBAH (p-hydroxy benzoic acid hydrazide) colorimetric assay [40], as previously reported [11].

2.4. HPAEC-PAD ANALYSIS

Oligogalacturonide profiles were obtained by anion exchange chromatography using a Dionex AL50 HPLC and a Dionex CarboPAC PA1 column coupled to a Dionex pulsed amperometric detector (PAD) as previously described [9].

2.5. OCTOMERIC GALACTURONIC ACID

The octamer of galacturonic acid $(GalA)_8$ was a kind gift of Stefan Eberhard at the CCRC. The LC buffers were made with acetic acid from J.T. Baker (Phillipsburgh, NJ) and acetonitrile from Fisher Scientific (Pittsburgh, PA). Buffer A consists of 94% H_2O and 6% acetic acid. Buffer B consists of 77% acetonitrile, 17% H_2O, and 6% acetic acid.

2.6. AMIDE EXCHANGE-MS WITH PEPSIN DIGESTION

Pepsin and D_2O were purchased from Sigma (St Louis, MO). Four trials were performed at pH 5using the D201E PG-II mutant (defined in this paper as PG-II'): PG-II' in H_2O (1), PG-II' in D_2O (2), PG-II' bound with $(GalA)_8$ in D_2O (3), and PG-II' bound with PGIP and then incubated with $(GalA)_8$ in D_2O (4). The PG-II' sample was approximately 1mg/mL in concentration, and 10μL aliquots were placed into 4 micro-centrifuge tubes. An equi-molar amount of PGIP was added to tube 4 and allowed to incubate at room temperature over night. Then a 100 molar excess of $(GalA)_8$ was added to tubes 3 and 4 and allowed to incubate at room temperature for 24hours. Subsequently, 10μL of H_2O was added to tube 1, and D_2O was added to tubes 2, 3 and 4 to reach a 50% D_2O concentration. The samples were left to exchange for 24 hours. At the end of the incubation, the exchange was quenched by cooling the sample and lowering the pH. The tubes were placed in an ice bath and approximately 1/10 the volume in each tube of .1M HCl was added to reach pH 2. 4μl of 1mg/mL pepsin was then added and the digestion progressed for 6 minutes. The solution was then analyzed by HPLC-MS. To ensure against any possible back exchange an ice bath was constructed to house the injection loop and reverse-phase HPLC column (25 mm long, 50 micron MAGIC C18 Bullet, Michrom Bioresources, Inc., Auburn, CA). A Micromass Q-Tof recorded spectra continuously during the LC gradient program, which ran linearly from 20%B to 65%B over 8 minutes.

2.7. UV-FLUORESCENCE

Using a Shimadzu Spectrofluorophotometer RF-5301 PC, emission spectra were generated of free PG-II' (1), PG-II' in the presence of oligosaccharide (2), and PG-II'/PGIP in the presence of oligosaccharide (3). Three 30 nM samples of PG-II' were prepared. An equal molar amount of PGIP was added to sample 3 and allowed to incubate overnight. A 100 molar excess of substrate was added to samples 2 and 3. All samples were left at room temperature for 24 hours. The samples were excited at 290nm and their maximum emission was recorded between 293-294 nm.

3. Results and discussion

3.1. INHIBITION/ACTIVATION SPECIFICITY STUDIES

We tested the interactions of PGIPs from bean with the constitutive PGs of *Aspergillus niger*, PG A and PG B, and compared these to the interactions of the same PGIPs with two inducible *A. niger* PGs, PG I and PG II (Table 1). While previous experiments have shown that bean PGIP isoforms inhibit PG I and PG II (although not to the same extent), data from reducing sugar assays show that both PG A and PG B were activated by native bean PGIP-1 at pH 5.0 (Table 1). Under some conditions, the activation was highly pronounced, exhibiting as high as a 30-fold increase (Table 1). Depending upon the chromatographic fraction chosen from across the ion-exchange elution profile during PGIP purification, native PGIP-2 either activated or inhibited both PG A and PG B (Table 1). These PGs and PGIPs were homogeneous as defined by SDS-PAGE, and/or assays for liberation of reducing sugar against PGI and *Fusarium moniliforme* PG. Therefore, these changes may be due to minor variations in the 'hypervariable' LRR region. These results were confirmed by Dionex profiles, and is represented by profiles of 16 hr digests of polygalacturonic acid by PG A in the presence or absence of bean PGIPs (Fig. 2).

Table 1. Relative reducing sugar activity* of *A. niger* PGs in the presence of bean PGIPs.

	PG A	PG B	PG I	PG II
"PGIP - 1"	1.5-32.05	1.0-1.8	0.01-0.03	0.9-0.94
"PGIP - 2"	0.7-8.8	0.8-16.3	0.01-0.02	0.01-0.02

*** Normalized for each PG to a value of 1.0 in the absence of PGIP.**
Substrate: 0.025% PGA in 50mM NaAc at pH 5.0
Values <1 = Inhibition
Values >1 = Activation

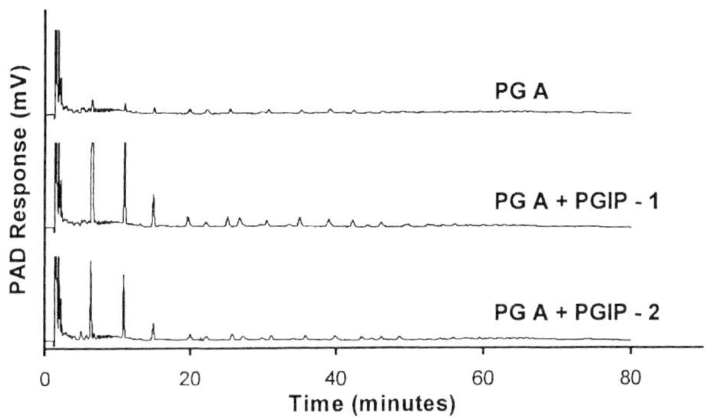

Figure 2. 16 hour digests of polygalacturonic acid by *A. niger* PG A in presence or absence of bean PGIPs at pH 5.0.

The PGIPs all interacted with our standard PGs as expected, i.e., the native PGIP-1 inhibited *A. niger* PG I, weakly inhibited PG II, but had no effect on the PG from *F. moniliforme*, while the native PGIP-2 inhibited *A. niger* PG I and PG II as well as the *F. moniliforme* PGs (data not shown).

Activation of the constitutive PGs proved to be sensitive to pH; according to both reducing sugar analysis and HPAEC, in the absence of PGIP both PG A and PG B increased in activity as the pH dropped to 4.24. However, as pH dropped below 4.5, PG A and PG B became inhibited by both the native bean PGIP isoforms, and at pH 4.24, both PG A and PG B were strongly inhibited by native PGIP-1 and PGIP-2 (Table 2). It is interesting to speculate that the interactions of PG A, PG B, and PGIP during pathogenesis represents a recent evolutionary development wherein the fungus produces these two PGs in response to the presence of host-plant PGIPs. Alternatively, perhaps the plant, by virtue of developing PGIPs capable of activation instead of inhibition, has enabled itself to more efficiently generate a steady-state concentration of biologically active oligogalacturonides, thus optimizing host defense reponsiveness. PGIPs with completely different ranges of activity may co-exist within a host, as indicated already for bean PGIP [29]. Thus, one can envision different PGIPs being called into play in different tissues, at different times, perhaps by intrusion from different pathogens. Progressive changes in the apoplastic environment at the site of host/pathogen interaction may alter the outcome of catalysis by different cell wall degrading enzymes produced by various pathogens on various hosts. In this regard, it appears that the EPG/PGIP system may be more complicated than initially suspected.

Table 2. Effect of pH on the activity of various PG/PGIP complexes at pH 5.0 or 4.2.

	"PGIP - 1"		"PGIP - 2"	
	pH 4.2	pH 5.0	pH 4.2	pH 5.0
PG A	0.18	32.05	0.10	8.82
PG B	0.95	1.64	0.15	16.26

*** Normalized for each PG to a value of 1.0 in the absence of PGIP.**
Substrate: 0.025% PGA in 50mM NaAc
Values <1 = Inhibition
Values >1 = Activation

3.2. STRUCTURAL CONFORMATION STUDIES

3.2.1. *PG-II/ polygalacturonic acid*

As part of an ongoing program studying the structure and function of pectin degrading enzymes, we wished to begin a study of the substrate binding sites of EPGs. Pectin is a major component of many plant cell walls, which serve as part of the primary defense against pathogens. In addition, the use of PDEs for industrial purposes has heightened the interest in understanding the interactions between the PDEs released by the pathogen and the cell wall carbohydrates. The study of protein-carbohydrate interactions has proven difficult if not impossible by traditional techniques, thus the use of hydrogen-deuterium exchange MS to study the binding of EPGs with an oligomer of galacturonic acid seemed highly appropriate.

Amide exchange-MS is typically interested in observing the rate at which amide hydrogens on a protein backbone exchange with hydrogen or deuterium in the solvent. Amide hydrogens are labile and will freely exchange with the protons in solution. When a protein is immersed in deuterated water, amide hydrogens will be replaced by deuterons. The protein is typically dissolved in D_2O or a mixture of D_2O and H_2O. Each exchanged proton results in an increase of 1 Dalton to the overall mass of the protein, a change which is easily monitored by mass spectrometry. Amide exchange-MS has been used in the past to take a look at the three-dimensional conformations of proteins. From the number of amino acid residues, it can be determined how many possible deuterons could be incorporated into the protein. The difference between this maximum possible incorporation and the experimental results indicates how many amide hydrogens are being protected from exchange with the solvent by the folding of the protein. Using ESI-MS the mass of the protein can be monitored over time, until the mass comes to an equilibrium. Therefore, it is possible to observe the acquisition of deuterium into the protein. [41-43]. This technique has been predominately used to study the folding and unfolding of proteins.

A recent variation of amide exchange-MS involves the quenching of the deuterium exchange. The spectra are no longer slices of a continually exchanging system, but a picture of a quickly exchanged system. After less than a minute, all readily available amide hydrogens will be able to exchange, at neutral pH. A quenching technique can extend the life time of the deuterium on the protein from 30-120 minutes which is long enough to allow for LC-MS analysis [44,45]. Not only can this method tell the number of hydrogens being protected, but by adding an enzymatic digestion step, the protected hydrogens can be located onto specific peptides (Fig. 3).

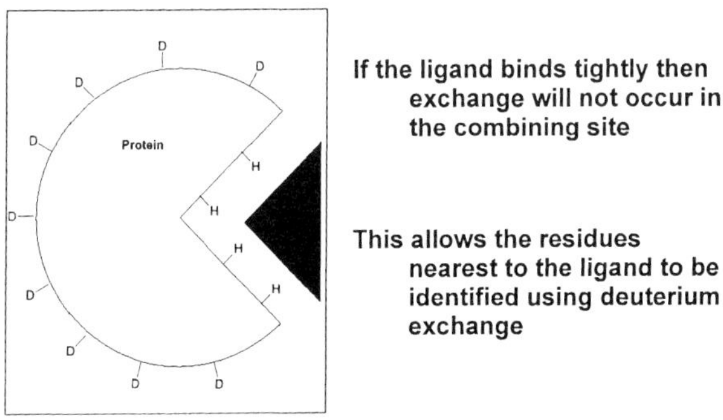

Figure 3. The effect of ligand binding.

What we report here is the novel application of hydrogen/deuterium (H/D) exchange-MS to locating regions of interaction between PG-II and polygalacturonic acid by looking for areas of decreased deuterium incorporation when the oligomer substrate is present.

The enzyme PG II from the fungus *A. niger* was chosen for the initial experiments. Several specific concerns regarding the system of choice needed to be addressed. First, the PG II is a hydrolase, and it was therefore necessary for a mutant to be prepared that maintained the binding capability without the hydrolysis activity. A number of mutants had been previously prepared and analyzed for activity and binding, and the D201E mutant (PG-II') was selected for meeting both these requirements [37]. A second consideration is that the active pH range for PG II lies between pH 4-5, far off the optimal pH for amide H/D exchange. The sample must therefore be allowed to exchange for longer than typical, pH neutral, protocols because exchange rates decrease with a decrease in pH. Finally, a large percent of the protein consists of β-pleated sheets, including the entire active site cleft. Because these sheets demonstrate extensive internal hydrogen bonding involving the amide hydrogens, it is unlikely that much deuterium incorporation will occur in this region. However, as will be seen, the large degree of internal hydrogen bonding proved to be advantageous.

Three trials were performed, PG-II' in H_2O, PG-II' in D_2O, and PG-II' with the oligosaccharide in D_2O. The samples were quenched, digested with pepsin, separated and detected by a LC-ESI mass spectrometer. The peptides from the control trial were identified by matching their masses to those of a computer generated peptic digest, and only those yielding unambiguous matches were used for analysis. Deuterium incorporation into peptides in the binding cleft with and without the substrate indicated that the few residues that had been deuterated were protected from exchange by the presence of the oligosaccharide (Fig. 4).

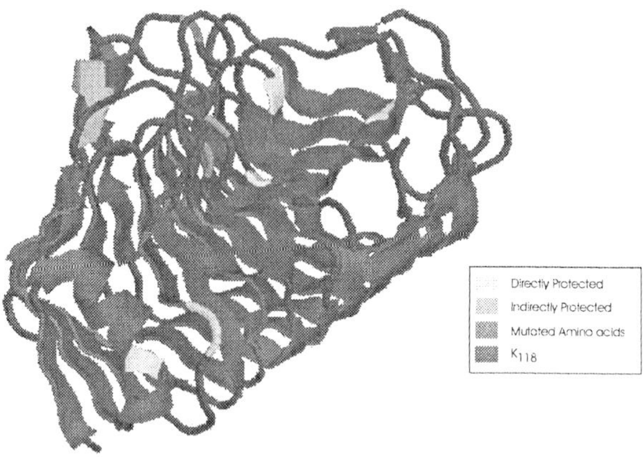

Figure 4. Amino acids of PG-II protected upon binding to (Gal A)$_8$.

(GalA)$_8$ was selected as substrate because this molecule would be the approximate length of the cleft. A substrate that is too long might provide non-specific protection to amino acids on sites outside the cleft, while a substrate that is too short would not indicate how much of the cleft is occupied by the substrate and may not participate in binding representative of PGA. This data suggests that the substrate lies somewhat linearly along the entire cleft.

As expected, due to the high degree of secondary structure, most of the identified peptides were related to regions within β-pleated sheets that incorporated little or no deuterium in the free PG-II'. The lack of incorporation when the protein is in the free state fits with the previously described understanding of this protein's secondary structure. Surprisingly, when the PG-II' was bound to the substrate, deuterium was incorporated into the β-sheets on the backside of the β-barrel (Fig. 5).

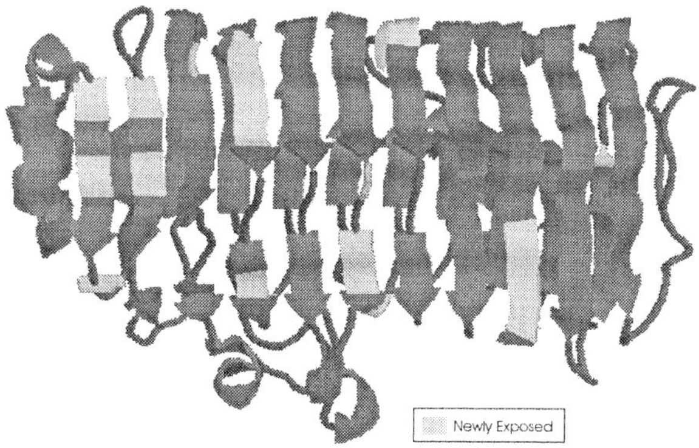

Figure 5. Amino acids exposed to deuterium exchange upon binding of PG-II' to (Gal A)$_8$.

The explanation for this is likely to be a conformational change in the protein, specifically, the disruption of these β-sheets on the backside of the molecule. The detection of a local increase in deuterium incorporation as a result of ligand binding far from the region of initial interest may point toward a possible mechanism for the activity of PG-II. PDEs often have lysine residues that stabilize the carboxylates on the carbohydrates. Two loops containing lysines hang predominantly over the binding cleft. If these two overhanging loops are required to "lean" into the cleft to allow the lysines to interact with and stabilize the substrate, the PG-II' may flex the β-sheets on the underside of the barrel to accommodate the necessary movement.

The PG-II'/oligosaccharide complex was also analyzed by UV-fluorescence. Fluorescence can be used to detect general changes in protein structures because fluorescence may be affected by interactions with solvent, neighboring amino acid sidechains, and other prosthetic groups. Tryptophan and tyrosine residues can both fluoresce, and if their microenvironment changes, a noticeable change in fluorescence intensity may be observed [46]. While exciting at 290 nm, so that only the tryptophans will fluoresce, an increase of intensity was observed in the presence of the oligosaccharide (Fig. 6).

The increase in intensity indicates the tryptophan residues are undergoing a perturbation in their microenvironment. Of the seven tryptophans, five are on the underside far from the binding cleft. The remaining two are on one of the loops described above. Whether or not the change in the environment of the tryptophans is caused by a conformational change of PG-II' or by the carbohydrate's direct contact with the residues and which tryptophans are most responsible for the fluorescence change can not be determined unambiguously at this point.

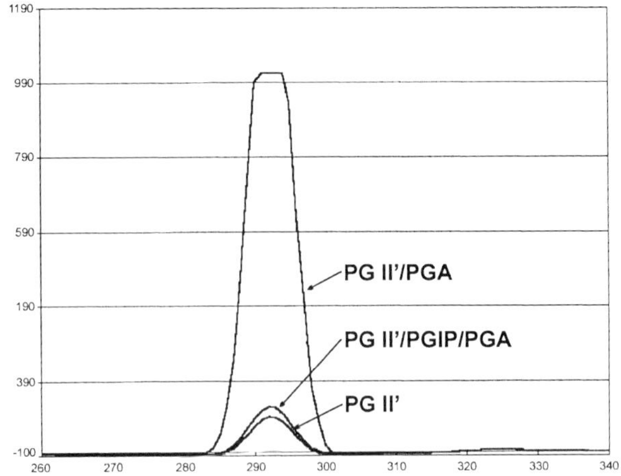

Figure 6. Fluorescence of the enzyme PG-II, the enzyme complexed with substrate, and the enzyme complexex with the substrate and PGIP.

3.2.2. *PG-II'/PGIP/PGA*

The presence of the PGIP caused a change in the pattern of deuterium incorporation. As described above, deuterium was not incorporated into the β-sheets of free PG-II', while in the presence of substrate, the β-sheets are apparently disrupted allowing deuterium to be incorporated allong the underside of the barrel. When PGIP is present in an PG-II'/PGIP/PGA complex, deuterium is no longer incorporated into the β-sheets (Fig. 7).

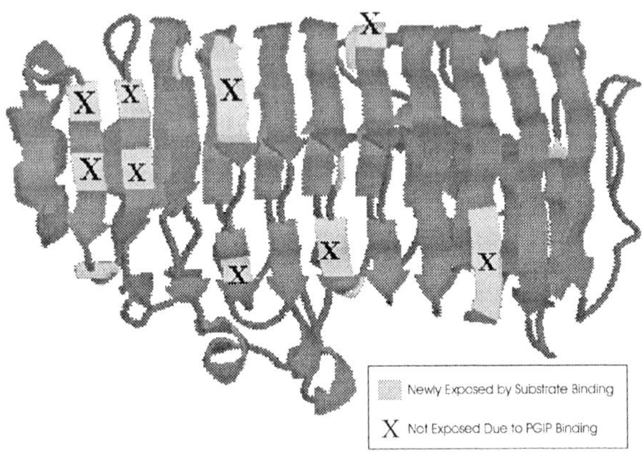

Figure 7. Regions of PG-II' protected from deuterium exchange by PGIP-2.

This protection from exchange may have two causes. The PGIP may be directly protecting this region from the solvent, or the inhibitor may be preventing the disruption of the β-sheets by the PGA.

A fluorescence experiment was performed to verify the binding of the 3 component system, PG-II'/PGA/PGIP. Figure 6 shows the fluorescence of the free PG-II', PG-II'/PGA, and PG-II'/PGA/PGIP. As described above, the fluorescence increases dramatically with the binding of the substrate supporting our previous findings of a conformational change. The presence of the PGIP has an equally remarkable effect on the level of fluorescence. The fluorescence data implies that the PGIP may be able to prevent the conformational changes caused by the presence of the substrate. The dramatic decrease in fluorescence caused by the presence of the PGIP clearly implicates the involvement of the inhibitor and indicates another change in the tryptophan environment. The similarity between this fluorescence intensity and that of the free PG-II' suggests that the inhibitor may be restraining the conformational changes that occurred in the EPG/oligosaccharide system. Thus, it is likely that the β-sheets are locked in their original conformation, preventing deuterium incorporation.

The use of H/D-MS also allowed us to begin to assess the site of interaction of PG-II' with PGIP. The inhibitor is, for unknown reasons, resistant to proteolysis, perhaps as a result of its three dimensional LRR structure. Differential peptide mapping suggests that the presence of the inhibitor is affecting the pepsin digest of the PG-II'. While pepsin is thought to be somewhat non-specific, cleaving at most hydrophobic residues, from experience the digests are very reproducible. The inhibitor caused a change in this consistent pattern of digestion. The expectation was that the inhibitor would block cleavage sites where it was bound. The observable effect would be a decrease in the population of small peptides that had cleavage sites in the binding region of the inhibitor, and a subsequent increase in population of larger peptides containing the protected cleavage sites. When comparing the dissapearing peptides with the newly produced peptides the findings were dramatic. They all occurred in the same region and the larger peptides contained the smaller peptides that had disappeared, indicating the region where the inhibitor was interacting with the PG-II'. Four residues (E_{95}, G_{104}, D_{110}, and I_{139}) that had consistently been cleavage sites in the PG-II' and PG-II'/PGA samples were now apparently protected from pepsin by the inhibitor. These residues are all located on the underside of the barrel, opposite to the binding site. This is consistent with reports of PGIP acting as a non-competitive inhibitor [16].

4. References

[1] Cervone, F., De Lorenzo, G., Caprari, C., Clark, A.J., Desiderio, A., Devoto, A., Leckie, F., Nuss, L., Salvi, G. and Toubart, P. (1993) The interaction between fungal polygalacturonase and plant cell wall PGIP (polygalacturonase-inhibiting protein), in B. Fritig and M. Legrand (eds.), *Mechanisms of Plant Defense Responses*, Kluwer Academic Publishers, Dordrecht, pp. 64-67.

[2] Hahn, M.G., Cheong, J.-J., Alba, R. and Côté, F. (1994) Characterization of hepta-β-glucoside elicitor binding protein(s) in soybean., in D.J. Bowles, P.M. Gilmartin, J.P. Knox, and G.G. Lunt (eds), *Molecular Botany: Signals and the Environment*, Portland Press Ltd., London, pp. 101-112.

[3] Carpita, N.C. and Gibeaut, D.M. (1993) Structural models of the primary cell wall in flowering plants: consistency of molecular structure with the physical properties of the walls during growth, *Plant J.* **3**, 1-30.

[4] Clay, R.P., Bergmann, C.W., and Fuller, M.S. (1997) Isolation and characterization of an endopolygalacturonase from *Cochliobolus sativus* and a cytological study of fungal penetration of barley, *Phytopathology* **87**, 1148-1159.

[5] O'Neill, M. Albersheim, P. and Darvill, A.G. (1990) The pectic polysaccharides of primary cell walls, in P.M. Dey (ed.) *Methods in Plant Biochemistry*, Vol. 2, Academic Press, London, pp. 415-441.

[6] Rose, J.K.C., O'Neill, M.A., Albersheim, P., and Darvill, A. (1999) Functions of the plant primary cell wall, in B. Ernst, G. Hart, P. Sinay (eds.), *Oligosaccharides in Chemistry and Biology, Vol. II Biology of Saccharide*, Wiley/VCH, Weinheim.

[7] Jones, T.M., Anderson, A.J. and Albersheim, P. (1972) Host-pathogen interactions IV. Studies on the polysaccharide-degrading enzymes secreted by *Fusarium oxysporum* f. sp. *lycopersici*, *Physiol. Plant Pathol.* **2**, 153-166.

[8] Mankarios, A.T. and Friend, J. (1980) Polysaccharide degrading enzymes of *Botrytis allii* and *Sclerotium cepivorum*. Enzyme production in culture and the effect of the enzymes on isolated onion cell walls, *Physiol. Plant Pathol.* **17**, 93-104.

[9] Cook, B.J., Clay, R.P., Bergmann, C.W., Albersheim, P., and Darvill, A.G. (1999) Fungal polygalacturonases exhibit different substrate degradation patterns and differ in their susceptibilities to polygalacturonase inhibiting proteins, *Mol. Plant Microbe Interact.* **12**, 703-711.

[10]Karr, A.L., Jr. and Albersheim, P. (1970) Polysaccharide-degrading enzymes are unable to attack plant cell walls without prior action by a "wall-modifying enzyme", *Plant Physiol.* **46**, 69-80.

[11]Darvill, A., Bergmann, C., Cervone, F., De Lorenzo, G., Ham, K., Spiro, M.D., York, W.S., and Albersheim, P. (1994.) Oligosaccharins involved in plant growth and host-pathogen interactions, *Biochem. Soc. Symp.* **60**, 89-94.

[12]Cervone, F., Hahn, M.G., De Lorenzo, G., Darvill, A., and Albersheim, P. (1989) Host-pathogen interactions XXXIII. A plant protein converts a fungal pathogenesis factor into an elicitor of plant defense responses, *Plant Physiol.* **90**, 542-548.

[13] Cervone, F., De Lorenzo, G., Aracri, B., Bellincampi, D., Caprari, C., Devoto, A., Leckie, F., Mattei, B., Nuss, L., and Salvi, G. (1996) The PGIP (polygalacturonase-inhibiting protein) family: Extracellular proteins specialized for recognition, in G. Stacey, B. Mullin, and P.M. Gresshoff (eds.), *Biology of Plant-Microbe Interactions*, International Society for Molecular Plant-Microbe Interactions, St. Paul, pp. 93-98.

[14]Centis, S., Guillas, I., Séjalon, N., Esquerré-Tugayé, M.T., and Dumas, B. (1997) Endopolygalacturonase genes from *Colletotrichum lindemuthianum*: Cloning of *CLPG2* and comparison of its expression to that of *CLPG1* during saprophytic and parasitic growth of the fungus, *Mol. Plant-Microbe Interact.* **10**, 769-775.

[15]Di Pietro, A. and Roncero, M.I.G. (1996) Endopolygalacturonase from *Fusarium oxysporum* f sp *lycopersici*: Purification, characterization, and production during infection of tomato plants, *Phytopathology* **86**, 1324-1330.

[16]Stotz, H.U., Bishop, J., Bergmann, C.W., Koch, M., Albersheim, P., Darvill, A., and Labavitch, J.M. (2000) Identification of target amino acids that affect interactions of fungal polygalaturonases and their plant inhibitors, *Physiol. Mol. Plant Pathol.* **56**, 117-130.

[17]Benen, J.A.E., Kester, H.C.M., and Visser, J. (1999) Kinetic characterization of *Aspergillus niger* N400 endopolygalacturonases I, II and C, *Eur. J. Biochem.* **259**, 577-585.

[18]Parenicova, L., Kester, H.C.M., Benen, J.A.E., and Visser, J. (2000) Characterization of a novel endopolygalacturonase from *Aspergillus niger* with unique kinetic properties, *FEBS Lett.* **467**, 333-336.

[19]Kester, H.C.M. and Visser, J. (1990) Purification and characterization of polygalacturonases produced by the hyphal fungus *Aspergillus niger*, *Biotech. Appl. Biochem.* **12**, 150-160.

[20]Parenicov , L., Benen, J.A., Kester, H.C.M., and Visser, J. (2000) *pga*A and *pga*B Encode two constitutively expressed endopolygalacturonases of *Aspergillus niger*, *Biochem. J.* **345**, 637-644.

[21] Benen, J.A.E., Kester, H.C.M., Parenicov , L., and Visser, J. (1996) Kinetics and mode of action of *Aspergillus niger* polygalacturonases, in J. Visser and A.G.J. Voragen (eds.), *Pectins and Pectinases* Elsevier Science, Amsterdam, pp. 221-230.

[22]Bussink, H.J.D., Buxton, F.P., and Visser, J. (1991) Expression and sequence comparison of the *Aspergillus niger* and *Aspergillus tubigenesis* genes encoding polygalacturonase II, *Curr. Genet.* **19**, 467-474.

[23]Wubben, J.P., Ten Have, A., van Kan, and Visser, J. (2000) Regulation of endopolygalacturonase gene expression in *Botrytis cinerea* by galacturonic acid, ambient pH and carbon catabolite repression, *Curr. Genet.* **37**, 152-157.

[24]Ramanathan, V., Simpson, C.G., Thgow, G., Iannetta, P.P.M., McNicol, R.J. and Williamson, B. (1997) cDNA cloning and expression of polygalacturonase-inhibiting proteins (PGIPs) from red raspberry (*Rubus ideus*), *J. Exp. Bot.* **48**, 1185-1193.

[25] Jones, D.A. and Jones, J.D.G. (1997) The role of leucine-rich repeat proteins in plant defences, *Adv. Bot. Res.* **24**, 89-167.

[26]Kobe, B. and Deisenhofer, J. (1994) The leucine-rich repeat: a versatile binding motif, *Trends Biol. Sci.* **19**, 415-421.
[27]Sharrock, K. and Labavitch, J.M. (1994) Polygalacturonase inhibitors of 'Bartlett' pear fruits: Differential effects of *Botrytis cinerea* polygalacturonase isozymes, and influence on products of fungal hydrolysis of pear cell walls and on ethylene induction in cell culture, *Physiol. Mol. Plant Pathol.* **45**, 305-319.
[28]Stotz, H.U., Contos, J.J.A., Powell, A.L.T., Bennett, A.B., and Labavitch, J.M. (1994) Structure and expression of an inhibitor of fungal polygalacturonases from tomato, *Plant Mol. Biol.* **25**, 607-617.
[29]Desiderio, A., Aracri, B., Leckie, F., Mattei, B., Salvi, G., Tigelaar, H.,Van Roekel, J.S.C., Baulcombe, D.C., Melchers, M.S., De Lorenzo, G., and Cervone, F. (1997) Polygalacturonase-inhibiting proteins (PGIPs) with different specificities are expressed in *Phaseolus vulgaris*, *Mol. Plant-Microbe Interact.* **10**, 852-860.
[30]De Lorenzo, G., Cervone, F., Bellincampi, D., Capodicasa , C., Caprari, C., Federici , L., Ferrari, S., Biuli, P., Laurenzi , M., Mattei, B., Raiola, A., Salvi, G., Sicilia, F., Vairo , D., and Zabotina, O. (1999) Structure-function studies on the leucine-rich repeat polygalacturonase-inhibiting protein (PGIP), Molecular Plant-Microbe Interactions 9th International Congress. Amsterdam.
[31]Pickersgill, R., Smith, D., Worboys , K., and Jenkins, J. (1998) Crystal structure of polygalacturonase from *Erwinia carotovora* ssp. *carotovora*, *J. Biol. Chem.* **273**, 24660-24664.
[32]Van Santen, Y., Benen, J.A.E., Schroer , K.-H., Kalk, K.H., Armand, S., Visser, J., and Dijkstra, W. (1999) 1.68-Å Crystal structure of endopolygalacturonase II from *Aspergillus niger* and identification of active site residues by site-directed mutagenesis, *J. Biol. Chem.* **274**, 30474-30480.
[33]Kajava, A.V. (1998) Structural diversity of leucine-rich repeat proteins, *J. Mol. Biol.* **277**, 519-527.
[34]Mattei, B., Bernalda, M., Luca Federici, S., Roepstorff, P., Cervone, F., and Boffi, A. (2001) Secondary structure and post-translational modifications of the leucine–rich repeat protein PGIP (Polygalacturonase-Inhibiting Protien) from *Phaseolus vulgaris*, *Biochemistry* **40**, 569-576.
[35]Leckie, F., Mattei, B., Capodicasa, C., Hemmings, A., Nuss, L., Aracri, B., De Lorenzo, G., and Cervone, F. (1999) The specificity of polygalacturonase-inhibiting protein (PGIP): a single amino acid substitution in the solvent-exposed β-strand/β-turn region of the leucine-rich repeats (LRRs) confers a new recognition capability, *EMBO J.* **18**, 2352-2363.
[36]Parenicov , L., Benen, J.A.E., Kester, H.C.M., and Visser, J. (1998) PgaE encodes a forth member of the endopolygalacturonase gene family from *Aspergillus niger*, *Eur. J. Biochem.* **251**, 72-80.
[37]Parenicov , L., Benen, J.A.E., Kester, H.C.M., and Visser, J. (2000) PgaA and pgaB encode two constitutively expressed endopolygalacturonases of *Aspergillus niger*, *Biochem. J.* **345**, 637-644.
[38]Armand, S., Wagemaker, M.J.M., Sanchez-Torres, P., Kester, H.C.M., van Santen, Y., Eijkstra, B.W., Visser, J., and Benen, J.A.E. (2000) The active site topology of *Aspergillus niger* endopolygalacturonase II as studied by site-directed mutagenesis, *J. Biol. Chem.* **275**, 691-696.
[39]Toubart, P., Desiderio, A., Salvi, G., Cervone, F., Daroda, L., De Lorenzo, G., Bergmann, C., Darvill, A.G. and Albersheim, P. (1992) Cloning and characterization of the gene encoding the polygalacturonase-inhibiting protein (PGIP) of *Phaseolus vulgaris* L, *The Plant J.* **2**, 367-373.
[40]Lever, M. (1972) A new reaction for colorimetric determination of carbohydrates, *Anal. Biochem.* **47**, 273-279.
[41]Katta, V. and Chait, B.T. (1993) Hydrogen/deuterium exchange electrospray ionization mass spectrometry: a method for probing protein conformational changes in solution, *J. Am. Chem. Soc.* **115**, 6317-6321.
[42]Zhang, Z.Q. and Smith, D.L. (1993) Determination of amide hydrogen-exchange by mass-spectrometry - a new tool for protein-structure elucidation, *Protein Sci.* **2**, 522-531.
[43]Liu, Y.Q. and Smith, D.L. (1994) Probing high-order structure of proteins by fast-atom-bombardment mass-spectrometry, *J. Am. Soc. Mass Spectom.* **5**, 19-28.
[44]Smith, D.L. and Zhang, Z.Q. (1994) Probing noncovalent structural features of proteins by mass-spectrometry, *Mass Spectrom. Rev.* **13**, 411-429.
[45]Smith, J.B., Liu, Y.Q., and Smith, D.L. (1996) Identification of possible regions of chaperone activity in lens alpha-crystallin, *Exp. Eye Res.* **63**, 125-127.
[46]Nichols, N.M. and Matthews, K.S. (2001) p53 Unfolding detected by CD but not by tryptophan fluorescence, *Biochem. Biophys. Res. Commun.* **288**, 111-115.

STRUCTURE-FUNCTION AND MOLECULAR STUDIES ON FUNGAL POLYGALACTURONASES AND THEIR INHIBITORS PGIPs

FELICE CERVONE

Dipartimento di Biologia Vegetale, Universita' di Roma "La Sapienza", Piazzale A. Moro, 00185 Roma, Italy.

E-mail: felice.cervone@uniroma1.it

Abstract

To invade a plant tissue, phytopathogenic fungi produce several cell wall degrading enzymes; among them, *endo*polygalacturonase (PG) catalyses the fragmentation and solubilisation of homogalacturonan. Polygalacturonase-inhibiting proteins (PGIPs), found in the cell wall of many plants, counteract fungal PGs by forming specific complexes with them and favouring the formation of elicitor-active oligogalacturonides. Against the many PGs produced by fungi, plants have evolved different PGIPs with specific recognition abilities. The genes encoding PGIPs are organized in families; the different members of each family may encode proteins with nearly identical characteristics but with different specificities. The members of a family also exhibit a different regulation, and are tipically induced by pathogen infection and stress-related signals. The recognition ability of PGIPs reside in their LRR (leucine-rich repeat) structure, where solvent exposed residues in the beta-strand/beta-turn motifs of the LRRs are determinants of specificity. Recently we have solved the crystal structure of PG from the phytopathogenic fungus *Fusarium moniliforme* (*Fm*PG) and studied the structural requirements of PG for interaction with PGIP.

1. Introduction

Fungal endopolygalacturonases (PGs) cause wall degradation and plant tissue maceration and also release from the plant cell walls oligogalacturonides (OG), which are elicitors of a variety of defense responses. Elicitor-active OGs are produced by the action of PGs if the enzyme activity is controlled by polygalacturonase-inhibiting proteins (PGIPs), extracellular plant proteins with inhibitory capabilities. As parasitic fungi need PGs for a successful invasion of the plant tissues, recognition of these enzymes by PGIPs is thought to be an efficient defense strategy devised for plant self-defense. Against the many PGs produced by fungi, plants have evolved different PGIPs with specific recognition abilities. PGIPs occur in a variety of dicotyledonous plants, as well as in the pectin-rich monocotyledonous plants onion and leek [1].

F. Voragen et al. (eds.), Advances in Pectin and Pectinase Research, 293-297.
© 2003 *Kluwer Academic Publishers. Printed in the Netherlands.*

PGIPs are typically effective against fungal PGs and ineffective against other pectic enzymes either from microbial or plant origin. Not only do PGIPs from different plant sources differ in their inhibitory activities, but also PGIPs from a single plant source inhibit, with different strength, PGs from different fungi or different PGs from the same fungus [2].

2. PGIPs in *Phaseolus vulgaris* and Arabidopsis

Pgip genes have been cloned from many plants, where they usually exist as gene families [1]. Typically, *pgip* genes code for protein products comprising a signal peptide for secretion and a mature polypeptide of 300-315 amino acids with several potential glycosylation sites. The mature PGIP is characterized by the presence of 10 leucine-rich repeat (LRR) elements matching the extracytoplasmic consensus GxIPxxLxxLxxLxxLxLxxNxLx also found in several *R* genes, which participate in gene-for-gene resistance [3,4].

In the genome of *P. vulgaris*, four *pgip* genes are clustered. The characterization of *Pvpgip1* and *Pvpgip2* provides a striking example of the structural diversification within a *pgip* gene family and its functional significance. The coding sequence of *Pvpgip1* and *Pvpgip2* differ by 26 nucleotides [5,6]. The 11 nonsynonymous nucleotide substitutions show a nonrandom distribution as they occur within or very close to the regions encoding the xxLxLxx motifs predicted to form solvent-exposed β-sheet/β-turn structures important for ligand binding. The genes *Pvpgip3* and *Pvpgip4* are closely related to each other (93.4% identity), but share only 79% identity with *Pvpgip1/Pvpgip2*.

Two Arabidopsis genes, *Atpgip1* and *Atpgip2,* which share 80% identity, are located 400 bp apart on chromosome 5.

We have expressed all the isolated *pgip* genes of bean and Arabidopsis either stably through genetic transformation or transiently through infection of *Nicotiana benthamiana* with a modified potato virus X (PVX) [6]. The products of *Pvpgip1* and *Pvpgip2,* expressed either in transgenic tomato and tobacco plants or in PVX-infected *N. benthamiana* plants, exhibit identical inhibitory activity and specificity irrespective of the plant used for expression. PvPGIP2 is the most efficient inhibitor, with the broadest recognition specificity, whereas PvPGIP3 inhibits PG from *Colletotrichum* sp. and *Stenocarpella maydis* but does not inhibit PG of *Fusarium moniliforme.* AtPGIP1 and AtPGIP2 separately expressed in transgenic Arabidopsis have been purified and shown to inhibit PG from *Colletotrichum* sp. and *Botrytis cinerea* but not PGs from *Aspergillus niger* or *F. moniliforme.*

The genes encoding PGIPs not only encode proteins with different specificity, but are also differentially regulated. Transcripts accumulate in suspension-cultured bean cells following addition of elicitors such as OGs and fungal glucan, and in *P. vulgaris* hypocotyls in response to wounding or treatment with salicylic acid [7]. In the interaction between *C. lindemuthianum* and *P.vulgaris,* rapid accumulation of mRNA at the infection site correlates with the appearance of the hypersensitive response in incompatible interactions [8,9]. RT-PCR analysis shows that *Pvpgip1, Pvpgip2* and *Pvpgip3* are regulated in a different manner. *Pvpgip2* transcripts are induced by wounding, salicylic acid, OGs and glucan, whereas *Pvpgip3* transcripts are induced by glucan and OGs, but not by wounding or salicylic acid; *Pvpgip1* responds to wounding only.

3. The PG-PGIP interaction.

Modeling studies based on the known structure of the porcine RNAse inhibitor (PRI) [10] suggest that PGIP exhibits a parallel stacking of β-strand/β-turns, forming a solvent-exposed surface; the protein assumes an arch-shaped protein fold resembling that of the β-helical structure [6]. A detailed analysis of the secondary structure of PvPGIP2 by far-UV CD and infrared spectroscopies coupled to constrained prediction methods indicates the presence of 12 alfa and 12 beta secondary structure segments. The protein consists of three domains, namely the central LRR region and two cysteine-rich flaking domains. Two N-linked oligosaccharides with a structure resembling the typical complex plant N-glycan are located on Asn 64 and Asn 141 [11].

The mature proteins encoded by *Pvpgip1* and *Pvpgip2* of *P. vulgaris* differ by only 8 amino acid residues located preferentially within or contiguous to the motif xxLxLxx. The two proteins exhibit distinct specificities: PvPGIP1 is not able to interact with PG of *F. moniliforme* and interacts with PG of *A. niger*; PvPGIP2 interacts with both. The single mutation of a lysine at position 253 of PvPGIP1 into the corresponding amino acid of PvPGIP2, a glutamine, is sufficient to confer to the protein the capacity of interacting with *F. moniliforme* PG [6]. These data provide evidence that variations in the solvent-exposed β-sheet/β-turn structure of PGIP have a functional significance and determine the discriminatory ability for a specific recognition of PG.

The residues of PG involved in the interaction with PGIP are also under investigation [12]. Recently, the 3-D structures of *F. moniliforme* PG has been elucidated. The enzyme exhibits, like other characterized pectic enzymes, a β-helix structure [13]. The enzyme has a substrate binding cleft characterized by a highly positive electrostatic potential for the polyanionic polygalacturonic acid. Several amino acids of *F. moniliforme* PG, located inside or in proximity of the cleft, have been mutated and their contribution to the formation of the complex with PvPGIP2 has been investigated. The enzyme forms multiple contacts with the inhibitor; Lys269 and Arg 267, located inside the active side cleft, and His 188, at the edge of the active site cleft, are among the residues directly involved in the formation of the complex. Since Lys 269 and Arg 267 putatively bind the substrate at subsites +1 and –1, they cannot easily mutate without affecting the enzyme activity, and therefore their involvement in the interaction with PGIP minimizes the possibility for the fungal PGs to escape recognition [14].

In order to locate the domain of PG recognised by the inhibitor, a combination of surface plasmon resonance (SPR) and matrix-assisted laser-desorption-ionization-time-of-flight mass spectrometry (MALDI-TOF-MS) was used. A peptide mixture was prepared by limited proteolysis of the native PG from *F. moniliforme* enzyme with Endoproteinase Lys-C. The peptide mixture was passed in flow over the sensor surface with immobilised PvPGIP-2, and over a reference flow-cell as a control. One peptide with *m/z* 6479.9 was specifically recovered from the flow-cell with immobilised PvPGIP-2. The peak at m/z 6479.9 corresponds to the PG fragment spanning from amino acid 181 to amino acid 244 comprising several residues which are conserved in all the known fungal PGs and form the active site of this class of enzymes [15]. Our isolation of the peptide 181-244 which has a strong capacity of interaction with PvPGIP-2 suggests that most if not all of the residues critical for the interaction are located in the active site of PG from *F. moniliforme* [16]. It now becomes feasible to pinpoint those residues by combining site-directed mutagenesis and SPR analysis.

4. Conclusions

PGIP, a widespread cell wall protein, exhibits features typical of the defence and pathogenesis-related proteins, and also belongs to the super-family of LRR proteins, which includes the LRR-based resistance gene products. The interaction between PGIPs and their ligands PGs is a unique model system to gain knowledge on the recognition specificity of LRR proteins in plants. Knowledge of the structural requirements that confer on PGIPs the ability of interacting specifically with their ligands can be exploited to devise *in vitro* mutagenesis-based strategies to obtain more efficient inhibitors, or inhibitors with novel recognition abilities. Improvement of resistance may therefore be obtained by transforming crop plants either with a set of different *pgip* genes with complementary recognition abilities, or with single genes with multiple specificities.

Acknowledgements

We thank the Armenise-Harvard Foundation and the Institute Pasteur-Fondazione Cenci Bolognetti for supporting our research.

References

1. De Lorenzo, G., D'Ovidio, R. and Cervone F. (2001). The role of polygacturonase-inhibiting proteins (PGIPs) in defense against pathogenic fungi. Annu.Rev.Phytopathol. **39**:313-335.

2. Desiderio, A., Aracri, B., Leckie, F., Mattei, B., Salvi, G., Tigelaar, H., Van Roekel, J.S.C., Baulcombe, D.C., Melchers, L.S., De Lorenzo, G. and Cervone, F. (1997) Polygalacturonase-inhibiting proteins (PGIPs) with different specificities are expressed in *Phaseolus vulgaris*. Mol.Plant-Microbe Interact. **10**:852-860.

3. De Lorenzo, G., Cervone, F., Bellincampi, D., Caprari, C., Clark, A.J., Desiderio, A., Devoto, A., Forrest, R., Leckie, F., Nuss, L. and Salvi G. (1994). Polygalacturonase, PGIP and oligogalacturonides in cell-cell communication. Biochem.Soc.Trans. **22**:396-399.

4. Jones, D. A. and Jones, J.D.G. (1997) The role of leucine-rich repeat proteins in plant defence. Adv.Bot.Res. **24**:89-166.

5. Toubart, P., Desiderio, A., Salvi, G., Cervone, F., Daroda, L., De Lorenzo, G., Bergmann, C., Darvill, A.G. and Albersheim., P. (1992) Cloning and characterization of the gene encoding the endopolygalacturonase-inhibiting protein (PGIP) of *Phaseolus vulgaris* L. Plant J. **2**:367-373.

6. Leckie, F., Mattei, B., Capodicasa, C., Hemmings, A., Nuss, L., Aracri, B., De Lorenzo, G. and Cervone, F. (1999) The specificity of polygalacturonase-inhibiting protein (PGIP): a single amino acid substitution in the solvent-exposed beta-strand/beta-turn region of the leucine-rich repeats (LRRs) confers a new recognition capability. EMBO J. **18**:2352-2363.

7. Bergmann, C., Ito, Y., Singer, D., Albersheim, P., Darvill, A.G., Benhamou, N., Nuss, L., Salvi, G., Cervone, F. and De Lorenzo, G. (1994) Polygalacturonase-inhibiting protein accumulates in *Phaseolus vulgaris* L. in response to wounding, elicitors, and fungal infection. Plant J. **5**:625-634.

8. Devoto, A., Clark, A.J., Nuss, L., Cervone, F. and De Lorenzo, G. (1997). Developmental and pathogen-induced accumulation of transcripts of poly-galacturonase-inhibiting protein in *Phaseolus vulgaris* L. Planta. **202**:284-292.

9. Nuss, L., Mahé, A., Clark, A.J., Grisvard, J., Dron, M., Cervone, F. and De Lorenzo, G. (1996) Differential accumulation of polygalacturonase-inhibiting protein (PGIP) mRNA in two near-isogenic lines of *Phaseolus vulgaris* L. upon infection with *Colletotrichum lindemuthianum*. Physiol.Mol.Plant Pathol. **48**:83-89.

10. Kobe, B. and Deisenhofer, J. (1995) A structural basis of the interactions between leucine-rich repeats and protein ligands. Nature **374**:183-186.

11. Mattei, B., Bernalda, M.S., Federici, L., Roepstorff, P., Cervone, F.and Boffi, A. (2001) Secondary structure and post-translational modifications of the leucine-rich repeat protein PGIP (polygalacturonase-inhibiting protein) from *Phaseolus vulgaris*. Biochem. **40**:569-576.

12. Caprari, C., Mattei, B., Basile, M.L., Salvi, G., Crescenzi, V., De Lorenzo, G. and Cervone, F. (1996) Mutagenesis of endopolygalacturonase from *Fusarium moniliforme*: Histidine residue 234 is critical for enzymatic and macerating activities and not for binding to polygalacturonase-inhibiting protein (PGIP). Mol.Plant-Microbe Interact. **9**:617-624.

13. Herron, S. R., Benen, J.A., Scavetta, R.D., Visser, J. and Jurnak F. (2000) Structure and function of pectic enzymes: virulence factors of plant pathogens. Proc.Natl.Acad.Sci.USA **97**:8762-8769.

14. Federici, L., Caprari, C., Mattei, B., Savino, C., Di Matteo, A., De Lorenzo, G., Cervone, F. and Tsernouglou, D. (2001) Structural requirements of endopolygalacturonase for the interaction with PGIP (polygalacturonase inhibiting protein). Proc.Natl.Acad.Sci.USA **98**:13425-13430.

15. Armand, S., Wagemaker,M.J., Sanchez-Torres, P., Kester, H.C.M., van Santen, Y.D.W., Visser, J. and Benen, J.A.E. (2000) The active site topology of *Aspergillus niger* endopolygalacturonase II as studied by site-directed mutagenesis. J.Biol.Chem. **275**:691-696.

16. Mattei, B., Cervone, F. ad Roepstorff, P. (2001) The interaction between endopolygalacturonase from *Fusarium moniliforme* and PGIP from *Phaseolus vulgaris* studied by surface plasmon resonance and mass spectrometry. Comp.Funct.Genomics **2**: 359-364.

RHAMNOGALACTURONAN ACETYLESTERASE, A MEMBER OF THE SGNH-HYDROLASE FAMILY

ANNE MØLGAARD
Centre for Crystallographic Studies, University of Copenhagen, Universitetsparken 5, 2100 Copenhagen, Denmark
E-mail: anne@ccs.ki.ku.dk

Abstract

Rhamnogalacturonan acetylesterase (RGAE) from *Aspergillus aculeatus* acts in synergy with rhamnogalacturonases on the enzymatic degradation of rhamnogalacturonan I (RG-I). RGAE is a member of the Carbohydrate Esterase Family 12 (CEF12), which also includes two pectin acetylesterases, a cephalosporin C deacetylase and a number of bacterial proteins of unknown function. The proteins within this family are similar in sequence and are expected to share the same fold. The carbohydrate esterase family 12 has been shown to be a subfamily of a family of hydrolases described by Dalrymple et al. (1997). The proteins in this family are characterized by having three blocks of conserved sequence, but have little or no sequence similarity outside these blocks.

The structure of RGAE was first determined at 1.55 Å resolution, and has recently been refined using synchrotron data to atomic resolution (1.12 Å). Although the overall fold of RGAE resembles the common α/β hydrolase fold, there are a number of differences, which has led to the identification and characterization of a new protein fold. This fold is shared with three other proteins so far, the serine esterase from *Streptomyces scabies* (SsEst), the platelet-activating factor acetylhydrolase from *Bos taurus* (PAF-AH) and the haemagglutinin-esterase-fusion glycoprotein from influenza C virus (HEF). All of these proteins are members of the hydrolase family described by Dalrymple et al. and a comparison of their common structures reveal four blocks of residues which are structurally conserved, and which contain all the catalytically important residues. One amino acid residue in each block is completely conserved, a serine, glycine, asparagine and histidine respectively, which has led to the name the SGNH-hydrolase family.

1. Introduction

1.1. RHAMNOGALACTURONAN I

One of the main components of pectin is rhamnogalacturonan I (RG-I). It is composed of alternating rhamnose (Rha) and galacturonic acid (GalUA) residues with the dimer repeat unit (1,2)-α-L-Rha-(1,4)-α-D-GalUA. Between *ca.* 20-80% of the rhamnosyl residues have arabinosyl- and galactosyl-rich sidechains attached to O4, and arabinan, galactan and two forms of arabinogalactan have also been found to be covalently attached to the backbone of RG-I (Albersheim et al., 1996). All this makes RG-I a

F. Voragen et al. (eds.), Advances in Pectin and Pectinase Research, 299-313.
© 2003 *Kluwer Academic Publishers. Printed in the Netherlands.*

highly branched or 'ramified' structure. The GalUA residues can be methylated at the carboxylic acid group, and can be acetylated at the O2 and/or the O3 position (Kouwijzer et al., 1996; Ishii, 1997).

Due to the structural complexity of RG-I, several enzymes are needed for its breakdown. The backbone of RG-I is degraded by rhamnogalacturonases, and it has been shown that their action is hindered by the acetyl groups on the GalUA residues. These enzymes are therefore found to work in synergy with a rhamnogalacturonan acetylesterase (RGAE), which deacetylates the substrate for subsequent degradation by rhamnogalacturonases (Schols et al., 1990; Searle-van Leeuwen et al., 1992; Kofod et al., 1994).

As a step toward an understanding of the degradation of RG-I at a molecular level, structures of three of the enzymes active toward RG-I have been determined. The first of these structures to be determined was the structure of RGase A from *Aspergillus aculeatus* (Petersen et al., 1997), an endo-acting enzyme which hydrolyses the glycosidic α-D-GalUA-(1,2)-α-L-Rha bond (giving it the systematic name α-D-galactopyranosyluronide-(1,2)-α-L-rhamnopyranosyl hydrolase (Mutter et al., 1996)). The fold of this enzyme is a right-handed parallel β helix, a fold which is also found in most of the pectinolytic enzymes that have been structurally investigated so far, namely the pectate lyases (Yoder et al., 1993; Pickersgill et al., 1994; Lietzke et al., 1996), the pectin lyases (Mayans et al., 1997; Vitali et al., 1998), polygalacturonases (Pickersgill et al., 1998; van Santen et al., 1999) and pectin methyl esterase (Jenkins et al., 2001). The first pectinase found not to have a right handed parallel β helix fold was RGAE from *Aspergillus aculeatus*, which has an α/β type fold. More recently, the structure of a β-(1,4)-galactanase from *Aspergillus aculeatus*, active on RG-I, was found to have a $(\beta/\alpha)_8$ barrel fold (Ryttersgaard et al., 2002).

1.2. THE α/β HYDROLASE AND OTHER ESTERASE FOLDS

Most neutral lipases and esterases whose structures are known adopt the α/β hydrolase fold. This fold was first characterized by Ollis et al. in 1992 to be shared between five apparently unrelated hydrolases, and since then the number of members has grown to about 50 in 1999 (Heikinheimo et al., 1999; Nardini and Dijkstra, 1999).

The α/β hydrolase database ESTHER (Cousin et al., 1998) (http://www.ensam.inra.fr/cholinesterase) currently lists 213 PDB entries (including redundant structures), with 27 structural families.

The 'canonical' α/β hydrolase fold was originally described as being an α/β structure with a central eight-stranded mostly parallel β sheet with strand β2 antiparallel to the rest (Fig.1). The topology of the strands is +1x, +2x, -1x, +2x, $(+1x)_3$ giving the strand order 12435678. All of the structurally characterized α/β hydrolase enzymes have a catalytic triad of a nucleophile (Ser, Cys or Asp), His and acid. The nucleophile is located in a sharp turn between β5 and the highly conserved α helix C. This motif is known as the nucleophile elbow motif, and it imposes a strain on the backbone of the nucleophile, causing it to acquire a strained ε conformation (Derewenda and Derewenda, 1991; Ollis et al., 1992). It was argued, that the benefits of this arrangement could be twofold. The placement of the nucleophile on an extremely sharp turn makes the relatively short sidechain of the serine more accessible to the substrate, and the positioning of the nucleophile with respect to the conserved helix C is optimal for the helix dipole to stabilize both the tetrahedral intermediate of the reaction and the ionized form of the nucleophile.

After many more structures became available, the original criteria for classifying a structure as a member of the α/β hydrolase family have been refined. The current necessary criteria according to Heikinheimo et al. (1999) and Nardini and Dijkstra (1999) are the presence of at least five parallel β strands, the order of nucleophile-histidine-acid of the catalytic triad residues and the presence of the nucleophile elbow motif.

There are some esterases which have a fold which is different from the α/β hydrolase fold. A very prominent exception is pectin methylesterase (PME), which not only does not have the α/β hydrolase fold, but also lacks the conventional catalytic nucleophile-His-acid triad, and seems to have a completely different mechanism of hydrolysis involving two aspartate residues and an arginine residue (Jenkins et al., 2001). Although the fold of RGAE, being an α/β structure, roughly resembles the α/β hydrolase fold, there are some important differences which will be described below, which justify the classification of the fold of RGAE as a unique fold.

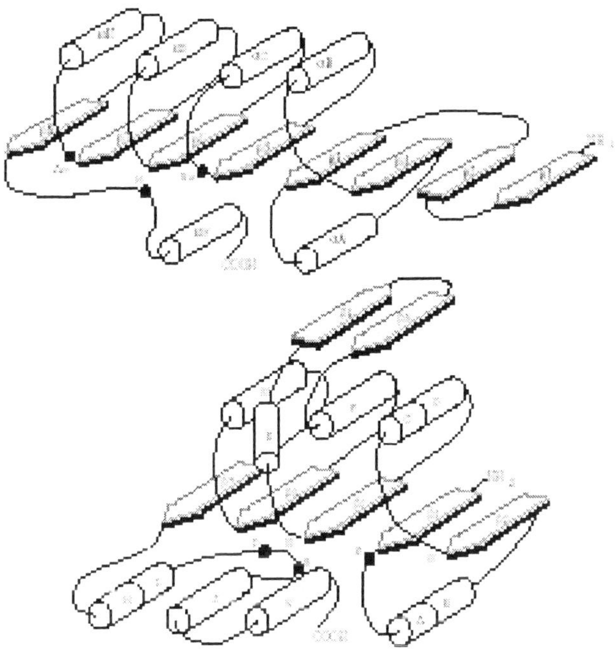

Figure 1. A schematical diagram of the 'canonical' α/β hydrolase fold (top) and the fold of RGAE (bottom). The location of the catalytically important residues is indicated in the diagrams.

1.3. PROTEINS RELATED TO RGAE BY SEQUENCE SIMILARITY

RGAE shows no obvious sequence homology to any protein with known three dimensional structure. Its closest homologues belong to the Carbohydrate Esterase Family 12 (CEF12), as defined by Coutinho and Henrissat (2000) (http://afmb.cnrs-mrs.fr/~pedro/CAZY/db.html), and the structure of RGAE is the first three-dimensional structure in this family. The classification of the families is based on sequence similarities, and the members within the same family are expected to share the same fold. There are currently nine members of CEF12, which in addition to RGAE from *Aspergillus aculeatus* includes two pectin acetylesterases, one from *Erwinia chrysanthemi* (PaeY) (Shevchik and Hugouvieux-Cotte-Pattat, 1997) and one from *Bacillus subtilis* (YxiM) (Bolvig et al., 2002), a cephalosporin C deacetylase from *Bacillus* sp. *KCCM10143* (ccd) (Choi et al., 2000) and five open reading frames (ORF), two from *Bacillus subtilis* (YesY and YesT), one from *Bacillus halodurans* (BH3963), one from *Streptomyces coelicolor* (SCI39.28c) and one from *Caulobacter crescentus* (CC0812). The members of the family are included in the multiple sequence alignment shown in Fig. 2 as the first nine sequences. The sequences were aligned using the program MULTALIGN (Barton and Sternberg, 1987; Barton, 1990). The alignment of the last three sequences in Fig. 2 is based on a structural alignment performed using the program DALI (Holm et al., 1992; Holm and Sander, 1994). The indicated PDB codes correspond to platelet-activating factor acetylhydrolase from *Bos taurus* (1wab), *Streptomyces scabies* serine esterase (1esc) and haemagglutinin-esterase-fusion glycoprotein from Influenza C virus (1flc). The alignment shown covers the entire sequence of the mature form of RGAE. The other sequences have been truncated, in the cases where they extend beyond this region.

2. RGAE - the enzyme

RGAE was first identified and purified from *Aspergillus aculeatus* (Searlevan Leeuwen et al., 1992) and a RGAE has also been identified in *Aspergillus niger* (Searle-van Leeuwen et al., 1996). The *A. aculeatus* RGAE was cloned and overexpressed in *Aspergillus oryzae* (Kauppinen et al., 1995). The recombinant enzyme consists of a single chain of 250 amino acid residues, including a 17 residue signal peptide, and the molecular mass of the mature enzyme was determined by SDS-PAGE to be 32-35 kDa. The discrepancy from the calculated molecular mass of 24.6 kDa was ascribed to glycosylation. Mass spectrometric measurements have confirmed this and have shown the glycosylation to be heterogeneous and to depend upon the exact method of purification (Kauppinen et al., 1995; Mølgaard et al., 1998). There are two putative N-glycosylation sites in the primary sequence, and X-ray crystallographic studies have shown both sites to be glycosylated. There was no evidence for O-glycosylation in the electron density maps (Mølgaard et al., 2000). The temperature and pH optima of the enzyme were found to be 40°C and 6.0, respectively (Searle-van Leeuwen et al., 1992; Kauppinen et al., 1995). The enzyme was tested for activity with different substrates, and it was found to be specific for RG-I, showing no activity against acetylated xylan, mannan, beet pectin, triacetin or acetylsalicylic acid (Searle-van Leeuwen et al., 1992; Kauppinen et al., 1995).

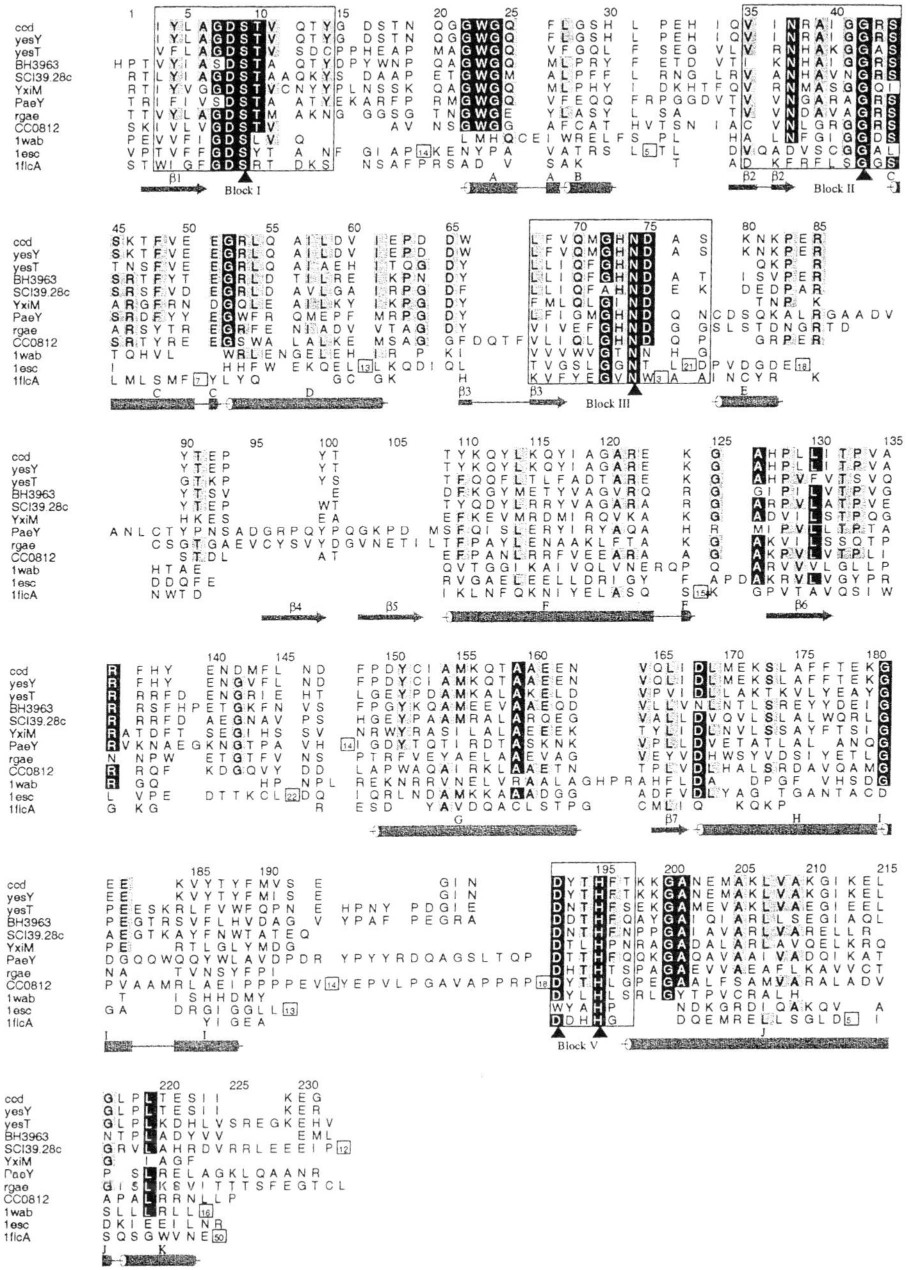

Figure 2. Multiple sequence alignment of the 9 CEF12 enzymes and 3 SGNH hydrolases. In places where residues have been omitted within the alignment, the number of omitted residues is indicated in a box. The numbering above the sequences and the secondary structure annotation correspond to RGAE. The conserved blocks of sequence are boxed and the catalytically important residues are indicated by black triangles.

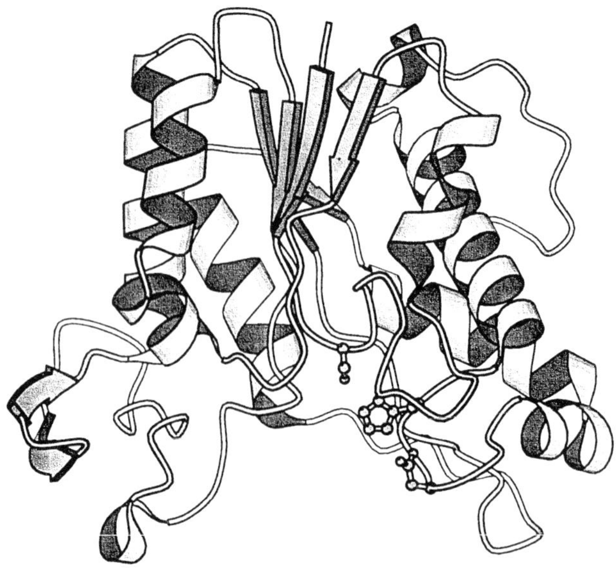

Figure 3. The overall structure of rhamnogalacturonan acetylesterase. The catalytic triad residues are shown in ball-and-stick representation.

3. RGAE - the overall structure

The purified recombinant protein was crystallized in two polymorphic modifications, a trigonal and an orthorhombic crystal form, despite being heterogeneously glycosylated (Mølgaard et al., 1998). The structure was initially determined to 1.55 Å resolution by Multiple Isomorphous Replacement for the orthorhombic crystal form grown with Li^+ or NH_4^+ sulfate as a precipitant (Mølgaard et al., 2000). Later, a synchrotron data set was obtained for the same crystal form, and the structure was refined using data to 1.12 Å resolution (Mølgaard and Larsen, 2002).

The enzyme folds into a three-layer α/β/α structure, with a central five-stranded parallel β sheet sandwiched between α helices (Fig.3). There is an insertion of a two-stranded antiparallel β sheet after the third β strand in the central sheet, as shown in the topology diagram in Fig. 1. The active site is located at a topological switchpoint, where Ser9, situated at the C-terminal end of β strand 1, forms a characteristic catalytic triad with His195 and Asp192 (Fig.1). A sulfate ion is found in the active site, where it connects four different regions of the protein structure through hydrogen bonds (Fig.4). One of its oxygen atoms occupies the oxyanion hole, forming hydrogen bonds to the mainchain amide groups of Ser9 and Gly42 and the sidechain amide of Asn74. These three amino acid residues are located at the carboxy terminal ends of the first three β strands in the central, parallel sheet (Fig.1). Another oxygen atom makes a hydrogen bond to the imidazole ring of His195. The active site is at the bottom of an open cleft making it readily accessible to a large substrate. There are no aromatic residues in the cleft which could make hydrophobic stacking interactions with the polysaccharide chain of the substrate as is often seen in carbohydrate active enzymes, but several arginine residues are positioned in suitable positions to interact with the negatively charged acid groups of the GalUA residues in the substrate. Arginine, aspartate and glutamate residues are not uncommonly found in the binding site for carbohydrates, as they can form bidentate interactions with adjacent hydroxyl groups on the sugar molecules (Taroni et al., 2000).

4. Structurally related proteins

A search for structures homologous to RGAE using the program DALI (Holm et al., 1992; Holm and Sander, 1994) found three structures, all of which are members of the family of hydrolases described by Dalrymple et al. (1997), namely the serine esterase from *Streptomyces scabies* (SsEst) (Wei et al., 1995), the α_1 subunit of the platelet-activating factor acetylhydrolase from *Bos taurus* (PAF-AH) (Ho et al., 1997) and more recently the esterase domain of haemagglutinin-esterase-fusion glycoprotein from Influenza C virus (HEF) (Rosenthal et al., 1998). The alignment of these enzymes with the members of CEF12 in Fig. 2 is based on a structural alignment with RGAE performed by DALI.

The structural alignment shows that in addition to the structural conservation of the central parallel five-stranded β sheet, there are four regions in the structures which also have conserved sequence (Fig.2), although the overall sequence identities are low (between 11-19%). These four regions correspond to blocks I, II, III and V of the five conserved blocks of sequence for a new lipolytic family of enzymes described by Upton and Buckley (1995) (Mølgaard et al., 2000). A structural alignment of the four proteins in the active site region, where these four blocks bring the catalytic residues close together is shown in Fig. 4.

Three of these blocks are also described by Dalrymple et al., 1997 to identify a family of hydrolases, which also has RGAE, SsEst and PAF-AH as members.

4.1. THE SGNH-HYDROLASE FOLD

When the alignment in Fig. 2 is compared with the alignments of the hydrolase families described by Upton and Buckley and Dalrymple et al. there is only one residue that is completely conserved in each of the four blocks; Ser, Gly, Asn and His. These four residues all play a central role in catalysis, as described below. We propose that the conservation of these four residues in four blocks of conserved sequence define a family of enzymes with members from as different organisms as *Bos taurus*, *Bacillus subtilis*, *Sacch. cerevisiae* (isoamyl acetate esterase (Fukuda et al., 1996), not included in Fig. 2) and *Aspergillus aculeatus*, and we found it appropriate to name this family the SGNH-hydrolase family after these four residues (Mølgaard et al., 2000).

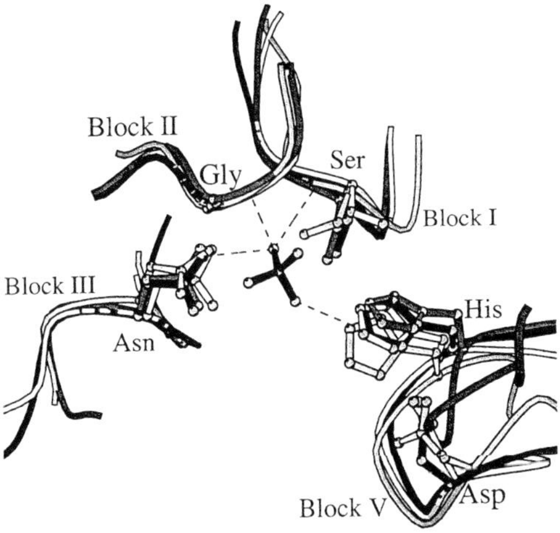

Figure 4. A structural alignment of RGAE (black), PAF-AH (white), SsEst (dark gray) and HEF (light gray) in the active site region, where the four blocks of conserved residues bring the active site residues close together. The sulfate ion in the figure is from the structure of RGAE.

4.1.1. Block I

As shown by Dalrymple et al., 1997, the characteristic sequence motif in this block is GDS, where the nucleophilic serine residue is completely conserved. As opposed to the α/β hydrolases, where the nucleophile is located on a nucleophile elbow motif and has a strained ε conformation, the nucleophilic serine in the SGNH hydrolases (Nu) is located on a type I β turn from Nu-1 to Nu+2 (in RGAE from Asp8 to Met11), with a completely unstrained conformation of the backbone (Fig. 5). The β turn is stabilized by hydrogen bonds from the backbone amides of Nu+1 and Nu+2 to the carbonyl group of Nu-1, leaving only room for a small residue at the Nu-2 position. In RGAE this residue is a glycine, and in all members of the family, this is a small residue. The also highly conserved aspartic acid (Nu-1) stabilizes the most flexible part of the β turn by forming a hydrogen bond between OD1 and the Nu+1 amide group. This gives the aspartic acid residue an unfavorable backbone conformation, which is conserved in the structures of SsEst and PAF-AH (Mølgaard and Larsen, 2002).

4.1.2. Block II

This block, which is structurally conserved in the four SGNH-hydrolase structures, was not mentioned by Dalrymple et al. (1997). There is only one completely conserved residue in this block, a glycine (Gly42 in RGAE), which is involved in hydrogen bonding to the oxyanion hole, through its backbone amide. This conserved glycine is located in the R3 position of a type II β turn, a position for which glycine residues are preferred, to avoid steric hindrance between the β carbon of the R3 residue with the carbonyl oxygen of the previous residue (Chou and Fasman, 1977).

4.1.3. Block III

This block, which is situated at the carboxy terminal end of β strand 3, contains a completely conserved asparagine (Asn74 in RGAE) as part of its fingerprint motif

GXND (Dalrymple et al., 1997). This asparagine donates a hydrogen bond from its sidechain amide to the oxyanion hole. The also highly conserved aspartic acid in this block, (Asp75 in RGAE) has a function similar to the aspartic acid in Block I, namely to stabilize the β turn by forming a hydrogen bond with the mainchain amide of the highly conserved glycine in the GXND motif, and it also participates in hydrogen bonds with residues from other parts of the structure, in RGAE with Asp87 and Ala45.

4.1.4. Block V

The fingerprint motif for this block is DXXHP (Dalrymple et al., 1997). The only completely conserved residue in this block is a histidine (in RGAE His195), which is the histidine member of the catalytic triad. The third member of the triad is located three residues upstream from the histidine, (Asp192 in RGAE), but is lacking in some members of the family. The serine esterase from *Streptomyces scabies* (SsEst) was indeed the first enzyme to be shown to have a catalytic dyad of serine and histidine, lacking the acid member of the conventional catalytic triad (Wei et al., 1995). The orientation of the histidine in SsEst is maintained by a hydrogen bond between the ND-H of the histidine and a mainchain oxygen. This can be clearly seen in Fig. 4, where the mainchain of SsEst takes the place of the sidechain oxygen of the aspartic acid in the catalytic triad in the three other structures. The *Vibrio parahaemolyticus* haemolysin (Taniguchi et al., 1986) and the *Streptomyces deastochromogenes* esterase (Tesch et al., 1996) also lack the block V aspartate, and could be expected to possess a catalytic dyad similar to that of SsEst.

5. Differences between the α/β hydrolase fold and the SGNH-hydrolase fold

5.1. THE NUCLEOPHILE

The most conserved feature of the α/β hydrolase fold is the nucleophile elbow motif. This very sharp bend, which connects β5 and the conserved α helix C gives the backbone of the nucleophile a strained ε conformation. In the Ramachandran plot shown in Fig. 5, one representative structure has been chosen from each of the 27 structural families defined in the ESTHER database of α/β hydrolases (see Table 1), and their nucleophile residues have been plotted using the program PROCHECK (Laskowski et al., 1993). Where there was more than one member in a family, a structure determined at a resolution of 2.0 Å or better was chosen. Also included in the plot are the nucleophilic residues of the SGNH hydrolases RGAE, SsEst, PAF-AH and HEF. It can clearly be seen that the nucleophiles of the α/β hydrolases cluster quite close together in the ε region of the Ramachandran plot, around $(\varphi,\psi) = (58°,-115°)$, whereas the nucleophiles of the SGNH hydrolases are found in the unstrained α helical region of the plot (-60°,-48°). The nucleophiles of the SGNH hydrolases RGAE, SsEst and PAF-AH are quite close together in the plot, whereas the nucleophile of HEF (1flc) seems to be an outlier, though it is still in the same general region of the plot. However, the absolute position of the nucleophile of HEF in this plot should be regarded with caution, as the structure was refined at 3.2 Å resolution, and the overall quality of the Ramachandran plot for this structure is poor with 11% outliers, according to the definition of Kleywegt and Jones, (1996). (A normal, better than 2.0 Å structure is expected to have less than 5% outliers.) This gives reason to expect that a higher resolution structure of HEF would be more similar to RGAE (1.55 Å), SsEst (2.10 Å) and PAF-AH (1.70 Å).

TABLE I. The α/β hydrolase structures in the Ramachandran plot

PDB	Protein name (resolution)
1ea5	*Torpedo californica* acetylcholinesterase (Harel et al., 2000) (1.80 Å)
1bs9	*Penicillium purpurogenum* Acetyl xylan esterase (Ghosh et al., 1999) (1.10 Å)
1hlg	Human gastric lipase (Roussel et al., 1999) (3.00 Å)
1dqy	*Mycobacterium tuberculosis* Antigen85c (Ronning et al., 2000) (1.83 Å)
1qlw	Bacterial Carboxylesterase (Bournc et al., 2000) (1.10 Å)
1qge	*Pseudomonas glumae* lipase (Lang et al., 1999) (1.70 Å)
1c4x	*Rhodococcus Sp.* Hydrolase Bphd Enzyme (Nandhagopal et al., 1997) (2.40 Å)
1qe3	*Bacillus subtilis* Para-Nitrobenzyl esterase (Spiller et al., 1999) (1.50 Å)
1wht	Wheat serine carboxypeptidase II (Bullock et al., 1994) (2.00 Å)
2bce	Bovine cholesterol esterase (Chen et al., 1998) (1.60 Å)
1agy	*Fusarium solani* cutinase (Nicolas et al., 1997) (1.15 Å)
1din	*Pseudomonas sp.* Dienelactone hydrolase (Pathak and Ollis, 1990) (1.80 Å)
1qo7	*Aspergillus niger* Epoxide hydrolase (Zou et al., 2000) (1.80 Å)
1cle	*Candida cylindracea* Cholesterol esterase (Ghosh et al., 1995) (2.00 Å)
1tca	*Candida antarctica* triacylglycerol lipase (Uppenberg et al., 1994) (1.55 Å)
1b6g	*Xanthobacter Autotrophicus* Haloalkane dehalogenase (Ridder et al., 1999) (1.15 Å)
1a7u	*Streptomyces aureofaciens* Chloroperoxidase T (Hofmann et al., 1998) (1.75 Å)
1jkm	*Bacillus subtilis* Brefeldin A esterase (Wei et al., 1999) (1.85 Å)
1qj4	*Hevea Brasiliensis* Hydroxynitrile lyase (Gruber et al., 1999) (1.10 Å)
1i6w	*Bacillus subtilis* lipase (Pouderoyen et al., 2001) (1.50 Å)
1fj2	Human acyl protein thioesterase 1 (Devedjiev et al., 2000) (1.50 Å)
1jfr	*Streptomyces exfoliatus* lipase (Wei et al., 1998) (1.90 Å)
1ei9	Bovine Palmitoyl thioesterase (Bellizzi III et al., 2000) (2.25 Å)
1bu8	Rat pancreatic lipase-related protein 2 (Roussel et al., 1998) (1.80 Å)
1c5t	*Sus scrofa* (mutant) Prolyl oligopeptidase (Fulop et al., 2000) (1.70 Å)
1qtr	*Serrata marcescens* Prolyl aminopeptidase (Yoshimoto et al., 1999) (2.32 Å)
1tht	*Vibrio harveyi* Myristoyl-ACP-specific thioesterase (Lawson et al., 1994) (2.10 Å)

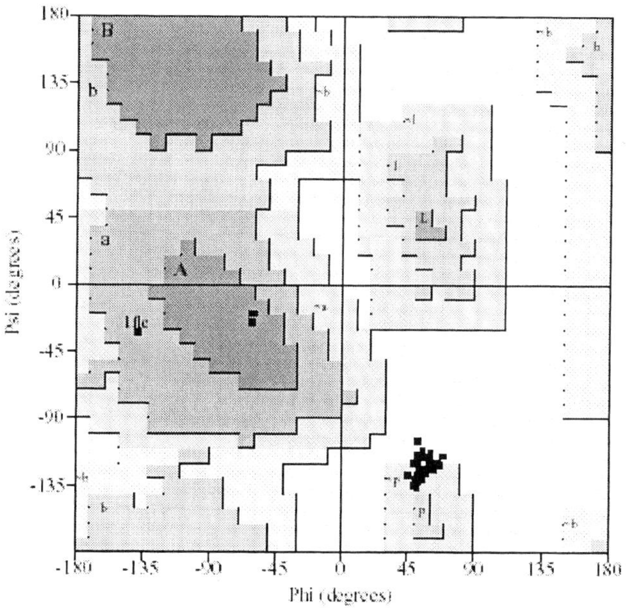

Figure 5. A Ramachandran plot showing the backbone conformation of the nucleophile residues in 27 representative α/β hydrolases and in RGAE (1deo), SsEst (1esc), PAF-AH (1wab) and HEF (1flc).

5.2. THE LOCATION OF THE CATALYTIC TRIAD RESIDUES

Similar to the α/β hydrolases, the SGNH hydrolases have a catalytic triad in the sequence order nucleophile-acid-histidine. However, the location of these residues with respect to the topology is different from that of the α/β hydrolases. In the α/β hydrolases, the nucleophile is invariably located after β strand 5, the acid is most often found after β strand 7, and the histidine is in a loop region following β strand 8. In the SGNH hydrolases, the nucleophile is on the other side of the topological switchpoint, on what corresponds to β strand 5, and the acid and histidine are placed in a loop region following β strand 7 separated by only two amino acid residues. In all the α/β hydrolase enzymes, the spacing between the acid and the histidine is much larger.

This different location of the catalytic triad residues leads to a different relative orientation of these residues with respect to the central parallel β sheet. In the α/β hydrolases, they align parallel to the central β sheet, but in the SGNH hydrolases, the alignment with the central sheet is almost perpendicular (Mølgaard et al., 2000).

6. Conclusion

The structure of rhamnogalacturonan acetylesterase from *Aspergillus aculeatus* was determined, and through a comparison with the structurally related enzymes platelet-activating factor acetylhydrolase from *Bos taurus*, *Streptomyces scabies* serine esterase and haemagglutinin-esterase-fusion glycoprotein from Influenza C virus, a new fold, different from the α/β hydrolase fold, was characterized. The four enzymes are members of a family of proteins which have four blocks of conserved sequence, with one completely conserved residue in each block, a serine, glycine, asparagine and histidine, respectively. These four blocks of residues have been shown to be also structurally conserved, and we expect the proteins within this family to have a common active site and a common fold. This family is named the SGNH-hydrolase family after the four conserved amino acid residues.

7. Acknowledgements

The author would like to thank Prof. Sine Larsen, Dr Leila Lo Leggio and Dr Renuka Kadirvelraj for critical reading of the manuscript. Support from the Danish National Research Foundation is gratefully acknowledged.

References

Albersheim, P., A. G. Darvill, M. A. O'Neill, H. A. Schols, and A. G. J. Voragen: 1996, `A Hypothesis: The Same Six Polysaccharides are Components of the Primary Cell Walls of All Higher Plants'. In: J. Visser and A. G. J. Voragen (eds.): *Pectins and Pectinases*, Vol. 14 of *Progress in Biotechnology*. Elsevier, pp. 47-55.

Barton, G. J.: 1990, `Protein Multiple Sequence Alignment and Flexible Pattern Matching'. *Methods Enzymol.* **183**, 403-428.

Barton, G. J. and M. J. E. Sternberg: 1987, `A strategy for the rapid multiple alignment of protein sequences. Confidence levels from tertiary structure comparisons.'. *J. Mol. Biol.* **198**, 327-337.

Bellizzi III, J. J., J. Widom, C. Kemp, J. Y. Lu, A. K. Das, S. L. Hofmann, and J. Clardy: 2000, `The Crystal Structure of Palmitoyl Protein Thioesterase 1 and the Molecular Basis of Infantile Neuronal Ceroid Lipofuscinosis'. *Proc.Natl.Acad.Sci.USA* **97**, 4573-4578.

Bolvig, P. U., M. Pauly, C. Orfila, H. V. Scheller, and K. Schnorr: 2002, `Sequence analysis and characterisation of a novel pectin acetyl esterase from Bacillus subtilis'. In: J. Visser and A. G. J. Voragen (eds.): *Pectins and Pectinases*. Dordrecht: Kluwer Academic Publishers.

Bourne, P. C., M. N. Isupov, and J. A. Littlechild: 2000, `The Atomic Resolution Structure of a Novel Bacterial Esterase'. *Structure Fold Des.* **8**, 143-151.

Bullock, T. L., B. Branchaud, and S. J. Remington: 1994, `Structure of the complex of L-benzylsuccinate with wheat serine carboxypeptidase II at 2.0 Å resolution'. *Biochemistry* **33**, 11127-11134.

Chen, J. C., L. J. Miercke, J. Krucinski, J. R. Starr, G. Saenz, X. Wang, C. A. Spilburg, L. G. Lange, J. L. Ellsworth, and R. M. Stroud: 1998, `Structure of bovine pancreatic cholesterol esterase at 1.6 Å: novel structural features involved in lipase activation'. *Biochemistry* **37**, 5107-5117.

Choi, D., Y. Kim, I. Chung, S. Lee, S. Kang, and K. Han: 2000, `Gene cloning and expression of a cephalosporin C deacetylase from *Bacillus sp. KCCM-10143* '. unpublished.

Chou, P. Y. and G. D. Fasman: 1977, `Beta-turns in proteins.'. *J. Mol. Biol.* **115**, 135-175.

Cousin, X., T. Hotelier, K. Giles, J. P. Toutant, and A. Chatonnet: 1998, `aCHEdb: the database system for ESTHER, the α/β fold family of proteins and the Cholinesterase gene server'. *Nuc. Acids Res.* **26**, 226-228.

Coutinho, P. M. and B. Henrissat: 2000, `Carbohydrate-active enzymes: an integrated database approach'. In: B. H. H. J. Gilbert, G. Davies and B. Svensson (eds.): *Recent Advances in Carbohydrate Bioengineering*. Cambridge: The Royal Society of Chemistry, pp. 3-12.

Dalrymple, B. P., D. H. Cybinski, I. Layton, C. S. McSweeney, G.-P. Xue, Y. J. Swadling, and J. B. Lowry: 1997, `Three Neocallimastix patriciarum esterases associated with the degradation of complex polysaccharides are members of a new family of hydrolases.'. *Microbiology* **143**, 2605-2614.

Derewenda, Z. and U. Derewenda: 1991, `Relationships among serine hydrolases: evidence for a common structural motif in triacylglyceride lipases and esterases.'. *Biochem. Cell Biol.* **69**, 842-851.

Devedjiev, Y., Z. Dauter, S. Kuznetsov, T. Jones, and Z. Derewenda: 2000, 'Crystal Structure of the Human Acyl Protein Thioesterase I from a Single X-Ray Data Set to 1.5 Å'. *Structure Fold Des.* **8**, 1137-1146.

Fukuda, K., O. Kuwahata, Y. Kiyokawa, T. Yanagiuchi, Y. Wakai, K. Kitamoto, Y. Inoue, and A. Kimura: 1996, 'Molecular cloning and nucleotide sequence of the isoamyl acetate-hydrolyzing esterase gene (EST2) from *Saccharomyces cerevisiae*'. *J.Ferment.Bioeng.* **82**, 8-15.

Fulop, V., Z. Szeltner, and L. Polgar: 2000, 'Catalysis of Serine Oligopeptidases is Controlled by a Gating Filter Mechanism'. *Embo Reports* **1**, 277.

Ghosh, D., M. Erman, M. Sawicki, P. Lala, D. R. Weeks, N. Li, W. Pangborn, D. J. Thiel, J. H., R. Gutierrez, and J. Eyzaguirre: 1999, 'Determination of a Protein Structure by Iodination: The Structure of Iodinated Acetylxylan Esterase'. *Acta Crystallogr. D* **55**, 779-784.

Ghosh, D., Z. Wawrzak, V. Z. Pletnev, N. Li, R. Kaiser, W. Pangborn, H. Jornvall, M. Erman, and W. L. Duax: 1995, 'Structure of uncomplexed and linoleate-bound *Candida cylindracea* cholesterol esterase.'. *Structure* **3**, 279-288.

Gruber, K., M. Gugganig, U. G. Wagner, and C. Kratky: 1999, 'Atomic resolution crystal structure of hydroxynitrile lyase from *Hevea brasiliensis*'. *Biol.Chem.* **380**, 993-1000.

Harel, M., M. Weik, I. Silman, and J. L. Sussman: 2000, 'Native Acetylcholinesterase (E.C. 3.1.1.7) From *Torpedo californica* at 1.8 Å Resolution'. To be published.

Heikinheimo, P., A. Goldman, C. Jeffries, and D. L. Ollis: 1999, 'Of barn owls and bankers: a lush variety of α/β hydrolases'. *Structure Fold Des.* **7**, R141-R146.

Ho, Y. S., L. Swenson, U. Derewenda, L. Serre, Y. Wei, Z. Dauter, M. Hattori, T. Adachi, J. Aoki, H. Arai, K. Inoue, and Z. S. Derewenda: 1997, 'Brain acetylhydrolase that inactivates platelet-activating factor is a G-protein-like trimer'. *Nature* **385**, 89-93.

Hofmann, B., S. Tolzer, I. Pelletier, J. Altenbuchner, K. H. van Pee, and H. J. Hecht: 1998, 'Structural investigation of the cofactor-free chloroperoxidases'. *J. Mol. Biol.* **279**, 889-900.

Holm, L., C. Ouzounis, C. Sander, G. Tuparev, and G. Vriend: 1992, 'A database of protein structure families with common folding motifs'. *Protein Science* **1**, 1691-1698.

Holm, L. and C. Sander: 1994, 'The FSSP database of structurally aligned protein fold families.'. *Nucleic Acids Res.* **22**, 3600-3609.

Ishii, T.: 1997, 'O-acetylated oligosaccharides from pectins of potato tuber cell walls'. *Plant Physiol.* **113**, 1265-1272.

Jenkins, J., O. Mayans, D. Smith, K. Worboys, and R. W. Pickersgill: 2001, 'Three-dimensional structure of *Erwinia chrysanthemi* pectin methylesterase reveals a novel esterase active site'. *J. Mol. Biol.* **305**, 951-960.

Kauppinen, S., S. Christgau, L. V. Kofod, T. Halkier, K. Dörreich, and H. Dalbøge: 1995, 'Molecular Cloning and Characterization of a Rhamnogalacturonan Acetylesterase from *Aspergillus aculeatus*'. *J.Biol.Chem.* **270**, 27172-27178.

Kleywegt, G. J. and T. A. Jones: 1996, 'Phi/Psi-chology: Ramachandran revisited'. *Structure* **4**, 1395-1400.

Kofod, L. V., S. Kauppinen, S. Christgau, L. N. Andersen, H. P. Heldt-Hansen, K. Dörreich, and H. Dalbøge: 1994, 'Cloning and characterization of two structurally and functionally divergent rhamnogalacturonases from *Aspergillus aculeatus*'. *J.Biol.Chem.* **269**, 29182-29189.

Kouwijzer, M., H. Schols, and S. P'erez: 1996, 'Acetylation of rhamnogalacturonan I and homogalacturonan: Theoretical calculations'. In: J. Visser and A. G. J. Voragen (eds.): *Pectins and Pectinases*, Vol. 14 of Progress in Biotechnology. Amsterdam: Elsevier, pp. 57-65.

Lang, D. A., P. Stadler, A. Kovacs, F. Paltauf, and B. W. Dijkstra: 1999, 'Structural and Kinetic Investigations of Enantiomeric Binding Mode of Subclass I Lipases from the Family of Pseudomonadaceae'. To be published.

Laskowski, R. A., M. W. MacArthur, D. S. Moss, and J. M. Thornton: 1993, 'Procheck - a program to check the stereochemical quality of protein structures'. *J.Appl.Cryst.* **26**, 283-291.

Lawson, D. M., U. Derewenda, L. Serre, S. Ferri, R. Szittner, Y. Wei, E. A. Meighen, and Z. S. Derewenda: 1994, 'Structure of a myristoyl-ACP-specific thioesterase from *Vibrio harveyi*'. *Biochemistry* **33**, 9382-9388.

Lietzke, S. E., R. D. Scavetta, M. D. Yoder, and F. Jurnak: 1996, 'The refined three-dimensional structure of pectate lyase E from *Erwinia chrysanthemi* at 2.2 Å resolution'. *Plant Physiol.* **111**, 73-92.

Mayans, O., M. Scott, I. Connerton, T. Gravesen, J. Benen, J. Visser, R. Pickersgill, and J. Jenkins: 1997, 'Two crystal structures of pectin lyase A from Aspergillus reveal a pH driven conformational change and striking divergence in the substrate-binding clefts of pectin and pectate lyases'. *Structure* **5**, 677-689.

Mølgaard, A., S. Kauppinen, and S. Larsen: 2000, 'Rhamnogalacturonan acetylesterase elucidates the structure and function of a new family of hydrolases'. *Structure Fold Des.* **8**, 373-383.

Mølgaard, A. and S. Larsen: 2002, 'A branched N-linked glycan at atomic resolution in the 1.12 Å structure of rhamnogalacturonan acetylesterase'. *Acta Crystallogr. D* **58**, 111-119.

Mølgaard, A., J. F. W. Petersen, S. Kauppinen, H. Dalbøge, A. H. Johnsen, J.-C. N. Poulsen, and S. Larsen: 1998, 'Crystallization and preliminary X-ray diffraction studies of the heterogeneously glycosylated enzyme rhamnogalacturonan acetylesterase from *Aspergillus aculeatus*'. *Acta Crystallogr. D* **54**, 1026-1029.

Mutter, M., I. J. Colquhoun, H. A. Schols, G. Beldman, and A. G. Voragen: 1996, 'Rhamnogalacturonase B from *Aspergillus aculeatus* is a rhamnogalacturonan α-L-rhamnopyranosyl-(1-4)-α-D-galactopyranosyluronide lyase.'. *Plant Physiol.* **110**, 73-79.

Nandhagopal, N., T. Senda, T. Hatta, A. Yamada, E. Masai, M. Fukuda, and Y. Mitsui: 1997, 'Three-Dimensional Structure of Microbial 2-Hydroxyl-6-Oxo-6-Phenylhexa-2,4- Dienoic Acid (Hpda) Hydrolase (Bphd Enzyme) from *Rhodococcus Sp.* Strain Rha1, in the Pcb Degradation Pathway'. *Proc.Jpn.Acad.Ser.B* **73**, 154.

Nardini, M. and B. W. Dijkstra: 1999, 'α/β Hydrolase fold enzymes: the family keeps growing'. *Curr.Opin.Struct.Biol.* **9**, 732-737.

Nicolas, A., C. Martinez, and C. Cambillau: 1997, 'The 1.15 Angstrom Refined Structure of Fsp Cutinase Compared to Other Members of Alpha/Beta Hydrolase Fold Family'. To be published.

Ollis, D. L., E. Cheah, M. Cygler, B. Dijkstra, F. Frolow, S. M. Franken, M. Harel, S. J. Remington, I. Silman, J. Schrag, J. L. Sussman, K. H. G. Verschueren, and A. Goldman: 1992, 'The α/β hydrolase fold'. *Protein Eng.* **5**, 197-211.

Pathak, D. and D. Ollis: 1990, 'Refined structure of dienelactone hydrolase at 1.8 Å'. *J. Mol. Biol.* **214**, 497-525.

Petersen, T. N., S. Kauppinen, and S. Larsen: 1997, 'The crystal structure of rhamnogalacturonase A from *Aspergillus aculeatus*: a right-handed parallel β helix'. *Structure* **5**, 533-544.

Pickersgill, R., J. Jenkins, G. Harris, W. Nasser, and J. Robert-Baudouy: 1994, 'The structure of *Bacillus subtilis* pectate lyase in complex with calcium'. *Nat. Struct. Biol.* **1**, 717-723.

Pickersgill, R., D. Smith, K. Worboys, and J. Jenkins: 1998, 'Crystal structure of Polygalacturonase from *Erwinia carotovora ssp. carotovora*'. *J.Biol.Chem.* **273**, 24660-24664.

Pouderoyen, G. v., T. Eggert, K.-E. Jaeger, and B. W. Dijkstra: 2001, 'The Crystal Structure of *Bacillus subtilis* Lipase: A Minimal α/β Hydrolase Fold Enzyme'. *J. Mol. Biol.* **309**, 215-226.

Ridder, I. S., H. J. Rozeboom, and B. W. Dijkstra: 1999, 'Haloalkane Dehalogenase from *Xanthobacter Autotrophicus* Gj10 Refined at 1.15 Å'. *Acta Crystallogr. D* **55**, 1273-1290.

Ronning, D. R., T. Klabunde, G. S. Besra, V. D. Vissa, J. T. Belisle, and J. C. Sacchettini: 2000, 'Crystal Structure of the Secreted Form of Antigen 85C Reveals Potential Targets for Mycobacterial Drugs and Vaccines'. *Nat. Struct. Biol.* **7**, 141-146.

Rosenthal, P. B., X. Zhang, F. Formanowski, W. Fitz, C.-H. Wong, H. Meier-Ewert, J. J. Skehel, and D. C. Wiley: 1998, 'Structure of the haemagglutinin-esterase-fusion glycoprotein of influenza C virus'. *Nature* **396**, 92-96.

Roussel, A., S. Canaan, M. P. Egloff, M. Riviere, L. Dupuis, R. Verger, and C. Cambillau: 1999, 'Crystal Structure of Human Gastric Lipase and Model of Lysosomal Acid Lipase, Two Lipolytic Enzymes of Medical Interest'. *J.Biol.Chem.* **274**, 16995-17002.

Roussel, A., Y. Yang, F. Ferrato, R. Verger, C. Cambillau, and M. Lowe: 1998, 'Structure and activity of rat pancreatic lipase-related protein 2.'. *J.Biol.Chem.* **273**, 32121-32128.

Ryttersgaard, C., L. Lo Leggio, P. M. Coutinho, B. Henrissat, and S. Larsen: 2002, '*Aspergillus aculeatus* β-1,4-galactanase: substrate recognition and relations to other glycoside hydrolases in clan GH-A'. submitted.

Schols, H. A., C. C. J. M. Geraeds, M. F. Searle-van Leeuwen, F. J. M. Kormelink, and A. G. J. Voragen: 1990, 'Rhamnogalacturonase: a novel enzyme that degrades the hairy regions of pectin'. *Carbohydr.Res.* **206**, 105-115.

Searle-van Leeuwen, M. J. F., L. A. M. van den Broek, H. A. Schols, G. Beldman, and A. G. J. Voragen: 1992, 'Rhamnogalacuronan acetylesterase: a novel enzyme from *Aspergillus aculeatus*, specific for the deacetylation of hairy (ramified) regions of pectins'. *Appl. Microbiol.Biotechnol.* **38**, 347-349.

Searle-van Leeuwen, M. J. F., J.-P. Vincken, D. Schipper, A. G. J. Voragen, and G. Beldman: 1996, 'Acetyl esterases of *Aspergillus niger*: purification and mode of action on pectins'. In: J. Visser and A. Voragen (eds.): *Pectins and pectinases*, Vol. 14 of *Progress in Biotechnology*. Amsterdam: Elsevier, pp. 793-798.

Shevchik, V. E. and N. Hugouvieux-Cotte-Pattat: 1997, 'Identification of a bacterial pectin acetyl esterase in *Erwinia chrysanthemi*'. *Mol.Microbiol.* **24**, 1285-1301.

Spiller, B., A. Gershenson, F. H. Arnold, and R. C. Stevens: 1999, 'A Structural View of Evolutionary Divergence'. *Proc.Natl.Acad.Sci.USA* **96**, 12305-12310.

Taniguchi, H., H. Hirano, S. Kubomura, K. Higashi, and Y. Mizuguchi: 1986, `Comparison of the nucleotide sequences of the genes for the thermostable direct hemolysin and the thermolabile hemolysin from *Vibrio parahaemolyticus*'. *Microb.Pathog.* **1**, 425-432.

Taroni, C., S. Jones, and J. M. Thornton: 2000, `Analysis and prediction of carbohydrate binding sites'. *Protein Eng.* **13**, 89-98.

Tesch, C., K. Nikoleit, V. Gnau, F. Gotz, and C. Bormann: 1996, `Biochemical and molecular characterization of the extracellular esterase from *Streptomyces diastatochromogenes*'. *J.Bacteriol.* **178**, 1858-1865.

Uppenberg, J., M. T. Hansen, S. Patkar, and T. A. Jones: 1994, `The sequence, crystal structure determination and refinement of two crystal forms of lipase B from *Candida antarctica*'. *Structure* **2**, 293-308.

Upton, C. and J. T. Buckley: 1995, `A new family of lipolytic enzymes?'. *Trends Biochem.Sci.* **20**, 178-179.

van Santen, Y., J. A. E. Benen, K. H. Schroter, K. H. Kalk, S. Armand, J. Visser, and B. W. Dijkstra: 1999, `1.68 Å crystal structure of endopolygalacturonase II from *Aspergillus niger* and identification of active site residues by site directed mutagenesis'. *J.Biol.Chem.* **274**, 30474-30480.

Vitali, J., B. Schick, H. C. M. Kester, J. Visser, and F. Jurnak: 1998, `The three-dimensional structure of *Aspergillus niger* pectin lyase B. at 1.7 Å resolution'. *Plant Physiol.* **116**, 69-80.

Wei, Y., J. A. Contreras, P. Sheffield, T. Osterlund, U. Derewenda, R. E. Kneusel, U. Matern, C. Holm, and Z. S. Derewenda: 1999, `Crystal structure of brefeldin A esterase, a bacterial homolog of the mammalian hormone-sensitive lipase'. *Nat. Struct. Biol.* **6**, 340-345.

Wei, Y., J. L. Schottel, U. Derewenda, L. Swenson, S. Patkar, and Z. S. Derewenda: 1995, `A novel variant of the catalytic triad in the *Streptomyces scabies* esterase.'. *Nat. Struct. Biol.* **2**, 218-223.

Wei, Y., L. Swenson, C. Castro, U. Derewenda, W. Minor, H. Arai, J. Aoki, K. Inoue, L. Servin-Gonzalez, and Z. S. Derewenda: 1998, `Structure of a microbial homologue of mammalian platelet-activating factor acetylhydrolases: *Streptomyces exfoliatus* lipase at 1.9 Å resolution'. *Structure* **6**, 511-519.

Yoder, M. D., N. T. Keen, and F. Jurnak: 1993, `New domain motif - the structure of pectate lyase C, a secreted plant virulence factor'. *Science* **260**, 1503-1507.

Yoshimoto, T., T. Kabashima, K. Uchikawa, T. Inoue, N. Tanaka, K. T. Nakamura, M. Tsuru, and K. Ito: 1999, `Crystal Structure of the Prolyl Aminopeptidase from *Serratia Marcescens*'. *J.Biochem.(Tokyo)* **126**, 559-565.

Zou, J.-Y., B. M. Hallberg, T. Bergfors, F. Oesch, M. Arand, S. L. Mowbray, and T. A. Jones: 2000, `Structure of *Aspergillus Niger* Epoxide Hydrolase at 1.8 Å Resolution: Implications for the Structure and Function of the Mammalian Microsomal Class of Epoxide Hydrolases'. *Structure Fold Des.* **8**, 111-122.

SEQUENCE ANALYSIS AND CHARACTERISATION OF A NOVEL PECTIN ACETYL ESTERASE FROM *BACILLUS SUBTILIS*

PERNILLE ULDALL BOLVIG[1], MARKUS PAULY[2,*], CAROLINE ORFILA[2], HENRIK VIBE SCHELLER[2], AND KIRK SCHNORR[1]
[1]Fungal Screening, Novozymes A/S, Smoermosevej 25, Building 1B1.05, DK-2880 Bagsværd, Denmark, [2]Dept.of Plant Biology, Royal Vet. and Agric. Univ., 40 Thorvaldsensvej, DK-1871, Frederiksberg C, Denmark
E-mail: PUS@novozymes.com

Abstract

A recombinant pectin acetyl esterase (YxiM) from *Bacillus subtilis* was characterised and the enzymatic properties compared to previously characterised pectin acetyl esterases (PAE's). The YxiM protein had a pH optimum of 8.0 and was stable above pH 4.0. The enzyme released acetate from pectin isolated from different sources and from various synthetic compounds. In addition, YxiM showed a preference for acetylated tobacco homogalacturonan oligomers. Determination of the kinetic constants of YxiM indicated that the enzyme had a higher affinity to acetylated substrates, but a lower specific activity towards these substrates than PaeY from *Erwinia chrysanthemi.* Pre-treatment of sugar beet pectin with pectin methyl esterase (PME) from *Aspergillus aculeatus* increased the ability of YxiM to release acetate and similarly, pre-treatment with YxiM increased the ability of PME to release methanol. This demonstrates that YxiM acts in synergy with PME on the modification of plant cell wall pectin.

1. Introduction

Pectins belong to a very complex class of polysaccharides [1]. Two structurally different regions can be distinguished in pectins: a "smooth" homogalacturonan (HG) region of α-(1,4) linked D-galacturonic acid residues, and a highly branched "hairy" or ramnified rhamnogalacturonan I (RG-I) region composed of a backbone of alternating α-(1,2) linked L-rhamnose and galacturonic acid residues [2-4]. RG-I may carry complex side chains such as arabinans, galactans and arabinogalactans primarily attached to the C-4 of the rhamnose residues [3,5-6]. The HG regions are generally devoid of side-chains. However, certain highly conserved structures known as rhamnogalacturonan II (RG-II) carry complex side-chains attached to a HG backbone.

The carboxyl groups of galacturonic acid can be methylesterified to various degrees and the 2- and 3- hydroxyl groups can be partially *O*-acetylated [2,8]. Pectins isolated from

[*] *Present address: Max-Planck Institute for Molecular Plant Physiology, Am Mühlenberg 1, 14476 Golm, Germany*

F. Voragen et al. (eds.), Advances in Pectin and Pectinase Research, 315-330.
© 2003 *Kluwer Academic Publishers. Printed in the Netherlands.*

different sources shows much variation in acetyl content. For example, pectins isolated from apple and citrus fruits have a low acetate content, whereas pectins isolated from potato tuber and sugar beet have a high acetyl content [1].

An important property of pectins is their ability to form gels in the presence of Ca^{2+} or in the presence of sugar and acid. Pectin is therefore used as a gelling agent in many different industrial applications. All of the above mentioned substitutions influence the gelling properties of pectins [9], and consequently their use in the food industry.

Enzymes can be used to specifically modify the structure and functionality of pectins. Pectin methyl esterase (PME) is already widely used to adjust the gelling properties of commercially available pectins. The acetyl esters also strongly affect the gelling properties of pectin [9,10]. However, only a few pectin acetyl esterases (PAE's) have been identified and characterised, and this limits the possibilities for specifically modifying the acetylation of commercial pectins. A rhamnogalacturonan acetyl esterase (RGAE) has been cloned from *Aspergillus aculeatus* [11] and it was shown to act in synergy with other hydrolytic pectic enzymes on rhamnogalacturonan. RGAE posses a specific activity on the removal of the acetyl groups on the hairy region of pectin [11-13]. Likewise the plant pathogen *Erwinia chrysanthemi* 3937 was shown to produce a range of pectic enzymes and recently a pectin acetyl esterase (PaeY) was found and characterised [14]. PaeY preferentially deacetylates the smooth regions of pectin.

Shevchik and Hugouvieux-Cotte-Pattat [14] compared the amino acid sequences of PaeY and other PAE's. Low sequence identity (25%) between PaeY and RGAE was found along the entire length of the protein. The most conserved region consist of 42 residues and shows 47% identity and 61% sequence similarity [14]. It was concluded that PaeY and RGAE define a new class of enzymes because these two proteins lack sequence similarity with other known pectinolytic enzymes or esterases.

Recently, Mølgaard *et al* [15] determined the crystal structure of RGAE. They found that RGAE folds into an $\alpha/\beta/\alpha$ structure. The active site is an open cleft containing a serine-histidine-aspartic acid catalytic triad. Ser9 forms the characteristic hydrogen-bonding pattern of the catalytic triad with the nearby His195 and Asp192. A sulphate ion is bound with one of its oxygen atoms occupying the oxyanion hole making hydrogen bonds to the backbone amides of Ser9, Gly42 and the side chain nitrogen in Asn74 [15].

Many proteins of unknown function show sequence homology to the existing characterised PAE's, and are therefore classified as possible PAE's in family 12 of the carbohydrate esterase families defined by Coutinho and Henrissat [16].

In this study the cloned, expressed and purified recombinant YxiM protein from *Bacillus subtilis,* classified as a member of family 12 of the carbohydrate esterases, was characterised. Like PaeY this enzyme was active on the smooth regions of pectin. The substrate specificity, pH optimum and stability of the enzyme will be discussed and compared to the previously characterised pectin acetyl esterases.

2. Materials and methods

Materials: Recombinant YxiM protein from *Bacillus subtilis* was cloned, expressed and purified as described in the patent: US 6,184,028 B1, inventors: Michael D. Thomas and Kimberly M. Brown [17]. Well-characterised pectin from sugar beet, potato, and apple were a kind gift from CP Kelco, Lille Skensved, Denmark, and tobacco homogalacturonan and rhamnogalacturonan [18] were kindly provided by M.A. O'Neill (Complex Carbohydrate Research Center and Dept. of Biochemistry and Molecular Biology, University of Georgia, Athens, GA, USA). Xyloglucan oligosaccharides derived from the extracellular polysaccharides of bean and tomato suspension cultured cells [19,20] were provided by William S.York, also Complex Carbohydrate Research Center. The plasmid pVY47 that was used for overproduction of PaeY was a kind gift from Nicole Hugouvieux-Cotte-Pattat *(Unité de Microbiologie et Génétique Composante INSA, Villeurbanne Cedex, France).*

Analytical methods: Uronic acid content was determined by the *m*-hydroxybiphenyl method [21]. Total carbohydrate content was determined by the phenol-sulphuric acid method [22]. Hydrolysis of HG and RG I preparations was in 2 M TFA for 2 h, at 120°C. Monosaccharide composition was determined by separation of monosaccharides by high performance anion exchange chromatography (HPAEC), by isocratic elution with dH_2O on a Dionex PA-10 column (4 mm), at a rate of 1.5 ml/min. Post-column addition of 0.125 M NaOH (0.5 ml/min) ensured pulsed amperometric detection (PAD). Acetic acid was determined with an acetic acid kit (Boehringer Mannheim) adapted to microtiter plates. Methanol was determined according to a method by Klavons and Bennett [23], adapted to microtiter plates.

Acetyl esterase plate assay: 1% agarose in 100 mM Tris-HCl pH 7.0 was melted and cooled to 60°C. Then Brilliant Green (4 μg/ml) and triacetin (50 mM) were added. Holes were punched in the agarose and 20 μl YxiM sample (1.9 mg/ml) was applied. A control sample of 20 μl *Bacillus subtilis* sample broth with empty vector was included. The plates were incubated for 4-6 h at 37°C. Acetyl esterase activity was seen by the appearance of a green circle around the holes.

pH optimum assay: The assay mixture (240 μl) containing 160 μl 100 mM citrate/phosphate buffer of varying pH, 80 μl substrate (1 mM *p*-nitrophenyl-acetate (pNP-Ac) in 5% EtOH), and 80 μl YxiM (1.9 μg/ml) was incubated in microtiter wells at room temperature for 40 min and the hydrolysis of pNP-Ac was monitored spectrophotometrically at 405 nm by the formation of *p*-nitrophenol. Assay mixtures without substrate or enzyme were included as controls.

pH stability assay: YxiM protein (0.12 μg) was incubated in various buffers of different pH (100 mM) for 4 h at RT. Ammonium formate was used for pH 2.5-6.5, MOPS buffer was used for pH 5.5 – 10. The acetyl esterase activity was tested using sugar beet pectin as a substrate as described below.

Determination of kinetic constants: The specific activity and Michaelis-Menten constant (K_m) were determined by incubating 80 µl YxiM (1.9 µg/ml), 160 µl 100 mM Tris-HCl, pH 7.0 and 80 µl pNP-Ac at various concentrations (37.5 µM – 30 mM) for 12 min at room temperature. The formation of *p*-nitrophenol was determined as described above. Specific activity is expressed in mmol of acetate liberated min^{-1} mg^{-1} protein. The kinetic constants were determined for sugar beet pectin by incubating 1 µl YxiM (0.019 µg/ml) with various concentrations of the pectin (10 – 170 µg) in 100 mM ammonium formate, pH 6.0 for 10 min. The release of acetate was determined.

Digestion of sugar beet pectin: The reaction mixture (50 µl) contained 100 µg sugar beet pectin in 50 mM buffer solution (sodium citrate buffer, pH 4.5; ammonium formate buffer, pH 6.0; or Tris-HCl buffer, pH 7.5). Enzymes were used in the following concentrations: Yxim from *Bacillus subtilis* (0.05 U, 1 unit releases 1 µmol of acetic acid per min from triacetin); PaeY from *Erwinia chrysanthemi* 3937 (4 µl crude supernatant per 100 µl), PME from *Aspergillus aculeatus* (0.05 U, 1 unit releases 1 µmol methanol per min from citrus pectin), RGAE from *Aspergillus aculeatus* (0.005 U, 1 unit releases 1 umol acetic acid per min from *p*-nitrophenyl acetate) and endopolygalacturonase (EPG) from *Aspergillus niger* (Megazyme, 0.5 U, 1 unit releases 1 µmol of reducing galacturonate per min from polygalacturonic acid). The mixtures were incubated for 2 h at 30°C, with continuous mixing by agitation. After the incubations, the mixtures were brought to room temperature and immediately assayed for acetic acid content, or for methanol content. Only when sequential enzymatic reactions were carried out, the first enzyme mixture was inactivated by boiling for 5 min. Non-enzymatic reactions were in either buffer or base (0.1 M NaOH). Base treatment was followed by neutralisation with 0.1 M HCl. The digestion of potato pectin and apple pectin was carried out as follows: the reaction mixture (100 µl) contained 0,5% potato pectin or apple pectin in 100 mM Tris-HCl buffer, and 8 µl YxiM (1.9 µg/ml) or PaeY (crude supernatant as described above) was added. The mixtures were incubated in temperature-controlled mixers (Eppendorf, Hamburg, Germany) over night at 30°C. Assay mixtures without substrate or enzyme were included as controls. The total release of acetate was determined by adding alkali instead of enzyme. After the incubations, the mixtures were assayed immediately for acetic acid content.

Digestion of tobacco homogalacturonan (HG) and rhamnogalacturonan I (RG-I) oligomer preparations: The reaction mixture (50 µl) contained 250 to 500 µg of either HG or RG-I preparations in 50 mM ammonium formate buffer, pH 6.0. Enzymatic and non-enzymatic incubations were as described for the digestion of sugar beet pectin.

Digestion of xylan and xyloglucan: The reaction mixture (100 µl) contained 0,5% wheat arabinoxylan (medium viscosity, Megazyme) or rye arabinoxylan (Megazyme) in 100 mM Tris-HCl buffer, and 8 µl YxiM (1.9 µg/ml) was added. The mixtures were incubated in temperature-controlled mixers (Eppendorf, Hamburg, Germany) for 3 h at 30°C. Assay mixtures without substrate or enzyme were included as controls. The total release of acetate was determined by adding alkali instead of enzyme. After the incubations, the mixtures were assayed immediately for acetic acid content.

105 μl assay mixture consisting of 10 μl enzyme, 5 μl 1M Tris-HCl, 20 μl tomato xyloglucan oligosaccharides or bean xyloglucan oligosaccharides (10 mg/ml) and 75 μl H_2O was incubated over night at room temperature, followed by determination of acetate release as described above.

3. Results

Multiple sequence alignment of proteins from family 12 of the carbohydrate esterases. The PAE's represent a group of enzymes of which only a few have been characterised by biochemical analysis. However, many proteins of unknown function show similarity to the existing characterised PAE's and are therefore categorised as possible PAE's. The multiple sequence alignment in figure 1 shows that the YxiM protein contains the five conserved catalytically active amino acids described by Mølgaard et al [15] as well as the conserved region of 42 residues as shown by Shevchik and Hugouvieux-Cotte-Pattat [14]. YxiM shows 34% identity to PaeY and 36% identity to RGAE within this conserved region. 19% sequence identity was found between YxiM and both RGAE and PaeY along the entire length of the protein.

Acetyl esterase plate assay. An acetyl esterase assay for use in agar plates was developed by incorporating Brilliant Green and triacetin into the plates. The YxiM sample was applied onto the plates, and they were incubated for 4-6 h at 37°C. Acetyl esterase activity was then shown as a green circle around the application wells as Brilliant Green acts as a pH indicator when acetate is released (figure **2**).

pH optimum assay. The pH optimum for YxiM was measured using pNP-Ac as substrate. YxiM had optimal activity at pH 8 (figure 3). The background level for the spontaneous deacetylation of pNP-Ac (substrate incubated without enzyme) was subtracted for all data points. The substrate was found to be unstable at pH values higher than pH 8 (data not shown).

```
                                                     ↓ Ser9
YESY      ---------------------------------MANHIYLAGDSTVQTYGDSTN--QGGW 25
ccd       ---------------------------------MANHIYLAGDSTVQTYGDSTN--QGGW 25
BH3963    DGQLT-IELTGKQPKINALT---IKERSKRNPAVHPTVYIASDSTAQTYDPYWNP-QAGW 211
YEST      ------------------------------MMKKPIQVFLAGDSTVSDCPPHEAP-MAGW 29
YXIM      DGQLN-LLVTEGKAGTAFTLSALKIKKLSDQPVTNRTIYVGGDSTVCNYYPLNSSKQAGW 202
RGAE      ------------------MKTAALAPLFFLPSALATTVYLAGDSTMAKNGGGSGT--NGW 40
PaeY      DIKFRPIAGLNASGSYEPLDLTSANQHRKALEQSHTRIFIVSDSTAATYEKARFP-RMGW 299
                                               ::: .***             **

                                ↓ Gly42                         ↓ Asn74
YESY      GQFLGSHLPEH--IQVINRAIGGRSSKTFVEEGRLQAILDVIEPDDWLFVQMGHNDASKN 83
ccd       GQFLGSHLPEH--IQVINRAIGGRSSKTFVEEGRLQAILDVIEPDDWLFVQMGHNDASKN 83
BH3963    GQMLPRYFETD--VTIKNHAIGGRSSRTFYTEGRLDTILREIKPNDYLLIQFGHNDATIS 269
YEST      GQVFGQLFSEG--VLVRNHAKGGASTNSFVEEGRLQAIAEHITQGDYLLIQFGHNDQ--- 84
YXIM      GQMLPHYIDKHT-FQVRNMASGGQIARGFRNDGQLEAILKYIKPGDYFMLQLGINDTN-- 259
RGAE      GEYLASYLSA----TVVNDAVAGRSARSYTREGRFENIADVVTAGDYVIVEFGHNDGGSL 96
PaeY      GQVFEQQFRPGGDVTVVNGARAGRSSRDFYYEGWFRQMEPFMRPGDYLFIGMGHNDQNCD 359
          *: :   :       : * * .*  :. :  :* :  :   :  .*:.:: :* **

YESY      ------------------------KPERYTEPYTTYKQYLKQYIAGAREKGAHPLLITPV 119
ccd       ------------------------KPERYTEPYTTYKQYLKQYIAGAREKGAHPLLITPV 119
BH3963    ------------------------VPERYTS-VEDFKGYMETYVAGVRQRGGIPILVTPV 304
YEST      ------------------------KP-RGTKPYSTFQQFLTLFADTAREKGAHPVFVTSV 119
YXIM      --------------------------PKHKESEAEFKEVMRDMIRQVKAKGADVILSTPQ 293
RGAE      STDNGR-----TDCSGTGAEVCYSVYDGVNETILTFPAYLENAAKLFTAKGAKVILSSQT 151
PaeY      SQKALRGAADVANLCTYPNSADGRPQYPQGKPDMSFQISLERYIRYAQAHRMIPVLLTPT 419
                                        .    :   :         :    :: :

YESY      AR-----------------FHYENGVFLNDFPDYCIAMKQTAAEENVQLIDLMEKSLAFF 162
ccd       AR-----------------FHYENDMFLNDFPDYCIAMKQTAAEENVQLIDLMEKSLAFF 162
BH3963    GRRS---------------FHPETGKFNVSFPGYKQAMEEVAAEQDVLLVNLNTLSREYY 349
YEST      QRR----------------RFDENGRIEHTLGEYPDAMKALAKELDVPVIDLLAKTKVLY 163
YXIM      GRAT---------------DFTSEGIHSSVNRWYRASILALAEEEKTYLIDLNVLSSAYF 338
RGAE      PNN-----------------PWETGTFVNSPTRFVEYAELAAEVAGVEYVDHWSYVDSIY 194
PaeY      ARVKNAEGKNGTPAVHSHLTKQNKAGGYAYIGDYTQTIRDTASKNKVPLLDVETATLALA 479
           .                    .          :       *    .  ::

                                    Asp192 ↓  ↓ His195
YESY      TEKGEEKVYTYFMIS--------------EGINDYTHFTKKGANEMAKLVAKGIKELGLP 208
ccd       TEKGEEKVYTYFMVS--------------EGINDYTHFTKKGANEMAKLVAKGIKELGLP 208
BH3963    DEIGPEGTRSVFLHVDAGVYPAF-----PEGRADDTHFQAYGAIQIARLLSEGIAQLNTP 404
YEST      EAYGPEESKRLFVWFQPNEHPNY-----PDGIEDNTHFSEKGAMEVAKLVAEGIEELGLP 218
YXIM      TSIGPERTLGLYMDG------------------DTLHPNRAGADALARLAVQELKRQGIA 380
RGAE      ETLGNATVNSYFPI-------------------DHTHTSPAGAEVVAEAFLKAVVCTGTS 235
PaeY      NQGDGQQWQQYWLAVDPDRYPYYRDQAGSLTQPDTTHFQQKGAQAVAAIVADQIKAT-PS 538
             .       :                     *  *    **  :*    . :     .
```

Figure 1 **CLUSTAL W (1.81) Multiple sequence alignment of proteins from family 12 of the carbohydrate esterases.** The five conserved catalytically important residues [15] are marked with arrows and the conserved region of 42 residues [14] is underlined. Sequences: YesY (SwissProt O31528), ccd (GenBank AF184175), BH3963 (GenBank 001520), YesT (SwissProt O31523), YxiM (SwissProt P42304), RGAE (SwissProt Q00017) and PaeY (SwissProt O32563). Identical or conserved residues in all sequences in the alignment are marked with *, conserved substitutions are marked : , and semi-conserved substitutions are marked •.

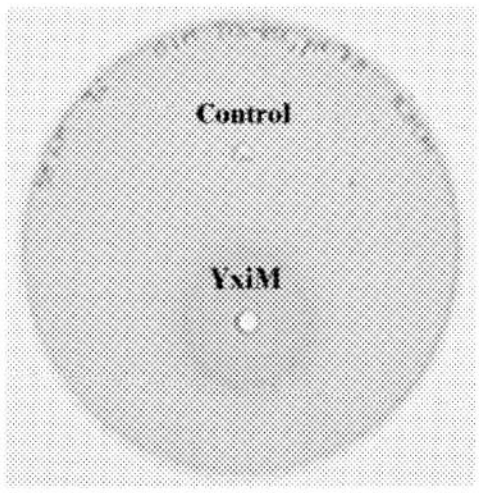

Figure 2 **Acetyl esterase plate assay.** Control: sample broth from *Bacillus subtilis* expressing the plasmid vector and YxiM: expressed and purified YxiM protein (1.9 mg/ml), were applied to the assay plate and incubated for 4-6 h at 37°C.

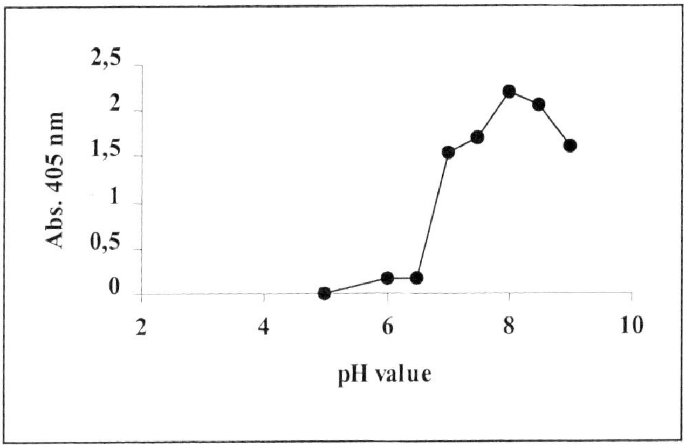

Figure 3 **pH optimum.** The assay mixture was incubated at room temperature for 40 min and the formation of the product *p*-nitrophenol was monitored spectrophotometri-cally at 405 nm.

pH stability assay. YxiM protein was incubated at different pH for 4 h to test its pH stability. After this incubation the enzyme activity was measured at pH 6.0 using sugar beet pectin as a substrate (Figure 4). YxiM remained active above pH 4.

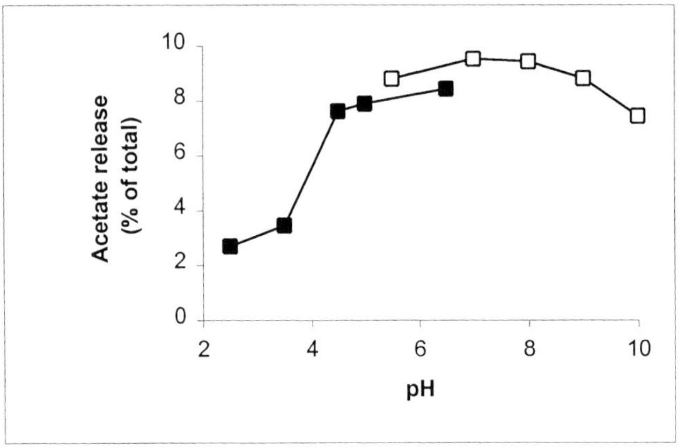

Figure 4 **pH stability.** The YxiM protein was incubated in buffers of varying pH for 4 h at room temperature, followed by determination of its activity using sugar beet pectin as substrate. Closed symbols - ammonium formate buffer, open symbols -MOPS buffer.

Determination of kinetic constants. The YxiM protein had higher affinity to pNP-Ac than PaeY as the K_m found in this study for YxiM was about five times lower than the K_m found previously for PaeY [14]. The specific activity for YxiM was about 10 fold lower than the corresponding value for PaeY (Table I).

Table I **Determination of kinetic constants**

	YxiM	PaeY[1]
K_m (mM pNP-Ac)	1.6	6
K_m (g/l sugar beet pectin)	12	25
Specific activity (mmol min^{-1} mg^{-1} protein, pNP-Ac)	154	1400
Specific activity ($mmol^{-1}$ mg^{-1} protein, sugar beet pectin)	0.3	1.2

[1]V.E. Shevchik and N. Hugouvieux-Cotte-Pattat (1997) [14]

Digestion of sugar beet pectin, potato pectin and apple pectin. The sugar beet pectin preparation used in this study had a degree of methylation (DM) of 66.1% and a degree of acetylation (DA) of 24.0 %. Figure 5 shows the percentage of total acetate groups released by Yxim or PaeY individually, or in combination with PME under different pH conditions. Yxim released ca. 9% of the total acetate groups from sugar beet pectin. The buffer pH had no significant effect on acetate release. PaeY released only about 3.5% of the total acetate groups, with a slightly higher release (up to 4.5% release) recorded at pH 6.0. The possible effects of the pH on the pectin substrate may interfere with any effects of the pH on enzyme activity. The inclusion of fungal PME in the reaction mixture increased the acetate release from sugar beet pectin, Yxim releasing 2 to 3 times, and PaeY 1 to 2 times more acetate than when the enzymes were used individually. Neither YxiM nor PaeY showed any detectable activity on apple pectin (results not shown). However, YxiM released ca. 24% of the total acetate groups from potato pectin (results not shown).

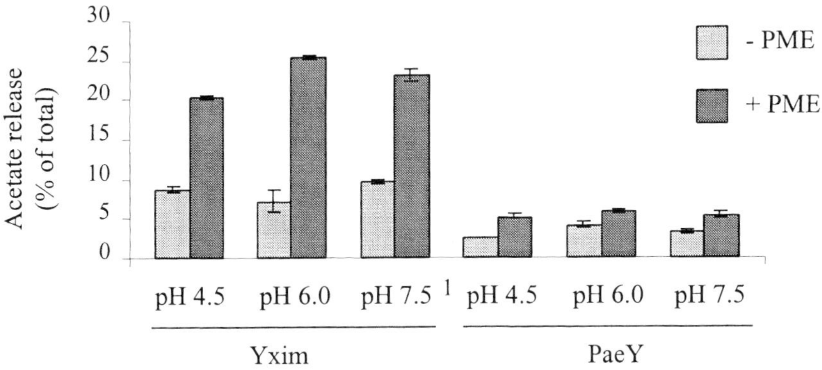

Figure 5 **Digestion of sugar beet pectin.** Sugar beet pectin (DA 24.0%) was incubated with either Yxim or PaeY individually, or in combination with PME, for 2 h at 30°C at different pH conditions. Acetate released was measured and compared to total acetate released by base treatment. Values shown are the means ± SD of three separate incubations.

Treatment of sugar beet pectin with various enzymes. Table II shows the effect of various enzyme treatments on the DM and DA of sugar beet pectin at pH 6.0. PME treatment resulted in the de-esterification of the sample down to a DM of 11%, and as expected had no effect on the DA. As mentioned above, Yxim treatment alone resulted in a decrease of the DA to 21.5%. However, when Yxim treatment was preceded by PME treatment, the DA further decreased to 18.2%. On the other hand, when Yxim treatment preceded PME treatment, the DM further decreased to 1.2 %. The treatment of sugar beet pectin simultaneously with both enzymes resulted in a DM of 0.9 % and a DA of 18.2 %. Simultaneous treatment with RGAE and PME resulted in a DA of 19.5%, and if followed by Yxim treatment, the DA could be further reduced to 17.9%. Inclusion of EPG in combination with PME in the reaction mixture did not increase the acetate release using either acetyl esterases (data not shown).

Table II **Effect of the treatment with various enzymes on the DM and DA of sugar beet pectin**

Treatment	DM (%)	DA (%)
No enzyme	66,1	24,0
PME	11,1	24,0
YxiM	66,1	21,5
PME then YixM	11,1	18,2
YixM then PME	1,2	19,2
PME and YixM	0,9	18,2
PME and RGAE	12,0	19,5
PME and RGAE then Yxim	12,0	17,9

Digestion of tobacco homogalacturonan (HG) and rhamnogalacturonan I (RG I) oligomer preparations. The polysaccharides in the HG preparation were composed mainly of uronic acid (98% (w/w)), with minor amounts of neutral sugars (3% (w/w)) which included Gal, Ara and Rha. The Rha:GalA ration was 0.1 (mol/mol). The HG preparation had a DM of 1.7% and a DA of 24.8%. Both acetyl esterases released a significant amount of acetate from the HG preparation at pH 6.0 (Figure 6). Yxim released ca. 40% and PaeY released ca. 67% of the total acetate from HG. Inclusion of PME in the reaction mixture had no effect on the acetate release using either acetyl esterases, presumably due to the low DM of the preparation. The polysaccharides present in the RG I preparation were composed of uronic acid (33% (w/w)) and neutral sugars (66% (w/w)) which included Gal, Ara and Rha. The Rha:GalA ration was 0.6 (mol/mol). The preparation had a DM of 1.1% and a DA of 118%. Yxim released ca. 35% of the total acetate from RG I, with a slight (2%) increase in the presence of PME (Figure 7). PaeY released ca. 9% of the acetate from RG I and a doubling of the acetate released was observed in the presence of PME. Treatment of the RG preparation with RGAE and PME simultaneously resulted in the release of ca. 65% of the acetate (data not shown). Subsequent treatment of RGAE de-acetylated material with YxiM did not result in further release of acetate from the preparation (data not shown).

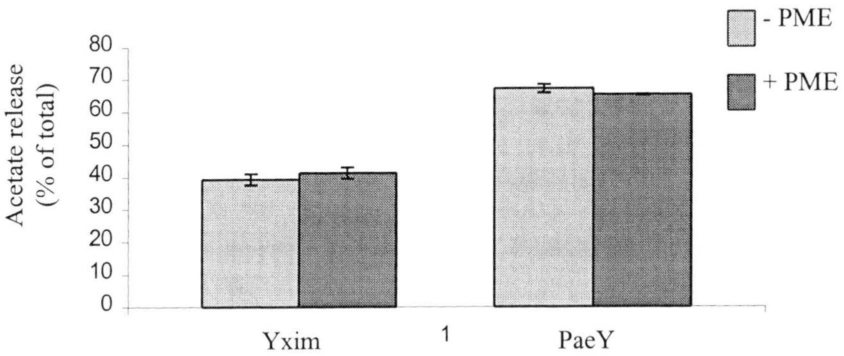

Figure 6 **Digestion of tobacco HG oligomers.** The HG preparation (DA 24.8%) was incubated with Yxim or PaeY individually, or in combination with PME, for 2 h at 30°C at pH 6.0. Acetate released was measured and compared to total acetate released by base treatment. Values shown are the means ± SD of three separate incubations.

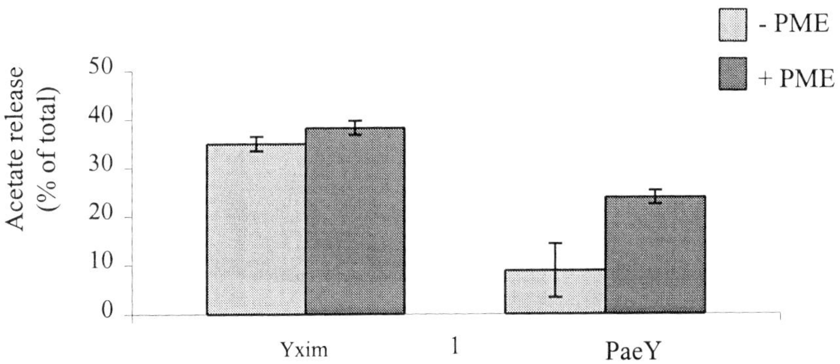

Figure 7 **Digestion of tobacco RG I oligomers.** The RG I preparation (DA 118%) was incubated with Yxim or PaeY individually, or in combination with PME, for 2 h at 30oC at pH 6.0. Acetate released was measured and compared to total acetate released by base treatment. Values shown are the means ± SD of three separate incubations.

Digestion of xyloglucan and xylan. Neither YxiM nor PaeY showed any detectable activity on xyloglucan oligosaccharides derived from tomato, but both enzymes released approximately 6% of the acetate groups on xyloglucan oligosaccharides from bean (results not shown). YxiM showed no significant activity on wheat arabinoxylan but released ca. 11% of the total acetate groups from rye arabinoxylan (results not shown).

4. Discussion

The sequence databases contain an enormous number of sequences of genes that encode proteins of unknown function. Often it is possible to use computational sequence tools to recognise similarity between a new sequence and a sequence about which something is already known. This allows predictions about structure and /or function of the new sequences [24].

The PAE's represent a group of enzymes of which only a few have been characterised by biochemical analysis. However, many proteins of unknown function show similarity to the existing characterised PAE's and are therefore categorised as possible PAE's. The hypothetical protein from *Bacillus subtilis*, named YxiM, shows homology with known acetyl esterases, RGAE from *Aspergillus aculeatus* and PaeY from *Erwinia chrysanthemi* [14, 15]. The YxiM protein contains the conserved region of 42 residues as found previously for PaeY and RGAE [14]. Furthermore, YxiM contains the five highly conserved catalytically important residues described by Mølgaard *et al* [15]. Like RGAE and PaeY, YxiM is classified as family 12 of the carbohydrate esterase families defined by Coutinho and Henrissat [16]. These data indicate structural relationship and support the theory that the YxiM gene encodes a protein specific for the de-acetylation of pectin.

The recombinant purified YxiM protein was tested for acetyl esterase activity. An acetyl esterase plate assay was developed by incorporating Brilliant Green and triacetin into the plates and the acetyl esterase activity of the enzyme was shown as a green circle around the application well. This plate assay is a very useful tool for a quick determination of acetyl esterase activity. The assay can also be used for the purpose of screening culture supernatants for activity.

When characterising the enzymatic properties of YxiM it was found that pH 8 was the optimal pH for the enzyme. However, pNP-Ac was not stable at pH levels higher than 8. As a comparison the optimal activity for RGAE was determined to be at pH 6 [13] and for PaeY, as for YxiM, it was found to be at pH 8 [14]. The YxiM protein was found to be stable over a large pH range as the activity was kept at the same level at pH 4-9. Determination of the kinetic constants of YxiM indicated that the enzyme had a higher affinity to acetylated substrates, but a lower specific activity towards these substrates than PaeY.

Sugar beet pectin has a high degree of acetylation and a high proportion of the acetyl groups have been shown to be carried by the homogalacturonan regions of the molecule [25]. Sugar beet pectin is thus a suitable substrate for studies on enzymes that de-acetylate pectin. Both YxiM and PaeY were found to be active on sugar beet pectin, however they only released a small proportion (9% and 3.5% respectively) of the total acetate groups. Shevchik and Hugouvieux-Cotte-Pattat showed that PaeY released 15% of the acetate groups from sugar beet pectin (degree of *O*-acetylation, 23%) and in combination with PME from *Erwinia chrysanthemi*, up to 40% of the acetate groups were released. This difference to the data presented here could be explained by the use of different batches and concentrations of PaeY. Also different types of PME were used. Furthermore, the incubation time in our experiment was only 2 hours at pH 4.5-7.5 as compared to 12 hours at pH 8.0. At elevated pH (8.0) a considerable amount of acetate is released chemically. The release of a small proportion of the total acetate groups present on the sugar beet pectin preparation could be explained by the high complexity of pectin preparations when compared to synthetic substrates as pNP-Ac and triacetin. Furthermore, the enzymes might only hydrolyse only one type of acetyl ester at either the *O*-2 and *O*-3 position of galacturonic acid, both of which have been found in native pectins [2,6]. NMR studies have indicated that a PAE purified from *Aspergillus niger* attacks only one type of acetyl esters of the polygalacturonan chain of pectin [26]. Pre-treatment of samples with PME increased the proportion of acetate released by both Yxim and PaeY, and could indicate that the accessibility of the acetyl esterases to the acetate groups was increased by de-methyl-esterification. The enzymes showed no affinity for apple pectin. As the acetyl groups in apple pectin have been shown to be present mainly in the hairy regions [27], the low affinity of the enzymes for apple pectin could indicate that none of the enzymes are able to de-acetylate native rhamnogalacturonan. This is in accordance with the data for PaeY of Shevchik and Hugouvieux-Cotte-Pattat [14]. Unlike YxiM and PaeY, RGAE was found to be specific for the de-acetylation of rhamnogalacturonan as the enzyme showed high activity on modified hairy (ramified) regions isolated from apples [12,13]. YxiM released a high proportion (24%) of the total acetate groups from potato pectin indicating that the activity towards potato pectin is higher than for sugar beet pectin.

The synergistic effect of YxiM and PME was further investigated by treatment of a sugar beet pectin preparation with various enzymes. Pre-treatment of the sample with PME increased the ability of Yxim to release acetate groups and similarly, pre-treatment of the samples with Yxim increased the ability of PME to release methanol groups. Even though the effect of YxiM activity on total DA is small, its activity may enhance the accessibility of PME to the methyl groups resulting in almost complete de-methyl-esterification, thereby improving the gelling properties of the pectin molecule. Enzymatic de-acetylation has been shown to improve the gelling ability of de-methyl-esterified sugar beet pectin in the presence of calcium [28]. In contrast to YxiM and PaeY, RGAE is not affected by treatment with PME [12,13]. Pre-treatment of sugar beet pectin with RGAE and PME released 19% of the total acetate, and subsequent treatment with Yxim released a further 10% of the acetate. This indicates that a subset of the acetate groups released by Yxim could not be removed by the action of RGAE. The acetate groups specifically removed by Yxim may have originated from the homogalacturonan domains of the pectin molecule. EPG treatment in combination with PME and Yxim, or PME and PaeY, did not increase the release of acetate from the sugar beet pectin sample at any of the pH values tested. This suggested that depolymerisation of pectin may not be an important factor for acetyl esterase activity (even though the enzymes appeared more active on oligomeric substrates than on polymeric substrates). However, the effect of EPG treatment on the extent of depolymerisation of the sample was not clear.

Both YxiM and PaeY showed a preference for the tobacco homogalacturonan oligomers. These data support the results of Shevchik and Hugovieux-Cotte-Pattat who determined the maximal release of acetate by PaeY on de-methyl-esterified oligomers of sugar beet pectin [14]. However, the low DA of the preparation (24.8%) may limit the activity of the enzymes. Surprisingly, both YxiM and PaeY were observed to be active on a tobacco rhamnogalacturonan preparation. YxiM seems to be more active than PaeY on this substrate, releasing 35% of the total acetate compared to a maximal release of 9% by PaeY. This was not expected, as we did not detect any activity on apple pectin. However, composition data on the RG I preparation revealed a more complex structure than could be anticipated for a pure acetylated RG I backbone and may also contain some (acetylated) homogalacturonan fragments. Nevertheless, no synergy was observed between RGAE and YxiM activities on the de-acetylation of the RG-I preparation. The results only confirm that pectin substrates may be very complex molecules and that comparison of specificity between different types of pectin molecules can be complicated. The use of more defined substrates, such as chemically acetylated homogalacturonan preparations [29] may prove useful for further characterisation of enzyme specificity. Furthermore, site-specificity of the different enzymes is an important issue to solve.

YxiM showed no detectable activity on wheat arabinoxylan, whereas the enzyme released 11% of the total acetate groups from rye arabinoxylan. The wheat arabinoxylan had been prepared with base extraction whereas a more gentle extraction procedure was used for the rye preparation. Hence, the wheat arabinoxylan may contain more tightly bound acetyl groups, which are less readily removed either by base treatment or by YxiM. The enzyme was not able to hydrolyse O-acetyl substituents present on the glucan backbone of xyloglucan, as they occur in e.g. tomato, but it was able to hydrolyse some of the O-acetyl substituents from xyloglucans, when these substituents are present on sidechain galactosyl-residues (e.g. bean xyloglucan). To explain the differences in affinity for these substrates further analysis has to be carried out.

The enzymatic properties of YxiM and the synergistic action with PME described here, suggest that this enzyme is important in improving the gelling properties of pectin. Like PaeY, this enzyme seems to have a preference for homogalacturonan oligomers whereas RGAE is specific for the hairy region of pectin. However, YxiM shows a broad substrate specificity and further analysis using more defined substrates and determination of site-specificity are needed to solve the exact specificity of the YxiM protein.

Acknowledgements

We thank Kimberly M. Brown and Michael D. Thomas for providing us with the overexpressed and purified recombinant YxiM protein, Nicole Hugouvieux-Cotte-Pattat for kindly providing the plasmid pVY47 that was used for overproduction of PaeY, and Malcolm O'Neill and William S. York for the gift of the oligosaccharide samples. We also thank CP Kelco A/S for the pectin samples and Alan Sloma for helpful discussions. This work was supported by a grant from the THOR (Technology by Highly Oriented Research) Program.

References

1. Schols, H. (1995) Structural characterization of pectic hairy regions isolated from apple cell walls. Thesis, Wageningen Agricultural University

2. Sakai, T., Sakamoto, T., Hallaert, J., and Pilnik, W. (1993) Pectin, Pectinases, and Protopectinases: Production, Properties, and applications. *Adv. Appl. Microbiol., 39, 213-293*

3. Hwang, J., Pyun, Y.R., and Kokini, J.L. (1993) Sidechains of pectins: some thoughts on their role in plant cell walls and foods. *Food Hydrocolloids,* 7(1), 39-53

4. De Vries, J.A., Rombouts, F.M., Voragen, A.G.J., and Pilnik, W. (1982) Enzymic degradation of apple pectin. *Carbohydr. Polym.* 2, 25-33

5. Thakur, B.R., Singh, R.K., and Handa, A.V. (1997) Chemistry and Uses of Pectin - A Review. *Critical Reviews in Food Science and Nutrition,* 37(1), 47-73

6. McCann, M.C. and Roberts, K. (1996) Plant cell wall architecture: the role of pectin. In *Pectins and Pectinases.* Visser, J., and Voragen, A.G.J. (eds). Elsevier, Amsterdam, 91-107

7. O'Neill, M., Albertsheim., P, and Darvill, A.(1990) The Pectic Polysaccharides of Primary Cell Walls. In *Methods in Plant Biochemistry.* P.M. Dey (ed), Vol. 2, *Carbohydrates,* Academic Press, London, 414-441

8.Axelos, M.A.V., and Thibault, J.F. (1991) The Chemistry of Low-Methoxyl Pectin Gelation. In *The Chemistry and Technology of Pectin.* Walter, R.H. (ed), Academic Press, London, 109-118.

9. Pippen, G.L., McCready, R.M., and Owens H.S. (1950) Gelation properties of partially acetylated pectins *J. Amer. Chem. Soc.,* 72, 813-616

10. Matthew, J.A., Howson, S.J., Keenan, M.J.H., and Belton, P.S. (1990) Improvement of the gelation properties of sugarbeet pectin following treatment with an enzyme preparation derived from *Aspergillus niger* - Comparison with a chemical modification. *Carbohydr. Polym.*,12, 295-306

11. Kofod, L., *et al.,* & Dalbøge H. (1994). Cloning and characterization of two structurally and functionally divergent rhamnogalacturonases from *Aspergillus aculeatus. J. Biol. Chem.* 269, 29182-29189.

12. Searle-van Leeuwen, M.J.F., vanden Broek, L.A.M., Schols, H.A., Beldman, G., and Voragen, A.G.J. (1992). Rhamnogalacturonan acetylesterase: a novel enzyme from *Aspergillus aculeatus,* specific for the deacetylation of hairy (rhamnified) regions of pectins. *Appl. Microbiol. Biotechnol.* 38, 347-349

13. Kappinen, S., Christgau, S., Kofod, L.V., Halkier, T., and Dalbøge, H. (1995) Molecular cloning and characterisation of a rhamnogalacturonan acetyl esterase from *Aspergillus aculeatus* – Synergism between rhamnogalacturonan degrading enzymes. *J. Biol.Chem.* 270 (45), 27172-27178

14. Shevchik, V.E., and Hugouvieux-Cotte-Pattat, N. (1997) Identification of a bacterial pectin acetyl esterase in *Erwinia chrysanthemi 3937 Mol. Microbiol.* 24(6), 1285-1301

15. Mølgaard, A., Kauppinen, S., and Larsen, S. (2000) Rhamnogalacturonan acetylesterase elucidates the structure and function of a new family hydrolases. *Structure* 8(4), 373-383

16. Coutinho, P.M. and Henrissat, B. (1999) Carbohydrate-Active Enzymes: an integrated database approach. In *Recent advances in Carbohydrate Bioengineering.* The Royal Society of Chemistry, Cambridge

17. Patent: US 6,184,028 B1, issued February 6, 2001 "Polypeptides having pectin acetyl esterase activity and nucleic acids encoding same", Assignee: Novo Nordisk Biotech, Inc., Davis, CA (USA)

18. Ishii, T., Matsunaga, T., Pellerin, P., O'Neill, M.A., Darvill, A., and Albersheim, P. (1999) The plant cell wall polysaccharide rhamnogalacturonan II self-assembles into a covalently cross-linked dimmer *J. Biol. Chem.* 274 (19), 13098-13104

19. Wilder, B.M. and Albersheim, P. (1973) The structure of plant cell walls IV: A structural comparison of the wall hemicellulose of cell suspension cultures of sycamore (*Acer pseudoplatanus)* and red kidney bean (*Phaseolus vulgaris*) *Plant Physiol.* 52, 889-893

20. York, W.S., Kolli, K., Orlando, R., Albersheim, P., and Darvill, A.G. (1996) The structures of arabinoxyloglucans produced by solanaceous plants. *Carbohydr. Res.* 285, 99-128

21. Blumenkrantz, N., Absoe-Hansen, G. (1973) New method for the quantitive determination of uronic acids. *Anal. Biochem.* 54, 484-489

22. Dubois, M., Gilles, J.K., Rebers, P.A., Smith, F. (1956) Colorimetric method for the determination of sugars and related substances. *Biochemistry* 28, 350-356

23. Klavons, J.A., and Bennett, R.D. (1986) Determination of methanol suing alcohol oxidase and its application to methyl ester content of pectins. *J. Agric. Food Chem.* 34, 597-599

24. Durbin, R., Eddy, S., Krogh, A., and Mitchison, G. (1998) Biological sequence analysis – probabilistic models of proteins and nucleic acids, Cambridge University Press

25. Rombouts, F.M., and Thibault, J.-F. (1986) Enzymatic and chemical degradation and the fine structure of pectins from sugar beet pulp. *Carbohydr. Res.* 154, 189-203

26. Searle-van Leeuwen, M.J.F., Vincken, J.-P., Schipper, D., Voragen, A.G.J., and Beldman, G. (1996) Acetyl esterases of *Aspergilus niger*: purification and mode of action on pectins. In *Pectins and Pectinases*. Visser, J., and Voragen, A.G.J. (eds). Elsevier, Amsterdam, 793-798

27. Schols, H.A., Geraeds, C.C.J.M, Searle-van Leeuwen, F.J.M., Kormelink, F.J.M., and Voragen, A.G.J. (1990). Rhamnogalacturonase: a novel enzyme that degrades the hairy region of pectin. *Carbohydr. Res.* 206, 105-115.

28. Oosterveld, A., Beldman, G., Searle-van Leeuwen, M.J.F., and Voragen, A.G.J. (2000) Effect of enzymatic de-acetylation of sugar beet pectin in the presence of calcium. *Carbohydr. Polym.* 43, 249-256

29. Renard, C.M.C.G., and Jarvis, M.C. (1999) Acetylation and methylation of homogalacturonanas 1: optimisation of the reaction and characterisation of the products. *Carbohydr. Polym.* 39, 201-207

THE TWO SECRETED PECTIN ACETYLESTERASES OF *ERWINIA CHRYSANTHEMI* 3937, PAEY AND PAEX

N. HUGOUVIEUX-COTTE-PATTAT and V. E. SHEVCHIK
Unité de Microbiologie et Génétique Composante INSA, UMR 5122 CNRS-INSA-UCB, Batiment Louis Pasteur, 11, Avenue Jean Capelle, F-69621 Villeurbanne Cedex, France
e-mail: cotte@insa-lyon.fr

Abstract

Erwinia chrysanthemi causes soft-rot diseases of various plants by enzymatic degradation of the pectin in plant cell walls. The structural complexity of pectin requires the combined action of several pectinases for its efficient breakdown. Part of the galacturonate residues of pectin are modified by methylesterification of the carboxyl function and/or acetylesterification of the hydroxyl function in C2 and/or C3 position. Esterases are necessary to cleave these methyl- or acetyl-esterifications. The *E. chrysanthemi* strain 3937 produces at least two pectin methylesterases (PemA and PemB) and two pectin acetylesterases (PaeY and PaeX). While PaeY is entirely secreted, PaeX is only partially secreted and found both in the extracellular medium and in the periplasm. The two pectin acetylesterases are secreted by the Out system responsible for the secretion of PemA and of eight endo-pectate lyases.

PaeY belongs to the family 12 of carbohydrate esterases which also includes the rhamnogalacturonan acetylesterase of *Aspergillus aculeatus*. PaeX is homologous to the C-terminal domain of the *Butyrivibrio fibriosolvens* xylanase XynB, to various proteins of unknown function and to a few bacterial esterases. PaeX contains the typical box (GxSxG) corresponding to the active site of the large family of serine hydrolases, like carboxylesterases, lipases and proteases. To characterize PaeY and PaeX, these proteins were purified after overproduction. Both enzymes release acetate from various synthetic substrates and from sugar-beet pectin. Their activity increased after previous depolymerization and demethylation of pectin, indicating that their preferred substrates are non-methylated oligogalacturonides. The three esterases PaeX, PaeY and PemA act in synergy with the endo-pectate lyases for the pectin breakdown process. Considering their different synergistic effects, it seems likely that the two *E. chrysanthemi* pectin acetylesterases, PaeX and PaeY, differ either in their specificity towards pectic substrate or in their reaction mechanism.

1. Introduction

Plant cell walls contain up to 7% O-bound acetyl groups by dry weight. These acetyl moieties are mostly linked to two plant polysaccharides, xylan and pectin (at the O-2 and/or O-3 position of sugars). Efficient breakdown of plant polysaccharides is

F. Voragen et al. (eds.), Advances in Pectin and Pectinase Research, 331-343.
© 2003 *Kluwer Academic Publishers. Printed in the Netherlands.*

achieved by the concerted action of several enzymes. The action of depolymerases is often precluded by the presence of esterifications whose removal is necessary prior to, or concomitantly with, the action of depolymerases. While several xylan acetylesterases have been characterized in a variety of fungi and bacteria, very few genetic and biochemical informations are available on pectin acetylesterases. Part of the galacturonate residues of pectin are modified by methyl-esterification at the carboxyl group and/or acetyl-esterification on the hydroxyl groups. The degrees of methylation and acetylation vary greatly, depending on the source of pectin. Many saprophytic and plant pathogenic microorganisms secrete enzymes involved in pectin degradation. Two microbial genes encoding acetylesterases active on pectic polymers have been identified: the *paeY* gene of *E. chrysanthemi* which encodes a pectin acetylesterase specific of linear regions (Shevchik and Hugouvieux-Cotte-Pattat, 1997) and a fungal gene which encodes a rhamnogalacturonan acetylesterase (RGAE), specific of the pectin ramified regions (Searle-van Leeuwen *et al.*, 1996).

E. chrysanthemi is a phytopathogenic enterobacterium which causes soft rot disease of various plants. This bacterium is capable of using pectin as its sole carbon source (Fig. 1). *E. chrysanthemi* accomplishes pectin degradation through the production of multiple secreted pectinases (Lojkowska *et al.*, 1995; Pissavin *et al.*, 1996; Shevchik *et al.*, 1997; Tardy *et al.*, 1997) which give rise to the formation of pectic oligomers (Fig. 1).

Endo-pectate lyases are the main causative agents of the symptom of maceration observed after plant infections (Collmer and Keen, 1986). *E. chrysanthemi* produces two pectin methylesterases, the secreted protein PemA, and the cell-linked outer membrane lipoprotein PemB (Shevchik *et al.*, 1996). Since endo-pectate lyases exhibit a reduced activity on methylated substrates (Tardy *et al.*, 1997), the depolymerization of pectin is favoured by the previous action of the pectin methylesterase PemA (Shevchik and Hugouvieux-Cotte-Pattat, 1997).

The main factor affecting the transcription of the *E. chrysanthemi* pectinase genes is induction in the presence of pectin (Hugouvieux-Cotte-Pattat *et al.*, 1996). Analysis of *E. chrysanthemi* culture supernatants revealed the presence of several acetylesterases whose synthesis is induced in the presence of pectin (Shevchik and Hugouvieux-Cotte-Pattat, 1997). This specific induction suggests that the physiological substrate of these esterases is pectin. Most of the *E. chrysanthemi* pectinase genes are organized in clusters on the bacterial chromosome. Analysis of one of these cluster allowed us to identify the first pectin acetylesterase, PaeY. Recently, we showed that another cluster of genes involved in pectin degradation encodes a second pectin acetylesterase, PaeX.

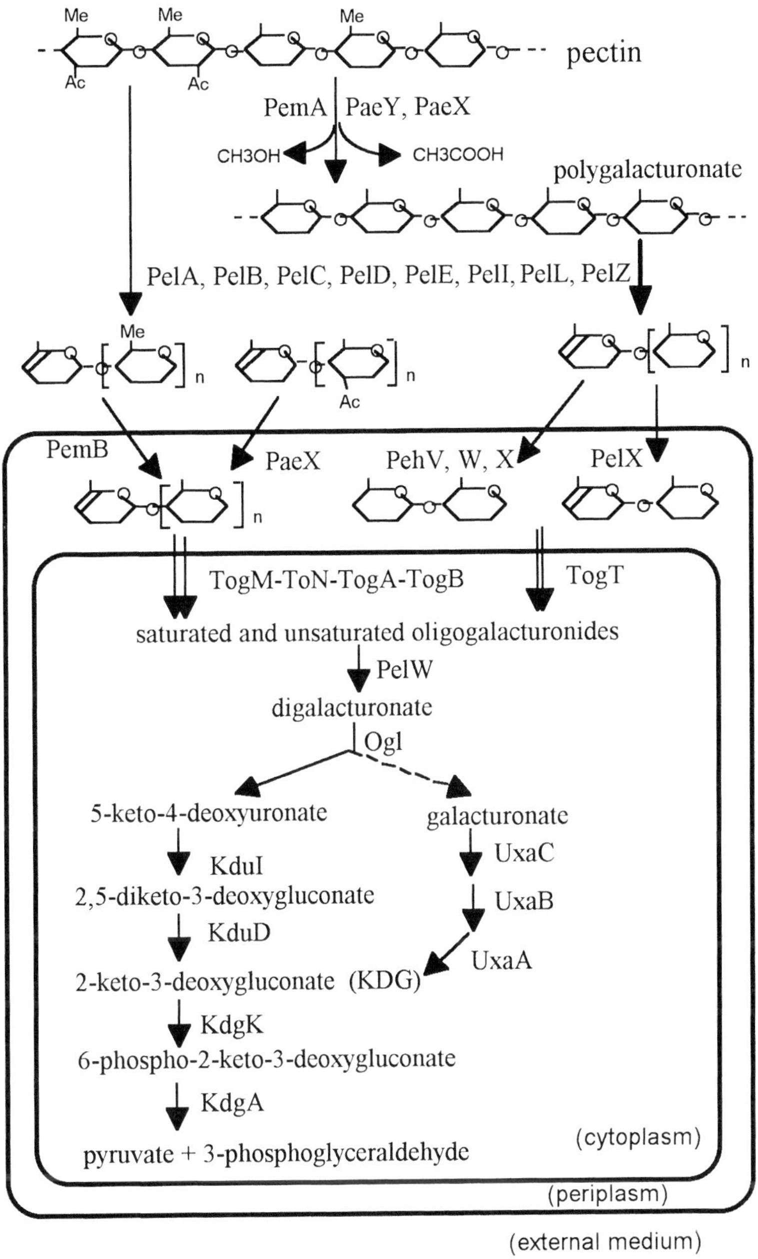

Figure 1. Pectin degradation by *Erwinia chrysanthemi.*
The different steps are catalyzed by the protein indicated below the corresponding arrow. Pem, pectin methylesterase; Pae, pectin acetylesterase; Pel, pectate lyases; Peh, polygalacturonase; Tog, oligogalacturonide transporter.

2. The Pectin Acetylesterase PaeY

2. 1. IDENTIFICATION OF THE *PAEY* GENE

On the basis of the induction in the presence of pectin, we isolated various mutations affecting pectin catabolism in the *E. chrysanthemi strain* 3937 (Hugouvieux-Cotte-Pattat and Robert-Baudouy, 1989). A large part of these mutations could not be assigned to an identified gene by simple phenotypic analysis. Analysis of two of these mutations revealed that they are localized in an unknown gene, situated in the cluster encoding the three pectate lyases, PelA, PelE, PelD and the pectin methylesterase PemA (Fig. 2). Since further studies revealed that this unknown gene encodes a pectin acetylesterase, we named it *paeY* (Fig. 2).

The nucleotide sequence of a 2.3 kb DNA fragment containing *paeY* was determined (accession number Y09828). Sequence analysis revealed the presence of a unique complete ORF which begins with an ATG codon at position 121 and ends with TAA at position 1767. The *paeY* gene encodes a 551 aminoacid protein. The 25 N-terminal aminoacids present the essential features of classical signal peptides. The precursor and potential mature forms of PaeY have a calculated molecular mass of 60 704 and 57 919 Da, respectively.

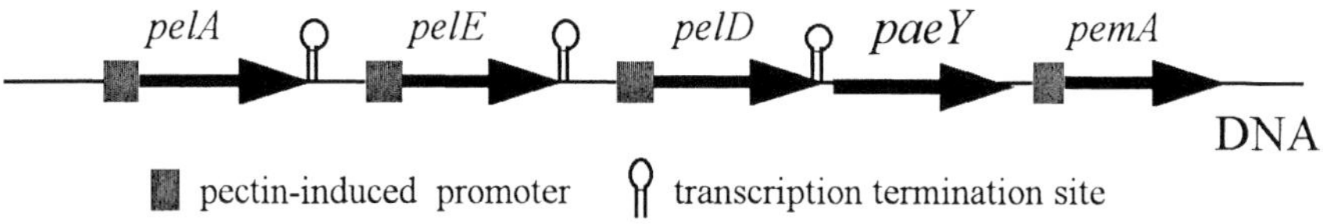

Figure 2. Genetic organisation of the region encoding the pectin acetylesterase PaeY.
The adjacent genes encode the following enzymes: *pelA*, *pelE* and *pelD*, endo-pectate lyases; *pemA*, pectin methylesterase.

2. 2. ANALYSIS OF THE PAEY AMINO ACID SEQUENCE

Comparison of the deduced amino acid sequence of PaeY with proteins present in databases gave the first indications on its function. Indeed, PaeY showed a low homology with the rhamnogalacturonan acetylesterase (RGAE) of *Aspergillus aculeatus* (GenBank accession number 1004217) (Kauppinen *et al.*, 1995) (Fig. 3).

RGAE is shorter than PaeY since it contains 250 residues, including a 17-aminoacid signal peptide. Homology is found along the entire length of RGAE, with a more conserved region of 42 residues in the C-terminal part of the proteins (residues 314 to 355 of PaeY). In addition, the hypothetical protein YxiM of *Bacillus subtilis* (GenBank accession number 665988) shows homology with PaeY and RGAE, with the same well-conserved region (Fig. 3).

A recent classification based on aminoacid sequence similarity distinguishes 12 families of carbohydrate esterases (Coutinho and Henrissat, 2000). PaeY of *E. chrysanthemi* and the RGAE of *Aspergillus aculeatus* belong to the family 12.

```
1                                     MKTAALAPLFFLPSALATTVYLAGDSTMAKNGGG--SGTNGW   40
224  GYNNNNGQASHTVLTDIKFRPIAGLNASGSYEPLDLTSANQHRKALEQSHTRIFIVSDSTAATYEKA-RFPRMGW  299
132  DGAEDTFQIP---VTDGQLNLLVTEGKAGTAFTLSALKIKKLSDQP-VTNRTIYVGGDSTVCNYYPLNSSKQAGW  203
                                                                  DST             GW

GEYLASYLS----ATVVNDAVAGRSARSYTREGRFENIADVVTAGDYVIVEFGHND----G----------GSLSTDNGR  102
GQVFEQQFRPGGDVTVVNGARAGRSSRDFYYEGWFRQMEPFMRPGDYLFIGMGHNDQNCDSQKALRGAADVANLCTYPNS  379
GQMLPHYIDKHT-FQVRNMASGGQIARGFRNDGQLEAILKYIKPGDYFMLQLGIND---------------TNPKHKESE  266
G              V N A  G   R     G           GDY     G ND

TDCSG--------TGAEVCYSVYDGVNETILTFPAYLENAAKLFTAKG---AKVILSSQTPNN--------------PWE  157
ADGRPQYPQGKPDMSFQISLERYIRYAQAHRMIPVLLTPTARVKNAEGKNGTPAVHSHLTKQNKAGGYAYIGDYTQTIRD  459
AEFKE--------VMRDMIRQVKAKGADVILSTP---QGRATDFTSEG------IHSSVNRWY-----------------  312
                                 P      A      G        S

TGTFVNSPTRFVEYAELAAEVAG--VEYVDHWSYVDSIYETLGNATVNSYFPIDHTHTSPAGAEVVAEAFLKAVVCTGTS  235
TASKNKVPLLDVETATLALANQGDGQQWQQYWLAVDPDRYPYYRDQAGSLTQPDTTHFQQKGAQAVAAIVADQIKATPSL  539
-R---ASILALAEEEKTYLIDLN--VLSSAYFTSIGP------ERTLGLYMDGDTLHPNRAGADALARLAVQELKRQGIA  380
            E                                        D  H    GA   A

LKSVLTTTSFEGTCL  250 RGAE Aspergilluz aculeatus
RELAGKLQAANR---  551 PaeY Erwinia chrysanthemi
GF-------------  382 YxiM  Bacillus subtilis
```

Figure 3. Comparison of the protein sequences of PaeY with its homologues. Conserved residues are indicated below the sequences.

2.3. CHARACTERIZATION OF THE PROTEIN PAEY

The *paeY* gene was cloned in a plasmid allowing overproduction of the recombinant protein (pVY47). Cell fractionation revealed that whole cell extracts of the *E. coli* strain BL21(DE3)/pVY47 contained both the precursor and the mature forms of PaeY (62 and 60 kDa) while the periplasmic fraction contained only the mature form (60 kDa) (data not shown). This result confirms that PaeY is an exported protein, presenting an N-terminal signal peptide efficiently cleaved in *E. coli.*

After optimization of the overproduction conditions, PaeY represented up to 50% of the periplasmic proteins. Purification of the mature PaeY protein consisted of the extraction of the periplasmic proteins from the recombinant *E. coli* cells, precipitation with ammonium sulfate and chromatography on a hydrophobic interaction column (Fig. 4).

Previously characterized pectin acetylesterases release acetyl groups from a range of synthetic substrates. We used two chromogenic substrates *p*-nitrophenyl acetate (pNP-acetate) and 5-bromo-6-chloro-3-indolyl acetate (X-acetate). Incubation of PaeY with pNP-acetate or X-acetate showed that PaeY is able to cleave these compounds, liberating the coloured product (Fig. 5). Triacetin is also a substrate of most acetylesterases. Incubation of PaeY with triacetin or sugar-beet pectin, followed by the detection of acetate liberation, clearly showed that PaeY is able to deacetylate these substrates (Fig. 5). PaeY removed about 15% of the acetyl groups of sugar-beet pectin.

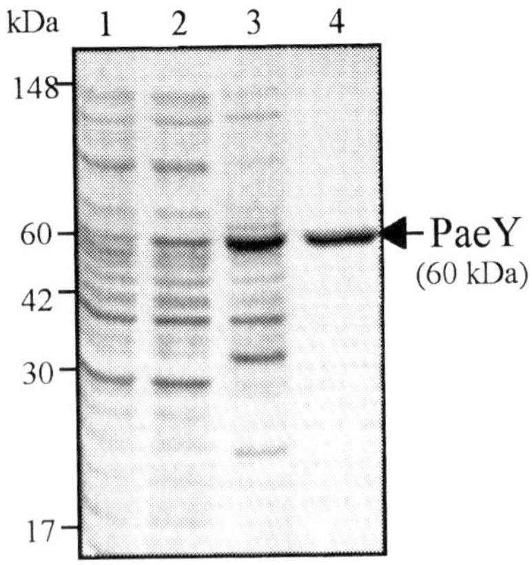

Figure 4. Purification of the PaeY protein.
The *E. coli* recombinant cells allowing overexpression of the *paeY* gene before (1) and after induction (2), periplasmic fraction (3) and PaeY purified after chromatography (4). The proteins are separated by SDS-PAGE and stained with Coomassie Blue G250.

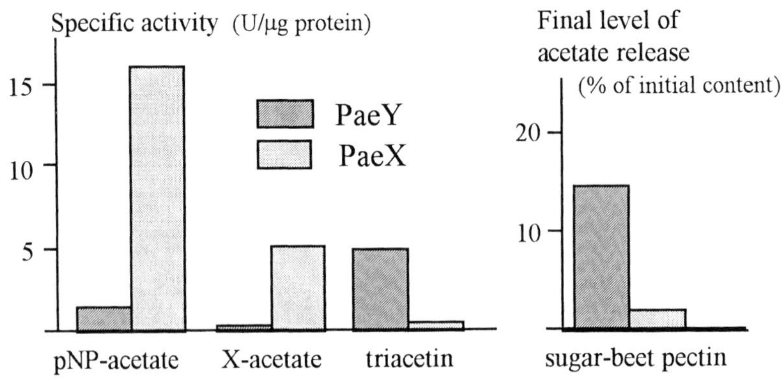

Figure 5. Acetylesterase activity of PaeX and PaeY on various substrates.
All assays were performed at 30°C. Hydrolysis of pNP-acetate (5 mM) in 0.1 M Na acetate buffer, pH 5.8, was directly monitored with a spectrophotometer at 400 nm. Cleavage of X-acetate (1 mM) was assayed in 0.1 M Tris-HCl buffer, pH 7.5; samples were removed at time intervals, added to 3 volumes of methanol and their absorbance was determined at 650 nm. Incubation with triacetin (50 mM) was performed in 0.1 M Tris-HCl buffer, pH 7.5; samples were removed at time intervals and, after heat inactivation, the concentration of acetate was determined using the acetic acid kit (Boehringer Mannheim).
Sugar-beet pectin was a gift from Copenhagen Pectin (mean molecular weight, 62 000; content in galacturonic acid residues, 66%; degree of methylation, 56%; degree of acetylation, 23%). To determine the final level of acetate release, incubation with sugar-beet pectin (10 mg/ml) was performed in 0.1 M Tris-HCl buffer, pH 7.5, for 12 hours and the concentration of acetate was determined using the acetic acid kit. In similar conditions, a negative control, i.e., an acetylesterase inactive on pectin, gave no acetate release.

3. The Pectin Acetylesterase PaeX

3. 1. IDENTIFICATION OF THE *PAEX* GENE

We recently identified the *E. chrysanthemi* genes involved in the uptake of oligogalacturonides resulting from extracellular pectin degradation, *togMNAB* and *kdgM* (Hugouvieux-Cotte-Pattat *et al.*, 2001) (Fig. 6).

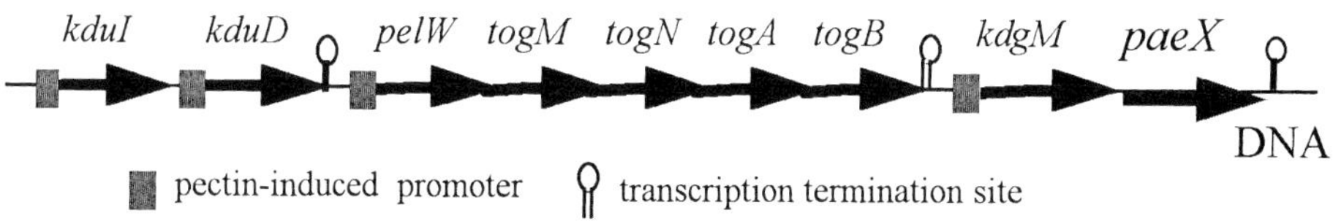

Figure 6. Genetic organisation of the region encoding the pectin acetylesterase PaeX.
The adjacent genes encode the following proteins: *kduI*, 5-keto-4-deoxyuronate isomerase; *kduD*, 2-keto-3-deoxygluconate oxidoreductase; *pelW*, exo-pectate lyase; *togMNAB*, the four elements of the ABC transporter involved in oligogalacturonide uptake; *kdgM*, an outer membrane porin.

To discover whether this region contains other genes involved in pectin degradation, we determined the nucleotide sequence of a 1.3 kb DNA fragment situated downstream of the *kdgM* gene (Fig. 6). Sequence analysis revealed the presence of a unique complete ORF which begins with an ATG codon at position 150 and ends with TAA at position 1116. This gene, named *paeX,* encodes a 322 aminoacid protein. The N-terminal sequence present the essential features of signal peptides but no typical cleavage site. In order to precise the position of the cleavage site, the N-terminal sequence of the purified mature PaeX protein (see below) was determined, indicating that the precursor form is cleaved after the alanine residue at position 21. The precursor and mature forms of PaeX have a predicted molecular mass of 34 940 and 32 716 Da, respectively.

3. 2. ANALYSIS OF THE PAEX AMINOACID SEQUENCE

Comparison of the deduced aminoacid sequence of PaeX with proteins present in data bases revealed homology with an hypothetical protein of *Caulobacter crescentus* (GenBank accession number AAK24284.1) (46% of identity over the entire length of the proteins) and with the C-terminal domain of the endo-1,4-β-xylanase XynB of *Butyrivibrio fibriosolvens* (37% of identity over 208 residues) (GenBank accession number X61495) (Fig. 7). Xylanases have often a modular organization, XynB is a large protein of 635 residues, supposed to contain two domains (Lin and Thomson, 1991). The N-terminal domain of XynB (residues 1 to 337) is homologous to xylanases (family 10 of glycosyl hydrolases) (Coutinho and Henrissat, 2000). In contrast, homology with PaeX is found in the XynB C-terminal domain of unknown function, from residue 405 to 609 (Fig. 7). Since enzymatic analysis revealed that PaeX has a pectin acetylesterase activity, the homology data indicate that the xylanase XynB is likely to be a bifunctional enzyme, with associated hydrolase and esterase activities.

In addition, other proteins showing similarity with PaeX were identified, including the heroin esterase Her of *Rhodococcus* sp. (24% of identity), the esterase EstB of *Thermotoga maritima* (23% of identity) and several hypothetical proteins from *C. crescentus* (34% of identity), *Deinococcus radiodurans* (34 and 27% of identity), *Lactococcus lactis* (33% of identity), and *Bacillus subtilis* (YtmA, 28% of identity).

```
                                                                          405  MLVIPGGGYC-MCCSHEGE  426
22  DTIFPIWPQGEAPGAATSSVQQQVVERSKDPTLPDRAVTGIRSPEITVYTPEKPNGTALLITPGGSYQRVVLDKEGS  100
40  TETMLLWPNG-APGGDGVDVEEKILERGDPKGLRDRAQVHTRVPKLVVFRPKTPNGAAVMLIPGGGYERVVLDKEGY  116
                                                                      PGG Y        EG

LAAMEFYNRGMNAFVLSYTTDITMSVPLHKQPLEDISRAVRFIRKNASKYNIDGKKLVIMGFSAGSHVCGSLAVHF-DDV  502
DLAPFFTRQGYTLFVMTYRMPGDGHQEGADAPLADAQRAIRTLRAHAAQWQIDPQRIGIMGFSAGGHVAASLGTRFAQTV  178
ETARWLSARGYTCFVLFYRLPGDGWAAGGDTPLQDAQRAVRLIRAQAAAMAFSADRVAVMGFSAGGHLAANLTTRFDVAT  195
  A    G  FV Y                PL D  RA R  R                MGFSAG H    L   F

KDNNPEYADISGRPDGVILSYPVITTGR-YTHADSVRTLLGANPTDEELTYFSLEKQVKDNTPPCFIWQTEEDSVVPVEN  542
YPAQDEIDHISARPDFMVLMYPVISMQENIAHAGSRKALIGSHPSDAQIQRYSAEKQVSAQTPPTFLVHAIDDPSVSVDN  219
YAPFDAADGLPAKPNLSALIYPVITLDPAFAHGGSRKQLLARQDTPERVAALSLEKQVREGLPPVFQLHAADDKAVPVEN  235
               P   L YPVI      H  S   L              S EKQV    PP F      D  V V N

SYLFANALREKKIPFAHYVFPRGFHGLTVANDEFFSGWSGGEYSMEQTMRARTAV              635  XynB
SLVMLAALRAHQIPAEIHLFEQGKHGFGIRGTVGLPAAIWPQLLDNWLTSLPLKKNTANQPDKK     322  PaeX
SLLMFSALKAKGVPAEMHIFEEGGHGFGLRFISGKPVAAWPGLFEAFAKRHGL                328  prot Cc
S    AL    P     F  G HG
```

Figure 7. Comparison of the protein sequences of PaeX with its homologues, XynB of *Butyrivibrio fibriosolvens* and a hypothetical protein of *Caulobacter crescentus* (prot Cc). Conserved residues are shown below the sequences and the residues supposed to be involved in the active site are underlined.

The most conserved region between PaeX, XynB and their homologues overlaps a sequence (GxSxG) typical of the active site of a large and diverse family of serine hydrolases, like carboxylesterases, lipases and proteases. The reaction mechanism of these enzymes involves a catalytic triad with serine, aspartic acid and histidine residues. Visual inspection of the sequence alignments revealed the presence of conserved aspartic acid and histidine in PaeX and homologous proteins. By analogy with the other serine hydrolases, S_{161} is the predicted serine active site residue of PaeX. The remaining two residues of the catalytic triad are probably D_{251} and H_{283} (Fig. 7).

The recent classification of esterases associated with the degradation of polysaccharides places PaeY and RGAE in the family 12 of carbohydrate esterases (Coutinho and Henrissat, 2000). In contrast, PaeX shows no significant homology with any members of the 12 defined families. Thus, PaeX and XynB appear to belong to a new family of carbohydrate esterases.

3.3. IDENTIFICATION OF THE PAEX PROTEIN

The *paeX* gene was inserted in a pT7 vector allowing specific expression of the cloned gene. Subcellular fractionation indicated that the precursor form of PaeX (32 kDa) is linked to the cells while the mature form (30 kDa) is found in the periplasmic fraction (data not shown). After optimization of the overproduction conditions, the mature PaeX protein was purified from the recombinant *E. coli* cells after fractionation and chromatography, first on a hydrophobic interaction column and then, on an ion exchange column (Fig. 8).

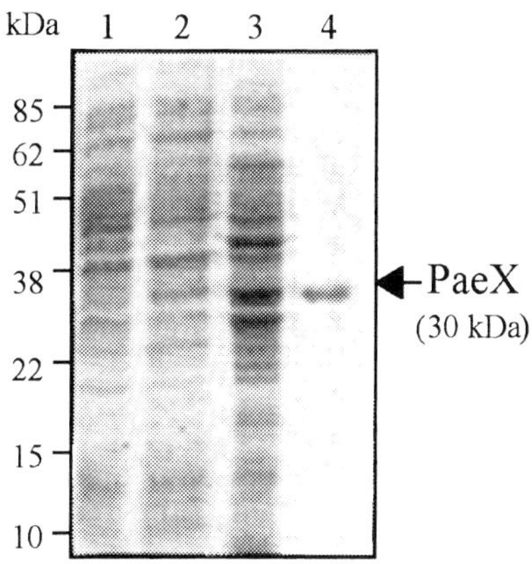

Figure 8. Purification of the PaeX protein.
The *E. coli* recombinant cells allowing overexpression of the *paeX* gene before (1) and after induction (2), soluble cell fraction (3) and PaeX purified after chromatography (4). The proteins are separated by SDS-PAGE and stained with Coomassie Blue G250.

The acetylesterase activity of PaeX was compared to that of PaeY, using various substrates (Fig. 5). Incubation of PaeX with the chromogenic substrates, pNP-acetate or X-acetate, showed that PaeX is able to cleave these compounds, liberating acetate and the coloured products. Incubation of PaeX with triacetin or sugar-beet pectin (23% acetylated), followed by the assay of acetate liberation, clearly showed that PaeX is able to deacetylate these substrates (Fig. 5). PaeX has a higher activity than PaeY with pNP-acetate or X-acetate as substrate but a lower activity than PaeY with triacetin or sugar-beet pectin. Indeed, PaeX removed only 3% of the acetyl groups of original sugar-beet pectin.

4. Detection of the Secreted *E. chrysanthemi* Acetylesterases

We developed a method for the direct detection of acetylesterases after isoelectrofocusing, using the synthetic substrate X-acetate (Shevchik and Hugouvieux-Cotte-Pattat, 1997) (Fig. 9). To analyse the different acetylesterase(s) produced by *E. chrysanthemi* in the presence of pectin, we used a *kdgK* mutant in which the expression of pectin-induced genes is strongly increased in the presence of pectin, polygalacturonate or galacturonate.

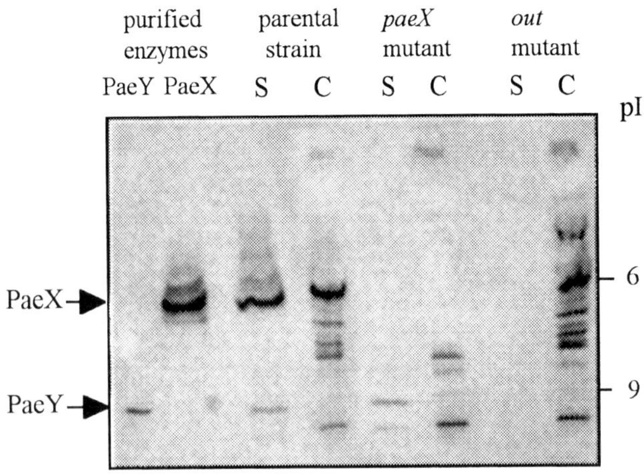

Figure 9. Electrofocusing followed by specific revelation of acetylesterase activity. The parental strain is the *kdgK* mutant A576, detection is performed with X-acetate. S: supernatant; C: cells.

A few coloured bands were observed in the supernatant of the *kdgK* mutant A576 grown in the presence of inducers (Fig. 9), No acetylesterase activity was detected in the supernatant of this strain grown in the absence of inducer (data not shown), indicating that these enzymes are produced only in the presence of pectic compounds.

Analysis of the purified proteins indicated that these bands correspond to the PaeX and PaeY proteins (Fig. 9). The apparent isoelectric point of PaeX and PaeY is about 6 and 9, respectively. Analysis of the *paeX* and *paeY* mutants confirmed these data, since the corresponding bands are absent from the respective mutant (Fig. 9 for *paeX*). PaeY is detected in the *E. chrysanthemi* culture supernatant but absent from the cells. In contrast, PaeX is found both in the supernatant and in the cells of strain A576. While PaeY is totally extracellular, PaeX appears to be only partially secreted.

The Out system is involved in the secretion of most of the *E. chrysanthemi* pectinases. We analysed an *out* mutant to determine whether this system is also involved in the secretion of PaeY and PaeX. No acetylesterase activity was detected in the supernatant of the *out* mutant, PaeX and PaeY are both retained in the periplasm (Fig. 8). Therefore, the two pectin acetylesterases are specifically secreted by the *E. chrysanthemi* Out system.

5. Synergy Between the *E. chrysanthemi* Pectinases

5.1. EFFECT OF DEMETHYLATION AND CLEAVAGE ON THE PECTIN ACETYLESTERASE ACTIVITY

We tested whether the action of other *E. chrysanthemi* pectinases, i.e., pectin methylesterases and pectate lyases, favours the action of the pectin acetylesterases PaeX and PaeY.

Both PaeY and PaeX act more efficiently after demethylation of pectin (pre-treatment with PemA) and even more efficiently, after both demethylation and cleavage of the pectic polymer (pre-treatment with PemA+PelB) (Fig. 10). Thus, PaeX and PaeY prefer demethylated pectic oligomers as substrate.

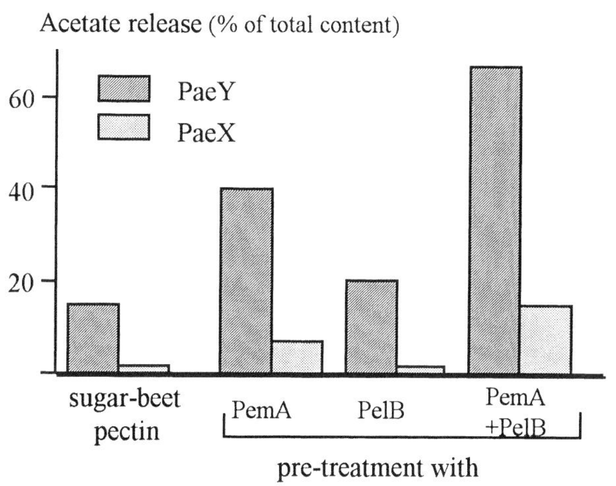

Figure 10. Acetyl esterase activity on sugar-beet pectin after pre-treatment with an endo-pectate lyase, PelB, or/and with a pectin methylesterase, PemA.

5.2. EFFECT OF DE-ESTERIFICATION ON THE PECTATE LYASE ACTIVITY

We also analysed whether the action of the extracellular endo-pectate lyases could be favoured by the deacetylation of pectin. We used the two *E. chrysanthemi* major pectate lyase isoenzymes PelD and PelE, and the two acetylesterases PaeX and PaeY (Fig. 11). Treatment of sugar beet pectin with any of the two acetylesterases led to a slight increase of the pectate lyase activity.

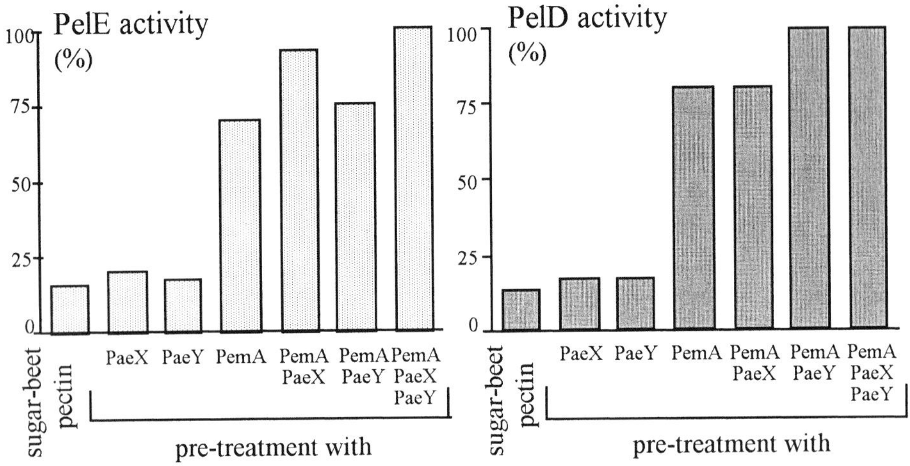

Figure 11. Pectate lyase activity of PelE and PelD on sugar-beet pectin after pre-treatment with the pectin esterases PaeX, PaeY and PemA

Since sugar beet pectin is also methyl-esterified (56%), we used the pectin methylesterase PemA to demethylate the substrate. The treatment of pectin with PemA strongly increased the activity of the pectate lyases. The combination of demethylation and deacetylation of the pectin led to a further improvement in pectate lyase activity. However, combined action of PemA with either PaeY or PaeX showed different effects, depending on the isoenzymes (Fig. 11). After PemA action, the PelE activity is further increased by PaeX action while the PelD activity is further increased by PaeY action. These differences in the increase in activity of the two pectate lyases could reflect a difference in the specificity (see below) of the two pectin acetylesterases towards the pectic substrates. Indeed, the existence of two isoenzymes produced by *E. chrysanthemi* rises the question of a putative difference in the specificity of each enzyme.

Two types of xylan acetylesterases were differentiated according to their substrate preference : either short oligomers or long-chain polymers (Christov and Prior, 1993). We have shown that both PaeX and PaeY prefer pectic oligomers.

Since the acetyl esters in pectin are O-2 or O-3 linked, an evident hypothesis is a specificity in the hydrolysis of only one type of esters. For instance, NMR spectrometry analysis indicated that the pectin acetylesterase of *Aspergillus niger* specifically attacks one type of acetyl ester of the pectic chain (Searle-van Leeuwen *et al.*, 1996). Such analysis could be performed to compare PaeY and PaeX.

Another type of specificity could be supposed with regards to the mechanism of de-esterification of these enzymes. For instance, pectin methylesterases act on pectin either by a processive mechanism, creating blocks of demethylated galacturonate, or by a multichain mechanism, giving a random demethylated polymer (Denes *et al.*, 2000).

Further studies of the pectin acetylesterases and of their reaction products will be necessary to clarify the type of specificity, towards substrates and/or mechanisms, that probably exists among these enzymes.

Acknowledgements

We gratefully acknowledge the members of this laboratory, particularly Sylvie Reverchon, Guy Condemine and William Nasser for helpful discussions. This work was supported by grants from the Centre National de la Recherche Scientifique and from the Ministère de l'Education Nationale et de la Recherche.

References

Christov, L. P., and Prior, B. A. à. l. (1993) Esterases of xylan-degrading microorganisms: production, properties, and significance. *Enzyme Microb Technol* **15**: 460-475.

Collmer, A., and Keen, N. T. (1986) The role of pectic enzymes in plant pathogenesis. *Annu Rev Phytopathol* **24**: 383-409.

Coutinho, P. M., and Henrissat, B. (2000) http:/afmb.cnrs.mrs.fr/~pedro/CAZY/.

Denes, J., Baron, A., Renard, C., and Drilleau, J. (2000) Different action patterns for pectinmethylesterase at pH7.0 and pH4.5. *Carbohydrate Research* **327**: 385-393.

Hugouvieux-Cotte-Pattat, N., Blot, N., and Reverchon, S. (2001) Identification of TogMNAB, an ABC tranporter which mediates the uptake of pectic oligomers in *Erwinia chrysanthemi* 3937. *Mol Microbiol*: in press.

Hugouvieux-Cotte-Pattat, N., Condemine, G., Nasser, W., and Reverchon, S. (1996) Regulation of pectinolysis in *Erwinia chrysanthemi*. *Annu Rev Microbiol* **50**: 213-257.

Hugouvieux-Cotte-Pattat, N., and Robert-Baudouy, J. (1989) Isolation of *Erwinia chrysanthemi* mutants altered in pectinolytic enzyme production. *Mol Microbiol* **3**: 1587-1597.

Kauppinen, S., Christgau, S., Kofod, L. V., Halkier, T., Dörreich, K., and Dalboge, H. (1995) Molecular cloning and characterization of a rhamnogalacturonan acetylesterase from *Aspergillus aculeatus*. *J Biol Chem* **270**: 27172-27178.

Lin, L.-L., and Thomson, J. A. (1991) Cloning, sequencing and expression of a gene encoding a 73 kDa xylanase enzyme from the rumen anaerobe *Butyrivibrio fibriosolvens* H17c. *Mol Gen Genet* **228**: 55-61.

Lojkowska, E., Masclaux, C., Boccara, M., Robert-Baudouy, J., and Hugouvieux-Cotte-Pattat, N. (1995) Characterization of the *pelL* gene encoding a novel pectate lyase of *Erwinia chrysanthemi* 3937. *Mol Microbiol* **16**: 1183-1195.

Pissavin, C., Robert-Baudouy, J., and Hugouvieux-Cotte-Pattat, N. (1996) Regulation of *pelZ*, a gene of the *pelBC* cluster encoding a new pectate lyase in *Erwinia chrysanthemi* 3937. *J Bacteriol* **178**: 7187-7196.

Searle-van Leeuwen, M. J. F., Vincken, J.-P., Schipper, D., Voragen, A. G. J., and Beldman, G. (1996) Acetyl esterases of *Aspergillus niger*: purification and mode of action on pectins. *Pectins and Pectinases*. Visser J. and Voragen A.G.J. Amsterdam, Elsevier: pp. 793-798.

Shevchik, V. E., Condemine, G., Hugouvieux-Cotte-Pattat, N., and Robert-Baudouy, J. (1996) Characterization of pectin methylesterase B, an outer membrane lipoprotein of *Erwinia chrysanthemi* 3937. *Mol Microbiol* **19**: 455-466.

Shevchik, V. E., and Hugouvieux-Cotte-Pattat, N. (1997) Identification of a bacterial pectin acetyl esterase in *Erwinia chrysanthemi*. *Mol Microbiol* **24**: 1285-1301.

Shevchik, V. E., Robert-Baudouy, J., and Hugouvieux-Cotte-Pattat, N. (1997) The pectate lyase PelI of *Erwinia chrysanthemi* belongs to a new family. *J Bacteriol* **179**: 7321-7330.

Tardy, F., Nasser, W., Robert-Baudouy, J., and Hugouvieux-Cotte-Pattat, N. (1997) Comparative analysis of the five major *Erwinia chrysanthemi* pectate lyases: enzyme characteristics and potential inhibitors. *J Bacteriol* **179**: 2503-2511.

DEVELOPMENT OF A VALENCIA ORANGE PECTIN METHYLESTERASE FOR GENERATING NOVEL PECTIN PRODUCTS

B.J. SAVARY[1], A.T. HOTCHKISS[1], M.L. FISHMAN[1], R.G. CAMERON[2], AND R.G. SHATTERS[3]
[1]Eastern Regional Research Center, USDA-ARS, 600 E. Mermaid Lane, Wyndmoor, PA 19038 (email: bsavary@arserrc.gov)
[2]Citrus and Subtropical Products Laboratory, USDA-ARS, 600 Ave. S NW, Winter Haven, FL 33883
[3]Horticultural Research Laboratory, USDA-ARS, 2001 S. Rock Road, Ft. Pierce, FL 34945

Abstract

We are interested in developing enzyme technologies for the preparation of novel specialty pectins and pectic materials from agricultural processing residues with new uses beyond current food applications. Enzymes can be used to modify efficiently and selectively pectin structure and to thereby change dependent functional properties. Recent work in our laboratories has focused on plant pectin methylesterases (PMEs), particularly those present in Valencia orange fruit. Our approach is to: 1) prepare highly pure individual isoenzymes, 2) correlate structural identity with biochemical and functional properties, 3) then apply the enzymes to well-defined pectin substrates, and 4) determine the structural and physical properties of the enzymatically modified pectins. We present here our progress to characterize the major salt-independent isoenzyme purified from Valencia orange peel and use it specifically to modify methylester patterns in citrus pectin.

1. Introduction

1.1. PECTIN UTILIZATION

Pectin is a major structural polysaccharide of cell walls in higher plants. The hydrocolloid properties of pectin are the basis for its primary use as a gelling and thickening agent in food applications. Annual consumption of pectin in the U.S. is now estimated to be about 9.5 million pounds, and about 100 million pounds world-wide. This represents a market value of at least \$50 million and \$500 million, respectively, with a growth of 5-6% annually. In the U.S., enormous quantities of pectin-rich residues are generated from citrus and sugar beet processing industries, from which about 8 billion pounds of pectin could potentially be produced annually. Despite the great abundance of these materials, no pectin is commercially produced domestically in the U.S. Such materials are used as low value animal feed or discarded. Assuming that high quality pectin could in fact be manufactured from this

F. Voragen et al. (eds.), Advances in Pectin and Pectinase Research, 345-361.
© 2003 *Kluwer Academic Publishers. Printed in the Netherlands.*

material, the potential production is in great excess of current food market demands. Therefore, to tap into this reservoir of underutilized biorenewable material, it is desirable to find new uses with industrial applications. Examples include biodegradable water-soluble films, bulking agents, coating agents, chelators, emulsifiers, viscosity modifiers and stabilizers. Such applications would generally represent high volume consumption with product values intermediate to current use in animal feed and food hydrocolloid applications. Development of new pectin-based products with unique functional properties will significantly advance this process. Such uses will also provide domestic substitutes for imported gums used in existing industrial applications.

1.2. THE DEGREE AND DISTRIBUTION OF METHYLESTERS IMPACT PECTIN FUNCTIONAL PROPERTIES

High molecular weight, high content of galacturonic acid, and high degree of esterification (DE) are critical structural properties of commercial pectins for their use as a gelling agent in food applications. The overall extent that galacturonic acids are methylesterified, the DE, dictate the conditions under which gels are formed and the gelling mechanism involved (Voragen et al., 1995; Rinaudo, 1996). The overall distribution of methyl esters, or conversely, free galacturonic acids, along the homogalacturonan chain are also critical for determining functional properties. The analysis of methyl ester distributions in pectin are addressed elsewhere in this volume in the chapter by Schols and Voragen. In gelling applications, pectins with DE > 50% require high soluble solids, typically sucrose (65%), and low pH (< 3.5) to gel, while pectins with DE < 50%, gelling requires polyvalent cations such as calcium. Calcium-mediated gelation of pectin may occur through formation of junction zones in which two antiparallel chains form cooperative interactions between blocks of at least seven un-esterified galacturonic acid residues on each pectin chain (Powell et al., 1982; Carpita and McCann, 2000). Calcium-mediated gelation of pectin may also occur through multiple parallel and antiparallel pectin chains given sufficient calcium is present. Some methylesterification can be tolerated in the multichain calcium-mediated association of pectin. Calcium gels with random ester distributions are stronger than gels with block ester distributions because of the homogeneous distribution of higher order interchain cross-linking prevalent in the random rather than block esterified pectins.

Upon initial extraction from plant tissues, commercial pectins possess a high DE (70-80%), which can be lowered by chemical de-esterification with acid, base, or ammonia (May, 1990; Voragen et al, 1995). Such treatments hydrolyze pectin methylesters randomly, but can also result in pectin depolymerization (Hills et al., 1949). Calcium sensitive pectins (CSPs) have been described in which pectins with DE > 50% can gel in the presence of calcium without the addition of sucrose as long as blocks of de-esterified pectin are present (Joye and Luzio, 2000). CSPs retain more water than low DE pectins (Joye and Luzio, 2000; Willats et al., 2001) making them softer and more commercially desirable for dietetic and many other food applications. A CSP, produced by limited treatment with the block-acting salt-dependent PME from navel orange peel, was able to stabilize casein in acidic milk-based beverage products (Glahn

1982; Christensen et al., 1997; Christensen et al., 1998). These products represent the fastest growing and highest value market for food grade pectin.

1.3. POTENTIAL FOR ENZYME TECHNOLOGIES

Enzymes have tremendous potential for creating new pectin bioproducts in an efficient and environmentally sustainable manner. Enzyme technologies have a range of advantages over conventional chemical approaches such as operation under mild temperature and pH conditions, high conversion efficiencies and substrate specificity, and eliminating exposure to, and disposal of, hazardous chemical catalysts. Carbohydrate modifying enzymes are now used extensively in a range of commercial processes (Table 1). For example, pectinases are increasingly important in preparation of fruit juices and processing of textiles (Kashyap et al., 2001). Pectinases can be used to study and selectively modify pectin structure (Williamson et al., 1998; Voragen et al., 2000), thereby providing a better understanding of fundamental relationships between pectin structure and functional properties. Using pectinases, a rational approach can be developed to generate pectins with improved performance and greater functionality beyond that possible with conventional chemical technologies. There are three requirements for rational application of an enzyme to pectin structural modification: 1) the enzyme must be catalytically pure, being free of contaminating pectinases, 2) the biochemical properties need to be well characterized, and 3) their action patterns on pectin substrates need to be established.

TABLE 1. Some enzymes used commercially in processing of carbohydrate substrates.

Starch and alcohol
Various amylases, pullulanase, glucose isomerase
Textiles; paper and pulp processing
Cellulases, xylanases, amylases, pectinases
Dairy, baking, and beverage products
β-galactosidase, glucose oxidase, amylases, pectinases

1.4. PROGRAM TO DEVELOP ORANGE PECTIN METHYLESTERASES

We are developing a library of highly pure and well-defined plant and microbial enzymes from both proprietary and commercial sources for use in our pectin-modification program. Recent work in our laboratories has focused on plant pectin methylesterases (PMEs) (Cameron et al. 1998; Savary, 2001), and details of current work are described elsewhere (Savary et al., 2002; Hotchkiss et al., 2002). PMEs specifically hydrolyze C6-carboxyl methylesters within homogalacturonan structural domains of pectin, releasing methanol from galacturonic acids. Plant PMEs generally are recognized to remove methylesters in a block-wise manner. Nevertheless it is possible some plant PMEs may act in a more random pattern. Block-wise de-esterification of pectin is only possible enzymatically. Such modifications can impact molecular flexibility and intermolecular interactions, subsequently influencing properties such as aggregation, chelation, and solubility. Our objective for this contribution is to summarize the current status of our progress to isolate and characterize the major PMEs from Valencia orange peel, principally the salt-independent isoenzyme, and to use it to modify methylesterification patterns in pectins present in fruit and vegetable processing residues. We also seek to highlight some of the recent advances in PME research relevant to our technological interest in this enzyme. We have focused on PMEs from Valencia orange peel since this is a rich source for producing these enzymes. A better understanding of their biochemical properties and action patterns will enable better manipulation of their activities. Further, since we are also interested in re-establishing pectin production in the U.S. from domestic orange peel, it is critical to understand how these endogenous enzymes may impact peel quality prior to pectin extraction.

For general reviews of PMEs, readers are directed to Rexová-Benková and Markovic (1976), Bordenave (1996) and Micheli (2001). Many new PMEs have been isolated and described in recent years. These can be readily located through public internet search databases such as ENTREZ (PubMed, GenBank and protein sequence searches) at the National Center for Biotechnology Information (www.ncbi.nlm.nih.gov/Entrez/) and the AGRICOLA database at the National Agricultural Library (www.nal.usda.gov/ag98/). Another valuable source is the set of EST sequence libraries, which are accessible through The Institute for Genomic Research website (www.tigr.org/tdb/tgi.shtml). For other recent reviews covering pectin and pectinases, readers are directed to Thakur et al. (1997), Williamson et al. (1998), Prade et al. (1999), Voragen et al. (2000), Kashyap et al. (2001) and Ridley et al. (2001).

2. Characterization of Orange Peel Pectin Methylesterase

2.1. MULTIPLE PECTIN METHYLESTERASES IN ORANGES

It is now well established that plants produce multiple forms of PMEs (Versteeg et al., 1978; Gaffé et al., 1994; Bordenave, 1996; Richard et al., 1996; Micheli, 2001). The diversity of PMEs in plants presumably reflects specialized functions in plant growth and development, and individual isoenzymes can be distinguished by tissue-specific expression patterns, nucleic acid sequence relationships, and their physical and biochemical properties. Evans and McHale (1968) first recognized that at least two major forms of PME occur in orange fruit tissues. Versteeg and coworkers (1978, 1980) subsequently described their properties in navel orange in detail. The PMEs were distinguished by relative binding strength on cation-exchange columns, pIs, kinetic properties, and their juice cloud destabilizing activities. Corresponding PMEs have been observed in lime (Evans and McHale, 1968), lemon (Macdonald et al., 1993), and grapefruit (Seymour et al., 1991; Cameron and Grohmann, 1995). We are conducting systematic evaluation of PMEs present in Valencia orange fruit (Savary et al., 2002; Cameron et al., 2002). We summarize our progress in subsequent sections, providing characterization of notable properties to establish unequivocal identity and behavior for subsequent application to modification of pectin methylesterification patterns.

We have classified the major PMEs isolated from Valencia orange into two groups based on the requirement for salt for enzyme activity (Savary et al., 2002). These two groups correspond to the navel orange PME-I and PME-II as designated by Versteeg et al. (1978). The distinguishing features between these two types, which are generally consistent with all the published literature on citrus PMEs, are summarized in Table 2. It is further notable that this classification extends to PMEs from other species, particularly those isolated from tomato fruit. We find orange PME types I and II correlate as orthologous homologues to the fruit-specific and vegetative ("ubiquitous") isoforms, respectively, in tomato (Gaffé et al., 1994; Warrilow and Jones, 1995; Savary, 2001).

Table 2. Properties distinguishing the two major PME types isolated from orange fruit tissues.

	Group I	**Group II**
Chromatography	Weak binding to cation exchangers (elute with ≤100 mM NaCl)	Moderate binding to cation exchangers (elute ≥200 mM)
Relative pI	Basic pI (about pH 9 or less)	Higher pI compared to PME-I (pIs >10)
Salt activation	Salt-independent	Salt-dependent
Tissue-specific expression	Fruit-specific?	Present in both fruit and vegetative tissues
Impact on juice cloud	Strong de-stabilizing activity	Weak de-stablizing activity
Kinetic rates	Higher Km compared to PME-II	Sensitive to polygalacturonate inhibition
De-esterification pattern	Block acting	Block acting
Amino acid sequence	Mature protein is not blocked	Blocked at terminal serine residue

2.2. PURIFICATION OF PECTIN METHYLESTERASES

We initially sought to use the fruit-specific PME from tomato for enzymatic de-esterification of pectin. The biochemistry, physiology, and genetics of this PME are well established in the literature. This enzyme was available as a commercial preparation, a convenience that has further promoted its wide use in many laboratories as a representative plant PME. However, preliminary studies with the preparation to modify pectin indicated undesired reduction of pectin polymer size. Subsequent characterization of this commercial preparation revealed significant contamination by the depolymerizing enzyme polygalacturonase, an acetylesterase, and a second PME isoenzyme with salt-dependent activity (Savary, 2001). The presence of these other enzymes may confound interpretations of action pattern for the fruit-specific PME (Grasdalen et al., 1996), and they would also interfere with our intent for specific modification of pectin methylester patterns.

To obtain a monocomponent (enzymatically pure) fruit-specific PME, a purification method based on perfusion chromatography media was developed to provide separations of this PME from contaminating pectinases (Savary, 2001). Application of perfusion chromatography (Afeyan et al., 1991) is a significant recent advancement in purification of PMEs because it provides high resolution separations with high flow rates. We were able to reduce routine chromatography runs from hours to minutes (Savary et al., 2002). Figure 1 provides a representative separation of Valencia orange PME-I by semipreparative cation-exchange chromatography. Similar separations have been performed using hydrophobic-interaction chromatography (Savary et al., 2001). Furthermore, we have similarly purified a range of enzymes by perfusion chromatography (Liao et al., 1998; Vivanco et al., 1999; Savary et al., 2001). Another approach reported recently for efficient purification of a PME was the use of a PME inhibitor protein in affinity chromatography (Denès et al., 2000a). The PME inhibitor protein was immobilized to an activated Sepharose matrix, then subsequently used to selectively isolate an apple PME from a crude fruit extract in a single step. The PME inhibitor protein does not seem to discriminate between PME types, since the apple PME purified behaves as a PME-II and the orange PME-I was used to originally isolate the inhibitor proteins. PME inhibitor proteins are presented elsewhere in this volume by Camardella and coworkers.

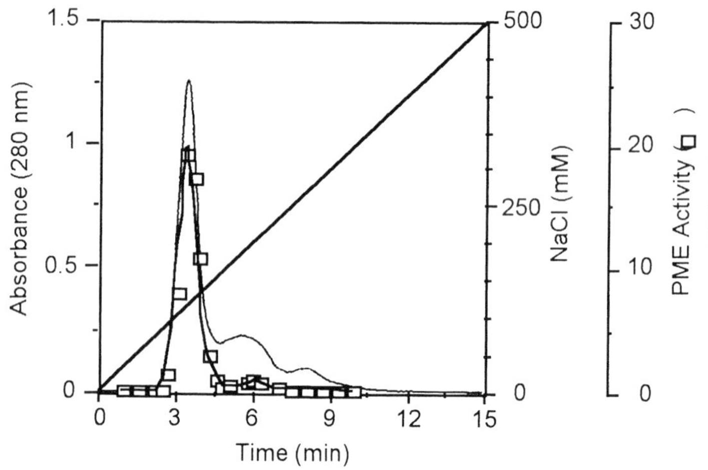

Figure 1. Representative separation of PME-I present in the commercial Valencia orange peel preparation using a Poros HS cation-exchange column (1.7 cm dia). Chromatography was performed at ca. 3 column volumes per min (50 ml/min).

Our research on plant PMEs shifted to the commercial Valencia orange peel PME because the manufacturer discontinued production of the tomato fruit PME preparation.

Our preliminary examination of the commercial preparation indicated equivalent properties with the tomato fruit-specific PME. Cameron and coworkers (1998) had recently separated several PME isoenzymes in Valencia orange peel using heparin-Sepharose media. Therefore, we began a detailed examination of the commercial orange PME to establish which isoenzyme it corresponded to. These studies (Savary et al., 2002) showed the commercial PME matched the peak 2 PME from Valencia peel (the PME-I) (Cameron et al., 1998) by identical elution properties on sulphopropyl- and heparin chromatography media in cation-exchange mode and by a hydrophobic interaction chromatography separation. Electrophoretic analysis further showed an identical relative molecular weight of 34,000 by SDS-PAGE (Figure 2A) and isoelectric point of 9.1 by IEF-PAGE, confirming equivalence of the commercial PME with Valencia peak 2 PME. The commercial enzyme preparation was also found to contain two smaller proteins (MW 27,000 and 8,000). These are derived from cleavage of the 34,0000 PME, which we determined by sequence homologies following partial sequencing of their N-termini (Savary et al., 2002). It is interesting that the fragments remained associated under native conditions and were fully active, since there were no differences in specific activities between purified commercial PME and the unfragmented peak 2 PME. Currently, we are further characterizing the peak 4 PME (PME-II) from Valencia orange fruit (Cameron et al., 1998; Cameron et al., 2002). Valencia PME-II is compared to PME-I in Figure 2B, migrating as a slightly smaller protein (MW about 33,500). The PME-II showed a much higher pI at 10.2 (Cameron et al. 2002). The relative elution conditions and pIs determined for Valencia peak 2 and peak 4 PMEs indicated they correspond, respectively, to PME-I and PME-II of navel orange PMEs (Versteeg et al., 1978; Christensen et al., 1998).

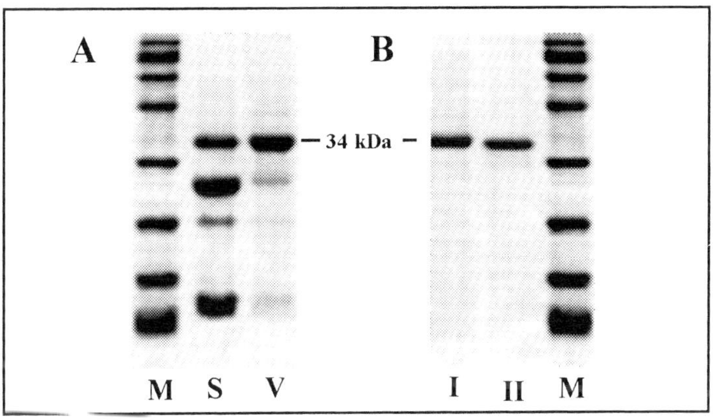

Figure 2. SDS-PAGE of purified Valencia orange PMEs. A, Comparison of the commercial PME (S, Sigma Chemical Co.) and Valencia peel PME-I (V) (peak 2 in Cameron et al., 1998). Fragments in S are 27,000 and 8,000. B, Comparison of PME-I and PME-II isolated from Valencia peel (peaks 2 and 4, respectively, in Cameron et al., 1998). Proteins are resolved on a 13.5% gel. Molecular weight markers (M) are broad range standards from Biorad.

To obtain further structural identification of the two Valencia PMEs, partial N-terminal amino acid sequences were obtained for both PME-I and PME-II (Savary et al., 2002; Cameron et al., 2002). The results from the sequencing studies are summarized in Figure 3. These results include a recent extension of the PME-I sequence. A partial internal peptide was also obtained for PME-I using the smaller fragment resolved electrophoretically from the commercial enzyme preparation. BLAST sequence searches using the N-terminal and internal peptides indicated the Valencia PME-I is new to the database. The Valencia PME-II was initially found to be blocked, like the salt-dependent PME-II previously sequenced from navel (Christensen et al., 1998) and Valencia orange (Nairn et al., 1998) and the tomato vegetative PME (Gaffé et al., 1997). However, treatment of protein blotted to PVDF membrane with trifluoroacetic acid (Wellner et al., 1990) provided sufficient cleavage of the block to determine the terminal six residues. The terminal residue was a serine, and the peptide aligned from this point with the other sequences reported for orange and tomato PME-IIs. It is interesting that the orange PME-I and PME-II showed sequencing behaviors similar to the tomato fruit-specific and vegetative PMEs, respectively. This suggests preservation of amino acid sequences or residues that direct final processing to form the mature protein. The data is organized in Figure 3 to reflect PME-I and -II as paralogous homologues, but establishing such phylogenetic relationships of orange PMEs awaits cloning and complete sequencing of PME-I.

```
Amino Terminal Peptide (Mature protein)                   Accession
Valencia PME-I          *1-vvpdvtvaad gsgnyltvaa vaaape
Tomato f-PME            *1-iianavvaqd gtgdyqtlae vaaapd   A25010
Tomato f-PME           230-iianavvaqd gtgdyqtlae avaaapd  S46527

Valencia PME-II      *1-s vtpnv
Navelina PME-II     †45-sss vtpnvvvaad gsgnfktvaa avaaapq  AAE44632
Valencia PECS1.2     44-sss vtpnvvvaad gsgnfktvaa avaaapq  AAB57668
Valencia PECS1.1    267-sss vtpnavvaad gsgnfktvaa avaaapq  AAB57667
Tomato PME-U        266-sst vtpdvvvaad gsgdyktvse avrkape  AAD09283

Internal Peptide
Valencia PME-I           *1-gsfetylgrp _krysrtv_m qsdl
Valencia CsPME2.1       408-ssvktflgrp wkqysrtvyi ktfl   AAB57699
Valencia CsPME1.1       482-gsfptylgrp wkeysrtvim qssi   AAB57667
Valencia CsPME1.2       188-gsfptylgrp wkeysrtvim qssi   AAB57668
Navelina PME-II         482-gsfptylgrp wkeysrtvim qssi ††AAE44632
Tomato PME-U            481-ksfptylgrp wkeysrtvim qssi   AAD09283
Tomato f-PME            442-kefptylgrp wkeysrtvvm esyl   S46527
```

[*]Sequences obtained directly from protein, others from cDNAs.
[†]Start of mature protein determined by protein sequencing.
[††]Accession corresponds to cDNA sequence in Christensen et al., 1998.

Figure 3. Sequence alignments of partial amino acid sequences obtained from Valencia orange PMEs and corresponding PMEs from orange and tomato. Accession numbers are given for location in ENTREZ protein database.

2.3. ACTIVITY PATTERNS OF PME ISOENZYMES

We associated distinctive salt-activation patterns with PME isoenzymes following analysis of the two PMEs isolated from the commercial tomato fruit enzyme preparation (Savary, 2001). We found similar activity patterns for PME-I and II from Valencia orange, which corresponded with the tomato fruit-specific and vegetative PMEs, respectively. These results are summarized in Figure 4. Type II PMEs exhibit an absolute requirement for supplemented salt (e.g., sodium chloride) for activity at neutral pH. In contrast, the type I PMEs remain highly active in the absence of supplemented salt at neutral pH. Furthermore, it remains active under mildly acidic conditions (pH 5) in the presence of salt, whereas the salt-dependent PME-II was completely inactive under these conditions.

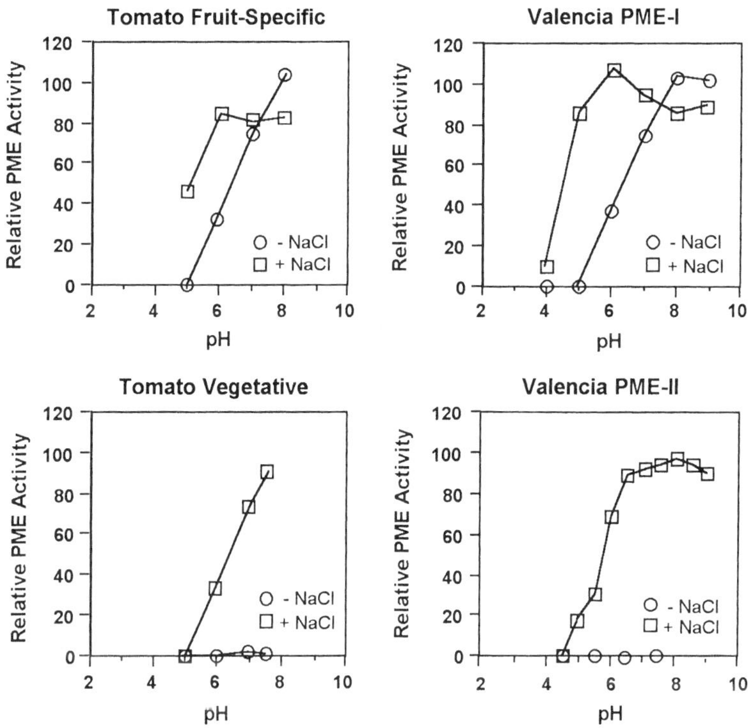

Figure 4. Salt activation patterns for tomato and orange PMEs. Partial pH optima activity curves were determined for both isoenzyme types in the presence or absence of sodium chloride.

Closer review of published reports on PMEs provided some data where two distinct isoenzymes purified from the same species showed such differences in salt-dependent pH activity curves (Warrilow and Jones, 1995; Bordenave et al., 1996). The salt-dependency of a plant PME, particularly those isolated from fruit tissues, may be a useful screen to initially determine PME type. The major PME isolated in apple was found to require supplemented salt for activity (Macdonald et al., 1996; Denès et al., 2000a). The pI was >pH 9, and had no effect on juice cloud stability, further properties consistent with PME-II of orange and tomato. Macdonald et al. (1996) had also detected a minor activity peak in apple fruit that preceded the major PME in elution from the cation-exchange column. They claimed it was this isoenzyme that promoted cloud clarification, which is suggestive of it being type 1 PME, but this second isoenzyme was not further characterized. In citrus, a third isoenzyme is distinguished for its tolerance to thermal inactivation at high temperature (Versteeg, 1980; Cameron et al., 1998). This isoenzyme has potent cloud destabilizing activity and is active without supplemented salts, indicating a relationship with PME-I. Detailed structural and biochemical analyses of this isoenzyme, however, have not been reported.

2.4. ENZYMATIC PECTIN DE-ESTERIFICATION PATTERNS

Plant PMEs are generally regarded to act in a block-wise pattern while microbial PMEs have been reported with both block-wise and random patterns (Rexová-Benková and Markovic, 1976; Ishii et al., 1979; Markovic and Kohn, 1984; Bordenave, 1996). Block-acting PMEs proceed linearly along the pectin backbone, starting with a free carboxyl group at the reducing end (Solms and Deuel, 1955; Lee & Macmillam, 1970) or nonreducing end (Limberg et al., 2000), producing a block of deesterified pectin. It is also possible that a random-acting plant PME may exist. Willats et al. (2001) recently reported immunolocalization evidence that non-blockwise sequences of pectin exist in the cell wall at the points of intercellular spaces. A phylogenetic analysis of a range of plant and microbial PMEs indicated a *Brassica napus* pollen-specific PME is structurally similar to phytopathogenic fungi and bacteria (Richard et al., 1996). Plant PMEs typically have pI and pH optima in the neutral to alkaline range, while microbial PMEs tend to cover a more acidic range. Markovic and Kohn (1984) suggested pI and pH optima may influence the mode of action by a PME. A plant PME with an acidic pI has been reported from jelly fig (Lin et al., 1989). The action pattern of jelly fig PME on pectin has not been described, but it would be interesting to know if this acidic plant PME de-esterifies pectin randomly at neutral pH.

NMR analysis has further refined the action patterns reported for plant PMEs (Grasdalen et al., 1996; Neiss et al., 1999). Block action patterns include single chain mode, where the enzyme acts processively along one pectin chain, and multiple-attack mode, where the enzyme attaches and moves along the pectin chain hydrolyzing only a limited number of methyl ester groups. Multiple-attack mode thereby produces short blocks of de-esterified pectin. Grasdalen et al. (1996) first reported the multiple attack PME action pattern on pectin using ^{1}H NMR. This direct analysis of an action pattern used the commercial tomato PME preparation (Sigma Chemical Co.) on a methylated polygalacturonic acid substrate (DE = 92%). However, it is now recognized that the commercial enzyme preparation contained a second PME isoenzyme as well as polygalacturonase (Savary, 2001). Therefore, it remains unclear if the smaller blocks of de-esterified pectin measured were principally due to the fruit-specific PME or the contaminating pectinases. The availability of both tomato isoenzymes in pure form warrants further investigation of their action patterns.

Recent reports based on NMR analysis have provided evidence that plant PME action pattern can differ depending on the pH. Studies with mung bean hypocotyl γ and β PME isoenzymes and an apple salt-dependent PME isoenzyme showed shifts from single chain mechanism to multiple-attack mechanism following a change from the isoenzymes' pH optimum to either a more basic or acidic pH (Catoire et al., 1998; Denès et al., 2000b). This suggests that the local pH is critical in determining the block size of de-esterified pectin produced by plant PMEs. Several reports of block-wise acting PMEs from orange have appeared (Powell et al. 1982; Markovic and Kohn, 1984; Schols et al., 1989) without specifying the isoenzyme type. Limberg et al. (2000) determined that the salt-dependent navel orange PME-II (Christensen et al., 1998) de-esterified methylated lime pectin (DE = 81%) in a block-wise pattern by enzymatic fingerprinting. We have also demonstrated that the salt-independent PME-I from Valencia orange acts in a block-wise manner by ^{1}H NMR and a calcium-sensitive pectin ratio assay (Hotchkiss et al., 2002; and discussed in Sec. 3).

2.5. ADVANCES IN MOLECULAR CHARACTERIZATION OF CITRUS PMEs

The first complete protein sequence obtained for a plant PME was reported for a tomato fruit-specific PME by Markovic and Jörnvall (1986). Numerous groups have produced corresponding cDNA and genomic clones to understand its role in fruit development (e.g., Pear et al., 1989; Turner et al., 1996). A tomato vegetative-type PME has also been cloned and shown to exist as a distinct single gene-copy isoform (Gaffe et al., 1997). Sequencing of the *Arabidopsis* genome has revealed as many as 67 different PMEs are present in that species (*Arabidopsis* Genome Initiative, 2000), indicating PMEs likely exist as large gene families in all higher plants. Two groups have now obtained genomic and cDNA clones of PMEs from Valencia and navel oranges, (Nairn et al. 1998; Christensen et al., 1998; respectively). Analysis of sequence homologies, activity properties (both are salt-dependent), and tissue expression patterns of the PMEs described in these two studies revealed all clones obtained commonly represent type-II PMEs. The Valencia PME clone that is homologous to the navel PME was found to be expressed in various tissues in addition to fruit tissue (Nairn et al., 1998). Tissue-specific accumulation patterns in fruit tissues were demonstrated by Christensen et al. for the navel orange PME-II (1998). Interestingly, expression-patterns for the second PME-II group (PECS2) isolated from Valencia orange were sensitive to ethylene induction and specific to abscission zones (Nairn et al., 1998). No clones have been reported from oranges that correspond to a PME-I isoenzyme. We suspect orange PME-I is a fruit-specific isoenzyme and that accumulation levels vary with maturity of fruit in patterns similar to accumulation of the PME isoenzymes in tomato fruit (Gaffé et al., 1994). Evidence for this is the wide variation in accumulation ratios reported between PME-I and PME-II (Evans and McHale, 1978; Versteeg et al., 1978; Cameron et al., 1998; Christensen et al., 1998).

A searchable database for screening citrus expressed sequence tag (EST) libraries is under development at the USDA-ARS Horticultural Research Laboratory (Ft. Pierce, FL). This database is intended to support programs targeting the genetic improvement of citrus crops. Currently it contains over 37,000 individual sequences with greater than 10,000 unique clones represented by at least one original sequence. Clones were obtained by conducting random one-pass sequencing of cDNA libraries created from various citrus tissues including flavedo, young whole fruit, root, flower, whole seedling and leaves. Progress is continuing to expand the tissue libraries. Each sequence obtained was compared to others in the database and those that shared sequence identity greater than 85% in overlap of at least 50 bases were combined into a single consensus. A consensus sequence (Figure 5) was derived from the internal peptide sequence from the salt-independent PME-I. This sequence contains residues that are highly conserved with all PMEs, including the previously sequenced salt-dependent PME-IIs. This sequence was therefore expected to not distinguish between PME isoenzyme types. The citrus EST library database was then searched for ESTs sharing greater than 80% homology. Fifteen different clones were thus obtained and further examined for identity. The library and distribution of sequence "hits" are shown in Table 3. All isolates were determined to encode members of the PECS1-group (PME-II) of PMEs (Nairn et al., 1998). Further searches will be conducted, particularly from additional libraries derived from albedo and late whole fruit, to isolate clones for the salt-independent PME. The lack of success for isolating this clone from the libraries produced so far is consistent with the hypothesis that the salt-independent PME is fruit-specific and expressed late in fruit development, like its functional homologue in tomato. Availability of a fruit-specific clone will provide a probe for characterizing the tissue-specific developmental expression patterns of this PME isoenzyme.

Table 3. Citrus EST Libraries screened for PME cDNA clones.

cDNA source	Total clones sequenced	Total PME clones matched
Trifoliate whole young fruit (Flying Dragon)	3,019	3
Sweet orange whole young fruit (Valencia)	6,854	4
Sweet orange whole seedli ng (Ridge Pineapple)	6,912	6
Trifoliate whole seedling (Large flower trifoliate)	6,912	1
Grapefruit young fruit flavedo (Marsh)	1,245	1
Sweet orange leaf (Valencia)	3,032	0
Trifoliate root (Swingle)	820	0
Sweet orange flower (Valencia)	2,110	0

Peptide	T	Y	L	G	R	P	W	K	-	Y	S	R	T	V	-	M
Sequence	ACG	TAC	CTC	GGC	AGG	CCC	TGG	AAG	NNN	TAC	TCG	AGG	ACG	GTG	NNN	ATG

Figure 5. Consensus sequence used to probe citrus EST libraries. This was derived from the PME-I internal peptide, but selected for its high conservation among plant PMEs.

3. Analysis of Pectins De-esterified with the Salt-Independent PME-I

3.1 CALCIUM-SENSITIVITY

We treated a non-calcium-sensitive pectin derived from lemon peel (DE = 76%) (Joye and Luzio, 2000) with the salt-independent PME-I from Valencia orange, reducing the total DE by just 6% (Hotchkiss et al., 2002). Examination of the treated pectin using ^{1}H-NMR showed new chemical shifts in the spectra at 4.80 - 4.83 ppm that were assigned to H_5 GGG blocks accordingly to Neiss et al. (1999). Further analysis of the same enzyme-treated pectins showed a dramatic shift to a calcium-sensitive pectin ratio values, indicating that sufficient blocks of de-esterified pectin had been formed to make these treated pectins sensitive to calcium. Together, these results clearly demonstrated the highly puried salt-independent orange PME-I in fact acts in a block-wise manner. Furthermore, they showed that calcium-sensitive pectin, with high methoxyl content, can be prepared with this isoenzyme.

3.2 DETERMINATION OF PECTIN MOLECULAR PROPERTIES

We subsequently used the block-wise acting salt-independent Valencia orange PME-I to produce a series of de-esterifed pectins using a rapid-set citrus pectin as the starting material (Hotchkiss et al., 2002). The pectins were analyzed using HPSEC with three detectors (multiangle lasar light-scattering, differential viscometer, and refractive index) (Fishman et al., 2000, 2001). This systems allows the measurement of weight-average molecular weight (M_w), the Z-average root mean square radius of gyration (R_{gz}), the Mark-Houwink exponent (a), and the intrinsic viscosity $[\eta]_w$. The R_{gz} is a measure of molecular size, a is a measure of shape, and $[\eta]_w$, is further a measure of molecular volume, provided solvent-molecule interactions do not change. These molecular properties are plotted as a function of degree of methyl-esterification (DE) of the treated pectins in Figure 6. Over the entire range of de-esterification, M_w, R_{gz}, and a remained essentially constant. The intrinsic viscosity $[\eta]_w$, remained relatively constant down to about 40% DE, but then decreased significantly below 40% DE. Since below 40% DE only $[\eta]_w$ decreased among molecular properties, we must assume this occurred as a result of changes in interactions between pectin and the aqueous solvent in which it is dissolved. Molecular volume could only change if R_{gz}, M_w or a changed. Normally, one might expect that increasing the percentage of polar groups on pectin (i.e., carboxyl groups) by de-esterification would increase interactions between pectin and polar water. This in turn would produce an increase in $[\eta]_w$. In previous studies (Fishman et al., 1993, 2000, 2001, and elsewhere in this volume) we have shown that part of pectin preparations exist as a network structure. The $[\eta]_w$ of these networks would partially depend on how well they drain solvent when in a flow field. If producing new carboxyl groups increased the cross-link density of the network by intramolecular hydrogen bonding, this might cause the network to be less free draining and thus decrease $[\eta]_w$.

The results with the salt-independent Valencia orange PME-I demonstrated that a plant PME can be used to produce a low methoxyl pectin without simultaneous depolymerization. Most previous reports on enzymatically de-esterified pectin showed extensive reduction in M_w and $[\eta]_w$. Recently, de-esterification of a lime pectin with a block-wise salt-dependent orange PME-II (Limberg et al., 2000) resulted in extensive reduction in M_w and $[\eta]_w$ (nearly 50%). The extent of M_w and $[\eta]_w$ reduction was similar to our results using an alkali hydrolysis without control of β-elimination depolymerization reaction (Hotchkiss et al., 2002). The PME-II enzyme preparation was reported to be free of contaminating pectinases (Christensen et al., 1998). It is not clear at this point if the difference in results between these two studies is due to the conditions for processing the treated pectins, properties of the pectins used, or somehow due to the enzymes themselves. These questions will be resolved by direct comparative treatments of a common pectin using both PME-I and II.

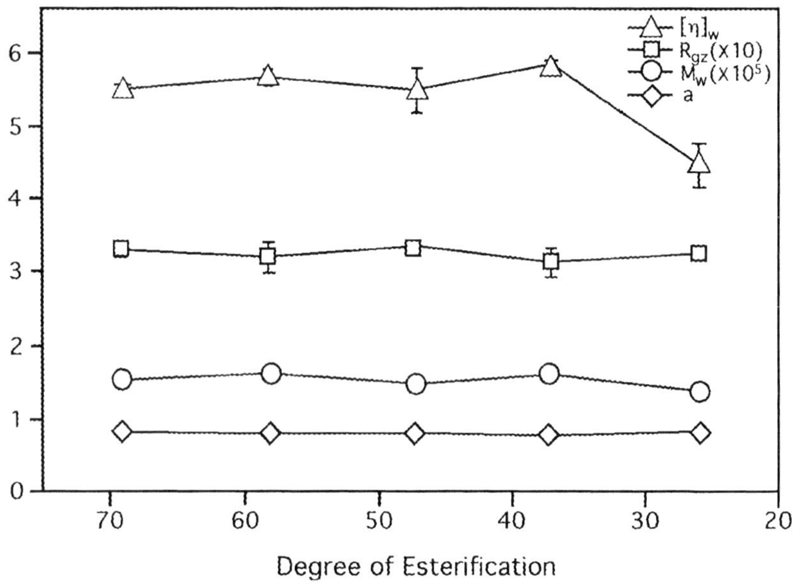

Figure 6. Molecular properties of pectins resulting from de-esterification using the salt-independent Valencia orange PME-I. The properties include weight average intrinsic viscosity, $[\eta]w$ (in dL/g), Z-average root mean square radius of gyration, R_{gz} (X 10 nm), weight average molar mass, M_W (X 10^5), and the Mark-Houwink exponent, a.

4. Conclusion

This work presented clear identification and further characterization of a salt-independent type I PME from Valencia orange. The range of biochemical and structural properties determined for it indicates it is an orthologous homologue to the fruit-specific PME of tomato. Further molecular characterization and demonstration of tissue-specific expression patterns are pending the isolation of a corresponding cDNA and genomic clone. We have demonstrated the potential industrial utility of this isoenzyme to generate novel pectins by preparing modified pectins with increased calcium sensitivity while preserving molecular weight.

Acknowledgements

We gratefully acknowledge the discussions of Kevin Hicks, Gary Luzio, and Jim Wade, and Michael Bausher for his efforts in developing the searchable citrus EST database, and we thank Rose Chau, André White, Pamela Rockwell-Warner, and Dave Coffin for their technical contributions. This research was partly supported through a Cooperative Research and Development Agreement with Hercules, Inc., Wilmington, DE.

References

Afeyan, N.B., Fulton, S.P., Regnier, F.E. (1991) Perfusion chromatography packing materials for proteins and peptides. *J. Chromatog.* **544**, 267-279.

Arabidopsis Genome Initiative, (2000) Analysis of the genome sequence of the flowering plant Arabidopsis thaliana. *Nature* **408**, 796-815.

Bordenave, M. (1996) Analysis of pectin methyl esterases. In *Plant Cell Wall Analysis*. H.F. Linskens, and J.F. Jackson (eds.), *Modern Methods of Plant Analysis*, Vol. 17, Springer-Verlag, Berlin, pp. 165-180.

Bordenave, M., Breton, C., Goldberg, R., Huet, J.-C., Perez, S., Pernollet, J.-C. (1996). Pectinmethylesterase isoforms from *Vigna radiata* hypocotyl cell walls: kinetic properties and molecular cloning of a cDNA encoding the most alkaline isoform. *Plant Mol. Biol.* **31**, 1039-1049.

Cameron, R. G., Baker, R.A., Grohmann, K. (1998) Multiple forms of pectinmethylesterase from citrus peel and their effects on juice cloud stability. *J. Food Sci.* **63**, 253-256.

Cameron, R.G., Grohmann, K. (1995) Partial purification and thermal characterization of pectinmethylesterase from red grapefruit finisher pulp. *J. Food Sci.* **60**, 821-825

Cameron, R.G., Savary, B.J., Hotchkiss, A.T., Baker, R.A., Grohmann, K. (2002). Separation and characterization of a salt-dependent pectin methylesterasefrom *Citrus sinensis* (var Valencia) fruit tissue. (Manuscript submitted for publication.)

Carpita, N., McCann, M. (2000) The cell wall. In *Biochemistry and Molecular Biology* of Plants, B. Buchanan, W. Gruissem, R. Jones (eds), American Society of Plant Physiologists, Rockville, p. 82.

Catoire, L., Pierron, M., Morvan, C., Hervé du Penhoat, C., Goldberg, R. (1998) Investigation of the action patterns of pectinmethylesterase isoforms through kinetic analyses and NMR spectroscopy. *J. Biol. Chem.* **273**, 33150-33156.

Christensen, T.M.I.E., Kreiberg, J.D., Thorsøe, H., Bucholt, H.C., Rasmussen, P., Mikkelsen, J.D. (1997). Process for stabilizing proteins in an acidic environment with a high-ester pectin. International patent,WO 97/03574.

Christensen, T.M.I.E., Nielson, J.E., Kreiberg, J.D., Rasmussen, P., Mikkelsen, J.D. (1998) Pectin methyl esterase from orange fruit: characterization and localization by *in-situ* hybridization and immunohistochemistry. *Planta* **206**, 493-503.

Denès, J. M., Baron, A., Drilleau, J.F. (2000a) Purification, properties and heat inactivation of pectin methylesterase from apple (cv Golden Delicious). *J. Sci. Food Agric.* **80**, 1503-1509.

Denès, J. M., Baron, A., Renard, C.M.G.C., Péan, C., Drilleau, J.F. (2000b) Different action patterns for apple pectin methylesterase at pH 7.0 and 4.5. *Carbohyd. Res.* **327**, 385-393.

Daas, P.J.H., Voragen, A.G.J., Schols, H.A. (2001) Study of the methyl ester distribution in pectin with endo-polygalactuonase and high-performance size-exclusion chromatography. *Biopolymers* **58**, 195-203.

Evans, R., McHale, D. (1978) Multiple forms of pectinesterase in limes and oranges. *Phytochem.* **17**, 1073-1075.

Fishman, M.L., Cooke, P., Hotchkiss, A.T., Damert, W. (1993) Progressive dissociation of pectin. *Carbohydr. Res.* ***248***, 303-316.

Fishman, M.L., Chau, H.K., Hoagland, P., Ayyad, K. (2000) Characterization of pectin flash extracted from orange albedo by microwave heating under pressure. *Carbohydr. Res.* **323**, 126-138.

Fishman ML, Chau HK, Kolpak F, Brady J. (2001) Solvent effects on the molecular properties of pectins. *J. Agric. Food Chem.* **49**, 4494-4501.

Gaffé, J., Tieman, D.M., Handa, A.K. (1994) Pectin methylesterase isoforms in tomato (*Lycopersicon esculentum*) tissues: effects of expression of a pectin methylesterase antisense gene, *Plant Physiol.* **105**, 199-203.

Gaffé, J., Tiznado, M.E., Handa, A.K. (1997) Characterization and functional expression of a ubiquitously expressed tomato pectin methylesterase. *Plant Physiol.* **114**, 1547-1556.

Glahn, P.E. (1982) Hydrocolloid stabilization of protein suspensions at low pH. *Prog. Food Nut. Sci.* **6**, 171-177.

Grasdalen, H., Andersen, A.K., Larsen, B. (1996) NMR spectroscopy studies of the action pattern of tomato pectinesterase: generation of block structure in pectin by a multiple-attack mechanism. *Carbohyd. Res.* **289**, 105-114.

Hall, L.N., Bird, C.R., Picton, S., Tucker, G.A., Seymour, G.B., Grierson, D. (1994) Molecular characterization of cDNA clones representing pectin esterase isozymes from tomato. *Plant Mol. Biol.* **25**, 313-318.

Hills, C.H., Mottern, H.H., Nutting, G.C., Speiser, R. (1949) Enzyme-demethylated pectinates and their gelation. *Food Technol.* **3**, 90-94.

Hotchkiss, A.T., Savary, B.J., Cameron, R.G. Chau, H.K., Brouillette, J., Fishman, M.L. (2002) Enzymatic modification of pectin to increase its calcium sensitivity while preserving its molecular weight. *J. Agric. Food Chem.* (Manuscript accepted for publication.)

Ishii, S., Kiho, K., Sugiyama, S., Sugimoto, H. (1979) Low-methoxyl pectin prepared by pectinesterase from *Aspergillus japonicus*. *J. Food Sci.* **44**, 611-614.

Joye, D.D., Luzio, G.A. (2000) Process for selective extraction of pectins from plant material by differential pH. *Carbohyd. Polymers* **43**, 337-342.

Kashyap, D.R., Vohra, P.K., Chopra, S., Tewari, R. (2001) Applications of pectinases in the commercial sector: a review. *Bioresource Technol.* 77, 215-227.

Lee, M., Macmillan, J.D. (1970) Mode of action of pectic enzymes. III. Site of initial action of tomato pectinesterase on highly esterified pectin. *Biochemistry* **9**, 1930-1934.

Liao, C.-H., Revear, L., Hotchkiss, A.T., Savary, B.J. (1999) Genetic and biochemical characterization of an exopolygalacturonase from *Yersinia enterocolitica*. *Can. J. Microbiol.* **45**, 396-403.

Limberg, G., Körner, R., Buchholt, H.C., Christensen, T.M.I.E., Roepstorff, P., Mikkelsen, J.D. (2000) Quantification of the amount of galacturonic acid residues in blocksequences in pectin homogalacturonan by enzymatic fingerprinting with exo- and endo-polygalacturonase II from *Aspergillus niger*, *Carbohyd. Res.* **327**, 321-332.

Lin, T.-P., Liu, C.-C., Chen, S.-W., Wang, W.-Y. (1989) Purification and characterization of pectinmethylesterase from *Ficus awkeotsang* Makino achenes. *Plant Physiol.* **91**, 1445-1453.

Macdonald, H.M., Evans, R. (1996) Purification and properties of apple pectinesterase. *J. Sci. Food Agric.* **80**, 321-36.

Macdonald, H.M., Evans, R., Spencer, W.J. (1993) Purification and properties of the major pectinesterases in lemon fruits (*Citrus limon*). *J. Sci. Food Agric.* **62**, 163-168.

Markovic, O., Jörnvall, H. (1986) Pectinesterase: the primary structure of the tomato enzyme. *Eur. J. Biochem.* **158**, 455-462.

Markovic, O., Kohn, R. (1984) Mode of pectin deesterification by *Trichoderma reesei* pectinesterase. *Experientia* **40**, 842-843.

May, C. D. (1990) Industrial pectins: sources, production and applications. *Carbohyd. Polym.* **12**, 79-99.

Micheli, F. (2001) Pectin methlyesterases: cell wall enzymes with important roles in plant physiology. *Trends Plant Sci.* **6**, 414-419.

Nairn, C.J., Lewandowski, D.J., Burns, J.K. (1998) Genetics and expression of two pectinesterase genes in Valencia orange. *Physiol. Plantarum* **102**, 226-235.

Neiss, T.G., Cheng, H.N., Daas, P.J.H., Schols, H.A. (1999). Compositional heterogeneity in pectic polysaccharides. In *Macromolecular Symposia*, Vol. 140, H.N. Cheng, G.L. Cote (eds.), Wiley-V C H Verlag Gmbh., Berlin, pp. 165-178.

Pear, J.R., Ridge, N., Rasmussen, R., Rose, R.E., Houck, C. (1989) Isolation and charactrization of a fruit-specific cDNA and the corresponding genomic clone from tomato. *Plant Mol. Biol.* 13, 639-651.

Powell, D.A., Morris, E.R., Gidley, M.J. Rees, D.A (1982) Conformations and interactions of pectins. II. Influence of residue sequence on chain association in calcium pectate gels. *J. Mol Biol.*, **155**, 517-531.

Rexová-Benková, L., Markovic, O. (1976) Pectic enzymes. *Adv. Carbohyd. Chem. Biochem.* 33, 323-385.

Richard, L., Qin, L.-X., Goldberg, R. (1996) Clustered genes within the genome of *Arabidopsis thaliana* encoding pectin methyleterase-like enzymes. *Gene* **170**, 207-211.

Ridley, B. L., O'Neill, M., Mohnen, D. (2001) Pectins: structure, biosynthesis, and oligogalacturonide-related signaling. *Phytochem.* **57**, 929-967.

Rinaldo, M. (1996) Physiochemical properties of pectins in solution and gel states. In *Pectins and Pectinases*, J. Visser, A.G.J. Voragen (eds.), Elsevier, New York, pp. 21-33.

Savary, B.J. (2001) Perfusion chromatography separation of the tomato fruit -specific pectin methylesterase from a semipurified commercial enzyme preparation. *Prep. Biochem. Biotech* **31**, 241-258.

Savary, B.J., Hicks, K.B., O'Connor, J.V. (2001) Hexose oxidase from *Chondrus crispus*: improved purification using perfusion chromatography. *Enzyme Microb. Tech.* **29**, 42-51.

Savary, B.J., Hotchkiss, A.T., Cameron, R.G. (2002) Characterization of a salt-independent pectin methylesterase from Valencia orange peel. (Manuscript submitted for publication.)

Seymour, T.A., Preston, J.F., Wicker, L., Lindsay, J.A., Marshall, M.R. (1991) Purification and properties of pectinesterases of Marsh White grapefruit pulp. *J. Agric. Food Chem.* **39**, 1080-1085.

Solms, J., Deuel, H. (1955) Über den mechanismus der enzymatischen verseifung von pektinstoffen. *Helv. Chim. Acta* **38**, 321-329.

Thakur, B.,R, Singh, R.K., Handa, A.K. (1997) Chemistry and uses of pectin. *Crit. Rev. Food Sc. Nutr.* **37**, 47-73.

Versteeg, C., Rombouts, F.M., Pilnik, W. (1978) Purification and some characteristics of two pectinesterase isoenzymes from orange. *Lebensm.-Wiss. u.-Technol.* **11**, 267-274.

Versteeg, C., Rombouts, F.M., Spaansen, C.H., Pilnik, W. (1980) Thermostability and orange juice cloud destabilizing properties of multiple pectinesterases from orange. *J. of Food Sci.* **45**, 969-971.

Vivanco, J.M., Savary, B.J., Flores, H.E (1999) Characterization of ribosome-inactivating proteins isolated from storage roots of *Mirabilis expansa*. *Plant Physiol.* **119**, 1447-1456.

Voragen, A.G.J., Daas, P.J.H., Schols, H.A. (2000) Enzymes as tools for structural studies. In *Bioactive Carbohydrate Polymers*, B.S. Paulsen (ed.), Kluwer Academic, pp. 129-145.

Voragen, A.G.J., Pilnik, W., Thibault, J.F., Axelos, M.A.V., Renard, C.M.G.C. (1995) Pectins. In *Food Polysaccharides*, Stephen, A.M. (ed.), Marcel Dekker Inc., NY, pp. 287-339.

Warrilow, A.G.S., Jones, M.G. (1995) Different forms of tomato pectinesterase have different kinetic properties. *Phytochem.* **39**, 277-282.

Willats, W.G.T., Orfila, C., Limberg, G., Buchholt, H.C., Van Alebeek, G.J.W.M., Voragen, A.G.J., Marcus, S.E., Christensen, T.M.I.E., Mikkelsen, J.D., Murray, B.S., Knox, J.P. (2001) Modulation of the degree and pattern of methyl-esterification of pectic homogalactuonan in plant cell walls. *J. Biol. Chem.* **276**, 19404-19413.

Williamson, G., Kroon, P.A., Faulds, C.B. (1998) Hairy plant polysaccharides: a close shave with microbial esterases. *Microbiology* **144**, 2011-2023.

STRUCTURE-FUNCTION OF A PROTEINACEOUS INHIBITOR OF PLANT PECTIN METHYLESTERASE

L. CAMARDELLA[1], A. GIOVANE[2], L. SERVILLO[2]
[1]*Institute of Protein Biochemistry and Enzymology (CNR), Via Marconi 10, 80125 Napoli (Italy); e-mail: camardel@dafne.ibpe.na.cnr.it.*
[2]*Department of Biochemistry and Biophysics, 2nd University of Naples, Via Costantinopoli 16, 80138 Napoli, (Italy)*

Abstract

A protein acting as a powerful inhibitor of plant pectin methylesterase has been isolated from kiwi fruit. The protein was purified by affinity chromatography on Sepharose-bound tomato pectin methylesterase. The purified protein has a molecular mass close to 16 kDa, as estimated by SDS-PAGE, whereas its molecular mass appears to be 25 kDa by gel filtration under native conditions. The inhibitor interacts with plant pectin methylesterase forming a 1:1 complex. The inhibitor primary structure, determined by direct protein analysis, comprises 152 amino acid residues; two disulphide bridges, connecting Cys9 and Cys18, and Cys74 and Cys114 have been identified. Amino acid sequence alignment indicates similarity with the inhibitors of invertase, and with the N-terminal non catalytic region of pectin methylesterase. Remarkably, in this alignment the four Cys residues involved in disulphide bridges in pectin methylesterase inhibitor are conserved.

1. Pectin and pectin methylesterase

Pectin methylesterase (PME, E.C. 3.1.1.11) is a carboxylic acid esterase and belongs to the group of hydrolases. The hydrolysis of methylester groups catalysed by the enzyme produces pectin with a lower methylation degree that can undergo further enzymatic cleavage by polygalacturonase. The two enzymes are synergically involved in the metabolism of plant cell wall. PME genes occur in multigene families and encode different isoforms. However the specific function and the mechanism of action of the different isoforms are far from being completely understood.

It is proposed that plant PMEs act in a blockwise fashion, producing stretches of contiguous unesterified galacturonic acid residues, whereas fungal PME act in a random fashion, resulting in the de-esterification of single galacturonic acid residues per enzyme/substrate interaction [1]. Cell wall pH is critical for PME activity [2, 3]. In fact, the enzyme action produces a local pH decrease, with a concomitant decrease of PME activity and increase of glycosidase and glycosyl transferase activities that participate to the cell-wall extension [4-6]. PME is thought to be an enzyme of physiological relevance in plant metabolism, being involved in all processes requiring remodelling of plant cell wall, such as fruit ripening and cell extension. It has been

F. Voragen et al. (eds.), Advances in Pectin and Pectinase Research, 363-372.
© 2003 *Kluwer Academic Publishers. Printed in the Netherlands.*

recently shown that the enzyme acts as host-cell receptor for the tobacco mosaic virus movement protein, and that this interaction is required for viral cell-to-cell movement [7, 8].

In industry, control of PME activity is important in processes concerning the production and storage of fruit juices and purees. It is well known that PME activity changes the texture of fruit products. For this reason storage at very low temperature or a high temperature pasteurisation process are required in order to inactivate the PME in industrial fruit products.

During the purification of PME from kiwi fruit we have found a powerful inhibitor of the enzyme. This finding may bring new insights in the control mechanism of PME in the plant cell. On the other hand, the inhibitor could be employed as a biotechnological tool in processes where the control of PME activity is critical.

2. Pectin methylesterase inhibitor (PMEI)

The purification procedure of PME from fruit generally starts with homogenisation in the presence of high salt concentration, since the enzyme is bound to the cell wall by electrostatic interaction. In this way the enzyme is released in the extraction buffer. Surprisingly, when we applied this procedure to the purification of PME from fully ripened kiwi fruit, no PME activity was found in the high salt extract. Moreover, the addition of this extract to PME solutions produced a dramatic loss of enzymatic activity. On the contrary, when the fruit was first homogenised in pure water and then the pellet obtained by centrifugation was again homogenised in a high salt buffer, PME activity was recovered. The addition of the aqueous supernatant to the high salt extract produced again a sharp loss of PME activity. These observations led us to suppose the presence of an inhibitor of PME. The protein nature of the inhibitor was confirmed by a number of tests. The full evidence that the inhibitor is a protein was obtained by purifying it to homogeneity.

2.1. PURIFICATION AND FUNCTIONAL CHARACTERISATION OF PMEI.

The inhibitor was purified from fully ripe kiwi fruit by protein purification procedure. The aqueous supernatant containing the inhibitor was subjected to ammonium sulphate precipitation, anion exchange chromatography on Q Sepharose, and gel filtration through Superose 12. The final specific activity of purified PMEI was 108.6 U/mg protein (the inhibition unit represents the minimum amount of the inhibitor that completely blocks 1 unit of PME after incubation for 10 min at 25°C), with a 21-fold purification and a yield of 56% [9]. Gel filtration of the native protein indicates a molecular mass of about 25 kDa, whereas SDS PAGE shows a doublet with a molecular mass close to 16 kDa. These data may indicate that the native protein has an elongated shape, which leads to a higher apparent molecular mass on the gel filtration column.

PMEI was tested against different amounts of PME purified from kiwi. No decrease of PME activity occurred when the inhibitor was incubated with an excess of the enzyme, even if the reaction was carried out at 37 °C for 20 h. The type of inhibition was determined by the Dixon plot, utilising preparations of purified kiwi PME and citrus pectin in the concentration range 0.1-0.3%. The inhibition was found to be competitive with an apparent $Ki=2.2x10^{-7}$ M. The inhibitor was not species-specific, since it inhibited PME from several plant sources, i.e. orange, apple, tomato, potato and banana. The inhibition was observed in the pH range from 7.5 to 4.0.

These data exclude the possibility that inhibition of PME occurs as a result of chemical modification of PME by the inhibitor. Most probably the inhibition occurs through the formation of an inactive 1:1 complex between the two proteins. Evidence of complex formation is given by isofocusing of a mixture of tomato PME and PMEI. PME and PMEI alone have different isoelectric points, 8.5 and 4.8, respectively. The mixture of the two proteins shows a band with an intermediate isoelectric point. Moreover, gel filtration experiments with a mixture of the two proteins shows a component which is eluted in a volume corresponding to a molecular mass higher than that of the individual proteins [10].

The high affinity of the inhibitor for plant PME allowed the development of a purification procedure based on affinity chromatography on PME-bound Sepharose column (PME purified from tomato fruit was used for coupling to CNBr-activated Sepharose). The inhibitor is retained by the affinity column at pH 7.5, and is eluted by increasing salt concentration and pH above 8.5 [10]. The purified protein was then used to raise antibodies against PMEI in rabbit. Since previous experiments demonstrated that in unripe fruit the PMEI activity was undetectable, the antibodies were utilised to reveal the presence of the inhibitor at various stages of fruit ripening. Kiwi fruits were collected from the same plant on October 1, 16, 30 and November 15. The fruits were homogenised in water and the proteins separated by SDS-PAGE. The gel was treated for Western-blot analysis using purified antibodies. In the unripe fruit a band at higher molecular mass (30 kDa) was recognised by antibodies, but no band corresponding to the active PMEI was detected. Instead, during the ripening, new antibody-recognised bands with lower molecular mass appeared, and PMEI activity became detectable. At the full ripening stage (November 15), a band with the same molecular mass of the purified PMEI was clearly evident, and the activity reached the maximum. These results suggest that PMEI is synthesised as an inactive precursor which is processed into the active protein during the ripening process [10].

2.2. AMINO ACID SEQUENCE OF PMEI

Amino terminal sequencing of PMEI indicated that it consists of a single polypeptide chain. However, both Glu and Ala residues were detected at the first position, indicating protein micro heterogeneity [11]. Isofocusing reveals two bands at pI 4.8 and 4.9, and SDS-PAGE invariably shows a doublet, with two bands with very close molecular mass (16 kDa). Bi-dimensional gel shows a complex pattern with several forms slightly differing by isoelectric point and/or molecular mass (Fig. 1). HPLC on a reverse phase column (Vydac C4) of the protein also reveals the presence of several forms, in variable amounts (Fig 2). Amino-terminal sequencing of each peak shows the same sequence, presenting both Glu and Ala at the first cycle.

The complete amino acid sequence of PMEI was obtained by direct protein chemistry methods, combining direct N-terminal sequencing of the entire protein, sequencing of peptides derived from acid cleavage of the two Asp-Pro bonds, and sets of peptides derived from trypsin, Asp-N and Lys-C endoproteinase digestions [12]. The total sequence comprises 152 amino acid residues, yielding a molecular mass of 16,277 Da, in good agreement with that obtained by SDS-PAGE. The sequence contains five Cys residues. Sequence analysis of fragments obtained after digestion of the protein alkylated without previous reduction identifies two disulphide bridges connecting Cys9 with Cys18, and Cys74 with Cys114, whereas Cys139 bears a free thiol group.

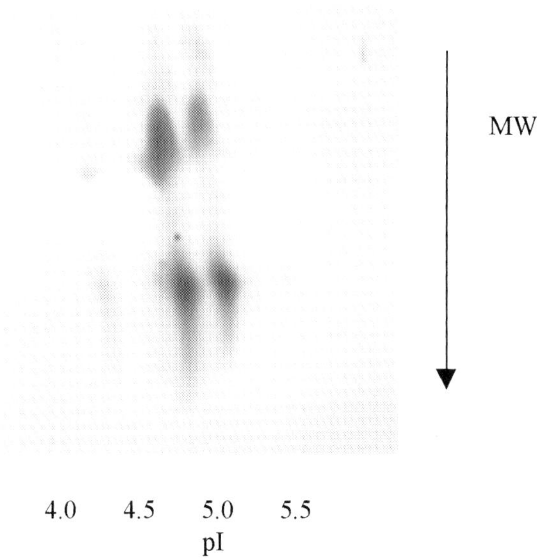

Figure 1. 2D-gel electrophoresis.

Protein micro heterogeneities are at the N-terminus (the polypeptide chain has a partially cleaved N-terminus with an additional Ala residue), and at five positions along the amino acid sequence (Ala/Ser56, Tyr/Phe78, Ser/Asn117, Asn/Asp123, and Val/Ile142) (Fig. 3). The presence of two forms, having or lacking the initial Ala residue, can be attributed to an incomplete processing of the protein, whereas the detection of different amino acid residues at defined positions along the polypeptide chain can be attributed to the presence of isoforms of the inhibitor in the kiwi fruit.

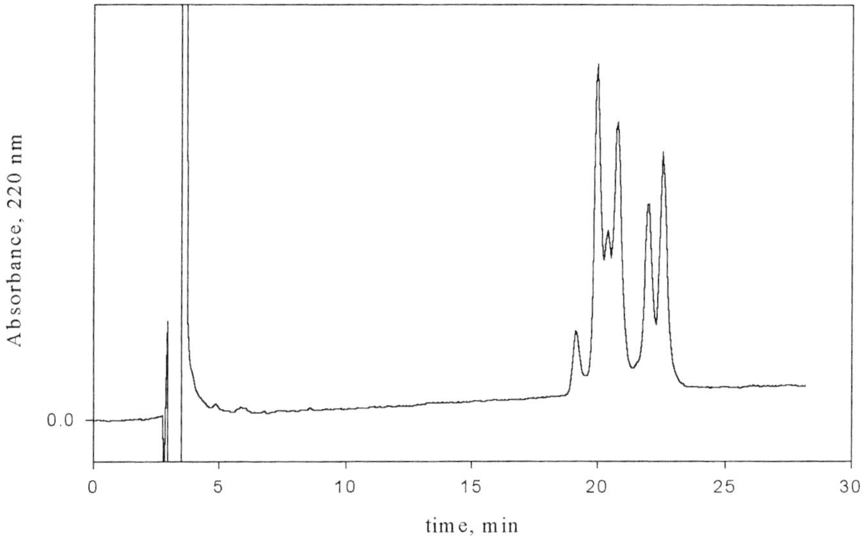

Figure 2. HPLC on a Vydac C4 column. The column is developed by increasing acetonitrile concentration from 40 to 60% in 0.1% trifluoracetic acid.

```
                    20                      40
-ENHLISEI**C**PKTRNPSL**C**LQALESDPRSASKDLKGLGQFS
A
                    60                      80
IDIAQASAKQTSKI IASLTNQATDPKLKGRYET**C**SENYAD
              S                     F
                     100                    120
AIDSLGQAKQFLTSGDYNSLNIYASAAFDGAGT**C**EDSFEG
                                     N
                     140        152
PPNIPTQLHQADLKLEDLCDIVLVISNLLPGS
 D                   I
```

Figure 3

2.3. SPECTRAL FEATURES OF PMEI

The structure of PMEI has been investigated by circular dichroism analysis. The CD spectrum of PMEI between 200 and 240 nm was characterised by two negative bands centered at 208 and 222 nm, indicative of a predominant of α-helical secondary structure (Fig. 4). The estimated percentage of secondary structure is 65.6, 4.2, 12, and 18.2% of α, β, β-turn and unordered structure, respectively. On the contrary the CD spectrum of tomato PME indicates a predominant β secondary structure. This observation may indicate a structural similarity of tomato PME with the recently determined three-dimensional structure of PME from *Erwinia chrisanthemi*. The latter enzyme shows the right-handed parallel β -helix structure, as seen in the pectinolytic enzymes pectate lyase, pectin lyase and polygalacturonase [13]. The mixture of PME and PMEI shows a CD spectrum that substantially coincides with the sum of the individual spectra, therefore it can be concluded that the secondary structure of the two proteins is not affected by their interaction.

The formation and the topography of the complex may be studied by fluorescence techniques. PME contains both tryptophan and tyrosine residues, whereas the inhibitor contains only tyrosine residues. The PMEI fluorescence spectrum after excitation at 275 nm shows the emission maximum at 305 nm, typical of fluorescence arising from tyrosyl residues. Therefore it is possible to study the interaction of the two proteins by following the fluorescence of tryptophan residues, which is specific for PME, both in the presence and in the absence of the inhibitor. The variation of fluorescence properties may depend on the formation of the complex.

2.4. AMINO ACID SEQUENCE ALIGNMENT.

Database search for homology with proteins of known function indicates similarity of PMEI with invertase inhibitors (INH). INHs are proteins about the same size as PMEI (17 kDa), have been isolated from plant tissues, and inhibit invertase of plant origin in a pH dependent manner with optimal inhibition at pH 5.0 [14]. The sequence of INH from tobacco (*Nicotiana tabacum*) cultured cells has been deduced from the sequence of the cDNA (Nt-inh1), and corresponds to the partial sequences obtained from the purified protein [15]. The presence of a signal peptide is in agreement with the apoplasmic location of the invertase inhibitor and indicates that the protein is synthesised as a precursor, which is processed through the secretory pathway. Several homologous sequences of INH have been detected either in *N. tabacum* or in other plants, including Arabidopsis; for some of them the invertase inhibitory activity of the recombinant protein has been demonstrated. For example, expression of Nt-inhh cDNA, encoding a putative homologue of INH, leads to inhibition of vacuolar invertase activity in transformed potato tubers, with substantial reduction of cold-induced hexose accumulation [16]. The comparison of PMEI and INH amino acid sequences show a low identity level (25-30%); few amino acid residues are conserved and, among these, the four Cys residues which are involved in two disulphide bridges in PMEI. Therefore, it is possible to hypothesise that the same disulphide bridges are present in INH; in fact it has been shown that INH is inhibited by treatment with a reducing agent [17].

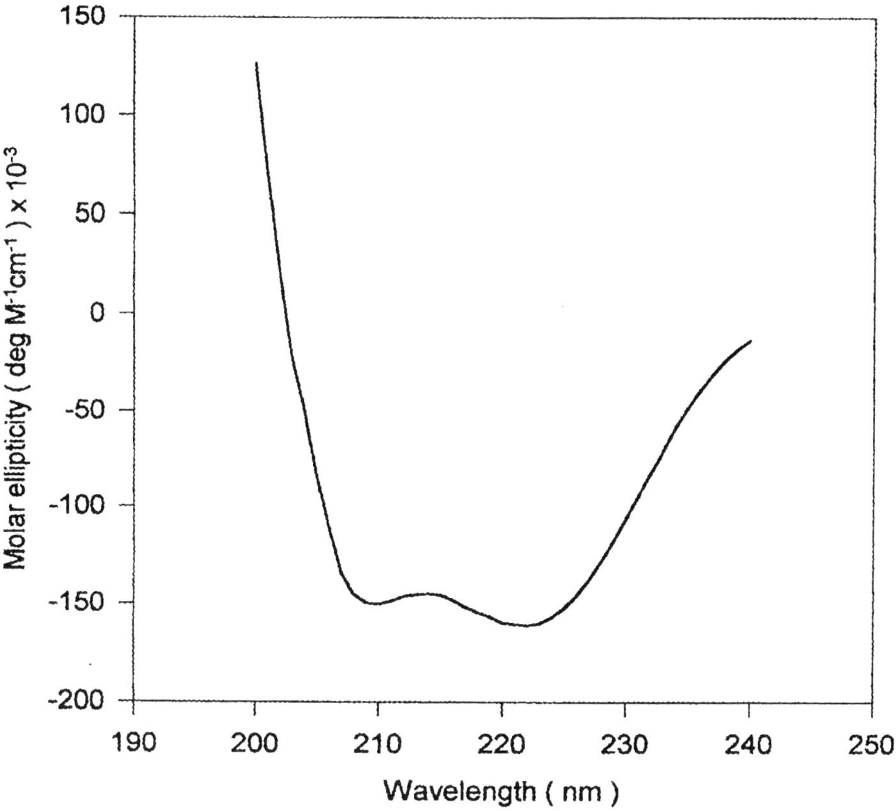

Figure 4. CD spectrum of PMEI

```
PMEI     ENHLISEICPKTRNPSLCLQALESDPRSASKDLKGLGQFSIDIAQASAKQTSKIIASLTNQATDPKLKGRYETCSENYAD
At1      QVADIKAICGKAKNQSFCTSYMKSNPKTSGADLQTLANITFGSAQTSASEGFRKIQSLVKTATNPTMKKAYTSCVQHYKS
At2      TSSEMSTICDKTLNPSFCLKFLNT--KFASPNLQALAKTTLDSTQARATQTLKKLQSIIDGGVDPRSKLAYRSCVDEYES
NtINH    ANNLVETTCKNTPNYQLCLKTLLSDKRSATGDITTLALIMVDAIKAKANQAAVTISKLRHSNPPAAWKGPLKNCAFSYKV
NtINHH   SNNIINTTCRATTNYPLCLTTLHSDPRTsgADLTTLGLVMVDAVKLKSIEIMKSIKKL--EKSNPELRLPLSQCYIVYYA
GRIP     STSFIKAKCSATRYPALCVQSLSVYASTIQKSPKQLAHTALTVSLAKA-QSAKAFVSKLGKIKGMKAREAVQDCLEEMND
AtPE1    ENHIVQAVCAPTDFKDTCVNSLMASPDSDDPDLIKLgkVTIKSINESLEKASGDIKAKADK--NPEAKGAFELCEKLMID
AtPE2    PSQAISKACELTRFPELCVDSLMDFPGSAASSSKDLIHVTVNMTLHHFSHALYSSASLSFVDMPPRARSAYDSCVELLDD
PvPE3    PAASLKAVCDTTRYPSSCFSSISSLPESNTTDPELLFKLSLRVA---IDELSSFPSK1ANAEQDARLQKAIDVCSSVFGD

PMEI     AIDSLGQAKQFLTSGDYNSLNIYASAAFDGAGTCEDSFEGPPNIPTQLHQADLKLEDLCDIVLVISNLLPGS
At1      AISSLNDAKQSLASGDGKGLNIKVSAAMEGPSTCEQDMADFKVDPSAVKNSG-DFQNICGIVLVISNMM---
At2      AIGNLEEAFEHLASGDGMGMNMKVSAALDGADTCLDDVKRLRSVDSSVVNNSKTIKNLCGIALVISNMLPRN
NtINH    ILtsLPEAIEALTKGDPKFAEDGMVGSSGDAQECEEYFKGSKSPFSALNIAVHELSDVGRAI--VRNLL---
NtINHH   VLHAVTVAVEALKRGPKFAENGMVDVAVE-AETCEFSFKYNGLVSPVSDMNKEIIELSSVAKSIIRMLL---
GRIP     SIDRLSRSVQE1SKGDFSNVDTWVSAALTDDNTCVDGFAGRALDGRVKASIRGWLVTVAQVTSNALALVNQF
AtPE1    AIDDLKKCMDHGFSVDQEDLRVWLSGSIAFQQTCMDSF-G--EIKSNLMQDMLKirELSsmVTRISTLIPNS
AtPE     SVDALSRALSSVVSSSAKPQDvwLSAALTNHDTCTEGFDGVDDGGVKDHMTA-ALQNLSELVSNCLAIFSAS
PvPE3    ALDRLNdaGRIASSASVSNVETWLSAALTDQDTCLDAVGELNSTAARGALQEIETAMRNSTEFASNSLAIVT
```

Figure 5. Alignment of PMEI sequence with the homologous regions of At1 (Q9LUV1), At2 (Q9LNF2), NtINH (O49908), NtINHH (O49603), GRIP (Q9M4H8), AtPE1 (Q9M7Y9), AtPE2 (Q9M3B0) and PvPE3 (Q43111). The accession numbers from the SWALL protein databank are in parentheses.

In addition to INH and INH homologues, database search identifies homology with a huge number of sequences sharing this pattern of Cys residues. These sequences mainly derive from plant gene sequencing and encode putative proteins of unknown function. An alignment of some representative sequences is shown in Fig 5. The sequences At1 and At2 showing the highest identity level are cDNA from Arabidopsis, encoding hypothetical proteins of 173 and 176 amino acid residues, respectively. The presence of a putative signal sequence may indicate an apoplasmic location; however, no indication is available about the protein function. Sequence GRIP (grape ripening-induced) is one of the cDNAs identified by differential screening in ripening grape berry and corresponds to a mRNA whose level increases at the onset of ripening [18]. Moreover, it is noticeable the presence, among the sequences homologous to PMEI, of the N-terminal non-catalytic domain of PME. The sequences encoding pectinesterase-like proteins from Arabidopsis (AtPE1 and AtPE2) and PME3 from *Phaseolus vulgaris* (PvPE3) are shown as examples. It is known that plant PME is encoded as a large size precursor of molecular mass >50 kDa, whereas the size of the mature protein present in the cell wall is in the range 32-42 kDa, corresponding to the C-terminal portion of the precursor. It is not known which is the site where the cleavage of the N-terminal extension occurs, and which is the function, if any, of the isolated N-terminal portion.

The most interesting observation coming from the comparison of PMEI and homologous amino acid sequences is the strong conservation of the four Cys residues that are involved in two disulphide bridges in PMEI. It is tempting to hypothesise that this arrangement of disulphide bridges is shared by the homologous proteins, constituting a common three-dimensional fold.

3. Conclusive remarks

A protein acting as a powerful inhibitor of PME has been isolated from kiwi fruit. The inhibitor is inactive against fungal PME, whereas it is active against plant PME, irrespective of the plant source assayed, forming with PME a 1:1 non-covalent complex. PMEI has been detected only in kiwi fruit. It is possible that it is transiently expressed in other plant and tissues, and that it plays a more general physiological role as regulator of PME activity. It is homologous to plant invertase inhibitors, which, in common with PMEI, modulate the activity of a plant enzyme of the carbohydrate metabolism located in the cell wall and involved in the control of plant development and in the defence against pathogen attack [19]. In addition, it is homologous to a number of putative proteins identified by cDNA and genome sequencing. All these proteins share a pattern of four Cys residues, which we demonstrated that are involved in two disulphide bridges in PMEI.

Most of the pectin modifying enzymes (pectate lyase, polygalacturonase, pectin methylesterase) share a novel structural fold, not observed before 1993, termed the parallel □helix [20]. This structure probably confers an "intrusive" shape to the protein, better adapted to the action in the cell wall environment. PMEI on the contrary has a mainly □-helix structure, and the same structure can be inferred for the homologues, on the basis of secondary structure predictions. Most of these homologous proteins are supposed to act in the cell wall context, either by directing the protein to the cell wall, as in the case of PME precursor, or interacting with cell wall enzymes, as in the case of PME inhibitor and invertase inhibitor. Does the observed disulphide knot confer a common, particular shape to these proteins? Elucidation of three-dimensional structure of PMEI or related proteins either by crystallographic or NMR analysis may give an answer to this question.

References

1. Willats, W.G.T., Orfila, C., Limberg, G., Buchholt, C., van Alebeek, G.W.M., Voragen, A.G.J., Marcus, S.E., Christensen, T.M.I.E., Mikkelsen, J.D., Murray, B.S. and Knox, J.P. (2001) Modulation of the degree and pattern of methyl-esterification of pectic homogalacturonan in plant cell walls, J. Biol. Chem. **276**, 19404-19413.
2. Denes, J.M., Baron, A., Renard, C.M., Pean, C. and Drilleau, J.F. (2000) Different action patterns of apple pectin methylesterase at pH 7.0 and 4.5, Carbohydr. Res. **327**, 385-393.
3. Goldberg, R., Pierron, M., Bordenave, M., Breton, C., Morvan, C. and du Penhoat, C.H. (2001) Control of mung bean pectinmethylesterase isoform activities. Influence of pH and carboxyl group distribution along the pectic chains, J. Biol. Chem. **276**, 8841-
4. Ricard , J. and Noat, G. (1986) Electrostatic effects and the dynamics of the enzyme reactions at the surface of plant cells. 1. A theory of the ionic control of a complex multi-enzyme system, Eur. J. Biochem., **155**, 183-190.
5. Moustacas, A.M., Nari, J., Diamantidis, G., Noat, G., Crasnier, N., Borel, M. and Ricard, J. (1986) Electrostatic effects and the dynamics of the enzyme reactions at the surface of plant cells. 2. The role of pectin methylesterase in the modulation of electrostatic effects in soybean cell walls, Eur. J. Biochem., **155**, 191-197.
6. Nari, J., Noat, G., Diamantidis, G., Woudstra, M. and Ricard, J. (1986) Electrostatic effects and the dynamics of the enzyme reactions at the surface of plant cells. 3. Iterplay between limited cell-wall autolysis, pectin methylesterase activity and electrostatic effects in soybean cell walls, Eur. J. Biochem., **155**, 199-202.

7. Dorokhov, Y.L., Makinen, K., Frolova, O.Y, Merits, A., Saarinen, J., Kalkkinen, N., Atabekov, J.G. and Saarma, M. (1999) A novel function for a ubiquitous plant enzyme pectin methylesterase: the host-cell receptor for the tobacco mosaic virus movement protein, FEBS Letters **461**, 223-228.
8. Chen, M.H., Sheng, J., Hind, J., Handa A.K. and Citovsky, V. (2000) Interaction between the tobacco mosaic virus movement protein and the host cell pectin methylesterase is required for viral cell-to-cell movement, EMBO J. **19**, 913-920.
9. Balestrieri, C., Castaldo, D., Giovane, A., Quagliuolo, L. and Servillo, L. (1990) A glycoprotein inhibitor of pectin methylesterase in kiwi fruit (*Actinidia chinensis*), Eur. J. Biochem. **193**, 183-187.
10. Giovane, A., Balestrieri, C., Quagliuolo, L., Castaldo, D. and Servillo, L. (1995) A glycoprotein inhibitor of pectin methylesterase in kiwi fruit. Purification by affinity chromatography and evidence of a ripening-related precursor, Eur. J. Biochem. **233**, 926-929.
11. Camardella, L., di Prisco, G., Giovane, A., Quagliuolo, L. and Balestrieri, C. (1994) Partial characterisation of pectin methylesterase inhibitor from kiwi fruit (*Actinidia chinensis*) purified by affinity chromatography, in L. Alberghina, L. Frontali and P. Sensi (eds) *ECB6: Proceedings of the 6th European Congress on Biotechnology* Elsevier Science B.V., Amsterdam, The Netherlands, pp. 635-638.
12. Camardella, L., Carratore, V., Ciardiello, M.A., Servillo, L., Balestrieri, C., and Giovane, A. (2000) Kiwi protein inhibitor of pectin methylesterase. Amino acid sequence and structural importance of two disulfide bridges, Eur. J. Biochem. **267**, 4561-4565.
13. Jenkins, J., Mayans, O., Smith, D., Worboys, K. and Pickersgill R.W. (2001) Three-dimensional structure of *Erwinia chrisanthemi* pectin methylesterase reveals a novel esterase active site, J. Mol. Biol. **305**, 951-960.
14. Pressey, R. (1994) Invertase inhibitor in tomato fruit, Phytochemistry **36**, 543-546.
15. Greiner, S., Krausgrill, S., and Rausch, T. (1998) Cloning of a tobacco apoplasmic invertase inhibitor. Proof of function of the recombinant protein and expression analysis during plant development, Plant Physiol. **116**, 733-742.
16. Greiner, S., Rausch, T., Sonnewald, U., and Herbers, K. (1999) Ectopic expression of a tobacco invertase inhibitor homolog prevents cold-induced sweetening of potato tubers, Nature Biotechnology **17**, 708-711.
17. Greiner, S., Koster, U., Lauer, K., Roskranz, H., Vogel R. and Rausch, T. (2000) Plant invertase inhibitor: expression in cell culture and during plant development, Aust. J. Plant Physiol. **27**, 807-814.
18. Davies, C. and Robinson, S.P. (2000) Differential screening indicates a dramatic change in mRNA profiles during grape berry ripening. Cloning and characterization of cDNAs encoding putative cell wall and stress response proteins, Plant Physiol. **122**, 803-812.
19. Sturm, A. and Guo-Qing, T. (1999) The sucrose cleaving enzymes of plants are crucial for development, growth and carbon partitioning, Trends in Plant Sciences **4**, 401-407.
20. Herron, S.R., Benen, J.A.E., Scavetta, R.D., Visser, J., and Jurnak, F. (2000) Structure and function of pectic enzymes: virulence factors of plant pathogens, PNAS **97**, 8762-8769.

DEGRADATION OF PECTINS WITH DIFFERENT NATURE, AMOUNT AND DISTRIBUTION OF THE SUBSTITUENTS BY ENDOPOLY-GALACTURONASE OF *FUSARIUM MONILIFORME*.

BONNIN E., LE GOFF A., AND THIBAULT J.-F.
Unité de Recherche sur les Polysaccharides, leurs Organisations et Interactions
INRA, BP 71627, 44316 Nantes Cedex 03, France
E-mail : bonnin@nantes.inra.fr

Abstract

The mode of action of the endopolygalacturonase from *Fusarium moniliforme* was studied towards a series of pectins with different amounts and distribution patterns of substituents. In homogalacturonan, the enzyme hydrolysed the linkages between two galacturonic acid residues according to a multi-chain attack mechanism, and the final products were monomer and dimer. The enzyme was able to degrade methylated pectins with a degree of methylation up to 70 and acetylated homogalacturonans with a degree of acetylation up to 38 to a lower extend but not xylogalacturonans.
When methylated pectins or acetylated homogalacturonan were used as substrates, the final percentage of hydrolysis decreased with increasing the degree of substitution. A blockwise distribution of the methyl-groups was more favourable to a higher velocity than a random one. In every case, the enzyme was shown to follow a multi-chain attack. The final products included methyl esterified oligogalacturonates. The enzyme seemed to degrade first the non-methylated area of the blockwise methylated substrate. The detailed structure of the oligomers was analysed by mass spectrometry, and showed that the enzyme was able to accomodate methylated or acetylated galacturonic acid residues in its active site.

1. Introduction

Fusarium moniliforme is a phytopathogenic fungus for a wide variety of plants. It causes foot or stalk rot and seedling blight in a wide variety of plant and especially in graminae such as maize (Nelson, 1992). It secretes four different isoforms of endopolygalacturonases (PG, Caprari *et al.*, 1993), that originate from an unique gene and differ only by their glycosylation pattern (Caprari *et al.*, 1996). The gene was cloned and expressed in *Saccharomyces cerevisiae* (Caprari *et al.*, 1996). Previous work about *Fusarium* PG deal with the purification of the 49 kDa isoform and its kinetic parameters and mode of action towards and GalA oligomers (Bonnin *et al.*, 2001a). The active site of this enzyme consists at least of 5 subsites, from -3 to +2.
However, native pectins in host cell walls are most often substituted. A general feature of galacturonans is their esterification with methanol and/or acetic acid (Voragen *et al.*,

F. Voragen et al. (eds.), Advances in Pectin and Pectinase Research, 373-384.
© 2003 *Kluwer Academic Publishers. Printed in the Netherlands.*

1995). They can also be substituted by xylose, as in the xylogalacturonans of pea hulls (Le Goff *et al.*, 2001) or apple (Schols *et al.*, 1995). For these reasons, it is of interest to study the behaviour of *Fusarium* PG towards substituted substrates.
In this study, homogalacturonan, acetylated homogalacturonan, xylogalacturonan and two series of methylated pectins with different distribution patterns of methyl ester groups were used as substrates and the mode of action of the enzyme was investigated by following the hydrolysis of the substrate and the nature of the products released.

2. Materiel and methods

2.1. SUBSTRATES

Homogalacturonan was extracted (Thibault *et al.*, 1993) from sugar beet pectin (degree of methylation DM 54, degree of acetylation DAc 16). Their molar mass was around 21,000 g/mol.
Two series of methylated pectins were given by Danisco-Cultor (Limberg *et al.*, 2000a). They were prepared from a lime pectin with DM 72, either with a plant PME from orange peels (P-series) or with a fungal PME from *Aspergillus niger* (F-series). Each of them is noted with a letter indicating the enzyme (P or F) and a number indicating the DM. The two series stretched from F11 to F76 and from P16 to P76, respectively. From what is known about the respective modes of action of these two PME, the P- and the F-series were expected to exhibit a blockwise and a more random distribution of the methyl groups, respectively.
For the extraction of acetylated homogalacturonan, sugar beet pulp was treated by extrusion-cooking (Micard and Thibault, 1999). Water-extractable pectins were then treated by a mixture of enzymes containing orange pectin methylesterase (Sigma Chemicals, L'Isle d'Abeau, France), fungal pectin methylesterase from *Aspergillus niger* (Novo Nordisk A/S, Copenhagen, DK), rhamnogalacturonan-hydrolase (Novo Nordisk A/S), endogalactanase and endoarabinanase previously purified in our lab from the commercial mixture Pectinex AR (Novo Nordisk A/, Bonnin *et al.*, 2001b). This mixture of enzyme was checked to be free of polygalacturonase. After centrifugation, the supernatant was loaded on a DEAE Sepharose CL6B column (14 x 2.6 cm) eluted with degassed 0.05 mol/L succinate buffer pH 4.5 and a linear gradient up to 0.4 mol/L NaCl in succinate buffer. The acetylated homogalacturonan eluted with 0,2 mol/L NaCl and had a DAc of 38. A controlled deacetylation in alkaline conditions led to two others substrates with DAc 28 and DAc 18.
Xylogalacturonan was extracted from pea hulls and had a degree of substitution by xylose of 70 (Le Goff *et al.*, 2001), meaning that 70% of GalA were substitued. Three other xylogalacturonans were obtained by a partial enzymic removal of xylose, leading to degrees of substitution of 58, 53, and 38.

2.2. ENZYME

Fusarium PG was purified to homogeneity from a transformed strain of *Saccharomyces cerevisiae* (Caprari *et al.*, 1996) as previously described (Bonnin *et al.*, 2001). The enzyme had a molar mass of 49 kDa and a specific activity of 1845 nkat/mg. It exhibited no lyase and methyl-esterase activities.

2.3. DEGRADATION OF POLYMERS BY FUSARIUM PG

Substrates were solubilized at 2 mg/mL in 50 mM acetate buffer pH 4. Each of them was hydrolysed by adding 8 pkat/mg substrate at t = 0 and 80 pkat/mg substrate at t = 48h.
Aliquots were withdrawn at intervals, the reaction was stopped by boiling the samples. The amount of reducing ends was quantified as previously described (Nelson, 1944, Sturgeon, 1990), allowing the calculation of the DP. The percentage of hydrolysis was calculated according to Thibault, 1983.

2.4. HIGH-PERFORMANCE CHROMATOGRAPHIES

High-performance size-exclusion chromatography (HPSEC) was performed on two columns in series (Shodex OH-Pack SB-804 HQ and SB-805 HQ) and was monitored using an on line Multi Angle Laser Light Scattering detector (Mini Dawn, Wyatt Technology Corp., USA) and differential refractometer (Erma 7512, Japan).
High-performance anion-exchange chromatography (HPAEC) was performed on a Dionex system equipped with a Carbopac PA1 column and a pulsed amperometric detection. The column was eluted with 0.5 mol/L NaOH, 1 mol/L sodium acetate and water as follows : initial conditions, 20/30/50; 30 min, 20/60/20; 30-34 min, 20/60/20; 35 min, 20/30/50. Borwin software (JMBS Developements, Grenoble, France) was used for data acquisition and processing. Monomer and oligomers of GalA up to DP 10 were used as standards.

2.5. MASS SPECTROMETRY

MALDI time-of-flight (TOF) mass spectra were acquired on a Bruker Reflex II spectrometer (Bruker Daltonik, Bremen, Germany). 2,4,6-trihydroxyacetophenone (THAP) was used as MALDI matrix in thin layer preparation as described in Limberg *et al.* (2000b).
Electrospray MS spectra were acquired on an Esquire quadripole ion trap (ESI-MS, Bruker-Franzen Analytik GmbH, Bremen, Germany) in the conditions described by Körner *et al.* (1998).

3. Results and Discussion

3.1. HOMOGALACTURONAN

The hydrolysis of homogalacturonan was followed by measuring the increase in reducing power and the percentage of hydrolysis was plotted versus time (Figure 1). The initial rate of hydrolysis was constant during the first 8 hours, where the percentage of hydrolysis reached 1.9%. Beyond this value, the reaction rate decreased. Further additions of enzyme led to an increase in hydrolysis, which finally reached 57%.

During the first steps of hydrolysis, the molar mass of homogalacturonan was lowered from 20,700 g/mol to 9,500 g/mol in 8 h. This rapid decrease of molar mass associated with a low percentage of hydrolysis (1.9%) is typical of an endo mode of action.

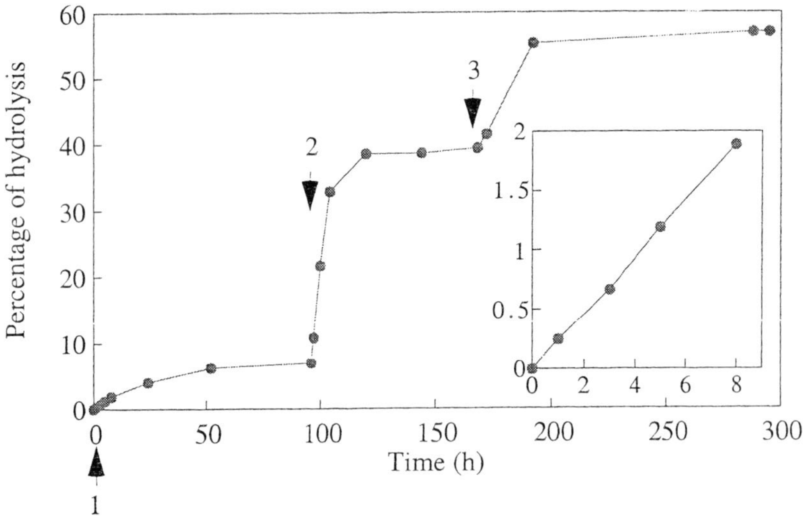

Figure 1 : Time course analysis of hydrolysis of homogalacturonan by *Fusarium* PG.
Arrows indicate the successive additions of enzyme :
1 = 8 pkat/mg substrate, 2 = 80 pkat/mg substrate, 3 = 6 nkat/mg substrate.
In inset : Percentage of hydrolysis during the first 8 h.

Figure 2 shows that the reciprocal of molar mass was linearly related to reaction time up to 8 h. According to Vink (1693), such a linear dependance was characteristic for a random hydrolysis of glycosidic linkages.

During the hydrolysis, the production of oligogalacturonates was followed by HPAEC (Figure 3).
No short oligomers were produced at the beginning of the hydrolysis, showing the multi-chain mode of attack. Further hydrolysis induced the release of oligomers with high dp that disappeared afterwards. At 20% hydrolysis, the tetra- and trimers were the major products. The final reaction products were obtained by adding larger amounts of enzyme and were monomer and dimer of GalA.
Fusarium PG was unable to degrade the dimer, and thus was similar to number of endoPG, such as several endoPGs from *A. niger* (Kester *et al*., 1999). In contrast, it was able to hydrolyze the trimer, which has not been seen for PG from *A. niger* (Pasculli *et al*., 1991).

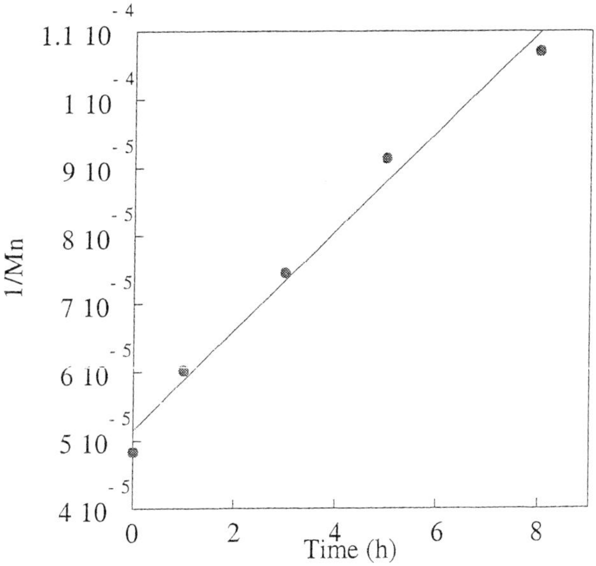

Figure 2 : Change in molar mass of homogalacturonan during the early stage of degradation (<2%) by *Fusarium* PG.

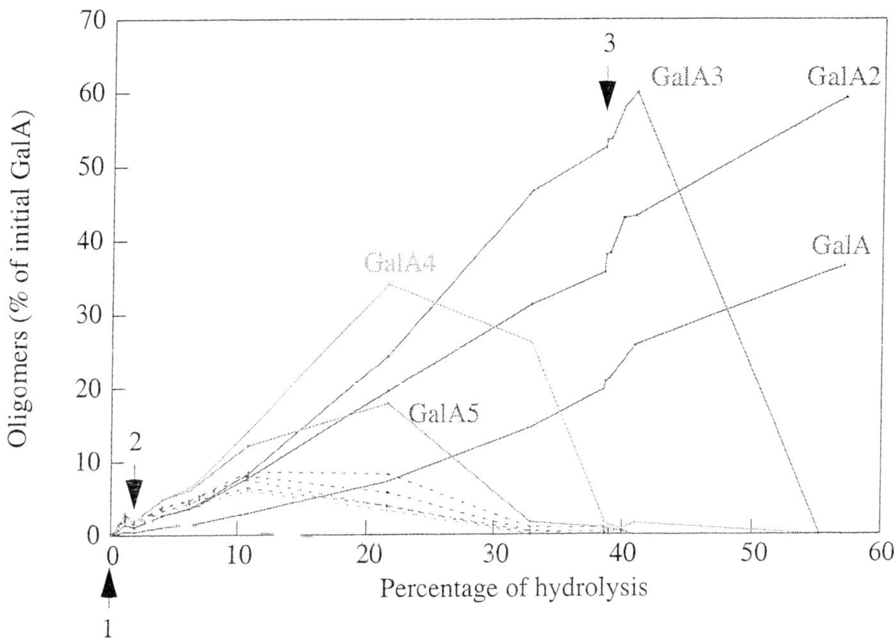

Figure 3 : Changes in oligomer concentration during the hydrolysis of homogalacturonan by *Fusarium* PG. The concentration in each oligomer is expressed as a percentage of initial galacturonic acid in the substrate. Arrows indicate the successive additions of enzyme:
1 = 8 pkat/mg substrate, 2 = 80 pkat/mg substrate, 3 = 6 nkat/mg substrate.
Dashed lines indicate the concentration of dp 6 to 10.

3.2. METHYLATED PECTINS

The F- and P-series of methylated pectins were hydrolysed by Fusarium PG in the same conditions as homogalacturonan. The enzyme was able to hydrolyse all the substrates but at different extents (Figure 4): the final percentage of hydrolysis decreased with increasing the DM. Moreover, for a similar DM between DM 15 and 50, the hydrolysis of the P-series was more complete than that of the F-series, suggesting that a blockwise distribution was more favourable than a random one to an efficient hydrolysis.

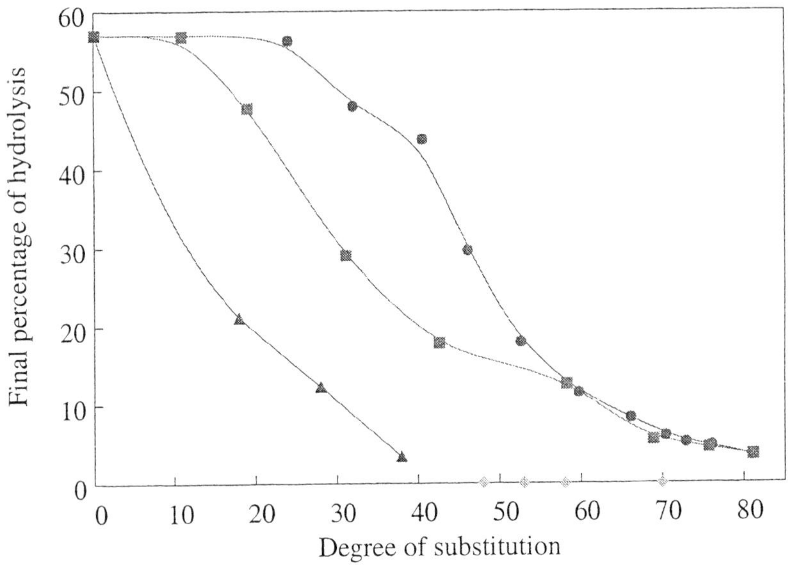

Figure 4 : Percentage of hydrolysis of homogalacturonan, methylated pectins from F-series (squares) and P-series (circles),acetylated homogalacturonans (triangles) and xylogalacturonans (lozenges) as a function of the degree of substitution of the substrate.

During the early steps of hydrolysis, the molar mass of pectins was followed and their reciprocal was plotted versus time for each series (Figure 5).

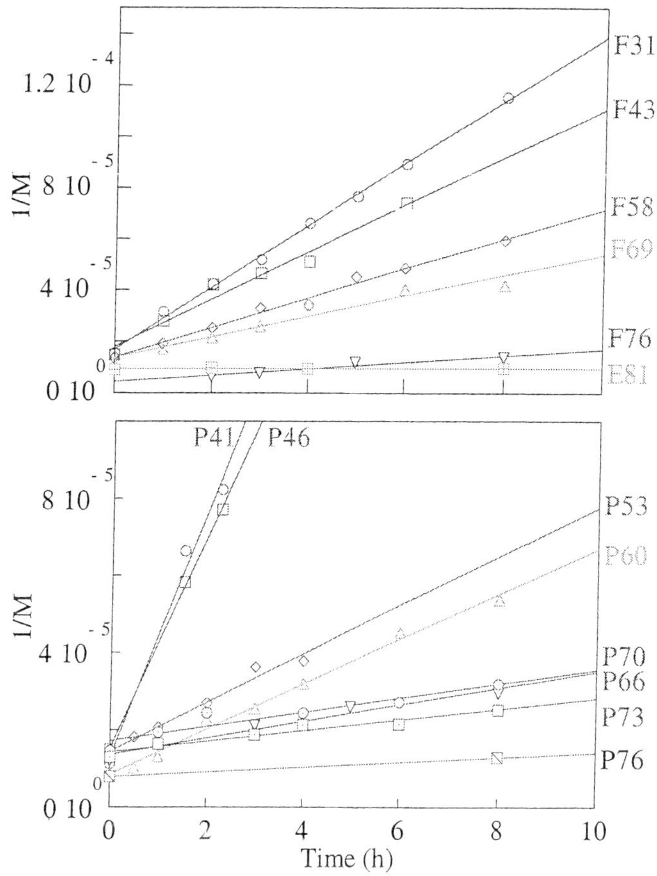

Figure 5 : Changes in molar masses of homogalacturonan and pectins of F- and P-series during the first step of their hydrolysis by *Fusarium* PG.

The reciprocal of molar mass varied linearly with time, demonstrating the same hydrolytic process as for homogalacturonan. The release of oligomers was followed for each substrate. Figure 6 shows the hydrolysis of P53 and F58. All the substrates of each series gave similar profiles. These two examples were chosen because they have about the same DM but different distribution pattern of the methyl groups.

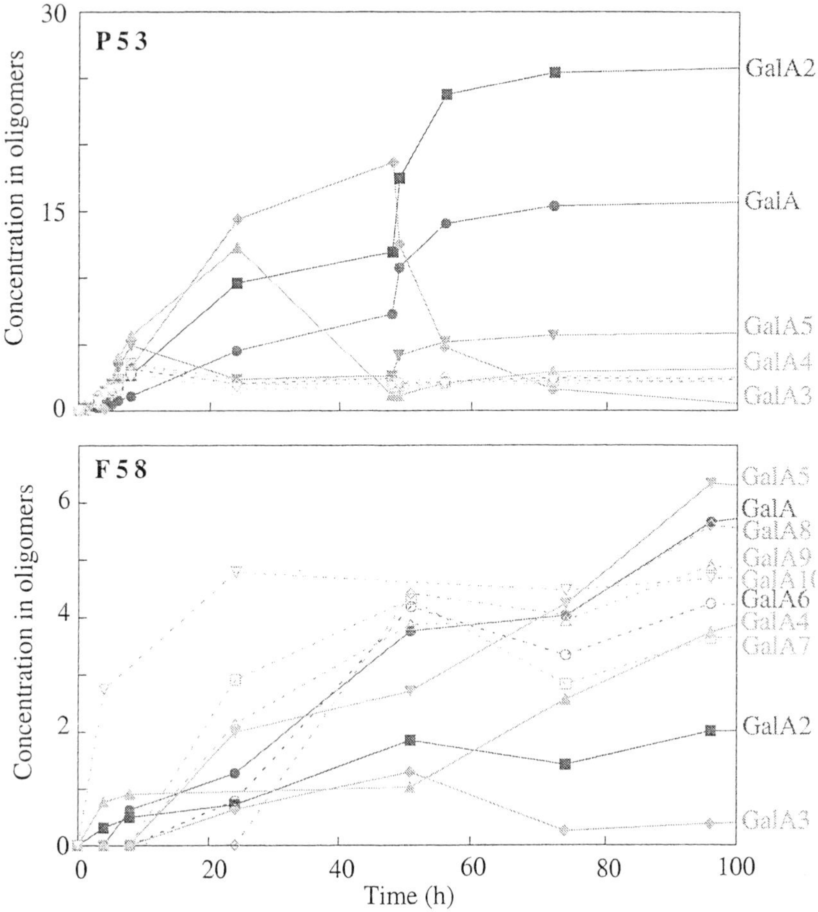

Figure 6 : Changes in oligomer concentration during the hydrolysis of P53 or F58 by *Fusarium* PG.

The concentration in each oligomer is expressed as a percentage of initial galacturonic acid in the substrate. Arrows indicate the successive additions of enzyme :
1 = 8 pkat/mg substrate, 2 = 80 pkat/mg substrate, 3 = 6 nkat/mg substrate.
Dashed lines indicate the concentration of dp 6 to 10.

From P53, no short oligomers were produced at the beginning of hydrolysis. When the hydrolysis was prolonged, the largest oligomers appeared and further disappeared but not completely. The final products were the dimer and the monomer together with low amounts of other DP.

As the pH conditions used for HPAEC induces a saponification of the oligomers, MALDI-TOF mass spectrometry was carried out on the digests to know the number of methyl groups on each oligomer. In the initial steps of hydrolysis, only methanol-free oligomers were released, suggesting that the enzyme degraded first the unsubstituted area of the pectin. After further degradation, more methylated oligmers appeared in the mixture, and in the final products, the largest oligomers were all methylated. As they were resistant to hydrolysis, that demonstrate that either the number of methyl groups or their location on the oligomer inhibited the enzymic activity.

The oligomer profile obtained from F58 differed from that obtained from P53 (Figure 6), as large oligomers were produced from the beginning of the reaction and were not further degraded, suggesting that they were methylated. Indeed, the MALDI-TOF MS experiments identified the same products as in the P53 digest but their proportion was completely different.

The oligomers present in the final mixture were analysed by ESI-IT MS, in order to locate the methoxyl groups on the oligomer. The analysis showed that the reducing and non-reducing ends of each oligomer were never methylated, demonstrating that the enzyme strictly required two free galacturonic acid residues on each side of the linkage to be broken. Moreover, the identified oligomers can be set in the active site of the enzyme (Table I). By setting the major form of each oligomer in the active site, we demonstrate that the subsites -3, -2 and +2 could accomodate a methylated galacturonic acid. The dimethylated tetramer 4/2 also showed that the subsite -3 and -2 could link methylated residues simultaneously. It would be of interest to confirm the ability of the subsites to accomodate methylated GalA by the mean of different methylated oligomers of GalA.

Due to their wide application fields in various industries, *Aspergillus niger* PG have been studied in detail. Among the seven enzymes already characterised, PG-I, -II, -C, and -E preferred non-methylated substrates (Benen *et al.*, 1999) while PG-A and PG-B preferred slightly methylated substrates (Parenicova *et al.*, 2000). The latter enzymes have a maximum activity for DM=22. Morever in our study, we showed that the mode of distribution of the methyl groups had a great effect on the efficiency of the hydrolysis.

Table I : Schematic representation of methylated products in the active site of *Fusarium* PG.

3/1 = DP 3 carrying 1 methyl group
-3 ... +2 = subsite of the enzyme
G = galacturonic acid residue
Me = methylated galacturonic acid residue
G* = reducing end

			-3	-2	-1	+ 1	+ 2			
3/1			G	Me	G*					
						G	Me	G*		

			-3	-2	-1	+ 1	+ 2			
4/1		G	Me	G	G*					
						G	Me	G	G*	
4/2		G	Me	Me	G*					
						G	Me	Me	G*	

			-3	-2	-1	+ 1	+ 2			
5/1	G	G	G	Me	G*					
	G	G	Me	G	G*					
						G	G	G	Me	G*
						G	G	Me	G	G*
5/2	G	G	Me	Me	G*					
						G	G	Me	Me	G*

3.3. ACETYLATED HOMOGALACTURONAN

Three acetylated substrates differing in their DAc were extracted from sugar beet pulp. When they were degraded by *Fusarium* PG, the final percentage of hydrolysis decreased with increasing the DAc (Figure 4). Thus, acetyl groups had the same effect as methoxyl groups in decreasing the efficiency of hydrolysis. By following the release of oligomers by HPAEC and their level of acetylation by mass spectrometry, it was obvious that large methylated oligomers were produced at the beginning of hydrolysis. In the final digest, a large panel of oligomers was present with different DP and different level of acetylation. Similarly to what was deduced from the production of methylated oligomers, the presence of monoacetylated trimer in the final products from acetylated homogalacturonan suggested that the active site of the enzyme could be combined with an acetylated galacturonic acid.

A negative effect of acetyl groups on the degradation of pectins has been reported in the past (Solms and Deuel, 1951). The same observation was done more recently on beet pectins degraded by polygalacturonases from *Aspergillus niger, Aspergillus aculeatus* or *Kluyveromyces fragilis* (Rombouts and Thibault, 1986, Pasculli *et al*., 1991).

3.4. XYLOGALACTURONAN

Four different xylogalacturonans with degrees of substitution between 48 and 70 were submitted to *Fusarium* PG. Their molar mass did not decrease, even if the incubation was prolonged up to 4 days. Thus, xylogalacturonan was not a substrate for *Fusarium* PG, at least if their degree of substitution is higher or equal to 48 (Figure 4).

4. Conclusion

When the final percentages of hydrolysis of all the substrates tested were gathered in an unique graph (Figure 4), it shows clearly that the number, the nature and the distribution pattern of the substituents had an effect on the ability of the enzyme to degrade the substrate. Indeed, the percentage of hydrolysis decreased with increasing the degree of substitution, whatever the substituent. A blockwise distribution of the substituents induced an easier degradation because the enzyme can attack first some unsubstituted area in the substrate. The comparison of the results obtained from methylated pectins and acetylated homogalacturonan demonstrated that the enzyme was more hindered by acetyl groups than by methoxyl groups, and this could be related to both the size and the respective position of the substituent on the galacturonic acid residue.

The differences in PG modes of action towards substituted substrates could be explained by the ability of their active site to link more or less substituted residues. As native pectins in the plant cell wall are more often substituted, it is of interest for a phytopathogenic fungus to produce polygalacturonases able to link substituted residues.

Acknowledgements

The authors thank Dr H.C. Buchholt from Danisco Cultor (Braband, DK) for kindly providing the methylated pectins, Dr C. Caprari (University of Roma, Italy) and Dr T. Christensen (Danisco Cultor, Copenhaguen, DK) for the access to *Fusarium* PG. We would like to thank also Dr R. Körner (University of Southern Denmark, Odense M, DK) and Dr van Alebeek (Wageningen Agricultural University, NL), for the mass spectrometry experiments.

Financial support from the EU (Contract N° ERB BIO4-CT96-0685) is gratefully acknowledged.

References

Benen J.A.E., Kester H.C.M. and Visser J. (1999) Kinetic characterization of *Aspergillus niger* N400 endopolygalacturonases I, II, and C. *Eur. J. Biochem.*, **259**, 577-585.

Bonnin E., Le Goff A., Körner R., van Alebeek G.-J.W.M., Christensen T.M.I.E., Voragen A.G.J., Roepstorff P., Caprari C., Thibault J.-F. (2001a) Study of the mode of action of endopolygalacturonase from *Fusarium moniliforme, Biophysica, Biochimica Acta*, in press.

Bonnin E., Dolo E., Le Goff A., Thibault J.-F. (2001b) Characterization of pectin subunits obtained by enzymic fractionation, submitted

Caprari A., Bergmann C., Migheli Q., Salvi G., Albersheim P., Darvill A., Cervone F. and De Lorenzo G. (1993) *Fusarium moniliforme* secretes four endopolygalacturonases derived from a single gene product. *Physiology Molecular Plant Pathology*, **43**, 453-462.

Caprari C., Mattei B., Basile M.L., Salvi G., Crescenzi V., De Lorenzo G. and Cervone F. (1996) Mutagenesis of endopolygalacturonase from *Fusarium moniliforme_*: Histidine residue 234 is critical for enzymatic and macerating activities and not for binding to polygalacturonase-inhibiting protein (PGIP). *Molecular Plant-Microbe Interactions*, **9**, 617-624.

Kester, H.C.M., Magaud, D., Roy, C., Anker, D., Doutheau, A., Shevchik, V., Hugouvieux-Cotte-Pattat, N., Benen, J.A.E. and Visser, J. (1999) *Journal of Biological Chemistry*, **274**, 37053-37059.

Körner R., Limberg G., Mikkelsen, J.D. and Roepstorff, P. (1998) Characterisation of enzymatic pectin digests by matrix-assisted laser desorption/ionization mass spectrometry. *Journal of Mass Spectrometry*, **33**, 836-842.

Le Goff A., Renard C.M.G.C., Bonnin E. and Thibault J.-F. (2001) Extraction, purification and chemical characterisation of xylogalacturonans from pea hulls. *Carbohydrate Polymers*, **45**, 325-334.

Limberg G., Körner R., Buchholt H.C., Christensen, T.M.I.E., Roepstorff P. and Mikkelsen J.D. (2000a) Analysis of different de-esterification mechanisms for pectin by enzymatic fingerprinting using endopectin lyase and endopolygalacturonase II from *Aspergillus niger*. *Carbohydrate Research*, **327**, 293-307.

Limberg G., Körner R., Buchholt H.C., Christensen, T.M.I.E., Roepstorff P. and Mikkelsen J.D. (2000b) Quantification of the amount of galacturonic acid residues in blocksequences in pectin homogalacturonan by enzymatic fingerprinting with exo- and endo-polygalacturonase II from *Aspergillus niger*. *Carbohydrate Research*, **327**, 321-332.

Micard, V. and Thibault, J.-F. (1999) Oxidative gelation of sugar-beet pectins : use of laccases and hydration properties of the cross-linked pectins. *Carbohydrate Polymers*, **39**, 265-273.

Nelson, N. (1944). A photometric adaptation of the Somogyi method for the determination of glucose. *Journal of Biology and Biochemistry*, **153**, 378-380.

Nelson P.E. (1992) Taxonomy and biology of *Fusarium moniliforme*. *Mycopathology*, **117**, 29-36.

Parenicova L., Benen J.A.E., Kester H.C.M., Visser J. (2000) pgaA and pgaB encode two constitutively expressed endopolygalacturonases of *Aspergillus niger*. *Biochemistry Journal*, **345**, 637-644.

Pasculli, R., Geraeds, C.,Voragen, A.G.J., Pilnik, W. (1991) Characterization of polygalacturonases from yeasts and fungi. *Lebensmittel Wissenchaft und Technologie*, **24**, 63-70.

Rombouts, F.M. and Thibault, J.-F. (1986) Enzymic and chemical degradation and the fine structure of pectins from sugar beet pulp. *Carbohydrate Research*, **154**, 189-203.

Schols, H.A., Bakx, E.J., Schipper, D., Voragen, A.G.J. (1995) A xylogalacturonan subunit present in the modified hairy regions of apple pectin. *Carbohydrate Research*, *279*, 265-279.

Solms J. and Deuel H. (1951) Untersuchungen an acetylierter Pektinsäure. *Helvetica Chimica Acta*, **34**, 2242-2249.

Sturgeon, R.J. (1990) Monosaccharides. In Methods in Plant Biochemistry (P.M.Dey, J.B. Harborne, eds), 2, pp. 1-37.

Thibault J.-F. (1983) *Aspergillus niger* endopolygalacturonase. 3- Action pattern on polygalacturonic acid. *Carbohydrate Polymers*, **3**, 259-272.

Thibault J.-F., Renard C.M.G.C., Axelos M.A.V., Roger P. and Crépeau M.-J. (1993) Studies of the length of homogalacturonic regions in pectins by acid hydrolysis. *Carbohydrate Research*, **238**, 271-286.

Vink, H. (1963) Degradation of some polymers in aqueous solutions *Makromolekulare Chemie* **67**, 105-123

Voragen, A. G. J., Thibault, J.-F., Axelos, M.A.V., Renard, C.M.G.C. & Pilnik W. (1995) Pectins, in A.M. Stephen (ed), Food polysaccharides and their applications, Marcel Dekker Inc., New York, pp. 287-339.

THE ROLE OF PECTINS IN PLANT TISSUE UPON STORAGE AND PROCESSING; ANALYSIS AND MATHEMATICAL MODELLING IN RELATION TO FIRMNESS

C. VAN DIJK* AND L.M.M. TIJSKENS
ATO.BV, P.O. Box 17, 6700 AA, Wageningen, The Netherlands
** email: c.vandijk@ato.wag-ur.nl*

Abstract

Preheating vegetables prior to processing (cooking or sterilisation) can result in a firmer product (Van Buren, 1979). It is assumed that this increase in firmness is due to the increase in total exerted activity of PME rendering the pectin in a form less susceptible to heat degradation. This formulation was expressed as a set of chemical reaction equations, which were converted into a set of differential equations and solved analytically for constant temperature. The solution was a triple exponential equation. The data were obtained by preheating carrot samples applying a range of temperature - time combinations followed by a standardised cooking. Analysis based on this model resulted in an explained part for the data of 95%.
Storage of tomatoes resulted in a temperature dependent softening of this fruit. This firmness decrease is assumed to be associated with the total exerted activity of endo-PG. For this reason the PG activity was monitored during storage of tomatoes harvested at two maturity stages. During prolonged storage at 3 °C the PG activity did virtually not change. Increasing the storage temperature resulted in a sudden, autocatalytic type of increase in PG activity. The time of onset of this increase decreased with increasing storage temperature. However, the total exerted PG activity could not be related to the decrease in observed firmness as was the case for peaches (Tijskens et al., 1999).
For potatoes the description of the attribute "firmness" is, in comparison to other vegetables, more complicated due to the contribution of starch. For this reason, the focus was put on the effect of storage on both the sensory perceived texture as well as the pectin moiety of the cell walls. It was observed that storage did not affect the sensory perceived texture, but significant changes were observed in the chemical composition of pectin upon storage. These observed changes were irrespective of cultivar, size and dry mater content. Preheating potatoes, followed by cooking also resulted into a more firm product, in accordance with previous observations for different vegetables. The magnitude of the sensory perceived firmness increase was larger at a lower dry matter content of the potato. A chemical analysis of the pectin moiety of the cell walls, as related to preheating and cooking, showed that, in contrast to storage, no large changes in the overall structure of the pectin could be observed. In case of preheating at 60 °C for 30 minutes, 80% of the pectin was retained in the cell wall material compared to the fresh material. In case of direct cooking (no preheating) this value was only 20%. Since the PME activity remained virtually constant during

F. Voragen et al. (eds.), Advances in Pectin and Pectinase Research, 385-401.
© 2003 *Kluwer Academic Publishers. Printed in the Netherlands.*

preheating at 60 °C, it is therefore concluded that also for potatoes tissue firming upon preheating followed by cooking is the consequence of the total exerted PME activity.

1. Introduction

In a simplified approach, the primary role of pectin in plant tissue can be compared with that of cement in a brick wall; gluing individual cells/bricks together thereby adding to the physical strength of the construction (Van Dijk and Tijskens, 2000). This metaphor can be worked out to a greater extend, but its main deficit is the dynamism of the systems involved. Brick walls are supposed to last at least last for decades with minimal changes during this period, resulting in a static construction in time. For fresh, edible plants, this situation is different since the genetically superimposed chain "growth, ripening and senescence" in general only lasts for a limited time. During this period, substantial changes are taking place in these food products. These dynamic changes are the consequence of a complex, mutually related and dependent, enzyme catalysed series of orchestrated events. This dynamism is somehow contained in the conserved and static DNA blueprint of the plant species under study.
With focus on pectin during the course of events described above, the pectin moiety within the plant is continuously being modified (Stolle-Smits et al., 1999). Due to these enzymatic based modifications not only the chemistry of pectin is affected, but also the physico-chemical properties and consequently its physiological role. With emphasis on post harvest processes like storage and heat treatments, the aim of our research was to obtain a better understanding of underlying processes, associated with pectin degradation, causing firmness loss (Van Dijk and Tijskens, 2000). In other words, can pectin degradation, either caused by enzymes or by heat induced β-eliminative degradation, be analysed and understood in such a way that this information is suitable to mathematically describe these processes and to relate them to changes in firmness?
On basis of their chemical nature pectins are substantially more prone to either enzyme-orchestrated modifications or chemical (β-elimination) degradation than the (hemi-) cellulose matrix. This information can be translates into two firmness components. The first component constitutes the variable part of firmness, assumed to be associated with the physico-chemical properties of the pectic matrix in the plant tissue. The second firmness component is related to a fixed firmness part, which is assumed to be associated with the physico-chemical properties of the cellulose and hemi-cellulose matrix. During either storage or processing it is assumed that firmness changes are only related to the variable part of the firmness. To work this hypothesis out into greater detail, information was gathered from a variety of edible plants or fruits, which could either be stored for a short (tomatoes) or long (potatoes) period, or are consumed after a heat process (potatoes, carrots). It has to be realised that, to obtain kinetic information, changes in time as function of either storage, or processing temperature have to be analysed. The information obtained was in all cases related to the physico-chemical properties of pectin, However, changes in the chemical structure of the pectin moiety of the product under study, were not necessarily the focus, rather than the enzymes and processes that caused these changes. This implied that the activity of pectolytic enzymes was measured and that the total exerted enzyme activity was modelled and related to changes in the firmness of the product. This approach appeared to be

successful for some agro-products like tomatoes, peaches, carrots, and green beans (Van Dijk and Tijskens, 2000). However, a major drawback is that the (chemical and kinetic) assumptions made to develop the models, still have to be validated.
For potatoes a different approach was followed in the quest to understand the role of pectin in this plant tissue. Here the overall pectin composition was analysed and related to conditions of storage and (pre-)heating. One of the reasons for a separate approach for potatoes is the dominant role of starch in the perceived texture of potatoes. Due to this role of starch, firmness models developed for e.g. carrots, were not valid for potatoes.

2. Materials and Methods

2.1. MATERIALS

2.1.1. Potatoes

Potato tubers (*Solanum tubersosum*), cv's. Nicola and Irene, representing two extremes regarding the sensory perceived texture after cooking (Van Marle, 1997) were stored after harvest at 6 °C using sprout inhibitor. One lot of potatoes was investigated immediately after harvesting (field-run (FR) potatoes) while the remainder was stored for either three months (first storage (FS) experiment) or six months (second storage (SS) experiment). Before analysis the lots were divided into three size categories: 30-45 mm (Small), 45-55 mm (Medium) and 55-65 mm (Large). Each size category was divided into three sub-categories (Low, Medium and High), relative to the dry matter content (see below). The dry matter (DryMa) distribution was determined for both cultivars and the three tuber sizes at each analysis time using under water weight (UWW) analyses of the individual potatoes (Von Scheele et al., 1935). For each distribution 25% of the potatoes with the lowest $DryMa_{UWW}$ and 25% of the potatoes with the highest $DryMa_{UWW}$ were sampled. These samples respectively represented the sub-categories with the Low and High dry matter content. The remainder of the batch (50%) represented the Medium dry matter sub-category. The $DryMa_{UWW}$ value of each sub-category was calculated as the average value.

2.1.2. Tomatoes

Tomatoes (*Lycopersicon esculentum* L., cv. Tradiro), grown in a glasshouse in Belgium, were either harvested at colour stage 6 or at colour stage 8. The colour stage was determined based on the colour card system used by both the Dutch and the Belgian Growers Associations (OECD, 1988). Throughout the remainder of this study colour stages 6 and 8 will be referred to as "Unripe" and "Ripe" respectively. The tomatoes were stored at respectively 3 °C (chilling injury temperature) , 12 °C (optimal storage temperature), 20 °C and 25 °C , all at a relative humidity of about 90%. From start of the storage experiment samples were withdrawn at regular intervals for further analyses. A sample consisted of 20 tomatoes, either Ripe or Unripe, stored at one of the above mentioned temperatures. In total 1400 individual tomatoes were analysed.

2.1.3. Carrots

Carrots, cv Amsterdamse Bak, were obtained from a local grower and, after arrival, directly submitted to a variety of preheating conditions, followed by cooking.

2.2. METHODS

2.2.1. Sensory based texture analysis of potatoes

The sensory perceived texture of cooked potatoes was determined according to Van Marle (1997) using eight mouthfeel (M) descriptors (firm-M, crumbly-M, moist-M, sticky-M, grainy-M, waxy-M, mealy-M, mashable-M).

2.2.2. Instrumentally based firmness measurements

For tomatoes non-destructive compression measurements were performed using a Universal Testing Machine, Instron 4301. A flat plate compression was applied at a speed of 0.02 $m.min^{-1}$ to a maximum force of 3N. The average value obtained for both the slope and the distance of twenty tomatoes, representing one temperature-time combination, was used as input value for further analysis.

For potatoes their firmness after cooking was assessed by measuring the force required to break the tissue using the "wedge method" as described by Vincent et al. (Vincent et al., 1991). The firmness measurements were performed using a Universal Testing Machine, Instron 4202, using a wedge with an angle of 10^0 at a rate of 4 $mm.min^{-1}$. The maximum force measured to break the tissue was used for further analysis of the data.

2.2.3. Cell wall analyses

Cell wall analysis was performed according to Selvendran et al. (1985), monosaccharide composition according to Stolle-Smits et al. (1997), and anhydro-uronic acids were determined as described by Ahmed and Labavitch (1977).

2.2.4. Enzyme analyses

Endo-polygalacturonase (PG; *EC 3.2.1.15*) activity was determined spectrophotometrically following derivatisation of the reaction product with UV-absorbing 2-cyanoacetamide as is described by Gross (1982). Pectin methyl esterase (PME; *EC 3.1.1.11*) activity was measured according to Hageman (1986).

2.2.5. Data Analysis

The data were analysed using the statistical program Unscrambler, version 6.1 (CAMO A/S, Norway) using Soft Independent Modelling of Class Analogy (SIMCA) analysis. SIMCA was used to distinguish between different classes of samples, in this case the sub-categories. Non-linear regression was performed with Genstat (Rothamsted, UK), model development with MapleV (Waterloo, Canada).

3. General considerations concerning texture

3.1. HIERARCHICAL ORDER OF TEXTURE

Considering the origin of texture three hierarchical related, interacting levels of abstraction can be discerned. The first, most basal level underlying texture refers to the molecular aspects of texture. Here the chemical nature of the biopolymers, with emphasis on polysaccharides, their physico-chemical properties and their interactions, which all together form the nature of the plant cell wall at the molecular level, is the key determinant. The second, cellular level refers to the type and intensity of the interactions between individual cells. At this level the key determinant of texture is the tissue "archestructure" (Van Dijk and Tijskens, 2000). This "archestructure" comprises aspects like the size and shape of individual cells and their organisation within the tissue, in combination with their interaction such as cell-cell adhesion, cell wall thickness etc. At the third and highest level actual texture measurements can be performed, either by sensory analysis, or rheological measurements. The way textural properties change depends on the change in the physico-chemical properties of the constituting bioplymers, and their interactions at the molecular level, affecting the cellular level, with emphasis on the cell-cell interactions and resultingly, the concomitant consequences for the organ level.

3.2. ORIGIN OF TEXTURE

Besides the hierarchical order in texture the origin of the texture of agricultural products is caused by a combination of physical forces originating from the following properties or processes (van Dijk & Tijskens, 2000):

- the turgor pressure inside living cells and the associated tissue turgor,
- special components within cells, with emphasis on starch,
- the cohesive forces within a cell; the physico-chemical properties of the cell wall,
- adhesive forces between cells: the physico-chemical properties of (middle lamellae) pectin,
- the overall structure and shape of individual cells,
- the overall structure and shape of the tissue.

3.3. PECTIN AND PECTIN EIGENVALUES

Pectin is composed of two distinguishable regions, a linear homogalacturonan region and a branched rhamno-galacturonan region (De Vries et al., 1982) The linear or 'smooth' regions consist of a backbone of galacturonic acid intermittent with insertions of rhamnose, which are responsible for 'kinks' in the linearity of the homo-galacturonic acid region. The galacturonic acid residues may either carry methyl, or acetyl groups. The branched or 'hairy' regions contain a high proportion of rhamnose residues with side chains of either single unit galactan, long arabinan, or arabinogalactan. In the middle lamella the 'smooth' regions predominate, while in the primary wall the proportion of hairy regions increases (Carpita and Gibeaut, 1993). A small percentage of the (neutral) sugars in the cell wall polysaccharides carry ferulic acids and related

phenolic groups, which may be cross-linked by the action of a peroxidase in the presence of H_2O_2 (Fry, 1986). These cross-linked structures can connect polysaccharides together forming intercellular bridges, responsible for thermal stability of the texture of e.g. Chinese water chestnut (Parker and Waldron, 1995).
Pectins comprise a whole family of rhamnogalacturonans, to which a varying amount of side chains are attached. These side chains vary in composition and size. Therefore it is worthwhile to add numerical values to the pectin one is dealing with, to have some basic clue about its structure. Besides the well known degree of methylation or acetylation both the linearity as well as the side chain extent can be encompassed into a numerical value. For this reason the ratio of galacturonic acid /rhamnose is considered to represent a measure for the linearity of the cell wall pectin and the ratio of galacturonic acid over the sum of the main pectic neutral sugars (NS=Ara+Rha+Gal) is considered to represent a measure for the side chain extent (Boeriu et al., 1998). This side chain extent either represents a continuous grading, ranging from a small number of relatively long (DP>) side chains to a large number of relatively small (DP<) side chains.

3.4. MATHEMATICAL MODELLING

Recently an overview has been presented about mathematical models and model levels (Van Dijk and Tijskens, 2000). The models presented were developed based on kinetic mechanisms which were assumed to underlie the observed phenomena, taking care of the well-known rules of chemical kinetics, including the temperature dependency of the rate constants according Arrhenius. The mathematical development and statistical analysis were carried out according to Tijskens *et al.* (1997). No transformations were applied to the data to prevent errors during the estimation (Ross, 1990). The data were analysed as one integral set using time and temperature simultaneously as explaining variables (Tijskens, 1994). To analyse the experimental data, analytical solutions of the model formulation at constant external conditions are required. These analytical solutions will be deduced from the differential equations, on basis of the kinetic mechanisms and are a logical consequence of the differential equations. The boundary conditions for the differential equations are defined by the experimental set-up.

4. Results and Discussion

Processing and storage are operations aimed at either bringing, or keeping the product quality at an optimal state; its optimal "fitness for use" by the consumer (Kramer and Twig, 1970; Steenkamp, 1989). After processing the product is brought into a "fixed" state were the product quality is rather independent of time. This situation is, in general, achieved by addition of substantial amounts of energy (heat or pressure) to the product during a relatively short period (minutes). As consequence the majority, if not all original biological activities of the raw material has been eliminated. Storage at the other hand aims at keeping the product quality as close to it's optimal "fitness for use" as possible. This is achieved by "enforced hibernation" of the product for a prolonged period by adapting the environmental conditions in a product specific way, in order to minimise the biological activity of the product.

Both for processing, as well as for storage, it is assumed that the firmness of the product consists of two parts; a variable and a fixed part. In other words:

$$\text{Firm} = \text{Firm}_{\text{fix}} + \text{Firm}_{\text{var}} \tag{1}$$

Contributing to the variable part of the firmness are for example the turgor and the part of the pectin susceptible for either enzymatic or heat induced degradation. Components adding to the fixed part of the firmness are for example the (chemically inert) cellulose-hemi-cellulose matrix and starch.

4.1. MATHEMATICAL MODELLING OF THE FIRMNESS AFTER PREHEATING FOLLOWED BY COOKING.

Using a verbal model to describe the events taking place during preheating followed by either cooking or sterilisation includes the following events. Preheating (T> 55 °C) first results in an irreversible membrane damage and a concomitant loss of cell turgor. Furthermore, at elevated temperatures the activity of the enzyme PME increases, thereby increasing both the rate and extent to which pectin becomes demethylated. A still further increase in temperature causes a progressive PME inactivation and concomitant decrease in exerted activity. Demethylated pectin is less susceptible to β-degradation. The intensity of β-degradation progressively increases at increasing temperatures. Due to this depolymerisation the contribution of pectin to both the strength of the cell wall, as well as the cell-cell adhesion is assumed to decrease. For this reason it is argued that the higher the total exerted PME activity, the less vulnerable pectin becomes to heat degradation, and the higher the value of the variable part of the firmness will remain.
Transferring this verbal model description into reaction mechanisms which describe these processes leads to:

$$\text{PME} \xrightarrow{k_d} \text{PME}_{na} \tag{2}$$

This equation describes the temperature dependent inactivation process of PME. The subscript na stands for "not active".

$$\text{(Methylated - Pectin)} \xrightarrow{k_s \cdot \text{PME}} \text{Pectine} + \text{Methanol} \tag{3}$$

This reaction equation describes the temperature dependent demethylation of pectin by PME. Both reactions 2 and 3 occur at blanching conditions. At cooking or sterilisation temperatures, all PME will be inactivated. In the following equation the temperature dependent β-degradation, occurring at cooking temperatures while negligible at blanching temperatures, is described :

$$(\text{Methylated - Pectine}) \xrightarrow{k\beta} (\text{Water Soluble Products}) \tag{4}$$

Deriving a set of differential equations from these underlying chemical reaction mechanisms and solving them for constant temperature, taking care of the proper boundary conditions, results in:

$$\begin{aligned} \text{Firm} &= \text{Firm}_{\text{fix}} + \text{Firm}_{\text{var}} \cdot \\ &\exp\{-k\beta \cdot t_c \cdot DE_0 \cdot \exp[k_s \cdot PME_0 \cdot (\exp[-k_d \cdot t_p] - 1)/k_d]\} \end{aligned} \tag{5}$$

This function describes that the observed firmness is the sum of a fixed and a variable firmness part. The contribution of the variable firmness part to the total firmness is expressed in the triple exponential function. The highest exponential function describes the effect of the preheating time (t_p) on the residual PME activity, where PME_0 represents this enzyme activity just before blanching. As can easily be seen the second exponential term describes the enzyme catalysed demethylation of pectin. The first exponent describes the β-degradation of pectin during the time of either cooking or sterilisation, t_c. For more detailed information about estimates of rate constants and activation energies see Van Dijk and Tijskens (2000). In Figure 1 a three dimensional

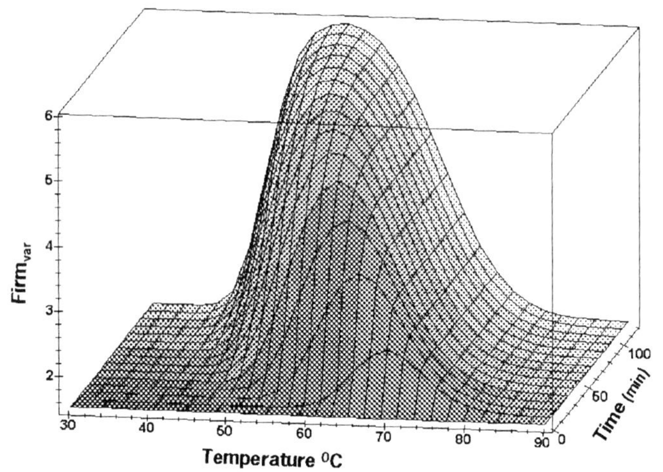

Figure 1 Three dimensional representation of the effect of preheating temperature and preheating time on the variable firmness, Firm_{var}, of carrots after cooking.

simulation is shown of the effects of preheating time and temperature, followed by cooking, on the measured firmness of 294 different carrot samples. From this figure it can be seen that both preheating below 45 °C, or above 75 °C has virtually no effect on the measured firmness. Below 45 °C the total exerted PME activity is too low to have

an effect on the degree of methylation and consequently on de intensity of the β-degradation of pectin during cooking or sterilisation. Above 75 °C PME is obviously already completely inactivated before the exerted activity can be observed. Between these two temperatures the observed increase in firmness is dependent on the time and temperature of preheating.
Here two remarks are at place. First, given the high explained part of the model (R^2 = 95.4%) this suggests the correctness of the assumptions made. Second, a decrease in the degree of pectin methylation is often difficult to observe after preheating and sterilisation (Stolle et al. (2000) , Binner et al. (2000)). It has however to be realised that small, local changes in the degree of methylation can have large consequences.

4.2. MATHEMATICAL MODELLING OF THE PG ACTIVITY OF TOMATOES DURING STORAGE.

Softening of the tomato fruit is the consequence of different processes active in product deterioration, such as cell wall and membrane breakdown, enzyme activation and inactivation etc. How these individual processes interact with one another is still largely unclear. In addition, it is already known for a long period that tomatoes are susceptible to chilling injury. This implies that below a temperature of about 12 °C the rate of deterioration increases with decreasing temperature.
Enzymes like PG and PME and other cell wall degrading enzymes often play an important role in the softening process. However, depending on the raw material, in some instances the exerted effects of these enzymes are hardly noticed. It becomes increasingly clear that the level of these enzymes, present in the individual fruits depend on growing and storage conditions. To obtain some systematic insight in the assumed relation between PG activity and tissue softening, tomatoes were stored at a four temperatures, of which one is known to cause chilling injury (3 °C) and the other three temperatures (12, 20 and 25 °C) are regarded safe temperatures as far as chilling injury is concerned. During storage of both Ripe and Unripe tomatoes, which lasted up to 4 weeks, both the softening of the tomato tissue was measured, as well as the PG activity. The firmness decay, which was also modelled (Van Dijk at al., 2001) is shown in Figure 2. From this figure it is obvious that the rate of the overall firmness decrease is minimal at12 °C. At 3 °C the decrease in firmness is slightly higher than at 12 °C. Apparently the gain obtained by lowering the overall biological activity caused by a decrease in temperature cannot compensate for the firmness loss caused by chilling injury. Increasing the temperature to either 20 °C or 25 °C causes a progressive increase in the deterioration of the firmness. A storage temperature of about 12 °C seems, in accordance with common practice, to be most optimal referring to a minimal change in firmness. What also can be observed, especially during storage at 20 °C and 25 °C, is that the firmness tends to extrapolate to a constant end value. This value, which is similar for both ripe and unripe tomatoes is assumed to represent the $Firm_{fix}$ value.

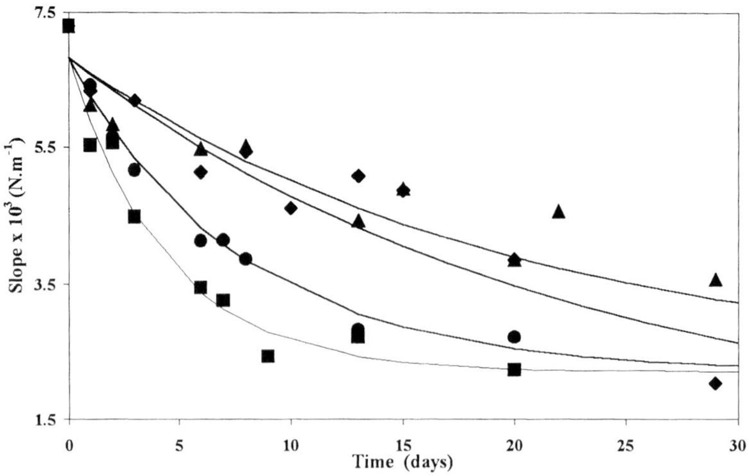

Figure 2 Effect of storage temperature on the firmness, expressed as compression slope, of unripe tomatoes stored at four different temperatures. Storage temperatures are ♦, 3 °C; ▲, 12 °C, •; 20 °C, ■, 25 °C. Drawn lines; simulation of the firmness according to Van Dijk et al. (20001)

At the beginning of the storage experiment, t=0, the firmness of the ripe tomatoes is lower than that of the unripe tomatoes (data not shown).
The measured PG activity for the same set of ripe and unripe tomatoes stored at the above mentioned temperatures is shown in Figure 3.

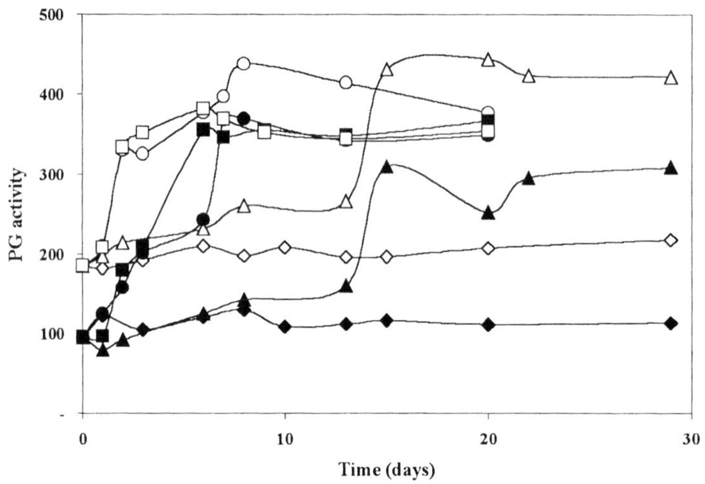

Figure 3 Effect of storage temperature on the PG activity of Ripe and Unripe tomatoes stored at four different temperatures. Storage temperatures are ♦, 3 °C; ▲, 12 °C, •; 20 °C, ■, 25 °C. Open symbols, Ripe tomatoes; closed symbols, Unripe tomatoes.

The following observations can be made with regard to the PG activity. At 3 °C the PG activity is virtually constant during storage. During storage at temperatures higher than 12 °C, a sudden increase in PG activity is observed, after which the activity remains constant at the attained higher activity level. The onset of this increase is faster in time at increasing temperatures, but the amplitude of this increase in activity is independent of the storage temperature. These observations are similar for the unripe compared with the ripe tomatoes.The magnitude of the increase is independent of ripeness stage, but the absolute level of PG is lower for the unripe compared with the ripe tomatoes. This information can be contained into the next chemical reaction mechanisms. Again the rate constants are assumed to obey Arrhenius' law.

$$PG_{pre} \xrightarrow{kf} PG \tag{6}$$

$$PG_{pre} + PG \xrightarrow{ka} 2PG \tag{7}$$

$$PG_{fix} = \text{constant} \tag{8}$$

Equation 6 describes the conversion of an inactive form of PG (=PG_{pre}) into the active form. Since the increase in PG activity is sigmoidal in shape this suggests an autocatalytic process, which is described in equation. 7. Deriving a set of differential equations from these mechanisms and solving them at constant conditions results in an (awkward) equation which is not transparent at all and will therefore not be presented. The explained part for this model, taking the PG activity measurements for the ripe and unripe tomatoes simultaneously and the differences in initial conditions into account, the explained part obtained for this model is R^2=88.5%. Given this reasonable value for the explained part the model presented rather accurately describes the observed PG activity in tomatoes.
However, at this stage a major problem is encountered. Previously, the relation between the measured PG activity and the decrease in firmness of peaches could be accurately modelled (Tijskens at al. 1999). At this moment it is still not possible to relate the measured or modelled PG activity to the observed firmness decrease in tomatoes. This suggests that for tomatoes additional, not yet understood mechanisms contribute to the softening process of the intact tomato fruit.

4.3. THE EFFECT OF STORAGE AND PROCESSING ON THE PECTIN MOIETY OF POTATOES.

Potatoes of cv's Nicola and Irene were divided into three size categories. From each size category a dry matter distribution was made, based on underwater weight analysis (Scheele et al., 1935). From this distribution 25% of the potatoes at the lower and 25% of the potatoes at the higher end of the distributions were further analysed. A pectin fractionation analysis was performed directly after harvest, and after three and six months of storage. In between the second and third harvest period potatoes were subjected to preheat treatments followed by cooking. These samples were also subjected to a pectin fractionation analysis to study these effects on the pectin composition. The procedure used to extract pectins from the Cell Wall Material (CWM) was designed to minimise both chemical (β-eliminative) as well as enzymatic degradation of pectins during their extraction and solubilisation in a state as close to their native form as possible (Selvendran et al., 1985). Distinguishable pectin fractions are obtained by a stepwise increase in the harshness of the extraction procedure. The pectic polysaccharides not connected to other cell wall polymers were extracted using the SDS-buffer during the preparation of the CWM and resulted into the Water Soluble Polymer (WSP) fraction. It seems reasonable to assume that, due to the ease of its extraction, the WSP does not contribute to the firmness of plant products. The amount of WSP is about 1 -1.6 times the amount of CWM. From the CWM the pectic polymers held in the cell wall by Ca^{2+} were solubilised by CDTA. (CDTA fraction). Dilute Na_2CO_3, subsequently extracted much of the CDTA-insoluble pectins at 4 °C (Carb 4 °C -fraction) and 20 °C (Carb 20 °C -fraction) respectively. The remainder is considered the Residue fraction. Of each of these fractions a sugar composition analysis was performed. The CDTA, Carb 4 °C, Carb 20 °C and Residue fractions amount for about 14, 12, 7 and 67% (w/w) of the CWM respectively.

The effect of storage on the sensory perceived texture of the potato cultivars, classified according to size and dry matter content, was analysed. In agreement with previous results (Van Marle et al., 1997) no effect of storage on the sensory perceived texture could be observed (data not shown). For the field run potatoes, as well as the potatoes stored for three and six months, a pectin fractionation study as well as a sugar compositional study was performed. Without going into detail it was shown that, with regard to both the side chain extent, as well as the linearity, no statistically significant differences could be observed neither between the cultivars studied, nor between sizes or dry matter content. However, significant effects of storage on both the linearity as well as side chain extent could be observed. In Figures 4A and 4B the linearity (Rha/UA) and side chain extent (NS/UA) of the water soluble polymers (WSP), the cell wall material (CWM) and sequentially extracted pectic fractions respectively are shown.

The overall change in the pectic composition upon storage is reflected in CWM. Upon storage the linearity and the side chain extent of the pectin of the cell walls of potatoes decreases (Fig 4A). These observed changes are especially reflected in the pectic fractions bound to the cell wall material, which can be extracted by carbonate (Carb 4 °C and Carb 20 °C), with an emphasis on the cold extracted pectic fraction. From Figure 4 it is obvious that upon storage of potatoes, large changes in the composition of pectin can be observed, which are not associated with changes in sensory perceived properties. In other words sensory perceived properties of potatoes are rather governed by the dry matter, respectively starch content, than by the contribution of pectin to the texture. The observed large changes in pectic composition upon storage are not reflected in any change contribution in sensory perception. Above, the consequences of preheating on the changes in firmness of carrots were described and modelled. In analogy, potatoes were preheated, cooled and steam cooked. It was observed that preheating at 60 °C, followed by cooking had a pronounced effect on the sensory perceived texture of potatoes.

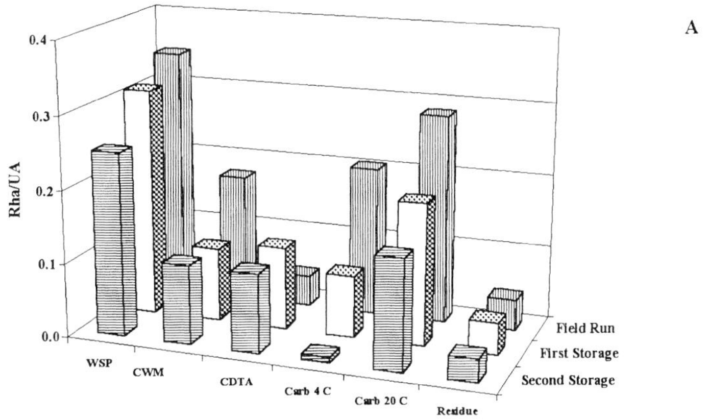

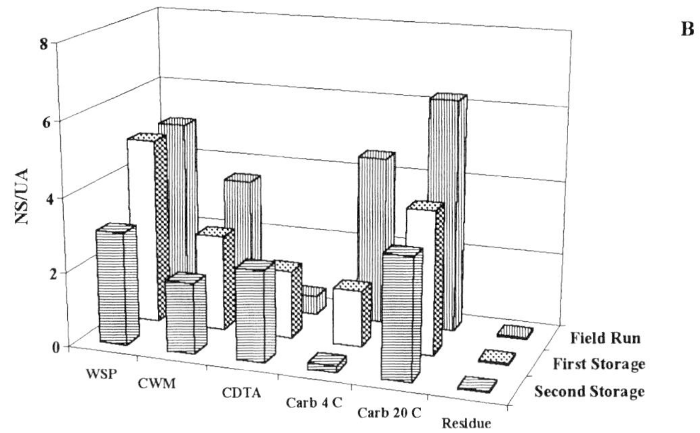

Figure 4 Effect of storage on the linearity (Rha/UA; Figure 4A) and side chain extent (NS/UA; Figure 4B) of WSP, CWM and sequentially extracted pectin fractions from potatoes directly after harvest and stored for three and six months.

In Figure 5 the effect of the preheating time on the sensory perceived firmness in relation to the Dry Matter content is shown. From this figure it is obvious that, without preheating, the potato firmness increases with decreasing DryMA content. After preheating for about one hour at 60 °C the sensory perceived firmness is maximal and has become independent of the dry matter content. In other words, preheating had a firming effect on potatoes. For this reason a pectin fractionation study was performed to analyse the potential changes in the composition of the pectin. Potato samples were either not preheated, preheated at 60 °C for 30 minutes, or preheated at 78 °C for 10 minutes, followed by cooking.

Figure 5 Effect of preheating time 60 °C on the sensory perceived firmness of potatoes after cooking, in

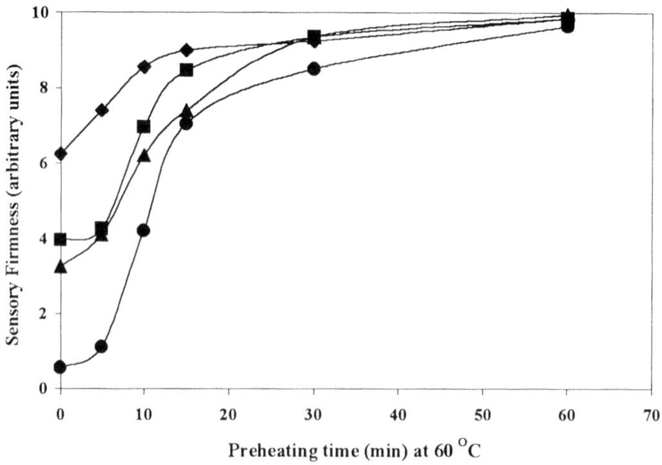

relation to the dry matter (DryMa) content of the potato. ♦, DryMa =14.7%; ■, DryMa =19.9%; ▲, DryMa =20.7%; ●; DryMa =25.9%

A pectin fractionation study, together with a sugar compositional study was performed. In Figure 6A the linearity and figure 6B the side chain extent of the WSP, the CWM and the sequentially extracted pectic fractions are presented. A general trend is that preheating results in both a small increase in linearity and a small decrease in side chain extent. This trend is, with some exceptions, observed in all sequentially extracted pectic fractions. In other words, the pectic compositional changes are substantially larger upon storage that upon preheating, the while the opposite is observed with regard to the sensory perceived firmness. Here it should be mentioned that, though preheating does not affect the pectin composition to a great extend, it does affect the amount of cell wall material. The amount of CWM of the potatoes directly cooked, or preheated at 78 °C, or at 60 °C, followed by cooking, relates as 1 to 2 to 4. In other words, preheating causes the pectin moiety of the cell wall to become substantially less susceptible to heat degradation, resulting in a fourfold increase in the amount of cell wall material after half an hour preheating at 60 °C. Again, the most probable explanation leads to the consequences of the PME activity. With regard to this enzymatic activity it was observed that the activity of this enzyme remained constant during preheating for one hour at 60 °C (data not shown).

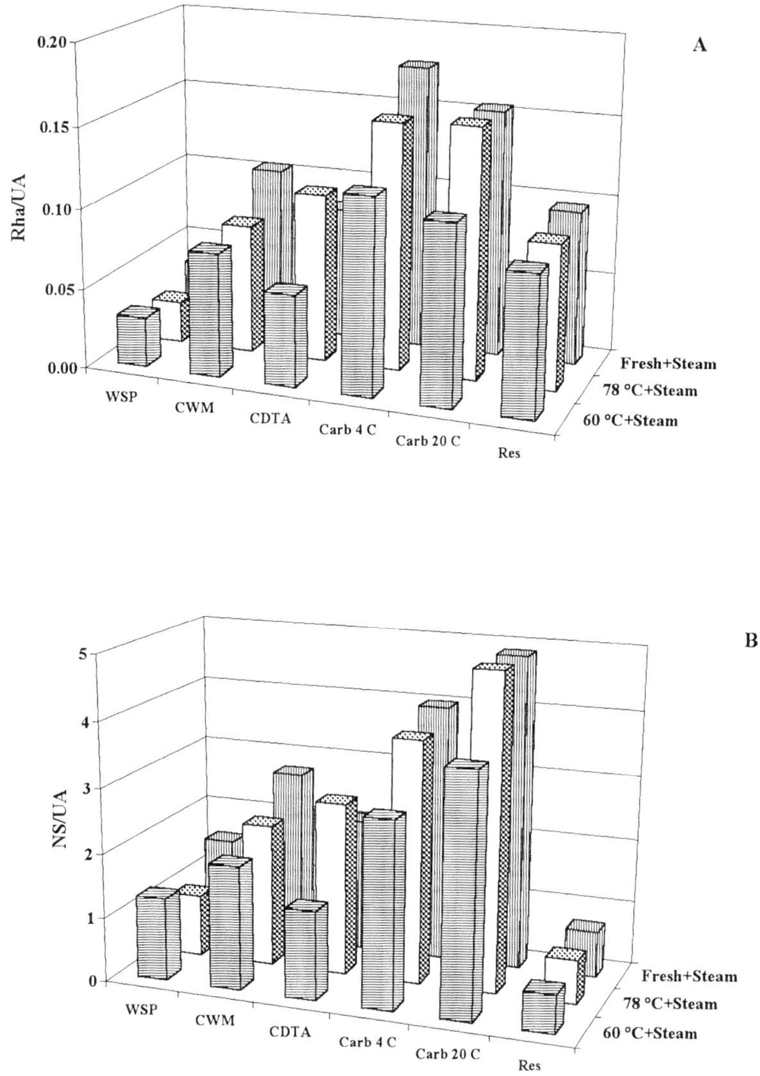

Figure 6 Effect of preheating on the linearity (Rha/UA; Figure 6A) and side chain extent (NS/UA; Figure 6B) of WSP, CWM and sequentially extracted pectin fractions from potatoes either directly cooked, preheated at 60 °C for 30 minutes, or at 78 °C for 10 minutes followed by cooking.

5. Conclusions

Summarising the results for potatoes it can be concluded, that storage is associated with substantial and significant changes in the chemical properties of pectin. These changes are not associated with changes in the sensory perceived texture. Preheating followed by cooking is not associated with changes in the pectin chemistry, but it is related with statistical significant changes in the sensory perceived texture and with an increase in the total amount of pectin in the cell wall material after cooking. This increase in cell wall material is ascribed to the consequences of the PME activity. This effect of the PME activity on the firmness was mathematically modelled, resulting into a triple exponential function describing the consequences of preheating temperature and time, followed by either cooking or sterilisation on the firmness of carrots. This model did not work for potatoes, probably due to it's starch.
Furthermore, the PG activity in tomatoes was also modelled on basis of an autocatalytic activation process.

Acknowledgements

The EU is acknowledged for their support of the research on tomatoes (FAIR CT95-302). Nestle is acknowledged for the financial contribution and joined efforts concerning the research on potatoes.

References

Ahmed, A.E.R. and Labavitch, J.M. (1977) A simplified method for accurate determination of cell wall uronide content, *J. Food Biochem.* **1**, 361-365.

Binner, S., Jardine, W.G., Renard, C.M.C.G. and Jarvis, M.C. (2000) Cell wall modifications during cooking of potatoes and sweet potatoes,. *J. Sci. Food Agric.* **80**, 216-218.

Boeriu, C., Stolle-Smits,T. and Van Dijk, C. (1998) Characterisation of cell wall pectines by Near Infrared Spectroscopy, *J. Near Infrared Spectrosc.* **6**, A299 - A303.

Carpita, N.C. and Gibeaut, D.M. (1993) Structural models of primary cell walls inf flowering plants: consistency of molecular structure with the physical properties of the walls during growth, *Plant Journal* **3**, 1-30.

Fry, S.C. (1986) Cross-linking of matrix polymers in growing cell walls of angiosperms, *Ann. Rev. Plant. Physiol.* **37**, 165-186.

Gross, K.C.A. (1982) A rapid and sensitive spectophotometric method for assaying polygalacturonase using 2-cyanoacetamide, *HortScience* **17**, 933-934.

Hagerman, A.E. and Austin, P.J. (1986) Continuous spectrophotometric assay for plant methyl esterase, *J. Agric. Food Chem.* **34**, 440-444.

Kramer A. and B.A. Twigg (1970) *Quality Control for the Food Industry*, AVI Pub. Co, Westport.

OECD, (1988). International standardisation of fruit and vegetables Tomatoes

Ross G.J.S. (1990). *Nonlinear Estimation.* Springer Verlag, New York.

Parker, M.L. and Waldron, K.W. (1985) Texture of Chinese Water Chestnut: Involvement of cell wall phenolics. *J. Sci. Food Agric.* **65**, 337-346

Selvendran, R.R., Stevens, B.J.H. and O'Neill, M.A. (1985) Developments in the isolation and analysis of cell walls from edible plants, in C.T. Brett and J.R. Hillman (eds.), *Society for experimental biology. Seminar series 28. Biochemistry of plant cell*; Cambridge University Press, London, pp. 39-78.

Steenkamp J.B.E.M.(1989) Product quality: an investigation into the concept and how it is perceived by consumers. Ph.D. Thesis, Agricultural University Wageningen, The Netherlands

Stolle-Smits, T., Beekhuizen, J.G., Dijk van, C., Voragen, A.G.J. and Recourt, K. (1997) Changes in pectice and hemicellolosic polymers of green beans (*Phaseolus vulgaris*) during industrial processing, *J. Agric. Food Chem.* **45**, 4790-4799.

Stolle-Smits, T., Donkers, C., Dijk van, C., Derksen, J.W.M. and Sassen, M.M.A. (1998) An electron microscopy study towards the texture of fresh, blanched and sterilised green bean pods (*Phaseolus vulgaris* L.), *Lebernsm. Wiss. Technol./Food Sci. and Technol.* **31**, 237-244

Stolle-Smits, T., Beekhuizen, J.G., Recourt, K., Voragen, A.G.J. and Van Dijk C. (2000) Preheating effects on the textural strength of canned green beans. 1. Cell wall chemistry, *J. Agric. Food Chem.* **48**, 5269-5277.

Tijskens L.M.M. (1994). Modelling colour of tomatoes. Advantage of multiple nonlinear regression. Proceedings COST94 Workshop *Post-Harvest Treatment of Fruit and Vegetables*, 14-15 September, Leuven, Belgium, pp. 175-185,

Tijskens, L.M.M.; Hertog, M.L.A.T.M.; Van Dijk, C. (1997) Generic modelling and practical applications, in B.M. Nicolaï and J. De Baerdemaeker (eds.*), Food quality modelling. Cost Action,* **915**, pp. 145-151.

Tijskens, L.M.M., Rodis. P.M., Hertog, M.L.A.T.M., Proxenia, N. and Van Dijk, C. (1999) Activity of pectin methyl esterase during blanching of peaches, *J. Food Engr.* **39**, 167-177.

Van Buren, J.P. (1979) The chemistry of texture in fruits and vegetables, *J. Texture Stud.* **10**, 1-23.

Van Dijk, C.and Tijskens, L.M.M. (2000) Mathematical modeling of enzymatic reactions as related to texture after storage and mild preheat treatments, in S.M Alzamore, M.S. Tapia, M.S. and A. Lopez-Malo (eds.), *Minimally processed fruits and vegetables,* Aspen Publishers Inc. Gaithersburg, Maryland. USA , pp. 127-152

Van Dijk, C., Stolle-Smits, T, Boeriu, C. and Tijskens, L.M.M. (2001) Fundamental modelling and NIR calibration of the firmness and moisture loss of batches of tomatoes (*cv. Tradiro*), Post Harvest Research, *to be submitted*

Van Marle, J.T., Van Vuurst- de Vries, R., Wilkinson, E.C. and Yuksel, D. (1997) Sensory evaluation of the texture of steam-cooked table potatoes, Potato Research **40**, 79-90

Vincent, J.F.V., Jeronimidis, G., Khan, A.A., Luyten, H. (1991) The wedge fracture test: A new method for measurement of food texture. *J. Texture Stud.* **22**, 45-57

Von Scheele, C., Svensson, G. and Rasmussen, J. (1935) Om bestämming an potatisens Stärelse och torrsubstanshalt med tiljhjäp av dess specifika vkt. Nord. Jordbr. Forsk. 11-37.

KINETIC INFORMATION ON THERMAL AND HIGH PRESSURE-TEMPERATURE INACTIVATION OF PECTINESTERASES

A. VAN LOEY, D. FACHIN, B. LY NGUYEN, I. VERLENT, and M. HENDRICKX
Katholieke Universiteit Leuven
Laboratory of Food Technology
Kasteelpark Arenberg 22, B-3001 Leuven, Belgium
E-mail: ann.vanloey@agr.kuleuven.ac.be

Abstract

Several key effects and advantages of high hydrostatic pressure have resulted in the current interest of the food industry to use this novel technology in food processing or preservation, replacing or complementing thermal processing. In this context, the current paper focuses on kinetic studies of the activity and stability of pectinesterase (PE) from oranges and tomatoes during thermal and high pressure processing. In addition, the influence of intrinsic product properties such as pH and Ca^{2+}-ions on the processing stability of PE is quantified.

1. Introduction

1.1. POTENTIAL FOOD APPLICATIONS OF HIGH PRESSURE

Where, in previous decades, the primary objective of food manufacturers was to produce safe shelf stable products and food quality was considered of secondary importance, throughout the 1980's to the present day, the food industry is driven by the consumer desire for high quality, minimally processed, additive free, shelf stable, convenient and safe food products. As a result, existing thermal preservation technologies are improved (e.g. better process control) or adjusted (e.g. aseptic processing, ohmic and microwave heating), or new physical food preservation methods are being introduced. In this context, the use of high hydrostatic pressure (HHP – up to 1000 MPa) receives plentiful interest from the food industry since it has been demonstrated experimentally that HHP at room temperature can inactivate vegetative microorganisms and some food deteriorative enzymes while it only slightly affects nutritional and sensorial food quality aspects (e.g. vitamin content, flavor and color) (Farr, 1990, Cheftel, 1991, Hoover, 1993, Knorr, 1993, Hayakawa *et al.*, 1994, Mozhaev *et al.*, 1994). Due to the extreme pressure stability of spores of some species, a combination of pressure with other, ideally synergistic, preservation technologies is required if pressure sterilization of low-acid foods is to be pursued. For this purpose, the effect of high pressure and elevated temperature on microbial as well as enzymatic inactivation is under investigation.

F. Voragen et al. (eds.), Advances in Pectin and Pectinase Research, 403-417.
© 2003 *Kluwer Academic Publishers. Printed in the Netherlands.*

Potential food applications of high pressure have been identified not only with respect to food preservation, but also for food processing. The latter include protein modification, tenderisation of meat products and texturisation of fish proteins, reaction enhancement or retardation with effect on bioconversions or metabolite production, and modification of enzyme functionality. Furthermore, by taking advantage of the effect of pressure on the solid-liquid phase diagram of water, potential applications are pressure assisted freezing, pressure assisted thawing and non frozen storage under pressure at subzero temperatures (Kalichevsky *et al.*, 1995).

1.2. THE EFFECT OF HIGH PRESSURE ON ENZYMES

Enzymes are a special class of proteins in which biological activity arises from an active site, brought together by the three dimensional configuration of the molecule. Even small changes in the active site can lead to a loss of enzymatic activity. Since protein denaturation is associated with conformational changes, it can change the functionality of the enzyme (e.g. increase or loss of biological activity, change in substrate specificity). Enzymes may be affected by pressure in several ways (Cheftel, 1992):

i. Pressurisation at room temperature may bring reversible or irreversible, partial or complete enzyme inactivation resulting from conformational changes in the protein structure, depending on the type of enzyme, the medium composition, the pressure, the temperature and the duration of processing. For multimeric enzymes pressure often enhances the dissociation into subunits, changing enzyme activity when the maintenance of the quaternary structure is required for proper functioning (Morild, 1981; Balny and Masson, 1993; Heremans, 1993). In general terms, the effect of pressure on enzyme (in)activation may be divided into two classes: in the low pressure range (100 to 300 MPa) reversible changes occur, while pressures above 400 MPa may induce irreversible inactivation. The two primary means by which enzyme inactivation occurs are alterations of intramolecular structures or conformational changes at the active site (Morild, 1981; Hoover *et al.*, 1989).

ii. Enzymatic conversions may be enhanced or inhibited by pressure, depending on the positive or negative reaction volume as reactions characterised by negative volume changes are favoured by application of high pressure. Reasons for pressure-induced changes in the catalytic rate may be classified into three main groups: (i) changes in enzyme-substrate interaction, (ii) changes in the reaction mechanism or (iii) effects of a particular rate-limiting step on the overall catalytic rate (Heremans, 1993; Mozhaev *et al.*, 1994).

iii. A macromolecular substrate (protein, starch) may become more or less sensitive to enzymatic depolymerisation or to modification once it has been pressurised.

iv. Intracellular enzymes may be released in extracellular fluids or cell cytoplasm provided the cell membrane or membranes of intracellular organelles are altered by pressure. This may facilitate enzyme-substrate reactions.

2. Pressure-temperature kinetic inactivation studies

2.1. MATERIALS

PE purified from orange peel [EC 3.1.1.11, Sigma (Bornem, Belgium), product P-5400] was purchased as a lyophilized powder. Orange PE dissolved in deionized water was used as a model system. The pH of this model system was 4.5. The influence of Ca^{2+} concentration was investigated by dissolving PE in 0.5, 1.0, and 1.5 M $CaCl_2$ solutions. The influence of pH on the inactivation of orange PE was investigated in a citric acid buffer (5.0 mM, pH 3.7). A citric acid buffer was chosen because of the relatively low negative ionization volume of this buffer and because of the occurrence of citric acid in natural products such as fruits and vegetables.

Tomato PE [EC 3.1.1.11, Sigma (Bornem, Belgium), product P-6763] was purchased as a lyophilized powder containing 145 units per mg solid. One unit was defined by Sigma as the release of 1 μequiv of acid from pectin within 1 min at a pH of 7.5 and 30°C. Tomato PE dissolved in deionized water was used as a model system. The pH of this solution was 5.2. To investigate the influence of Ca^{2+} ions a 60.0 mM and 1.0 M $CaCl_2$ solution was used. To study the influence of pH on the inactivation of tomato PE, the enzyme was dissolved in a 5.0 mM citric acid buffer with pH-values from 3.5 to 4.5, which is relevant for tomato products.

2.2. METHODOLOGY

2.2.1 Isothermal treatment

Isothermal inactivation experiments were performed in a water bath with temperature control. To ensure isothermal heating, the enzyme solution was enclosed in glass capillary tubes. After preset time intervals, the capillaries were withdrawn from the water bath and immediately cooled in ice water to stop thermal inactivation. The residual activity of PE was measured after 10–120 min storage in ice water as described in 2.2.3. During storage, no reactivation of the enzyme was observed.

2.2.2 Isobaric-isothermal treatment

The kinetic parameter values for the pressure-temperature inactivation of PE were determined on the basis of isobaric-isothermal inactivation experiments and determination of the residual enzyme activity. For this purpose, a laboratory pilot scale, multivessel high-pressure equipment (Resato, Roden, The Netherlands) was used.
The apparatus allows pressurization up to 1000 MPa in combination with temperatures ranging from –20 to 100°C. High pressure is generated using a pressure intensifier in the central pressure circuit. The pressure medium is a glycol-oil mixture (TR15, Resato). The temperature is controlled by a thermostated mantle, which surrounds each vessel and is connected to a cryostat. This apparatus is suited for kinetic studies, since eight individual vessels (volume = 8 mL, diameter = 10 mm, length = 100 mm) can be subjected to the same pressure level and the same temperature level.
Isobaric-isothermal treatments were performed as follows; flexible micro tubes [0.3 mL, Elkay (Overijse, Belgium)], filled with the solution to be treated were enclosed in the pressure vessels, already equilibrated at a preset temperature. Pressure was built up

slowly to minimize adiabatic heating. After reaching the desired pressure, the individual vessels were isolated and the pressure was maintained in the vessels until the valves were opened. Based on previous research (Weemaes et al., 1997), an equilibration period of 2 min to allow temperature to evolve to its desired value (input value) was taken into account. Starting the time course of the experiment ("zero point") after this equilibration period, the process could be considered as an isobaric-isothermal treatment. At this moment, one pressure vessel was decompressed and the activity of the corresponding enzyme sample was considered as the blank (A_0). The other seven vessels, each containing one enzyme sample, were then decompressed after preset time intervals. After pressure release, the samples were immediately cooled in ice water. The residual activity was measured after 10–120 min of storage in ice water as described in 2.2.3. During storage no reactivation of the enzyme was observed.

2.2.3 PE activity assay

PE activity was determined titrimetrically at pH 7.0 and 22°C. The reaction mixture in the standard assay method consisted of 250 μl of PE sample and 30 mL of a 0.35% apple pectin solution [70-75% esterification, Fluka (Bornem, Belgium)] containing 125 mM NaCl. Before injection of the enzyme solution, the pectin solution was adjusted to pH 7.0. During hydrolysis at 22°C, the pH was maintained at 7.0 by addition of 0.01 N NaOH using an automatic pH-stat titrator (Metrohm). Every 15 sec the consumption of 0.01 N NaOH was recorded during the 10 min reaction period. The PE activity is proportional to the rate of NaOH consumption ($\Delta V_{NaOH}/\Delta t$). It was confirmed that the commercially available pectin was not contaminated with PE, i.e. without addition of PE no acid formation was observed during the assay. With the assay described above, duplicates or triplicates could be measured with less than 5% error.

2.3. DATA-ANALYSIS

For the analysis of the inactivation data we proceed in two steps: At first it is investigated what happens with the enzyme as a function of treatment time, or in other words what is the rate of enzyme inactivation. So the first step is the selection of an adequate inactivation model. In a second step, it is investigated how the reaction rate changes with the processing variables pressure and temperature, or in other words the selection of pressure and temperature coefficient models.

2.3.1 Identification of an Inactivation Rate Equation

PE inactivation due to a thermal or high pressure-temperature treatment could usually be described by the following inactivation models:

First order model Thermal or pressure inactivation of enzymes can often be described by a first order reaction. Under isothermal-isobaric conditions a first order reaction can be expressed as Eqn. 1, which can be linearized by a logarithmic transformation, yielding Eqn. 2.

$$A/A_0 = \exp(-kt) \qquad (1)$$

$$\ln(A/A_0) = -kt \qquad (2)$$

where A is enzymatic activity at time t; A_0 is initial enzymatic activity; t is treatment time; and k is the first-order inactivation rate constant. In case of a first order inactivation model, the rate constant at a given temperature and pressure was estimated by linear regression analysis of $\ln(A/A_0)$ versus inactivation time at constant inactivation temperature and pressure.

Distinct isozyme model Enzymes characterized by several isozymes such as orange PE can often be subdivided into two (or more) fractions with different processing stability, e.g. one more thermal (pressure) resistant than the other and both inactivating according to a first-order decay kinetic model. For constant extrinsic (e.g. pressure, temperature) and intrinsic factors and assuming that the inactivation of both fractions is independent of each other, the inactivation can be modeled according to Eqn. 3.

$$A/A_0 = (1-\alpha)\exp(-k_l t) + \alpha\exp(-k_s t) \tag{3}$$

where α is the stable enzyme fraction, $1-\alpha$ the labile enzyme fraction, k the first-order inactivation rate constant where the subscripts s and l for k denote thermostable and thermolabile respectively. By plotting the residual activity after different time intervals versus time, the inactivation rate constant of the labile fraction (k_l-value), the inactivation rate constant of the stable fraction (k_s-value), and the activity of both fractions can be estimated using nonlinear regression analysis.

Fractional conversion model Fractional conversion refers to a (first-order) inactivation process that takes into account a nonzero activity upon prolonged heating and/or pressurizing ($=A_\infty$). A fractional conversion model can be expressed mathematically as Eqn. 4.

$$A = A_\infty + (A_0 - A_\infty)\exp(-kt) \tag{4}$$

This relation is valid in the temperature and/or pressure domain where only the labile enzyme fraction inactivates whereas the activity of the stable fraction (A_∞-value) does not change with respect to time. By plotting residual activity after different time intervals versus time, the inactivation rate constant (k-value) and the remaining activity after prolonged treatment (A_∞-value) can be estimated using nonlinear regression analysis.

2.3.2 Determination of a Temperature and Pressure coefficient model

Temperature dependence model The temperature dependence of the rate constant k can usually be expressed by an activation energy, E_a, as indicated in the Arrhenius relationship (Eqn. 5).

$$k = k_{Tref}\exp\left(\frac{E_a}{R_g}\left(\frac{1}{T_{ref}} - \frac{1}{T}\right)\right) \tag{5}$$

where k_{Tref} is the inactivation rate constant at T_{ref}, T_{ref} the reference temperature, E_a the activation energy and R_g the universal gas constant (R_g=8.314 J/Kmol). Eqn. (5) can be linearized which allows the activation energy at a certain pressure to be estimated based on linear regression analysis of the natural logarithm of k versus the reciprocal of the absolute temperature

Pressure dependence model The pressure dependence of the rate constant can usually be expressed by an activation volume (V_a), as indicated in the Eyring equation (Eqn. 6).

$$k = k_{Pref} \exp\left(\frac{-V_a}{R_g T}(P - P_{ref})\right) \quad (6)$$

where k_{Pref} is the inactivation rate constant at P_{ref}, P_{ref} the reference pressure, V_a the activation volume at a certain temperature, T the absolute temperature and R_g the universal gas constant (R_g=8.314cm³MPa/Kmol). Also the Eyring equation can be linearized by a logarithmic transformation. Hence, plotting the natural logarithm of the rate constant in function of pressure, the activation volume at a certain temperature can be derived from the slope of the regression line.

2.4 Results And Discussion

2.4.1 Thermal inactivation of orange PE

Thermal inactivation of commercially available orange PE (1 mg of solid/mL deionized water) was studied in the temperature domain 60-82°C. As two different fractions were observed, a heat labile fraction and a heat stable fraction, a biphasic model was used to estimate the kinetic inactivation parameters (see Table 1). For temperatures up to 70°C, the inactivation rate constant of the labile fraction was estimated, whereas the stable fraction did not inactivate. At higher temperatures (75-82°C), the inactivation rate constant of the stable fraction was estimated. In this temperature domain, the labile fraction inactivated too fast to estimate accurately the decline of activity. The inactivation rate constant of the labile fraction was found to be less temperature sensitive than the inactivation rate constant of the stable fraction. This confirms literature data (Wicker and Temelli, 1988; Versteeg *et al.*, 1980, Van den Broeck *et al.*, 1999a). The stable fraction did not inactivate for temperatures up to 70°C, and could be considered as a constant fraction. Five percent of the initial activity was estimated to be thermostable. In the temperature domain 60-70°C, where only the heat labile fraction inactivated, also a fractional conversion model was used to estimate the inactivation parameters (see Table 1). In Figure 1, the model fitting for thermal inactivation of commercial orange PE is depicted. This figure clearly illustrates that the stable fraction is independent of the temperature level applied. The inactivation parameters estimated by both models, i.e. the inactivation rate constant of the thermolabile fraction, the percentage of thermostable PE and the activation energy for the labile fraction, correspond very well.

To investigate the influence of Ca^{2+}-ions, naturally present in orange juice, on the inactivation kinetics, commercial orange PE was dissolved in 0.5 M, 1.0 M and 1.5 M $CaCl_2$ solution. Inactivation temperatures for the thermolabile PE fraction varied from 45 to 58°C. In all cases, thermal inactivation data of the thermolabile PE fraction could be modeled with a fractional conversion model. By comparing the thermal inactivation parameters of orange PE in the different systems, it could be deduced that the presence of Ca^{2+} ions decrease the thermal stability of orange PE. In case of 1.0 and 1.5 M $CaCl_2$ solution, a small decreased temperature sensitivity of the inactivation rate constant was noticed. Although orange PE is less thermostable in the presence of Ca^{2+} ions, the percentage of stable PE after treatment is higher. It is worthwhile mentioning that, in

presence of Ca^{2+} ions, a decreased initial activity of PE was observed, which agrees with literature indications (Van Buren et al., 1962, Alonso *et al.*, 1997).
With regard to the thermostable PE frcation, the presence of Ca^{2+}-ions clearly reduced the inactivation rate of the thermostable PE fraction, whereas the activation energy of the stable PE fraction was only slightly influenced by addition of Ca^{2+}-ions.

Table 1
Kinetic parameter estimates for isothermal inactivation at atmospheric pressure of orange PE (1 mg/mL) in distilled water.

	Biphasic model	Fractional conversion
Individual analysis of the heat labile fraction		
$k_{60°C}$ (min^{-1})	0.110 ± 0.013	0.103 ± 0.004
$k_{63°C}$ (min^{-1})	0.311 ± 0.014	0.314 ± 0.008
$k_{65°C}$ (min^{-1})	0.610 ± 0.106	0.637 ± 0.064
$k_{67°C}$ (min^{-1})	1.896 ± 0.128	1.766 ± 0.095
$k_{70°C}$ (min^{-1})	4.189 ± 0.491	3.848 ± 0.376
E_a (kJ/mole)	357.4 ± 21.7	353.1 ± 17.7
Individual analysis of the heat stable fraction		
$k_{75°C}$ (min^{-1})	0.037 ± 0.004	
$k_{77°C}$ (min^{-1})	0.112 ± 0.021	
$k_{80°C}$ (min^{-1})	0.551 ± 0.041	
$k_{82°C}$ (min^{-1})	1.234 ± 0.473	
E_a (kJ/mole)	518.2 ± 19.3	

The influence of pH on the isothermal inactivation of orange PE was investigated by using a 5.0 mM citric acid buffer (pH 3.7). A comparison of the kinetic parameters for thermal inactivation of the thermolabile PE fraction reveals that commercial orange PE is inactivated more rapidly and at lower temperatures in citric acid buffer (pH 3.7) than in water (pH 4.5). It can also be concluded for the thermolabile fraction of commercial orange PE that the inactivation rate constant in citric acid buffer (pH 3.7) is significantly less temperature sensitive than the inactivation rate constant in water. For the thermostable PE fraction, treatment in citric acid buffer (pH 3.7) also accelerated thermal inactivation of the stable fraction as compared to the inactivation in water (pH 4.5), while the activation energy for thermal inactivation of PE in both media was similar.

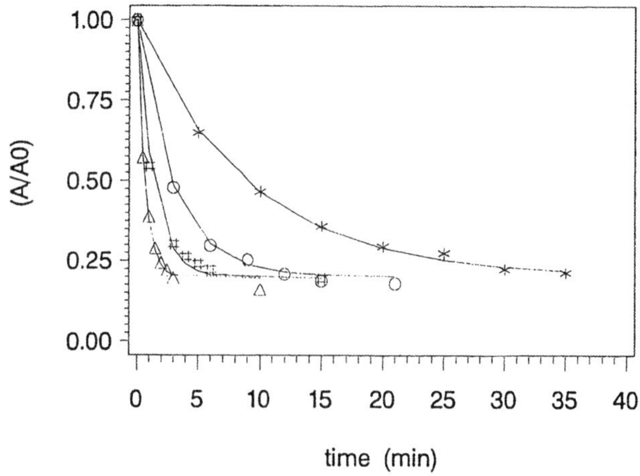

Figure 1

Thermal inactivation curves of commercial orange PE (1.5 mg solid/mL), thermally treated at 60 (*), 63 (o), 65 (#) and 67 (Δ)°C. A_0 and A are initial orange PE activity and activity after thermal treatment respectively.

2.4.2 Pressure-temperature inactivation of orange PE

A complete kinetic data-set for the combined pressure-temperature inactivation of orange PE was gathered (Van den Broeck *et al.*, 1999b, 2000a). The pressure-temperature inactivation of commercially available orange PE (0.4 mg solid/mL) was investigated in deionized water. Pressures from 0.1 to 900 MPa were applied combined with temperatures from 15 to 65°C. In Figure 2 the inactivation of commercial orange PE is illustrated for different combinations of pressure and temperature At all pressure/temperature combinations studied no complete inactivation of orange PE was attained. Only the labile fraction was inactivated whereas the activity of the stable fraction remained unchanged with respect to time. Nonlinear regression analysis per temperature level, according to the fractional conversion model was used to fit the inactivation data.

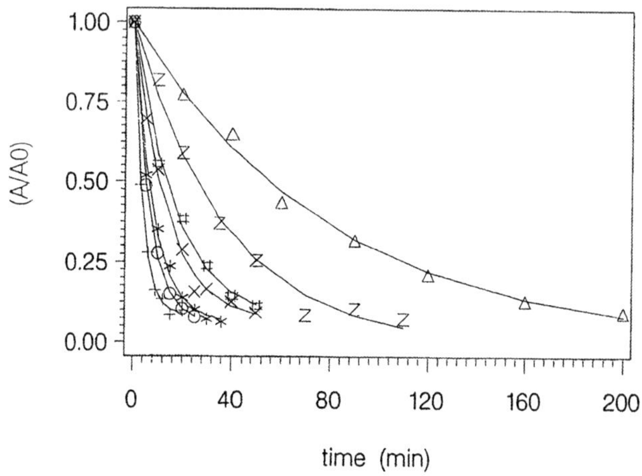

Figure 2

Combined pressure-temperature inactivation of commercial orange PE at pressure-temperature combinations (MPa/°C): (#) 0.1/57; (O) 0.1/60; (Δ) 100/57; (z) 150/60; (×) 400/60; (+) 800/35; (*) 850/20.

The pressure-stable PE fraction was independent of the pressure level applied and contributed to about 5% of the total PE activity. Because no complete inactivation of the enzyme was attained after temperature (see 2.4.1) and after pressure treatment, we assume that the PE isozymes with high heat resistance may also show high pressure resistance. However, to elucidate the relation between the pressure and temperature resistance of this fraction, more experiments are needed after purification of the individual PE isozymes.

For orange PE a pressure increase to 700-900 MPa at room temperature was required to inactivate the labile fraction. At elevated pressure and temperature, a synergistic effect of pressure and temperature on the inactivation of the labile fraction has been observed: increasing pressure at a fixed temperature increases the inactivation rate. Likewise, increasing temperature at a fixed pressure leads to increased inactivation rates. In the high temperature (T≥57°C) and low pressure (P≤300 MPa) region, on the contrary, a clearly antagonistic effect of pressure and temperature was observed. Increasing pressure caused a decrease of the inactivation rate constant. A similar antagonistic effect was observed for polyphenoloxidase (Weemaes *et al.*, 1998) and tomato PE (see 2.4.4). Based on the estimated *k*-values from isothermal as well as from isobaric-isothermal inactivation experiments, a pressure-temperature kinetic diagram (iso-rate contour) for commercial orange PE was constructed (Figure 3). The lines in this two-dimensional diagram represent combinations of pressure and temperature resulting in the same inactivation rate constant of the labile fraction. The synergistic effect of pressure and temperature in the high-pressure region (P≥350 MPa) and the antagonistic effect of pressure and temperature in the low pressure-high temperature domain can also be deduced from the shape of this kinetic diagram. Such iso-rate contour diagrams are extremely useful for adequate process design.

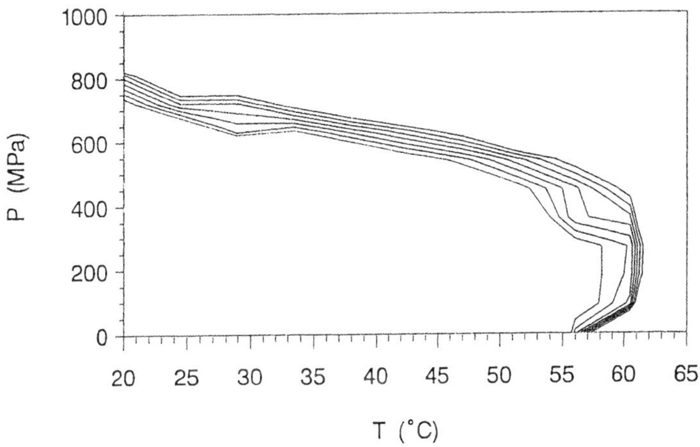

Figure 3

Pressure-temperature kinetic diagram for the P/T-inactivation of commercial orange PE in deionized water. The inner and outer line represent P/T combinations for which *k* equals 0.02 and 0.08 min^{-1} respectively

The influence of pH on the isobaric-isothermal inactivation of commercial orange PE was investigated in a 5.0 mM citric acid buffer (pH 3.7). Inactivation data of the pressure labile PE fraction could be modeled with the fractional conversion model. A

comparison of the kinetic parameters for the inactivation in water and in buffer at elevated pressure revealed that the enzyme is less pressure stable in an acid medium. The pressure sensitivity of the inactivation rate constant k was almost not influenced. The percentage pressure stable fraction remained the same, whether orange PE inactivated in water or in an acid medium.
The influence of Ca^{2+} ions on the pressure inactivation of commercial orange PE at room temperature was investigated in an 0.5, 1.0, and 1.5 M $CaCl_2$ solution. A comparison of the kinetic data indicates that an increased pressure inactivation was observed for the lower molarities (0.5 and 1.0 M) as compared to PE inactivation in the absence of Ca^{2+} ions, whereas in the presence of a 1.5 M $CaCl_2$ solution, a decreased inactivation was noticed. This is in contrast to thermal treatment, where an increased inactivation was observed for each of the Ca^{2+} concentrations studied (see 2.4.1). A slight increase in the pressure sensitivity of the inactivation rate constant k was observed, as compared to the inactivation in absence of Ca^{2+} ions. Notwithstanding the altered stability and sensitivity of orange PE, the remaining active fraction remained nearly the same: 4–5% of the original activity was found to be active after pressure treatment in the absence as well as in the presence of Ca^{2+} ions.

2.4.3 Thermal inactivation of tomato PE

Isothermal inactivation of tomato PE in deionized water (0.4 mg solid/mL) was studied in a temperature range from 57 to 65°C. Inactivation could be accurately described by a first-order kinetic model as can be seen from a plot of log(A/A_0) versus time (Figure 4) (Table 2).

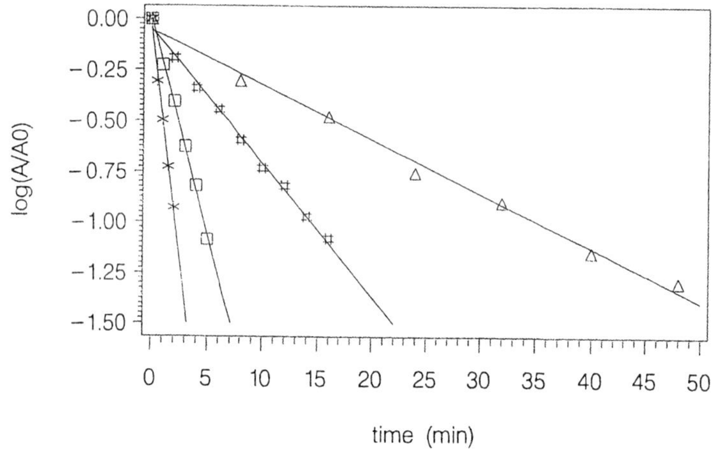

Figure 4
Thermal inactivation curves of commercial tomato PE (0.4 mg solid/mL deionized water), thermally treated at 57 (Δ), 60 (#), 63 (□) and 65 (∗) °C. A_0 and A are initial tomato PE activity and activity after thermal treatment respectively.

The influence of pH on the thermal inactivation of commercial tomato PE was investigated by heating the enzyme in a citric acid buffer (pH 3.5, pH 3.8 and pH 4.5). By comparing the inactivation in the buffer systems and in deionized water, it can be stated that tomato PE is more thermostable in buffer than in deionized water.

Comparing the different buffer systems mutually points out that tomato PE is more thermostable at higher pH values. In other words, acidification of the medium decreases the thermal stability. Moreover, an increased thermosensitivity of the rate constant, expressed by the Ea-value, is observed for inactivation in a citric acid buffer. These findings are consistent with earlier reported work of Rothschild et al. (1975), Eagerman and Rouse (1976) and Nath and Ranganna (1977) who reported that increasing the pH necessitated increasing the amount of heat treatment in order to inactivate PE.

Table 2
k- and E_a-values for the thermal inactivation of tomato PE in water and in citric acid buffers pH 4.5, 3.8 and 3.5.

	Water	Citric acid buffer pH 4.5	Citric acid buffer pH 3.8	Citric acid buffer pH 3.5
Inactivation rate constant (min^{-1})				
57°C	0.062±0.002			
60°C	0.152±0.004			
63°C	0.489±0.012			0.050±0.001
65°C	1.047±0.050	0.028±0.002	0.053±0.003	0.110±0.005
67°C		0.074±0.008	0.107±0.009	0.255±0.011
68.5°C		0.139±0.006	0.215±0.009	
70°C		0.327±0.009	0.441±0.017	
E_a-value (kJ/mole)				
	331.0 ± 15.5	468.1±18.2	410.2±24.1	388.3 ± 7.2

To investigate the influence of Ca^{2+} ions on the thermal inactivation of tomato PE, a 60.0 mM and 1.0 M $CaCl_2$ solution was used. Again, a first-order thermal inactivation was observed. By comparing the estimated kinetic parameters obtained for the inactivation of tomato PE in presence of and without Ca^{2+} ions, a clear distinction has to be made between the lower (60.0 mM) and higher molarity (1.0 M). In presence of a high concentration of Ca^{2+} ions an enhanced inactivation and a reduced temperature sensitivity of the *k*-value was noted. For the inactivation in presence of a 60.0 mM $CaCl_2$ solution the inactivation rate clearly reduced and a decreased temperature sensitivity of the *k*- value was noted.

2.4.4 Pressure-temperature inactivation of tomato PE

The influence of pressure on thermal inactivation of tomato PE (0.4 mg solid/ mL deionized water) was investigated at two different temperatures; a temperature of 60°C whereby tomato PE inactivates at atmospheric pressure, and a temperature of 40°C whereby no thermal inactivation of tomato PE was observed. The pressure range studied, varied from 100 MPa to 900 MPa. At 40°C, no inactivation of the enzyme was observed for pressures up to 700 MPa. A further increase in pressure inactivated tomato PE, which points to a small synergistic effect of pressure and temperature at 40°C.

At 60°C on the other hand, an outspoken antagonistic effect of pressure and temperature were observed, i.e. thermal inactivation of tomato PE was strongly counteracted by pressures in the range 100-500 MPa. The antagonistic effect is illustrated in Figure 5. Increasing pressure from atmospheric pressure to 500 MPa, suppressed inactivation. A further increase in pressure to 850 MPa slightly increased suppressed inactivation. A further increase in pressure to 850 MPa slightly increased the inactivation, but even at 900 MPa and 60°C ($k_{60°C,900\ MPa}$=0.077 min^{-1}), was the inactivation still slower as compared to the same treatment at atmospheric pressure ($k_{60°C}$=0.154 min^{-1}). It can thus be concluded that, although tomato PE is rather heat labile it is very pressure resistant.

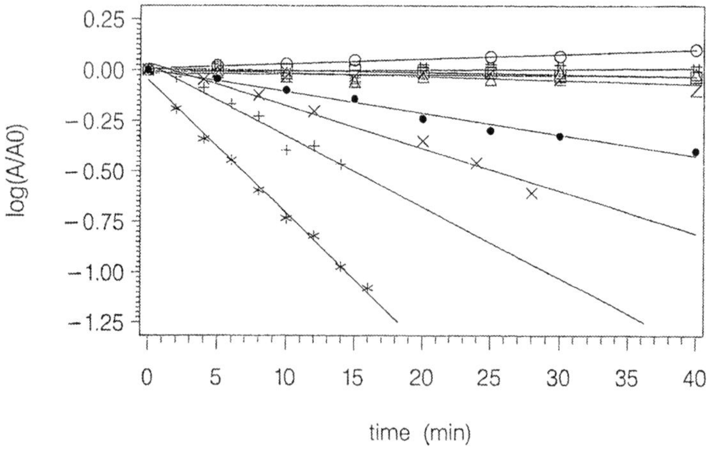

Figure 5
Combined pressure-temperature inactivation curves of commercial tomato PE (0.4 mg solid/mL deionized water), treated at 60°C in combination with pressures of 100 MPa (O), 200 MPa (#), 300 MPa (□), 500 MPa (Δ), 700 MPa (z), 800 MPa (•), 850 MPa (×), 900 MPa (+) and atmospheric pressure (*). A_0 and A are initial tomato PE activity and activity after treatment respectively.

A study was carried out in order to find out whether pH has an influence on the pressure-temperature inactivation of tomato PE and more specifically on the antagonistic effect of pressure and temperature. As for thermal inactivation a 5.0 mM citric acid buffer pH 4.5, 3.8 and 3.5 was used. As for the inactivation in deionized water an antagonistic effect of pressure and temperature was observed in citric acid buffer. However, the antagonistic effect is less pronounced in buffer than in deionized water. Comparison of the different buffered systems mutually points out that at lower pH values, higher pressures are needed to obtain a similar inactivation rate constant as at atmospheric pressure, i.e. a stronger antagonistic effect of pressure and temperature is observed at the lowest pH value. This can be explained by the fact that a decrease of pH from 4.5 to 3.5 has more influence on thermal inactivation than on pressure-temperature inactivation. As related to the $\Delta V^{\neq}$-value, the lowest pressure sensitivity of the *k*- value was observed for the inactivation of tomato PE in buffer pH 3.5. Concerning the stability of tomato PE in citric acid buffer towards temperature and/or pressure, it can be stated that tomato PE is less stable towards temperature and pressure at lower pH values, while the *k*-value is more thermosensitive and pressure sensitive at higher pH values.

Besides the influence of pH also the influence of Ca^{2+} ions on the pressure-temperature inactivation of tomato PE was investigated. It could be concluded that pressure-temperature inactivation enhances in presence of a 1.0 M $CaCl_2$ solution, whereas it decreases in a 60.0 mM $CaCl_2$ solution. A similar behavior was observed for thermal inactivation. For both Ca^{2+}-concentrations investigated, a pressure increase to 700 MPa was necessary to slowly inactivate tomato PE. A further increase in pressure from 700 to 900 MPa accelerated the inactivation process. However at 900 MPa was the inactivation still slower than at atmospheric pressure. At lower temperatures the antagonistic effect of pressure and temperature was no longer noted. A similar behavior was found for the inactivation in absence of Ca^{2+} ions.

2.4.5 Effect of Pressure and Temperature on tomato PE activity

As tomato PE is sensitive to temperature but resistant to pressure, it was decided to study the catalytic activity of tomato PE under combined pressure and/or temperature treatment (Van den Broeck *et al.*, 2000b). Therefore, in contrast to the work described in above, the enzyme together with its substrate was pressure and/or temperature treated. The enzymatic activity during thermal or combined pressure-temperature treatment was determined by measuring the release of methanol as function of time colorimetrically based on a method of Klavons and Bennett (1986).

The effect of increasing pressure on PE-catalyzed hydrolysis of pectin was investigated at the optimal pH for enzymatic reaction observed at atmospheric pressure, *i.e.* pH 7.2. An enzyme concentration of 0.1 mg solid/mL and a pectin concentration of 0.4% was used. The inactivation study (see 2.4.4) revealed that pressures from 100 to 500 MPa protected PE from thermal inactivation at temperatures higher than 40°C. In Figure 6, it is clearly illustrated that the enzyme is not only protected from thermal inactivation but even remains active for pressures up to 300 MPa.

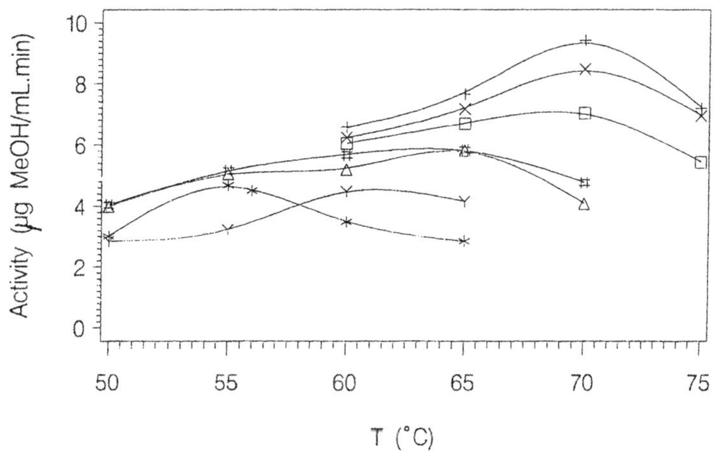

Figure 6

Enzyme activity as function of temperature at atmospheric pressure (*), 100 MPa (Δ), 200 MPa (#), 300 MPa (Y),100 MPa and 60.0 mM Ca^{2+} (□), 200 MPa and 60.0 mM Ca^{2+} (+), 300 MPa and 60.0 mM Ca^{2+} (x).

As at atmospheric pressure, the reaction rate at elevated pressure is dependent of temperature. The optimal temperature observed at elevated pressure shifted to higher values (respectively 60-65°C and 70°C in absence and presence of Ca^{2+} ions) as compared to atmospheric pressure (55°C). At 100 and 200 MPa, a higher enzymatic activity was observed in the temperature domain 50-70°C, as compared to atmospheric pressure, whereas at 300 MPa, only at temperatures above 60°C, a higher enzymatic activity was noted. As a consequence of these two tendencies (*i.e.* an increase in optimal temperature and acceleration of the reaction under pressure), the catalytic activity of PE at temperatures higher than 60°C, is significantly greater at pressures up to 300 MPa as compared to atmospheric pressure. Presence of Ca^{2+} ions even increased the enzymatic activity at elevated pressure up to 300 MPa and shifted the optimal temperature to 70°C (Figure 6).

3. Conclusion

For orange PE, a thermo- and pressure labile, and a thermo- and pressure stable PE fraction has been observed. An increased Ca^{2+}-concentration causes sensitization of orange PE to temperature, whereas the influence on pressure stability of PE was concentration dependent. Acidification decreases the thermal and pressure stability of orange PE.
For tomato PE, a single heat labile tomato PE fraction was observed as opposed to orange PE. Although heat labile, tomato PE shows extreme pressure stability. Temperature and pressure act antagonistically on tomato PE inactivation: pressures up to 500 MPa suppress thermal inactivation of tomato PE. It has been shown that tomato PE in presence of pectin maintains its activity under pressure. By application of low pressure (up to 300 MPa) its activity can even be enhanced. Treatments that increase PE activity may be interesting for improving the texture of processed fruits and vegetables. It may be possible to design treatments that minimize unwanted degradation by PG and maximize PE activity, possibly leading to tissue firming (Eshtiaghi *et al.*, 1994; Knorr, 1995). A recent patent described the use of high pressure to produce canned tomatoes with improved texture by activation of PE and inactivation of PG (Wilding and Woolner, 1997).

References

Alonso, J.; Canet, W.; Rodriguez, T. 1997. Thermal and calcium pretreatment affects texture, pectinesterase and pectic substances of frozen sweet cherries. J. Food Sci., 62, 511-515.

Balny, C. and Masson, P. 1993. Effects of high pressure on proteins. Food Reviews International, 9(4), 611-628.

Cheftel, J.-C. 1991. Applications des hautes pressions en technologie alimentaire. Industries Alimentaires et Agricoles, 108(3), 141-153.

Cheftel, J.-C. 1992. Effects of high hydrostatic pressure on food constituents: an overview. in: 'High Pressure and Biotechnology, vol. 24', eds. Balny, C., Hayashi, R., Heremans, K. and Masson, P., John Libbey Eurotext, Montrouge, pp. 195-209.

Eagerman, B. A.; Rouse, A. H. 1976. Heat inactivation temperature-time relationships for pectinesterase inactivation in citrus juices. J. Food Sci., 41, 1396-1397.

Eshtiaghi, M. N.; Stute, R.; Knorr, D. 1994. High-pressure and freezing pretreatment effects on drying, rehydration, texture and color of green beans, carrots and potatoes. J. Food Sci., 59, 1168-1170.

Farr, D. 1990. High pressure technology in the food industry. Trends in Food Science and Technology, 1(1), 14-16.

Hayakawa, I., Kanno, T., Tomita, K. and Fujio, Y. 1994. Application of high pressure for spore inactivation and protein denaturation. Journal of Food Science, 59(1), 159-163.

Heremans, K. 1993. The behaviour of proteins under pressure. in: 'High pressure chemistry, biochemistry and material science', eds. Winter, R. and Jonas, J., Kluwer Academic Publ., Dordrecht, pp. 443-469.
Hoover, D.G. 1993. Pressure effects on biological systems. Food Technology, 47(6), 150-155.
Hoover, D.G., Metrick, C., Papineau, A.M., Farkas, D.F. and Knorr, D. 1989. Biological effects of high hydrostatic pressure on food microorganisms. Food Technology, 43(3), 99-107.
Kalichevsky, M.T., Knorr, D. and Lillford, P.J. 1995. Potential food applications of high-pressure effects on ice-water transitions. Trends in Food Science and Technology, 6, 253-259.
Klavons, J. A.; Bennett, R. D. 1986. Determination of methanol using alcohol oxidase and its application to methyl ester content of pectins. J. Agric. Food Chem., 34, 597-599.
Knorr, D. 1993. Effects of high hydrostatic pressure on food safety and quality. Food Technology, 47(6), 156-161.
Knorr, D. 1995. High pressure effects on plant derived foods. In High pressure processing of foods; Ledward, D. A., Johnston, D. E., Earnshaw, R. G., Hasting, A. P. M., Eds.; Nottingham University Press: Nottingham, U.K., pp 123-125.
Morild, E. 1981. The theory of pressure effects on enzymes. Advances in Protein Chemistry, 34, 93-166.
Mozhaev, V.V., Heremans, K., Frank, J., Masson, P. and Balny, C. 1994. Exploiting the effects of high hydrostatic pressure in biotechnological applications. Trends in Biotechnology, 12(12), 493-501.
Nath, N.; Ranganna, S. 1977. Time/temperature relationship for thermal inactivation of pectinesterase in mandarin orange (Citrus reticulata Blanco) juice. J. Food Technol., 12, 411-419.
Rothschild, G.; Van Vliet, C.; Karsenty A. 1975. Pasteurization conditions for juices and comminuted products of Israeli citrus fruits. J. Food Technol., 10, 29-38.
Van Buren, J. P.; Moyer, J. C.; Robinson, W. B. 1962. Pectin methylesterase in snap beans. J. Food Sci., 27, 291-294.
Van den Broeck, I., Ludikhuyze, L. R., Van Loey, A. M., Hendrickx, M. E. 2000a. Inactivation of orange pectinesterase by combined high pressure and temperature treatments: a kinetic study. J. Agric. Food Chem., 48(5), 1960-1970.
Van den Broeck, I., Ludikhuyze, L. R., Van Loey, A. M., Weemaes, C. A., Hendrickx, M. E. 1999b. Thermal and combined pressure-temperature inactivation of orange pectinesterase: influence of pH and additives. J. Agric. Food Chem. 47 (7), 2950-2958.
Van den Broeck, I., Ludikhuyze, L. R., Weemaes, C. A., Van Loey, A. M., Hendrickx, M. E. 2000b. Effect of temperature and/or pressure on tomato pectinesterase activity. J. Agric. Food Chem, 48(2), 551-558.
Van den Broeck, I., Ludikhuyze, L., Weemaes, C., Van Loey, A., Hendrickx, M. 1999a. Thermal inactivation kinetics of pectinesterase extracted from oranges. J. Food Process. Preserv., 23, 391-406.
Versteeg, C.; Rombouts, F. M.; Spaansen, C. H.; Pilnik, W. 1980. Thermostability and orange juice cloud destabilizing properties of multiple pectinesterases from orange. J. Food Sci., 45, 969-971, 998.
Weemaes, C.; Ludikhuyze, L.; Van den Broeck, I.; Hendrickx, M. E. 1998. Effect of pH on pressure and thermal inactivation of avocado polyphenoloxidase: a kinetic study. J. Agric. Food Chem., 46, 2785-2792.
Weemaes, C.A., De Cordt, S.V., Ludikhuyze, L.R., Van den Broeck, I., Hendrickx, M.E. and Tobback, P.P. 1997. Influence of pH, benzoic acid, EDTA and glutathione on the pressure and/or temperature inactivation kinetics of mushroom polyphenoloxidase. Biotechnology progress, 13, 25-32.
Wicker, L.; Temelli, F. 1988. Heat inactivation of pectinesterase in orange juice pulp. J. Food Sci., 53 (1), 162-164.
Wilding, P. and Woolner, E. 1997. Process for the preparation of a food product. Patent WO 97/3859.

CHARACTERISATION OF PECTINS EXTRACTED FROM FRESH SUGAR-BEET ROOTS UNDER DIFFERENT CONDITIONS USING AN EXPERIMENTAL DESIGN

S. LEVIGNE, M.-C. RALET, J.-F. THIBAULT
Unité de Recherche sur les Polysaccharides, leurs Organisations et Interactions
Institut National de la Recherche Agronomique B.P. 71627
F-44316 Nantes Cedex 03 France

Abstract

An experimental design was constructed to study the influence of different parameters of acidic extraction on yield, chemical composition and intrinsic viscosity of pectins from fresh sugar-beet roots. pH, and, to a lower extent, time and temperature of extraction, affected greatly the yield and features of extracted pectins whereas the type of acid had no effect. Influences of extraction conditions on pectins characteristics were quantified and second order models were built.

1. Introduction

Industry traditionally uses citrus peels and apple pomace as raw materials for pectin production but alternative sources are currently investigated (Voragen *et al.*, 1995). In Europe, more than two million tons of sugar-beet pulp are generated annually by the sugar refining industry. It's availability, high content in pectins (20-25%) and low cost, make sugar-beet pulp a potential source of pectins (May, 1990). Sugar-beet pectins, although they exhibit the same main structural characteristics than citrus or apple pectins, have some distinctive features:

(i) their molar mass is relatively low (Axelos and Thibault, 1991);

(ii) they carry some acetyl groups on the O2 and/or O3 of the galacturonic acid residues (Keenan *et al.*, 1985);

(iii) they contain a high proportion of neutral sugars side chains, especially arabinan side-chains (Guillon and Thibault, 1988);

(iv) the presence of phenolic acids, mostly ferulic acid, esterifying arabinose and galactose residues, has been demonstrated (Guillon *et al.*, 1989; Ralet *et al.*, 1994; Colqhoun *et al.*, 1994).

The high acetyl content, together with the high arabinose content and the relatively low molar mass of sugar-beet pectins limit their use in the traditional applications of pectin - gel formation with acid/sugar or Ca^{2+} ions (Pippen *et al.*, 1950). The presence of ferulic acid ester-linked to the arabinose and galactose residues of the side-chains can however

F. Voragen et al. (eds.), Advances in Pectin and Pectinase Research, 419-430.
© 2003 *Kluwer Academic Publishers. Printed in the Netherlands.*

be used for oxidative cross-linking of pectins, leading to gel formation (Rombouts and Thibault, 1986; Oosterveld *et al.*, 2000, 2001).
Sugar-beet contains low amounts of water- and calcium chelating agents-extractable pectins (Rombouts and Thibault, 1986; Renard and Thibault, 1993; Oosterveld *et al.*, 1996). The highest amount of pectin is generally obtained by hot acid extraction (Rombouts and Thibault, 1986; Arslan, 1995), which is also the most convenient approach for industry. Temperature, pH, time, and the type of acid used were reported to affect the amount as well as the chemical characteristics of pectins extracted from sugar-beet pulp (Michel *et al*, 1985; Phatak *et al.*, 1988). The effect of the different extraction parameters was however not quantified.
In this work, our aim was to study the influence of several parameters of extraction (pH, time, temperature and type of acid) on yield, chemical composition (galacturonic acid, neutral sugars, methanol, acetic acid and ferulic acid contents), and intrinsic viscosity of sugar-beet pectins. An experimental central composite design was constructed and we report here on the quantification of extraction parameters on pectin characteristics and on the fitting of experimental values with second order models.

2. Experimental

2.1 MATERIAL

Fresh sugar beet roots were sampled in the experiment station of INRA in Mons (France) during the 1999 campaign. The roots were washed with tap water, manually peeled and rasped in a kitchen grinder. 1kg of rasped roots was immersed in 96% boiling ethanol (3l) for 20 min. The slurry was filtered through G3 sintered glass and the insoluble material was left under magnetic stirring for 30 min with 70 % ethanol (2l) and filtered again. This step was repeated until the filtrate gave a negative reaction in phenol sulphuric acid test (Dubois *et al.*, 1956). The residue was dried by solvent exchange (ethanol, aceton) and left one night at 40°C to give AIR (alcohol insoluble residue).

2.2 EXPERIMENTAL DESIGN

A factorial composite design was used to determine the effect of the four extraction variables (pH, temperature, time and nature of the acid) on the characteristics of the extracted pectins. The experimental design was constructed from a full two-level factorial design for four factors. The variables were standardised and coded as levels (-1, +1) (Table 1). The coefficients were tested for significance using Student's t-test at a significance level of P=0.05. The estimated regression equation was tested for the adequacy of fit using the Fisher F-test at a significance level of P= 0.05.
Central points were performed in duplicate for both acids. The quadratic effects were determined by six "star" points for HCl. The model coefficients reflected the linear, quadratic and interactive effects. The regression equations given in "Results and Discussion" only give the most influent factors.

TABLE 1. Full two-level[a] factorial design (Yates matrix)

Extraction	pH X_1	Temperature (°C) X_2	Time (min) X_3	Acid X_4
A	-1	-1	-1	-1
B	+1	-1	-1	-1
C	-1	+1	-1	-1
D	+1	+1	-1	-1
E	-1	-1	+1	-1
F	+1	-1	+1	-1
G	-1	+1	+1	-1
H	+1	+1	+1	-1
I	-1	-1	-1	+1
J	+1	-1	-1	+1
K	-1	+1	-1	+1
L	+1	+1	-1	+1
M	-1	-1	+1	+1
N	+1	-1	+1	+1
O	-1	+1	+1	+1
P	+1	+1	+1	+1
CP1[b]	0	0	0	-1
CP2	0	0	0	+1
SP1[c]	-1	0	0	-1
SP2	+1	0	0	-1
SP3	0	-1	0	-1
SP4	0	+1	0	-1
SP5	0	0	-1	-1
SP6	0	0	+1	-1

[a] The lower and upper level (-1, +1) of the variables correspond to 1 and 3; 75°C and 95°C; 30 min and 90 min; and HCl and HNO_3 for pH, temperature, time and type of acid, respectively.
[b] Central points (performed in duplicate for each acid).
[c] Star points (performed only with HCl).

2.3 ACIDIC EXTRACTIONS

AIR (1g) was stirred in an acid solution (30 ml) in a glass reactor bearing a reflux condenser. The temperature was regulated using an oil bath. The extraction was carried out 3 times. The residue was separated by filtration through G3 sintered glass.

Supernatants were pooled, adjusted to pH 4.5 with 2M NaOH, concentrated under vacuum at 35°C, and extensively dialysed against distilled water. The extract was freeze-dried, stored in a vacuum oven at 40°C and weighed. The required pH (1 to 3) of the acid solution, the temperature (75 to 95°C) and the time of extraction (30 to 90 min) were set according to the different values of the composite design.

2.4 ANALYTICAL

The galacturonic acid content was determined by the automated *m*-hydroxybiphenyl method (Thibault, 1979) and by the method of Ahmed and Labavitch (1977) in the soluble and in the insoluble materials, respectively.

Individual neutral sugars were analysed as their alditol acetate derivatives (Blakeney *et al.*, 1983) by gas chromatography after hydrolysis by 1M H_2SO_4 at 100°C for 2 and 6h. Six hours of hydrolysis were needed for a total rhamnose release (Micard *et al.*, 1996), but some degradation of arabinose and galactose could occur in this condition. A prehydrolysis step in 72% H_2SO_4 (30 min, 25°C) was used in the case of insoluble materials (Saeman *et al.*, 1954).

Methanol and acetic acid were released by alkaline deesterification (0.5 M NaOH, 1 h, 4°C), separated and quantified by HPLC on C18 Superspher (4 mM sulphuric acid, 0.7 ml/min, 25°C). Isopropanol was added as internal standard (Levigne *et al.*, 2001a). Degree of methylation (DM) and degree of acetylation (DA) were calculated as the molar ratio of methanol and acetic acid to galacturonic acid, respectively.

Phenolic compounds were determined by HPLC after saponification and extraction. The pectins were saponified by 2M NaOH at 35°C under argon during 30 min in the dark. After an internal standard (*o*-coumaric acid) has been added, the solution was neutralised with 2N HCl. Phenolic compounds were extracted with ether. The ether phase was evaporated at 40°C, 1 ml of methanol/H_2O (1:1v/v) was added and samples (20 μl) were injected on an HPLC system equipped with a C18 column (Purospher, Merk). Gradient elution was performed using methanol/acetic acid (1:0.01) (A) and H_2O/acetic acid (1:0.01) (B) at 0.7ml/min at 25°C: (0min, A = 20%; 20min, A= 60%; 21min, A= 80%; 30min, A= 80%; 31min, A= 20%). Phenolic compounds were detected at 320 nm. Response factors were determined relatively to *o*-coumaric acid (RF = 0.54 for ferulic acid).

2.5 INTRINSIC VISCOSITY

Pectins were solubilised in 50mM $NaNO_3$ solution (3mg/ml) at room temperature for 3h under magnetic stirring and filtered over a 0.45 μm Minisart RC15 Sartorius membrane. The solution was injected at 25°C on a high-performance size exclusion chromatography (HPSEC) system constituted of one Shodex SB-G pre-column followed by two Shodex OH-pak SB HQ 804 and 805 columns eluted at 0.7ml/min with 50mM $NaNO_3$, containing 0.02% NaN_3 as a preservative. On-line intrinsic viscosity determinations were performed at room temperature using a differential refractometer (ERC 7517A) (dn/dc=0.146 ml/g) and a differential viscometer (T-50A, Viscotek).

3. Results and discussion

3.1 COMPOSITION OF THE CELL WALL MATERIAL

42 mg of AIR were recovered per g of fresh sugar-beet. AIR was mainly composed of carbohydrates (650 mg/g). Galacturonic acid (200 mg/g), glucose (188 mg/g), arabinose (172 mg/g) and galactose (45 mg/g) were the main sugars detected together with minor amounts of rhamnose, xylose, mannose and fucose. Degrees of methylation and acetylation of 50 and 58, respectively, were calculated. Ferulic acid (8.2 mg/g) was detected. AIR composition was in close agreement with previously published data (Renard and Thibault, 1993; Micard *et al.*, 1994; Micard *et al.*, 1997).

3.2 CHARACTERISATION OF EXTRACTED PECTINS

Water-soluble pectins were shown to represent only a very minor amount of AIR (50 mg/g) (Levigne *et al.*, 2001b). Acid extractions were therefore directly carried out on AIR.

A first one level full experimental design, giving access to first order models, was constructed (Table 1). The statistical analysis revealed that the type of acid used (HCl or HNO_3) had no effect on the different characteristics of extracted pectins. P-value observed for each characteristic was indeed far over 0.05 (data not shown). Furthermore, the first order models obtained were unable to fit accurately with the experimental values. Six extra HCl extractions (star points) were added to the first one level full experimental design to give access to second order models (Table 1).

3.2.1 Yield

Yield varied from 23 to 362 mg of dialysed extract per g of AIR. The simplified regression equation derived for "yield" was calculated as:

$$\text{Yield (mg/g)} = 257.2 - 126.6\, X_1 + 30.6\, X_2 + 21.2\, X_3 + 43.1\, X_1^2 + 48.8\, X_2^2 - 71.8\, X_3^2 \quad (1)$$

With X_1 = pH; X_2 = temperature and X_3 = time of extraction (quoted +1, -1).

pH was the main significant parameter with increasing yield when decreasing pH. Yield ranging from 23 to 138, 87 to 362, and 280 to 354 mg/g AIR, were indeed observed at pH 3, 2 and 1, respectively. The quadratic effects of time of extraction, and to a lower extent, of temperature were also quite important. The complete second order model fitted experimental values with a R-squared of 82%. The response surface model obtained by selecting the two more important parameters of extraction (pH and time of extraction) is shown in Figure 1.

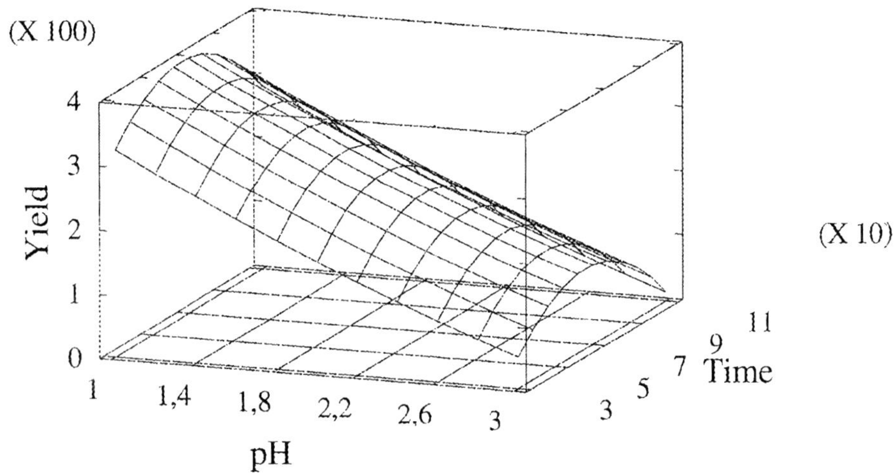

Figure 1. "Yield" (mg/g of AIR) response surface model according to pH and time (min) of extraction

Yields obtained at pH 1 were much higher than that observed on sugar-beet pulp for similar extraction conditions (172 to 207 mg/g; Michel *et al.*, 1985). The different treatments applied to the beet during the sucrose extraction process may modify the solubility of polysaccharides.

3.2.2 Sugar composition

Galacturonic acid content varied from 295 to 528 mg/g of extract but most of the values were in the range 400-450 mg/g. A very slight influence of pH on galacturonic acid content of the extract was evidenced, higher galacturonic acid contents being observed at pH 3. When taking extraction yields into account, extractions were shown to be more efficient at low pHs; 5 to 27%, 13 to 67% and 41 to 81% of the galacturonic acid initially present in the AIR were extracted at pH 3, 2 and 1, respectively.

Rhamnose content varied from 11 to 54 mg/g of extract. The simplified regression equation (whole model: R-squared = 92%) calculated for "rhamnose content" was:

$$\text{Rha (mg/g)} = 25.7 - 13.1\ X_1 + 11.0\ X_1^2 - 7.0\ X_3^2 \qquad (2)$$

pH and its quadratic effect affected very significantly rhamnose content while quadratic effect of the extraction time was also significant, but to a lesser extent (Figure 2).

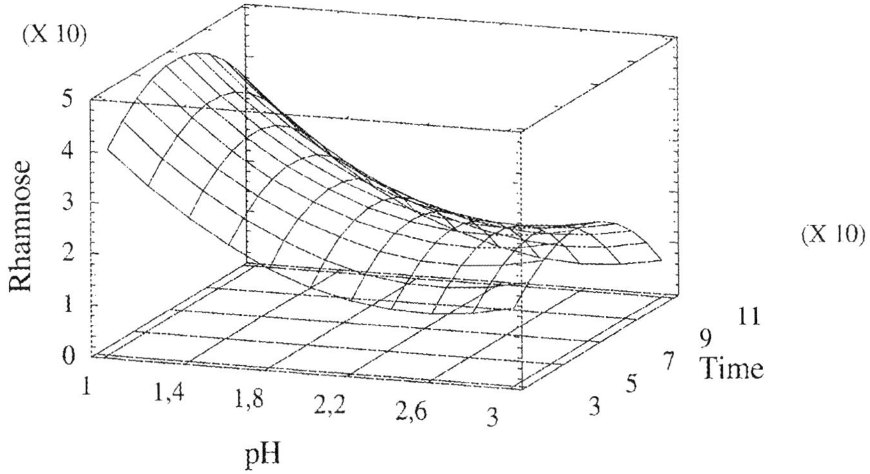

Figure 2. "Rhamnose" (mg/g of extract) response surface model according to pH and time (min) of extraction

Galactose content varied from 27 to 95 mg/g of extract The simplified regression equation (whole model: R-squared = 91%) calculated for "galactose content" was:

$$\text{Gal (mg/g)} = 51.1 - 23.7\, X_1 + 20.5\, X_1^2 - 9.5\, X_2^2 - 11.5\, X_3^2 \quad (3)$$

pH and quadratic effects of all factors affected significantly galactose content.
Arabinose content varied widely (from 3 to 477 mg/g of extract). The simplified regression equation (whole model: R-squared = 85%) calculated for "arabinose content" was:

$$\text{Ara (mg/g)} = 363.7 - 216.0\, X_1^2 + 66.0\, X_1X_2 \quad (4)$$

The quadratic effect of pH was by far the most important on arabinose content.
The highest contents in rhamnose and galactose were observed at pH 1 while the highest contents in arabinose were observed at pH 2.
The extraction conditions affected greatly the yield and the sugar composition of sugar-beet pectins. Among the extraction factors studied, pH was always the most significant. Different pectin "species" could be extracted according to pH. At pH 3, low yields of pectins particularly rich in homogalacturonic regions (GalA/Rha = 18.3 to 34.2 mol/mol) were recovered. At pH 2, higher yields of pectins richer in rhamnogalacturonic regions (GalA/Rha = 10 to 21.5 mol/mol) were obtained. Finally, at pH 1, very high yields of pectins particularly rich in rhamnogalacturonic regions (GalA/Rha = 5.4 to 13 mol/mol) were recovered. Whatever the pH, Gal/Rha ratio remained constant at 1.9 (± 0.3). Ara/Rha ratio remained constant at 16.3 (± 4) for pectins extracted at pH 2 and 3 and decreased drastically for pectins extracted at pH 1 (Figure 3). Degradation of arabinan side-chains probably occurred at this pH, the arabinofuranosyl linkages being particularly acid labile.

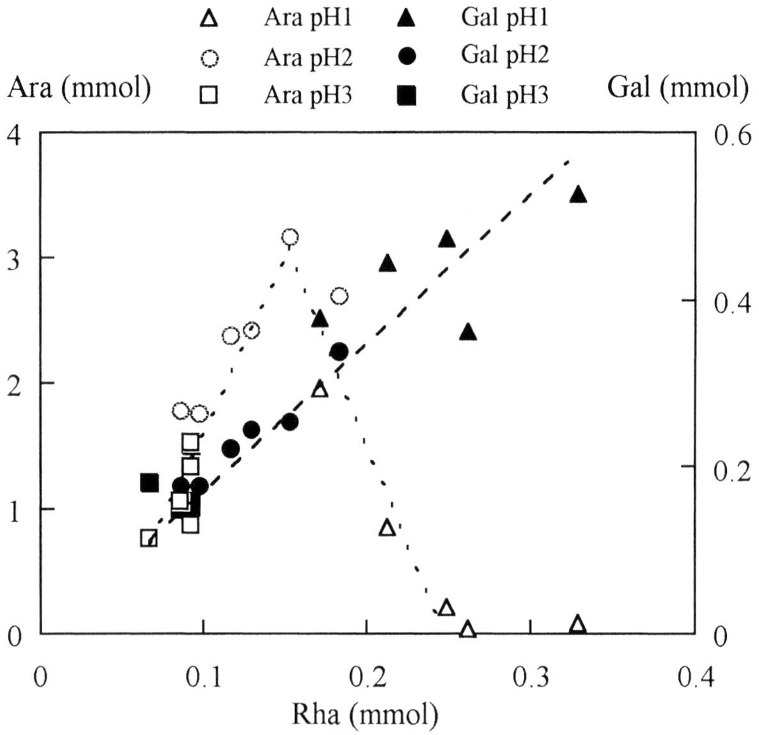

Figure 3. Galactose and arabinose content (mmol) versus rhamnose content (mmol)

3.2.3 Substitution

The degree of methylation (DM) varied from 42 to 94. Central point values (DM = 54 ± 6) were close to those determined by Michel *et al.* (1985) for pectins extracted from sugar-beet pulp under similar conditions. The data fitted the following simplified second order model (whole model: R-squared = 87%):

$$DM = 51.5 - 4.2\, X_2 - 11.1\, X_3 - 5.6\, X_1X_3 + 6.8\, X_1^2 + 10.8\, X_3^2 \qquad (5)$$

The time of extraction and its quadratic effect were the main significant parameters. Low yields of very highly methylated pectins were extracted with short extraction times, especially at high pHs, in agreement with previously reported data (Joye and Luzio, 2000).

The degree of acetylation (DAc) varied from 6 to 43. The simplified regression equation derived for DAc (whole model: R-squared = 88%) was:

$$DAc = 29.8 + 5.9\, X_1 - 5.4\, X_2 - 4.35\, X_3 - 8.98\, X_1^2 + 6.5\, X_2^2 + 4.3\, X_3^2 \qquad (6)$$

showing that the quadratic effect of pH and temperature of extraction were the main significant parameters. The highest DAc values were observed at pH 2 and low temperature.

DM and DAc did not vary in the same way, which suggests that:

(i) either the methyl and acetyl groups are not distributed similarly over the pectic molecules;

(ii) either the susceptibility to acid hydrolysis of the two types of ester linkages is not the same.

The acid-extracted pectins contained also some ferulic acid residues and ferulic acid content varied from 3 to 14 mg/g of extract. The data fitted the following simplified second order model (whole model: R-squared = 88%):

$$\text{FerAc (mg/g)} = 10.82 - 1.32\ X_1 + 0.60\ X_3 + 0.79\ X_1X_2 - 4.86\ X_1^2 \qquad (7)$$

pH and its quadratic effect were the main parameters influencing ferulic acid content. The highest contents in ferulic acid were observed at pH 2 (Figure 4).

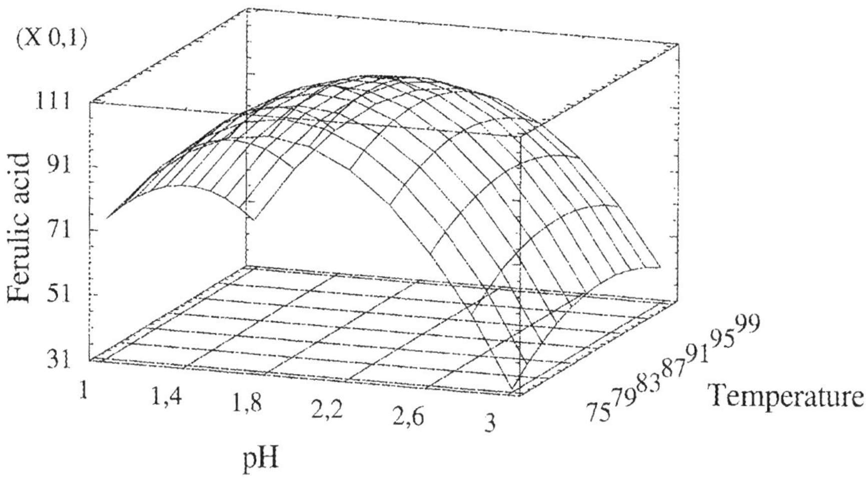

Figure 4. "Ferulic acid" (mg/g of extract) response surface model according to pH and temperature (°C)

At this pH, the highest contents in neutral sugar side-chains (arabinose + galactose), which to ferulic acid is linked (Guillon *et al.*, 1989; Ralet *et al.*, 1994) were indeed observed. Very high amounts of arabinose (close to 100%) and around 30 to 35% of galactose initially present in AIR were extracted at pH 2 for average time and temperature of extraction. In these conditions, 44 to 52% of the ferulic acid initially present in AIR could be extracted.

3.2.4 *Intrinsic viscosity*

Intrinsic viscosity varied from 172 to 493 mL/g. The simplified regression equation (whole model: R-squared = 84%) was:

$$\text{Int. Visc.} = 350.8 - 34.2\,X_1 - 42.1\,X_2 + 36.1\,X_1X_2 - 103.5\,X_2^2 \qquad (8)$$

pH, its quadratic effect, temperature and the interaction pH-temperature affected intrinsic viscosity. The response surface model obtained by selecting pH and temperature is shown in Figure 5. Except for extremely harsh conditions, the intrinsic viscosity values observed are higher than those observed for pectins extracted from pulp (Rombouts and Thibault, 1986; Arslan, 1995), showing the importance of the starting material.

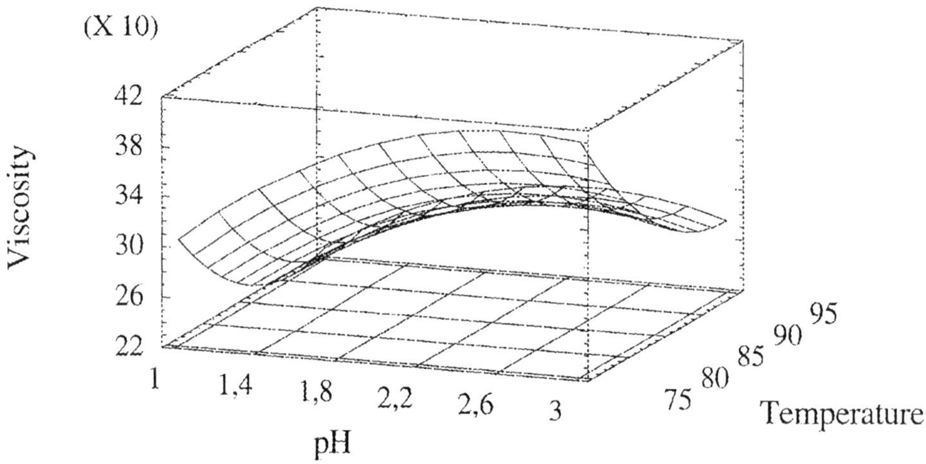

Figure 5. "Intrinsic viscosity" (mL/g) response surface model according to pH and temperature (°C)

4. Conclusion

The effect of experimental conditions on the feature of acid-extracted pectins from fresh sugar-beet AIR was investigated. pH, temperature and time of extraction, affected markedly the extraction yield, the chemical composition and the intrinsic viscosity of pectins. The type of acid used (HCl or HNO_3) had no significant influence on none of these characteristics. Pectins with peculiar features may be obtained:

(i) High amounts of pectins with low DAc, short arabinan side-chains and fair intrinsic viscosity values could be extracted at pH 1, 85 to 95°C and 60 to 90 min. These pectins could eventually gel with calcium or with high sugar concentrations in acidic conditions, depending on the DM.

(ii) Some pectins particularly rich in ferulic acid and of high intrinsic viscosity could be recovered with good yields at pH 2, 85°C and 60 min. These last could be particularly suitable for oxidative gelation.

5. Acknowledgements

The research reported in this paper was supported by the European Union within the Europectin QLK3-1999-00089 Program.

6. References

Ahmed, A. R., and Labavitch, J. M. A. (1977) A simplified method for accurate determination of cell wall polyuronides content, *J. Food Biochem.* **1**, 361-365.
Arslan, N. (1995) Extraction of pectin from sugar beet pulp and intrinsic viscosity-molecular weight relationship of pectin solutions, *J. Food Sci. and Technol.* **32**, 381-385.
Axelos, M.A.V., and Thibault, J.-F. (1991) Influence of the substituents of the carboxyl groups and of the rhamnose content on the solution properties and flexibility of pectins, *Int. J. Biol. Macromol.* **13**, 77-82.
Blakeney, A. B., Harris, P. J., Henry, R. J., and Stone, B. A. (1983) A simple and rapid preparation of alditol acetates for monosaccharide analysis, *Carbohydr. Res.* **113**, 291-299.
Colquhoun, I. J., Ralet, M.-C., Thibault, J.-F., Faulds, C. V., and Williamson, G. (1994) Structure identification of feruloylated oligosaccharides from sugar beet pulp by NMR spectroscopy, *Carbohydr. Res.* **263**, 243-256.
Dubois, M., Gilles, K. A., Hamilton, J. K., Rebers, P. A., and Smith, F. (1956) Colorimetric method for determination of sugars and related substances, *Anal. Chem.* **28**, 350-356.
Guillon, F., and Thibault, J.-F. (1988) Further characterization of acid- and alkali-soluble pectins from sugar beet pulp, *Lebensm. Wiss. Technol.* **21**, 198-205.
Guillon, F., Thibault, J.-F., Rombouts, F.M., Voragen, A.G.J., and Pilnik, W. (1989) Enzymic hydrolysis of the "hairy" fragments of sugar beet pectins, *Carbohydr. Res.* **190**, 97-108.
Joye, D. D., and Luzio, G. A. (2000) Process for selective extraction of pectins from plant material by differential pH, *Carbohydr. Res.* **43**, 337-342.
Keenan, M.H.J., Belton, P.S., Matthew, J.A., and Howson, S.J. (1985) A 13C-nmr study of sugar-beet pectin, *Carbohydr. Res.* **138**, 168-170.
Levigne, S., Thomas, M., Ralet, M.-C., Quemener, B., and Thibault, J.-F. (2001a) Determination of the degrees of methylation and acetylation of pectins using a C18 column and internal standards, *Food Hydrocoll.* Submitted.
Levigne, S., Ralet, M.-C., and Thibault, J.-F. (2001b) Characterisation of pectins extracted from fresh sugar-beet under different conditions using an experimental design, *Carbohydr. Polym.* Submitted.
May, C. (1990) Industrial pectins: sources, production and applications, *Carbohydr. Polym.* **12**, 79-99.
Micard, V., Renard, C. M. G. C., and Thibault, J.-F. (1994) Studies on enzymic release of ferulic acid from sugar beet pulp, *Lebensm. Wiss. Technol.* **27**, 59-66.
Micard, V., Renard, C. M. G. C., and Thibault, J.-F. (1996) Enzymatic saccharification of sugar beet pulp, *Enzyme Microbial Technol.* **19**, 162-170.
Micard, V., Renard, C. M. G. C., and Thibault, J.-F. (1997) Dehydrodiferulic acids from sugar beet pulp, *Phytochem.* **44**, 1365-1368.
Michel F., Thibault, J.-F., Mercier, C., Heitz, F., and Pouillaude, F. (1985) Extraction and characterisation of pectins from sugar beet pulp, *J. Food Sci.* **50**, 1499-1502.
Oosterveld, A., Beldman, G., Schols H.A., and Voragen, A.G.J. (1996) Arabinose and ferulic acid rich pectic polysaccharides extracted from sugar beet pulp, *Carbohydr. Res.* **288**, 143-153.
Oosterveld, A., Beldman, G., Schols H.A., and Voragen, A.G.J. (2000) Characterization of arabinose and ferulic acid rich pectic polysaccharides and hemicelluloses from sugar beet pulp, *Carbohydr. Res.* **328**, 185-197.
Oosterveld, A., Pol, I.E., Beldman, G., and Voragen, A.G.J. (2001) Isolation of feruloylated arabinans and rhamnogalacturonans from sugar bet pulp and their gel forming ability by oxidative cross-linking, *Carbohydr. Polym.* **44**, 9-17.

Phatak, L., Chang, K. C., and Brown, G. (1988) Isolation and characterisation of pectin in sugar beet pulp, *J. Food Sci.* **53**, 830-833.

Pippen, E. L., McCready, R. M., and Owens, H.S. (1950) Gelation properties of partially acetylated pectins, *J. Amer. Chem. Soc.* **72**, 813-816.

Ralet, M.-C., Thibault, J.-F., Faulds, C. B., and Williamson, G. (1994) Isolation and purification of feruloyted oligosaccharides from cell walls of sugar beet pulp, *Carbohydr. Res.* **263**, 227-241.

Renard, C. M. G. C., and Thibault, J.-F. (1993) Structure and properties of apple and sugar beet pectins extracted by chelating agents, *Carbohydr. Res.* **244**, 99-114.

Rombouts, F.M., and Thibault, J.-F. (1986) Feruloyated pectic substances from sugar beet pulp, *Carbohydr. Res.* **154**, 177-188.

Saeman, J. F., Moore, W. E., Mitchell, R. L., and Millet, M. A. (1954) Techniques for the determination of pulp constituents by quantitative paper chromatography, *Tappi* **37**, 336-343.

Thibault, J.-F. (1979) Automatisation du dosage des substances pectiques par la méthode au métahydroxydiphényle, *Lebensm. Wiss. Technol.* **12**, 247-251.

Voragen, A. G. J., Pilnik, W., Thibault, J.-F., Axelos, M. A. V., and Renard, C. M. G. C. (1995) Pectins, Food Polysaccharides. Stephen A.M. and Dea Y. (Eds), Marcel Dekker (London), chap. 10, pp 287-339.

SOL-GEL TRANSITIONS OF HIGH-ESTERIFIED PECTINS AND THEIR MOLECULAR STRUCTURE

S. NEIDHART, C. HANNAK, K. GIERSCHNER
Hohenheim University, Institute of Food Technology
Garbenstrasse 25, 70599 Stuttgart, Germany
neidhasy@uni-hohenheim.de

Abstract

Molecular structure, molecular heterogeneity and gelling properties of eleven high-esterified pectins with average degrees of esterification (DE) between 50% and 80% were investigated. For rheological studies in the small-strain-oscillation mode, gelation of sugar-acid-model systems was induced by cooling at constant rates from 93°C to usually 20°C. Junction zones of such physical gel networks are known to be mainly stabilized by hydrogen bonds besides hydrophobic interactions. However, in systems with high gelation temperatures far from 30°C, adverse thermal conditions for junction zone formation must be overcompensated by other effects reducing the mobility of pectin molecules, such as macromolecular entanglements intensified by long side-chains of the rhamnogalacturonan-backbone.

Gelation temperature of a pectin in a defined medium was dependent on the pectin concentration and the thermal conditions during cooling. Therefore, rather the gelling system than the respective pectin is characterized by the gelation temperature. The presented results contradict the prospective evaluation of gelation temperatures (or times) on the basis of DE. Obviously, the esterified galacturonic acid units are not causally related to the induction of gelation. This conclusion is mainly supported by the following observations: For the apple pectins studied, a correlation between DE and gelation temperature was only found in the DE-range above 70%. However, this was limited to systems with pectin concentrations exceeding 0.4 g pectin/100g gel. At concentrations below 0.175 g pectin/100g gel, "slow set" pectins with medium DE displayed higher gelation temperatures than "rapid set" and "ultra-rapid set" pectins with very high DE. Since side-chains are reduced along with DE during pectin production, gelation observed at high temperatures, i.e. under adverse thermal conditions for the formation of hydrogen bonds, would be only inducible by macromolecular entanglements, provided that the gelling systems involve both high pectin concentrations and very high DE. By analogy, calcium-ions, being a component of pectin preparations in concentrations even below the saturation-point of the respective unesterified carboxylic groups, resulted in an increased gelation temperature and seemed to cause higher gelation temperatures in gelling systems of constant galacturonic acid (AUA) content for citrus pectin and co-extracted apple/citrus pectins than comparable apple pectins. As a conclusion, the possibilities for hydrogen bonding primarily determine induction of gelation through the amounts of unesterified AUA units and actual thermal conditions, whereas adverse conditions for junction zone

F. Voragen et al. (eds.), Advances in Pectin and Pectinase Research, 431-448.
© 2003 *Kluwer Academic Publishers. Printed in the Netherlands.*

formation may be largely faced by reducing mobility of pectin molecules through the effects of long side-chains and interactions between calcium and unesterified galacturonan units. Therefore, the complex interactions contributing to sugar-acid-gelation of high-esterified pectins, including hydrophobic interactions, concur in different ways to junction zone formation during sol-gel transition and junction zone stabilization in the resulting gels, respectively.

1. Introduction

By definition, at least 50% of the carboxylic groups of galacturonic acid units (AUA) in high-esterified (HM) pectins are esterified with methanol. The gelling properties of HM-pectins are used in acid systems with low water activity such as jams and jellies. Recently, both extrinsic prerequisites for self-association of HM-pectin chains have been thoroughly discussed by Evageliou et al. (2000). The complex influence of sodium, potassium, calcium, and magnesium ions on the gelling behaviour of HM-pectins and on viscoelastic properties of their gels was previously reported (Neidhart et al., 1996). By increase in setting temperature, calcium ions were shown to induce marked association of all HM-pectins investigated, although formation of a superimposed calcium-pectate gel was not possible (Neidhart, 1997).

Prospective evaluation of the gelling properties with respect to molecular structure of HM-pectins is largely based on their average degree of esterification (DE) (table 1), which has been traditionally determined by the pectin producers based on subsequent titration of unesterified and esterified AUA units before and after their chemical de-esterification. Although fine structure of commercially available pectins has been the subject of various studies (Kravtchenko et al., 1993; Daas et al., 2001), influence of pectin structure on sugar-acid gelation was mostly discussed considering molecular weight, degree of esterification and distribution of methoxyl groups only (Ikkala, 1986; Crandall and Wicker, 1986). Due to their complex fine structure and molecular heterogeneity (figure 1), chemically similar pectins often differ in their rheological behaviour during gelation or may not follow the expected behaviour in practice. Gels produced with citrus pectins were generally described to be more stiff than gels from comparable apple pectins (May, 1990).

Table 1. Setting times of citrus pectins according to Ikkala (1986)

Pectin	Setting time [s]	DE [%]
Extra Slow Set (ESS)	260-330	58-62
Slow Set (SS)	170-225	63-67
Medium Rapid Set (MRS)	120-160	67-71
Rapid Set (RS)	<110	71-75

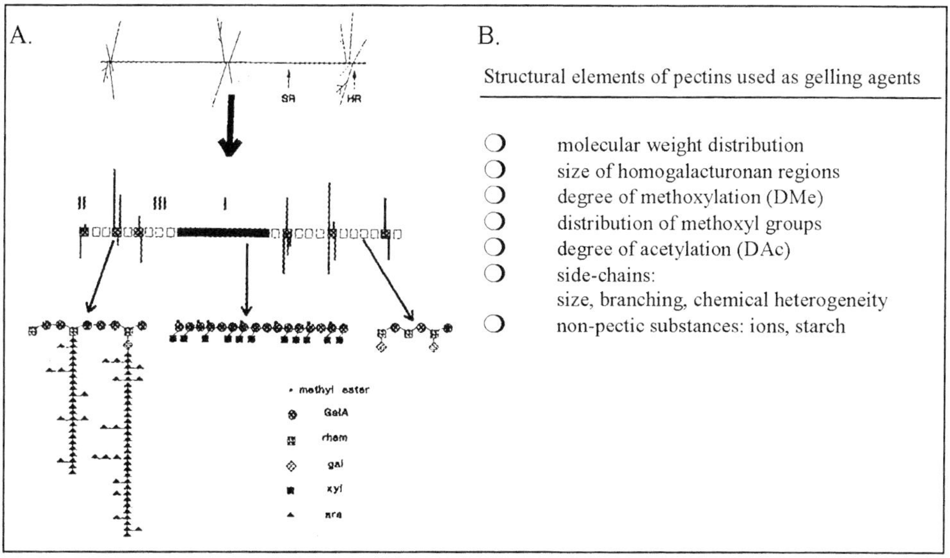

Figure 1. Structural elements (B.) considered in the present study and derived from the fine structure of pectins described by (A.) the model of Schols and Voragen (1996). SR = smooth region, HR = hairy region

The junction zones in these physical gel networks are largely stabilized by intermolecular hydrogen bonds besides hydrophobic interactions involving the methyl ester groups in the pectins (Oakenfull and Scott, 1984; Evageliou et al., 2000). Whilst hydrogen bonds are destabilized by increasing temperature, hydrophobic association becomes more stable. A complex temperature dependence of the gel properties results from the concurrence of both types of interactions.

This study was aimed at a more detailed insight into the structure-function relation of HM-pectins focussing on the mechanisms involved in self-association during sol-gel transition. Unfortunately, single parameters of the fine structure cannot be individually modified by either laboratory methods nor industrial processes. Therefore, both gelling properties and fine structure of eleven selected pectins including molecular heterogeneity according to the elements in figure 1 were investigated.

2. Experimental

2.1. CHARACTERIZATION OF THE PECTIN SAMPLES

Five medium-esterified pectins (50% < DE < 68%) and six pectins of higher DE (68% < DE < 80%) were selected among unstandardized commercial pectins or were produced for reasons of comparison (table 2). In this way, botanical origin (apple, citrus), pectin production technologies, co-extraction of apple and citrus waste and special molecular features such as DE and molecular weight were considered.

Anhydrogalacturonic acid content (AUA) was determined colorimetrically with m-hydroxydiphenol, while quantitative analysis of neutral sugars was realized gaschromatographically as their alditol acetates. Additionally, starch content was determined spectrophotometrically by means of the respective enzymatic assay (Boehringer, Mannheim). In addition to the titrimetrical determination of the degree of esterification (DE), degree of methoxylation (DMe) and acetylation (DAc) were calculated from HPLC analysis of methanol and acetate, relating both to the AUA content. Cation analysis of the pectin ash was performed by atomic absorption spectrometry. Molecular weight distribution resulted from analytical size exclusion chromatography (HPSEC) with RI detection. For determination of intrinsic viscosity according to Kravtchenko and Pilnik (1990), a micro-KPG-Ostwald-viscosimeter was used. Molecular heterogeneity of five pectins with DE $\geq$ 70% was investigated after preparative SEC-fractionation. Additionally, intramolecular distribution of methoxyl groups of eight selected samples was studied by preparative SEC after β-eliminative pectin degradation at pH 5.4 and 120°C for 5 hours. For further estimation of intramolecular rhamnose distribution of a citrus pectin (DE 70%) and four apple pectins (DE 57%, 70%, 74%, and 78%), preparative SEC was performed after partial acid hydrolysis. A more detailed description of the analytical procedures is reported by Hannak (1997).

2.2. GELLING SYSTEMS

For rheological characterization, pectin/sucrose gels were produced at 65.1 $\pm$ 0.3 °Brix soluble solid substance and pH 3.04 $\pm$ 0.02 in a potassium acetate-lactic acid buffer (0.25 mol / 100g gel) with pectin contents (c_P) of 0.1, 0.15, 0.175, 0.25, 0.4, 0.55, and 0.85 g / 100g gel, respectively (Neidhart, 1997). Constant soluble solid substance was maintained in the samples by adapting the sucrose concentrations (61.93 - 62.68g / 100g gel) to c_P. The buffer corresponded in components and pH with the more concentrated one (0.03 mol / 100g gel) used for breaking strength analysis of standard gels with the Herbstreith-Pektinometer (table 2).

Preparation of gels was performed as previously described (Neidhart et al., 1996), carefully dispersing the dry pectin premixed with a small portion of the sucrose in the cold diluted buffer. Soluble solid substance was adjusted by cooking the mixture to a precalculated net weight of 170g under stirring. Subsequently, the sample was poured into the hot measuring device (see 2.3) of the rheometer.

2.3. RHEOLOGICAL EXPERIMENTS

Dynamic rheological measurements were performed using a Bohlin CS controlled-stress rheometer with improved angle resolution, equipped with a Searle cylinder device (diameter of the fixed cylinder 25 mm) for pectin contents $c_P \geq 0.4$ g / 100g gel or a modified double gap device (diameter of the fixed cylinders 40 and 50 mm, respectively) for $c_P \leq 0.4$ g / 100g gel considering the viscosity range before setting of the gels (Neidhart, 1997). In order to avoid water loss and incrustation of the sample surface, a humid atmosphere over the free sample surface was provided in the modified double gap device (Neidhart et al., 1996) or the free sample surface in the Searle cylinder was covered with hot low-viscous liquid paraffin, respectively.

If not otherwise stated, gel formation was monitored by measuring the storage (G') and the loss (G'') modulus as well as the related phase angle δ = arc tan (G''/G') at a fixed frequency (1 Hz) as a function of time during cooling with constant rate (-1.047 ± 0.009 K/min) from 93 to 20 °C and subsequent thermostatting at 20°C for 1 hour (figure 2A). The strain amplitude was $\gamma = 0.015$, but for higher pectin contents ($c_P \geq 0.55\%$), $\gamma = 0.0015$ had to be used to ensure precise measurements in the linear-viscoelastic range and at the required resolution in all cases. In order to characterize the final gel, a mechanical spectrum, i.e. a frequency sweep, was subsequently performed in the range of 0.005 to 10 Hz at the same strain amplitude and 20°C (e.g. figure 9). The linear-viscoelastic range of each gel at 20°C was checked by a subsequent strain sweep at a frequency of 1 Hz as the last part of this cycle. Temperature control was achieved with an adjustable water bath allowing the application of temperature gradients, when cylindrical devices are used as in this case.

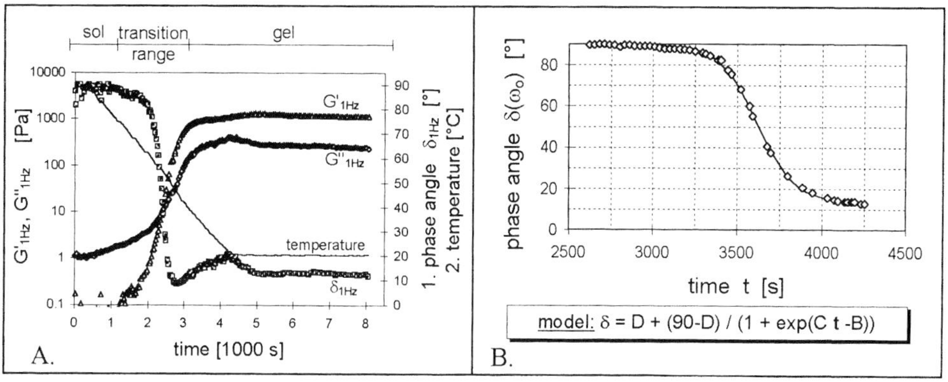

Figure 2. Rheological monitoring of gelation (A.) and modelling of the sol-gel transition range (B.)

The phase angle-time curves in the sol-gel transition range were modeled according to figure 2B (Neidhart, 1997). R^2 ranged from 0.956 to 0.9998 with a median at 0.997. The curves in the transition range were characterized by their turning-points with the time $t_W = B/C$ and the phase angle $\delta_W(1\text{Hz}) = 45° + D/2$. For each curve, time and corresponding temperature of the turning-point, of the crossover-point of the moduli ($\delta_{co}(1\text{Hz}) = 45°$), and of the pseudo gel point defined by $\delta_{crit}(1\text{Hz}) = 75°$ were calculated. As deduced in 3.2, setting time and temperature were detected by the pseudo gel point defined by $\delta_{crit} = 75°$, since all techniques based on the rheological gel point definition of Winter and Chambon (1986) could not be applied with the used rheometer.

Gel strength of the final gels at 20°C was deduced from their mechanical spectra. Because of the negligible frequency dependence of both moduli G'(ω) and G''(ω) in the low frequency range (figure 9), gel strength was defined by G' at 0.01 Hz. Non-linear regression analysis of the dependence of setting temperature and gel strength, respectively, on pectin concentration was performed as described by Neidhart (1997).

Table 2. Characteristics of the HM-pectin samples

Pectin labels: A = apple pectin, C = citrus pectin, AAC or ACC = co-extraction of apple and citrus waste; the figure indicates the respective degree of esterification (DE) determined by titration. TS = dry matter, NaAc = sodium acetate, HAc = acetic acid, HPE = Herbstreith-Pektinometer units, n.d. = not detectable.

	High-esterified pectins						Medium-esterified pectins				
Pectin	A78	A77	A74	C70	A70	AMr68	A57	AVf53	AAC56	ACC52	AACp56
DE [%]	77.8	77.3	74.0	70.4	69.5	68.4	56.8	52.7	56.2	51.8	56.0
AUA [g/100g TS]	70.77	69.75	70.26	72.33	73.40	72.05	79.23	79.98	68.30	72.56	66.08
DMe [%]	70.60	69.46	70.60	65.26	67.46	58.87	54.62	53.77	52.34	53.37	53.62
DAc [%]	5.68	4.71	2.60	1.77	3.15	3.11	0.92	1.23	1.39	0.95	1.27
Starch [g/100g TS]	1.44	1.74	2.06	0.34	1.68	2.52	1.76	1.66	11.92	3.31	10.76
Ash [g/100g TS]	2.94	2.21	1.48	8.09	1.82	6.98	2.57	2.21	5.45	6.00	8.82
Neutral sugars [mol / 100 mol AUA]:											
Rhamnose	3.40	3.72	3.42	3.03	3.26	3.77	2.96	3.40	3.31	3.37	3.81
Arabinose	16.53	14.54	7.72	8.31	6.54	5.34	2.36	1.79	2.00	1.85	1.64
Galactose	7.27	7.52	8.62	8.08	10.77	6.58	5.40	7.45	8.39	6.80	8.96
Xylose	2.78	2.71	2.39	0.28	2.37	1.66	2.26	2.33	1.45	1.25	1.74
Fucose	0.27	0.28	0.23	0.18	0.19	0.14	0.15	0.18	0.12	0.12	0.13
Mannose	0.15	0.13	0.10	0.06	0.17	0.13	0.15	0.14	0.14	0.12	0.13
Glucose (not from starch)	1.83	2.07	3.09	0.43	1.59	0.32	1.80	2.62	0.08	1.58	2.18
Intrinsic viscosity [η] [dl/g], in (0.4M NaAc/HAc+0.025M Na_2SO_4), pH 3.0:											
	6.3	6.1	8.4	5.8	5.0	2.6	4.5	4.0	4.7	4.3	4.6
Molecular weight distribution, in (0.4M NaAc/HAc+0.025M Na_2SO_4), pH 3.0, calibration based on dextran:											
Mw (x 10^4)	219	194	220	169	171	67.3	121	164	137	122	125
Mn (x 10^4)	14.1	10.5	10.3	14.4	14.3	8.29	35.0	35.0	39.0	35.4	31.0
D	15.6	18.5	21.3	11.7	11.9	8.1	9.4	14.0	10.3	9.3	12.0
Gel strength [°USA-Sag]	167	182	209	171	183	n.d.	225	181	180	219	150
Breaking strength [HPE]	460	710	1275	730	1040	n.d.	1230	720	660	1090	430

3. Results and discussion

3.1. FINE STRUCTURE OF THE PECTINS INVESTIGATED

Main characteristics of the pectins studied are summarized in table 2. Similarly to earlier reports (Voragen et al., 1986, Kujawski and Tuszynski, 1987), the sum of DMe and DAc reflected the degree of total esterification (DE) of the AUA units (eq. 1), which is accessible by titration (see 1.), and DAc exponentially increased with rising DE (eq. 2).

$$(DMe + DAc) = 0.8772 \cdot DE + 6.6351 \qquad R^2 = 0.935 \qquad (1)$$

$$DAc = 0.0465 \cdot \exp(0.0584 \cdot DE) \qquad R^2 = 0.846 \qquad (2)$$

Due to its enzymatically reduced molecular weight, pectin AMr68 did not form measurable gels, when generally accepted standard procedures for gel strength and breaking strength were used (table 2). With the exception of AMr68, differences in molecular weight of the other pectin samples were rather low. Constant rhamnose-AUA ratios indicated only small variations of the avarage homogalacturonan size. It should be noted, however, that arabinose content [in g/100g dry matter] exponentially decreased with declining DE (eq. 3) among the selected samples, indicating increasing degradation of arabinans during acid de-esterification, since glycosidic bonds between furanose monomers are especially prone to acid hydrolysis (Aspinall, 1982). Therefore, marked differences in the side-chain pattern could be expected among the pectins (table 3).

$$Ara = 0.0131 \cdot \exp(0.084 \cdot DE) \qquad R^2 = 0.945 \qquad (3)$$

Based on the pectin model suggested by Schols and Voragen (1996) (figure 1A), the assumptions on the neutral sugar distribution described in table 3 were made in order to estimate roughly number and size of side-chains from preparative SEC data (table 3). For this simple estimation, the following arbitrary assumptions were made: (1) 75% of the rhamnose units are linked to side-chains of arabinose or galactose, (2) a constant ratio of 1:2 for arabinan and galactose/galactan side-chains, (3) 75% of the galactose units form monomer side-chains according to figure 1 and 25% long (unbranched) side-chains (Neidhart, 1997). Whilst in HM-pectins of highest DE, arabinan side-chains of up to 10 monomers might be possible (table 3), arabinan side-chains of medium-esterified pectins should be reduced to only two or three monomers as a result of unavoidable hydrolysis during pectin production. According to these assumptions, galactose contents of the pectins would allow non-monomer galactan side-chains of 3.6 to 6.6 monomers irrespective of DE. Therefore, size of side-chains should strongly decrease with DE and long side-chains can be only assumed to be present in the group of higher esterified pectins (table 3).

In spite of similar AUA content and DE, the pectins A70 and C70 markedly differed in their standard gel properties (table 2) and in molecular heterogeneity as revealed by preparative SEC (Hannak, 1997). Intermolecular distribution of DMe was completely different for both pectins. Their β-eliminative degradation patterns indicated the existence of some non-methoxylated galacturonan sequences in pectin C70 compared to the apple pectins of this DE range, possibly in accordance with its higher ash and especially calcium content observed for pectins produced using citrus waste (table 2).

Table 3. Assumptions on side-chain structure of pectins based on the pectin model in figure 1 and the composition of the SEC fractions. Concentrations of saccharides in mmol /100g dry matter

Assumptions on the *number* of side-chains (*NSC*) [mmol/100g dry matter]:			
NSC_{Ara}	= [Rha] / 4	NSC_{Xyl}	= [Xyl]
$NSC_{Gal\text{-}long}$	= [Rha] / 8	$NSC_{Gal\text{-}mono}$	= 3 [Rha] / 8
Assumptions on the *length* (degree of condensation) of side-chains (*LSC*):			
LSC_{Ara}	= [Ara] / NSC_{Ara}	LSC_{Xyl}	= 1 for [Xyl] ≤ [AUA]
$LSC_{Gal\text{-}long}$	= 0.25 [Gal] / $NSC_{Gal\text{-}long}$	$LSC_{Gal\text{-}mono}$	= 1

pectin samples	A78	A77	A74	A70	C70
mean number of side-chains [mmol/100g dry matter]:					
NSC_{Ara}	1.73	2.53	3.36	2.33	2.18
$NSC_{Gal\text{-}long}$	0.87	1.26	1.68	1.16	1.09
$NSC_{Gal\text{-}mono}$	2.60	3.79	5.03	3.49	3.27
NSC_{Xyl}	4.49	9.43	7.69	5.71	2.13
total	9.69	17.01	17.76	12.70	8.66
mean length of side-chains:					
LSC_{Ara}	10.08	9.89	6.22	5.87	5.90
$LSC_{Gal\text{-}long}$	5.25	5.10	4.05	4.35	2.35

[Ara] = arabinose, [Gal] = galactose, [AUA] = galacturonic acid, [Rha] = rhamnose, [Xyl] = xylose

All in all, significant differences of the distribution patterns of various structural elements were found on a descriptive level for those pectins, which were rather similar with respect to their average chemical characteristics (Hannak, 1997).

3.2. EXTRINSIC EFFECTS ON GELATION OF HM-PECTIN SYSTEMS AND THEIR SIGNIFICANCE FOR RHEOLOGICAL GEL POINT DETECTION

The rheological gel point definition (Winter and Chambon, 1986) implies that the frequency dependence of both moduli follows a power law with the same exponent *n* resulting in a phase angle δ_{GP} independent of frequency:

$$G'(\omega) \sim G''(\omega) \sim \omega^{n} \tag{4}$$

This power law behaviour at the gel point had been shown to be valid for chemical gels cross-linked by covalent bonds (Scanlan and Winter, 1991) as well as physical gels of biopolymers (Cuvelier and Launay, 1990; Michon et al., 1993; da Silva and Conçalves, 1994). However, neither this exponent *n* nor resulting δ_{GP} are universal parameters, but dependent on the nature of the network, i.e. on type and concentration of the biopolymer (Cuvelier and Launay, 1990). Since sol-gel transition of the pectin-sugar systems is limited to an extremely short period characterized by a very sharp decrease of the phase angle (figures 2A and 4), gel point detection would require multi-frequency gelation experiments for each sample in order to detect δ_{GP} by its frequency independence. Considering the available measurement technique, a suitable mode of gel point detection had to be deduced from the character of the $\delta(\omega_o,t)$-curves in the transition range revealing important extrinsic influences on pectin-sugar-acid gelation.

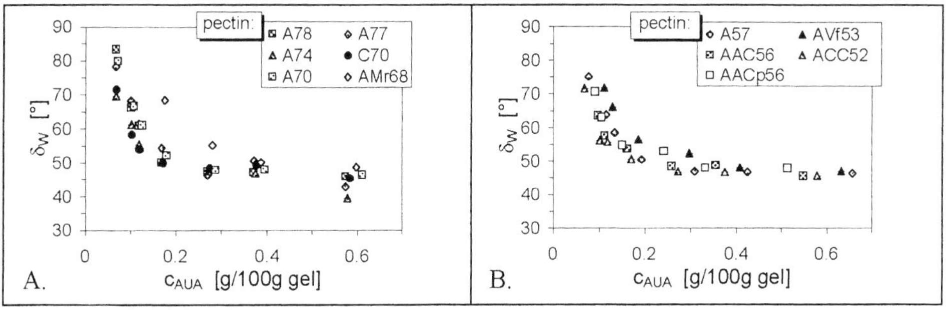

Figure 3. Phase angle $\delta_W(\omega_o,t_W)$ at the turning point of the gelation curves $\delta(\omega_o,t)$ in the sol-gel transition range of samples with various pectin concentrations expressed as galacturonic acid content (c_{AUA}); f = 1 Hz.

The character of the $\delta(\omega_o,t)$-curves during sol-gel transition strongly depended on type and concentration of pectin as shown by the phase angle $\delta_W(\omega_o)$ of their turning-points (figure 3). Only at higher pectin concentrations ($c_P \geq 0.4\%$), $\delta_W(\omega_o)$ was near to the crossover of $G'(\omega_o)$ and $G''(\omega_o)$. Altogether, the introduction of the pseudo gel point defined by $\delta_{crit}(\omega_o) = 75°C$ enabled a largely precise detection of setting time and temperature for all samples in such single-frequency gelation experiments compared to the use of either the crossover point ($\delta(\omega_o) = 45°$) or the calculated turning-point.

Moreover, the $\delta(\omega_o,t)$-curves during sol-gel transition depended on the cooling rate and especially on the applied frequency (figure 4A) supporting the definition of the pseudo gel point by the rather high phase angle of 75°. However, both parameters influenced gel point detection, but did not significantly affect the resulting gel properties (figure 4B). Setting temperature linearly decreased with increasing cooling rate (Neidhart, 1997). The actual setting temperature characterized rather the individual process applied to a gelling pectin system, than the pectin itself.

Cooling of samples below 20°C resulted in retardation of gel ageing (figure 5). This is in accordance with other approaches (Evageliou et al., 2000; Oakenfull and Scott, 1984; da Silva et al., 1995) suggesting a broad temperature range at approx. 30°C for maximum gelation rate.

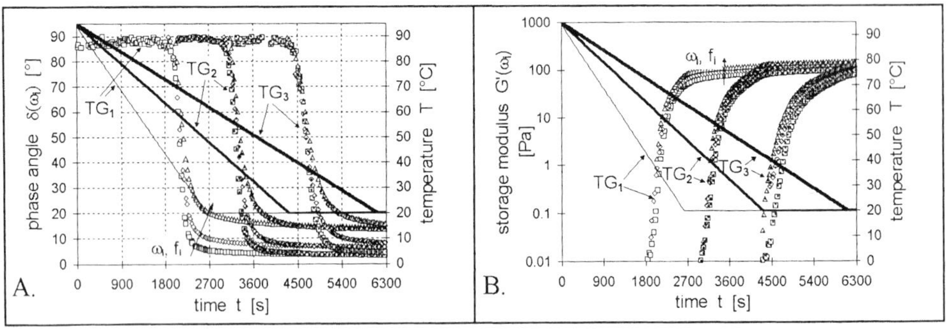

Figure 4. Influence of cooling rate and frequency on sol-gel transition (A.) and final gel strength (B.) of HM-pectin systems. Strain γ = 0.015. Cooling rates [K/min]: TG_1 = -1.69, TG_2 = -1.04, TG_3 = -0.734. Lines: temperature. Symbols: phase angle and storage modulus, resp., at f_1 = 0.2 Hz (square), f_2 = 1 Hz (rhomb) or f_3 = 5 Hz (triangle)

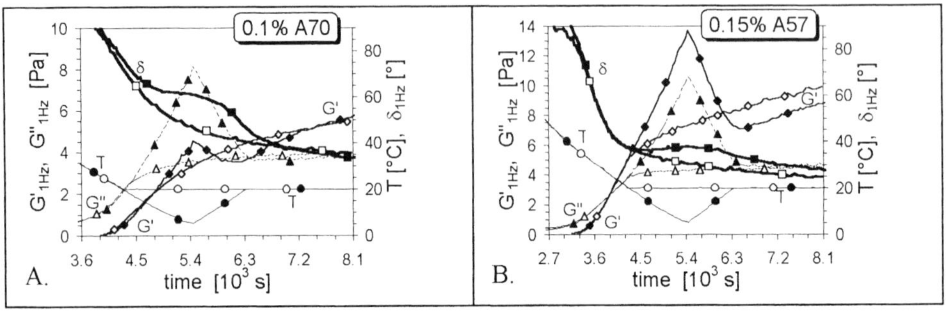

Figure 5. Effect of cooling with -1 K/min to 5°C (filled symbols) and to 20°C (open symbols), resp., prior to ageing of HM-pectin samples at 20°C. Circle = temperature, square = δ_{1Hz}, rhomb = G'_{1Hz}, triangle = G''_{1Hz}

Kinetic models describing the effect of biopolymer concentration on gelation rate assume that gelation rate is inversely proportional to setting time at constant temperatures (Oakenfull and Scott, 1986; Ross-Murphy, 1991). Although the gelation experiments of the present study were not performed at constant temperatures, but at constant cooling rate, the dependence of the gelation rates on pectin concentration was studied by analogy to those models (figure 6). As implied by the kinetic model (Ross-Murphy, 1991; da Silva et al., 1995), power law approximation of the data was possible for the medium-esterified pectins only (R^2 = 0.952 up to 0.997), but resulted in very low exponents n' = 0.18 up to 0.32 due to simultaneous cooling, whilst da Silva et al. (1995) reported n' = 3.6 for gelation of pectin systems with DMe = 64% at 20°C. The medium-esterified pectins studied set at temperatures between 30 and 63°C depending on type and concentration of pectin (figure 7B). For the higher esterified pectins setting in a broad temperature range (figure 7A), decreasing pectin concentration resulted in an increasingly weaker decrease of gelation rate (figure 6). All in all, it became obvious that gelation rate increased, when temperature was sinking in the investigated range (93°C to 20°C). Pectin concentration and temperature showed opposed effects on gelation rate.

Obviously, the supercooling effect shown for the sol-gel transition (figure 4A) depended on non-stationary thermal conditions during cooling on the one hand, and the broad optimal temperature range around 30°C for the formation of junction zones on the other hand.

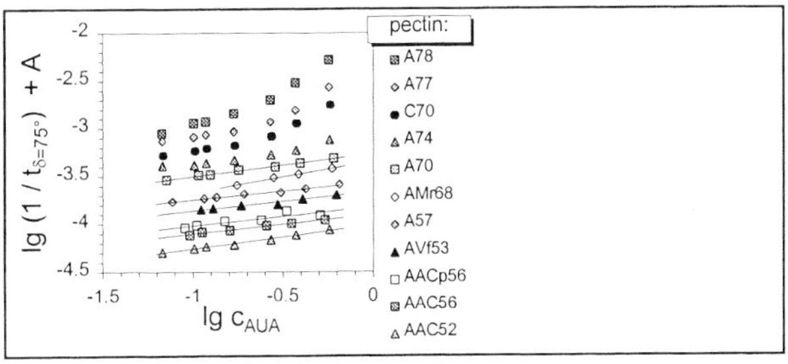

Figure 6. Gelation rate calculated as reciprocal setting time $t_{\delta=75°}$ of HM-pectin systems (pectin content expressed as AUA content, c_{AUA}). A = arbitrary shift factors on the ordinate

3.3. INTRINSIC EFFECTS ON GELATION OF HM-PECTINS AS REVEALED BY THEIR CONCENTRATION DEPENDENCE

The optimum temperature of gelation rate, i.e. junction zone formation, around 30°C seemed to be the same for all pectins investigated (figure 6). This could be ascribed to the conclusion of other studies that at least two types of interaction are involved in intermolecular association of high-esterified pectins, one of which is destabilized by increasing temperature, while the other becomes more stable, corresponding to hydrogen bonding and hydrophobic association, respectively (Oakenfull and Scott, 1984, Evageliou et al., 2000). However, the pectins strongly varied in the concentration dependence of setting time and temperature (figure 7). This should be related to the contribution of interactions enabled by the individual fine structure of the pectin. Setting temperature is not a fixed characteristic of the respective pectin, but characterizes the corresponding gelling system. In samples with high setting temperatures far from 30°C, adverse thermal conditions for junction zone formation must be overcompensated by other effects reducing the mobility of pectin molecules.

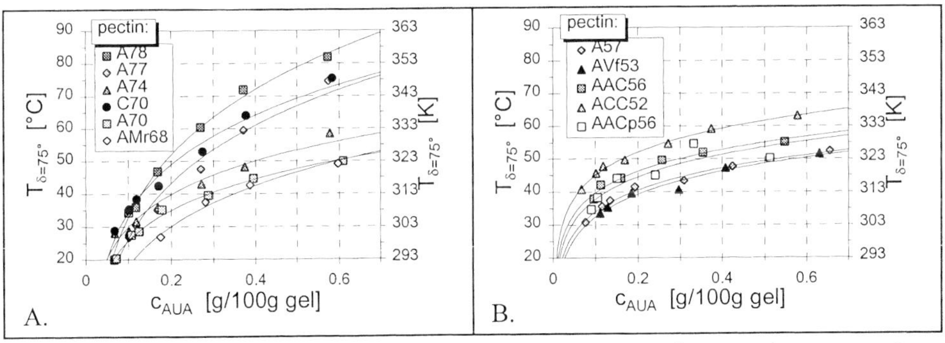

Figure 7. Setting temperature $T_{\delta=75°}$ of HM-pectins at various pectin concentrations expressed as AUA content (c_{AUA}). Lines: regression curves with $T_{crit.} = T_{\delta=75°}\,[K] = a\,c_{AUA}^{B}$

Dependence of setting temperature on the AUA content of the gels could be modeled by an empirical power law (R^2 = 0.962 up to 0.995) valid for the temperature range between 20 and 93°C (figure 7):

$$T_{crit.} = a\, c_{AUA}^{\,B} \qquad \text{with } T_{crit.} = T_{\delta=75°} \qquad (5)$$

The temperature T_W at the turning point of the gelation curves similarly depended on the AUA content of the gels (R^2 = 0.953 up to 0.997).

According to figure 7, several observations contradict a simple correlation between setting temperature and pectin methoxylation. Altough the methoxylated AUA units are involved in intermolecular hydrophobic interactions and junction zone stabilization, they do not seem to be causally related to the induction of gelation. At least, the following results are opposed to the prospective evaluation of setting temperatures (or times) on the basis of DE. Among the apple pectins, "rapid set" and "ultra-rapid set" pectins (figure 7A) only showed high setting temperatures at high pectin concentrations (c_P = 0.85 g / 100g gel, $c_{AUA} \approx$ 0.6 g / 100g gel). A correlation between setting temperature and DE, which was, however, approximately the sum of DMe and DAc (see 3.1), was only found in the DE range above 70% (lines in figure 8), and even this was limited to high pectin concentrations above c_P = 0.4 g / 100 g gel, i.e. $c_{AUA} \approx$ 0.25 - 0.3 g / 100g gel (figure 7A). At very low pectin contents (c_P = 0.1 - 0.175 g / 100 g gel), most "slow set" pectins with medium DE set at even higher temperatures (figure 7B) compared to "rapid set" and "ultra-rapid set" pectins. It should be noted, however, that gel strength at these low pectin contents was lower than required for food (Neidhart, 1997). The citrus pectin C70 and co-extracted apple/citrus pectins (open symbols in figure 8) always set at higher temperatures than comparable apple pectins (figure 7). Finally, gelling behaviour of both "medium rapid set" pectins, A70 and C70, completely differed from each other (figure 7A). Indeed, setting temperature of A70 was peculiarly low at each concentration.

In figure 8, the parameters a and B of equation 5 have been related to DE summarizing these findings. Similarly, Barfod and Pedersen (1990) had found three ranges with different dependence of setting temperature on DE for lime pectin ($c_P \approx$ 0.25% at 65°Brix). However, setting temperature most strongly increased with DE in the range 63% < DE < 70%, whilst it rised only slightly at DE ≥ 70% and was independent of DE at DE < 63%. In samples with 70°Brix soluble solids, the region of the strongest effect of DE was shifted to lower DE.

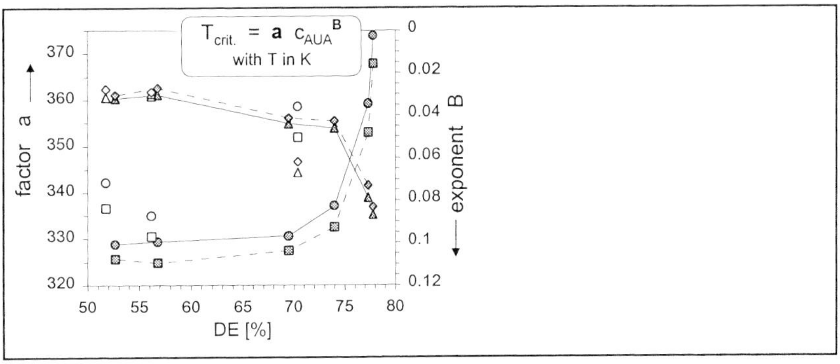

Figure 8. Evaluation of the effect of esterification on the setting temperature $T_{crit.}$ (DE = degree of esterification determined by titration). Setting temperature $T_{crit.}$ is indicated both as the pseudo gel point temperature $T_{\delta=75°}$ and the turning-point temperature T_W. Filled symbols = apple pectins, open symbols = C70, ACC52, AAC56. Factor *a* calculated for $T_{crit.} = T_{\delta=75°}$ (circle) and $T_{crit.} = T_W$ (square), exponent *B* calculated for $T_{crit.} = T_{\delta=75°}$ (triangle) and $T_{crit.} = T_W$ (rhomb)

The mechanical spectra of the final gels (figure 9) permit an insight into the nature of the networks, at least on a qualitative level. At all pectin concentrations, the mechanical spectra showed the typical characteristics of physical gels by the storage modulus G'(ω) being largely independent of frequency and, at least at low frequencies, by G'(ω) highly dominating G''(ω). At low pectin concentrations up to c_P = 0.25 g / 100 g gel, the basic characteristic of the mechanical spectra was the same for all pectins studied (figure 9A/B). However, at higher pectin concentrations ($c_P \geq$ 0.4 g / 100 g gel), this remained unchanged for medium-esterified pectins only (figure 9D) and pectin AMr68 reduced in its molecular weight (figure 9C), whilst the higher esterified pectins revealed another characteristic of their mechanical gel spectra in spite of similar values of the moduli (figure 9C). In the latter case, a second plateau of G'(ω) appeared in outlines associated with the occurence of a minimum of tan δ(ω) = G''(ω)/G'(ω) in the frequency range investigated. According to the marked increase of G'(ω) at frequencies f ≥ 1 Hz, mobility of pectin molecules was more limited in these gels compared to the gels of medium-esterified pectins, obviously as a result of additional entanglements of the macromolecules.

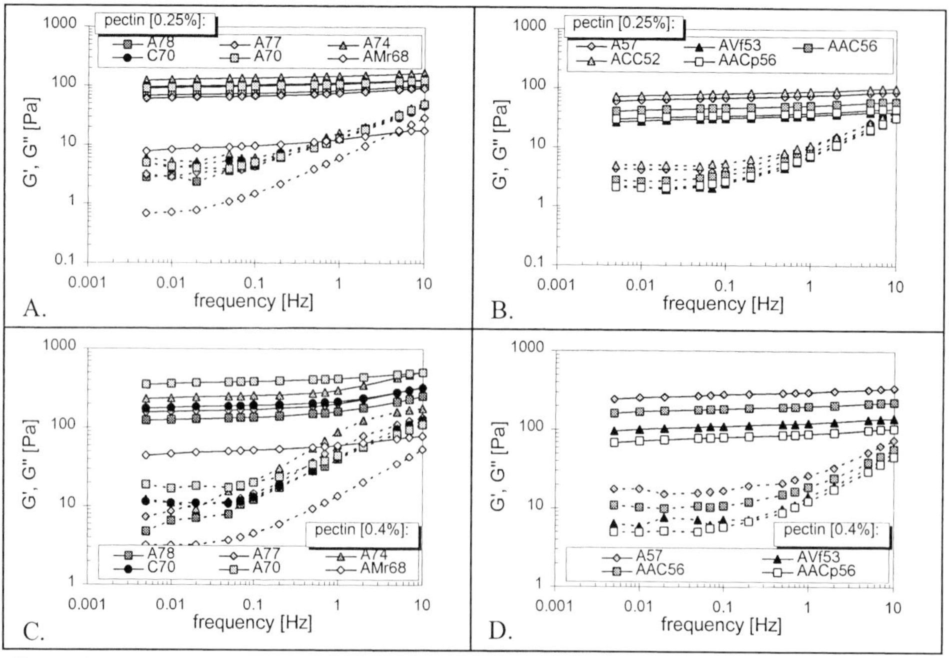

Figure 9. Frequency dependence of the final gels with 0.25% and 0.4% pectin, respectively, showing characteristic differences among the two groups of high-ersterified pectins (left) and medium-esterified pectins (right) at pectin contents $c_p \geq 0.4\%$. —— storage modulus G'(ω), - - - - loss modulus G''(ω)

Such macromolecular entanglements are detectable by their influence on viscoelasticity in the upper frequency range (f ≥ approx. 10 Hz) (Clark and Ross-Murphy, 1987). They especially occur in concentrated polymer systems with long, branched side-chains (Hwang and Kokini, 1992). Longer side-chains were likely for the higher esterified pectins only (3.1), which showed the deviating mechanical gel spectrum in figure 9C.

At the low frequency of 0.01 Hz, both moduli of the gels were found in the plateau zone (figure 9) and an apparent equilibrium relaxation modulus G_e could be deduced from G'(ω) of this zone characterizing network stability. Irrespective of pectin type, the loss modulus $G''_{0.01Hz}$ in the plateau zone could be expressed as a general empirical function of the corresponding storage modulus $G'_{0.01Hz}$ by

$$\lg[G''(\omega_o)] = a + b\,\lg[G'(\omega_o)] + c\,(\lg[G'(\omega_o)])^2 \tag{6}$$

with $f_o = \omega_o/2\pi = 0.01$Hz, $a = -0.85141$, $b = 0.61676$, and $c = 0.08552$ (figure 10A). Therefore, the mechanisms which are detectable at this frequency and which underlie stability and relaxation behaviour of the junction zones were identical for all eleven pectins, i.e. for both medium-esterified and higher esterified ones. Obviously, molecular fine structure only defines the pectin concentration, at which a certain stability of junction zones defined by $G'_{0.01Hz}$ is achieved in the gel network. Investigations on the possible influence of solvents on this general function between both gel moduli exceeded the scope of the present study and were not yet performed.

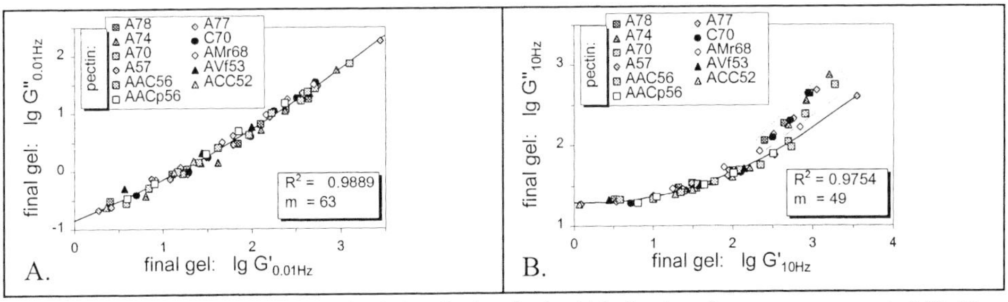

Figure 10. The rheological moduli of the final gels in (A.) the low-frequency range, at 0.01 Hz, and (B.) the high-frequency range, at 10 Hz. Regression analysis with *m* data sets irrespective of type and content of pectin. The broken-line frame in fig. 9B marks data sets deviating from the regression curve

As already suggested by figure 9C/D, specific mechanisms of individual pectins influencing relaxation behaviour of the gels could be only detected at higher frequencies $f \geq 1$Hz (figure 10B), provided that high gel strength was achieved by high pectin concentrations. These specific effects became evident in deviations from the general function between both moduli at frequencies above 1 Hz. $G''(\omega_o)$ was much higher than predicted by the general function as shown for 10 Hz in figure 10B, indicating more restricted mobility of chain segments in the network. By intensifying macromolecular entanglements, long side-chains of pectins may thus cause gel relaxation behaviour specific for the respective pectin.

Both gel strength ($G'_{0.01Hz}$) and pseudo gel point temperature ($T_{\delta=75°}$) were related to pectin concentration (c_{AUA}) by empirical power law functions (Neidhart, 1997). However, the influence of c_{AUA} was not the same, and for gel strength, the power laws even included a critical concentration (Neidhart, 1997). There was no general, pectin-independent correlation between $G'_{0.01Hz}$ of the gels and the setting temperature $T_{\delta=75°}$ (figure 11). Therefore, both properties could not be influenced by pectin fine structure in the same way.

At very low pectin contents (0.1-0.175%), medium-esterified pectins (figure 11 B) formed much softer gels than higher esterified pectins (figure 11A), but set at much higher temperatures. In contrast to gel strength, pseudo gel point temperature $T_{\delta=75°}$ strongly increased with pectin concentrations above 0.25 g/100g gel (A70: 0.4 g/100g gel), but only in systems of higher esterified pectins (figure 11 A). However, exactly for those samples and concentration range, the mechanical spectra of the gels (figure 9C) had revealed the deviations from the general function between $G''(\omega_o)$ and $G'(\omega_o)$ at $f \geq$ 1 Hz (figure 10B), which had been explained by macromolecular entanglements.

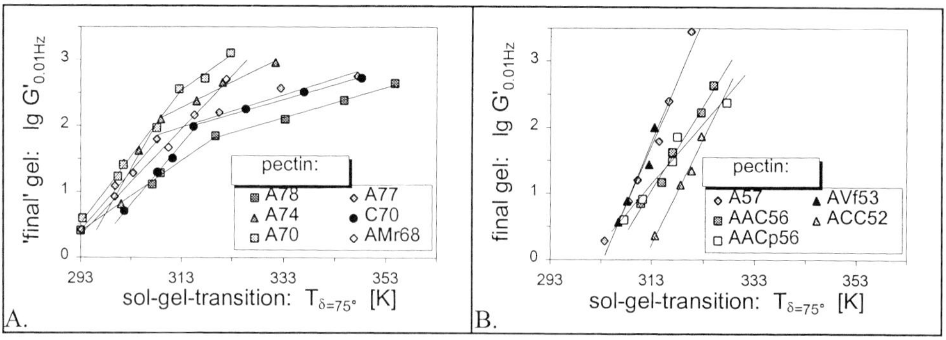

Figure 11. Setting temperature and gel strength of HM-pectin systems at various pectin concentrations

From these observations, it may be concluded that junction zone formation during induction of gelation at high temperatures is facilitated by intensified immobilization of pectin chains by entanglements related to long side-chains in the fine structure of certain pectins.

3.4. CONCLUSION

The comparative studies on the concentration dependence of both mechanical gel spectra and the setting temperature gave evidence that gelation of HM-pectins (DE > 50%) is primarily induced depending on the possibilities for intermolecular formation of hydrogen bonds, i.e. the amounts of unesterified AUA units and actual thermal conditions. Adverse conditions may be faced by reducing mobility of pectin molecules through the effects of long side-chains, as described in 3.3, and interactions between calcium and unesterified galacturonan units (Neidhart, 1997). This conclusion is not necessarily contradictory to the known mechanisms involved in junction zone stabilization based on hydrogen bonding and hydrophobic interactions (Evageliou et al., 2000). However, considering the lacking correlation between setting temperature and DMe or DE, hydrophobic interactions may be more decisive for thermal stability of the junction zones existing in the gels than for intermolecular association during sol-gel transition. The idea of the types of interactions contributing to sugar-acid gelation of HM-pectins should be extended as outlined above.

4. Acknowledgement

The authors thank Herbstreith & Fox KG, Neuenbürg / Germany, for their manifold support, Prof. Dr. M. Kuhn, Hohenheim University, who made the rheometer available for this study, and Mrs Carola Bolay for her assistance in the chemical analyses. Part of this work was financially supported by the "Bundesministerium für Forschung und Technologie" (BMFT).

5. References

Aspinall, G.O. (1982) Chemical characterization and structure determination of polysaccharides. In G.O. Aspinall (ed.), *The polysaccharides*, Vol. 1, Academic Press, New York, pp. 36-131

Barfod, N.M. and Pedersen, K.S. (1990) Determining the setting temperature of high-methoxyl pectin gels. *Food Technology (Chicago)* **44**(4), 139-141,148

Clark, A.H. and Ross-Murphy, S.B. (1987) Structural and mechanical properties of biopolymer gels, *Adv. Polym. Sci.* **83**, 57-192

Crandall, P.G. and Wicker, L. (1986) Pectin internal gel strength: Theory, measurement, and methodology. In M.L. Fishman, J.J. Jen (eds.), *Chemistry and function of pectins*, ACS Symp. Ser. 310, Washington, D.C, pp. 88-102

Cuvelier, G. and Launay, B. (1990) Frequency dependence of viscoelastic properties of some physical gels near the gel point, *Makromol. Chem., Macromol. Symp.* **40**, 23-31

Daas, P.J.H., Voragen, A.G.J., and Schols, H.A. (2001) Study of the methyl ester distribution in pectin with *endo*-polygalacturonase and high-performance size-exclusion chromatography, *Biopoly.* **58**, 195-203

da Silva, J.A.L. and Gonçalves, M.P. (1994) Rheological study into the ageing process of high methoxyl pectin / sucrose aqueous gels, *Carbohydr. Polym.* **24**, 235-245

da Silva, J.A.L., Gonçalves, M.P., and Rao, M.A. (1995) Kinetics and thermal behaviour of the structure formation process in HMP / sucrose gelation, *Int. J. Biol. Macromol.* **17**(1), 25-32

Evageliou, V., Richardson, R.K., and Morris, E.R. (2000) Effect of pH, sugar type and thermal annealing on high-methoxy pectin gels, *Carbohydr. Polym.* **42**, 245-259

Hannak, C. (1997) Einflüsse der chemischen Zusammensetzung sowie besonders der Verteilung der Molekulargewichte und der freien Carboxylgruppen auf die Geliereigenschaften von ausgewählten Pektinen, PhD thesis, Hohenheim University, Germany

Hwang, J. and Kokini, J.L. (1992) Contribution of the side branches to rheological properties of pectins, *Carbohydr. Polym.* **19**, 41-50

Ikkala, P. (1986) Characterization of pectins. A different way of looking at pectins. In G.O. Phillips, D.J. Wedlock, P.A. Williams (eds.), *Gums and stabilisers for the food industry 3*, Elsevier Applied Science, London, pp. 253-267

Kravtchenko, T.P. and Pilnik, W. (1990) A simplified method for the determination of the intrinsic viscosity of pectin solutions by classical viscosimetry. In G.O. Phillips, P.A. Williams, D.J. Wedlock, (eds.), *Gums and Stabilizers for the Food Industry 5*, IRL Press, Oxford, pp. 281-285

Kravtchenko, T.P., Penci, M., Voragen, A.G.J., and Pilnik, W. (1993) Enzymic and chemical degradation of some industrial pectins, *Carbohydr. Polym.* **20**(3), 195-205

Kujawski, M. and Tuszynski, T. (1987) A comparison of some methods for the determination of methoxyl groups in commercial pectin preparations, *Die Nahrung* **31**, 233-238

May, C.D. (1990) Industrial pectins: Source, production, and application, *Carbohydr. Polym.* **12**, 79-99

Michon, C., Cuvelier, G., and Launay, B. (1993) Concentration dependence of the critical viscoelastic properties of gelatin at the gel point, *Rheol. Acta* **32**, 94-103

Neidhart, S. (1997) Einfluß unterschiedlicher Strukturelemente von Pektinen auf das rheologische Verhalten entsprechender Gele, PhD thesis, Hohenheim University, Germany

Neidhart, S., Hannak, C., and Gierschner, K. (1996) Investigations of the influence of various cations on the rheological properties of high-esterified pectin gels. In J. Visser, A.G.J. Voragen (eds.), *Pectins and pectinases. Progress in Biotechnology,* Vol.14, Elsevier Science, Amsterdam, pp. 583-590

Oakenfull, D. and Scott, A. (1984) Hydrophobic interaction in the gelation of high methoxyl pectins. *J.Food Sci.* **49**, 1093-1098

Oakenfull, D. and Scott, A. (1986) New approaches to the investigation of food gels. In G.O. Phillips, D.J. Wedlock, P.A. Williams (eds.), *Gums and stabilisers for the food industry 3*, Elsevier Science Publishers, New York, pp. 465-475

Ross-Murphy, S.B. (1991) Concentration dependence of gelation time. In E. Dickinson (ed.), *Food polymers, gels and colloids,* Royal Society of Chemistry, London, pp. 357-368

Scanlan, J.C. and Winter, H.H. (1991) Composition dependence of the viscoelasticity of end-linking poly(dimethylsiloxane) at the gel point, *Macromolecules* **24**, 47-54

Schols, H.A. and Voragen, A.G.J. (1996) Complex pectins: structure elucidation using enzymes. In J. Visser, A.G.J. Voragen (eds.), *Pectins and pectinases. Progress in Biotechnology,* Vol.14, Elsevier Science, Amsterdam, pp. 3-19

Voragen, A.G.J., Schols, H.A., and Pilnik, W. (1986) Determination of the degree of methylation and acetylation of pectins by h.p.l.c., *Food Hydrocolloids* **1**(1), 65-70

Winter, H.H. and Chambon, F. (1986) Analysis of linear viscoelasticity of a crosslinking polymer at the gel point, *J. Rheol.* **30**, 367-382

RHEOLOGICAL CHARACTERIZATION OF GUM AND JELLY PRODUCTS

H.-U. ENDRESS, F. MATTES
Herbstreith & Fox KG Pektin-Fabrik Neuenbürg
Turnstraße 37, 75305 Neuenbürg, Germany
dr.endress@herbstreith-fox.de, f.mattes@herbstreith-fox.de

Abstract

Beside colour and flavour the texture profile of confectionery products influence their properties. Texture defines shape and more important the chewing character of a wide range of products like gummy articles or jelly fruits. It is therefore necessary to determine texture not only for development work but also for quality assurance. This work shows the possibilities of using texture analysis and oscillating rheometer to determine texture of confectionery jellies with pectin and influences of selected recipe parameters, i.e. pectin raw material, degree of esterification, type of retardator on the texture profile. With a texture analyser beside firmness the breaking distance to determine the brittleness of the manufactured samples was used and differences between the samples measured. The breaking distance could be used to judge the brittleness by comparing the finished products. To come to objective results the measurement of storage modulus (G') and loss modulus (G") with an oscillating rheometer was used.
The tests show that for getting an evident texture profile a minimum gel firmness of more than 750 PE was needed. It was then seen that the used retardator had a big influence on the texture. In comparison to sodium citrate with sodium tartrate products having increased firmness but more brittle texture were obtained. Gels made with citrus pectin showed increased brittleness compared to products manufactured using apple pectin but having the same firmness. This behaviour is also valid if the degree of esterification is increased form 60 % to 65 %. From these measurements it was concluded that for jelly fruits with firm and brittle texture citrus pectin will be an appropriate pectin with sodium tartrate as retardator. For soft jellies apple pectin with sodium citrate as retardator should be used. Products with a gummy character can be manufactured with apple pectin having a high degree of esterification in addition with sodium tartrate.

1. Introduction

The confectionery sector comprises a wide range of products, which are substantially different in appearance, texture and other sensory aspects. Very important ingredients which the food industry uses to give their products uniqueness are gelling agents. Gelling agents give the products a certain appearance, which is beside colour the first impression consumers get from the products. During consumption the texture of the

F. Voragen et al. (eds.), Advances in Pectin and Pectinase Research, 449-465.
© 2003 *Kluwer Academic Publishers. Printed in the Netherlands.*

product will have a major influence on the taste of the product. Besides flavour release, chewing properties are very important for the consumers' acceptance of the product. For example jelly fruits manufactured with pectin or agar agar will give the finished product a typical visco-elastic texture. Products with agar agar having a dry and brittle texture can be distinguished from pectin based products with a viscous texture. Pastilles manufactured with gum arabic have a hard chewy texture, whereas for Turkish Delight manufactured with modified starch or pectin a soft viscous texture is required by the consumer. Very famous are gummy products like jelly bears or jelly babies manufactured with gelatine, modified starch or pectin and combinations of these ingredients. Other products can be aerated either with air or nitrogenium like Zefir, Angel Kisses or Marshmallows. All products have in common that gelling agents are used to give the product a defined appearance and texture.

The textural properties can be easily described by using sensory tests and are therefore often used in the product development or quality control of these products. But as a subjective analysis this kind of analysis is very difficult to carry out. It needs a trained panel to obtain good reproducible results. Nowadays development work has to be done much quicker and the production needs quick decisions so that objective methods to characterise the manufactured products are an advantage. To characterise textural properties of liquids or solids rheological tests have been successfully introduced in the food industry not to replace sensory tests, but as an additional analysis. Sensory tests are still the best method to get an overall impression of a product. With rheometers supported by computers very small but decisive differences of textural properties of confectionery products can be seen, giving the product development quick support in their work.

2. Pectin as a Gelling Agent for Confectionery Products

Every gelling agent influences the product's texture because of the different structure network built as a backbone to bind and to give the necessary firmness. Also within the group of pectins different gelling behaviour is seen, influenced by the raw material for the pectin production. Pectins extracted from pomace show the forming of viscous gels with high mechanical loading capacity because of a regular distribution of methoxyl groups. The methoxyl groups of citrus pectins have a block by block distribution giving the molecule a high affinity to ions, especially divalent ions. The molecule structure of citrus pectins leads to forming of elastic or brittle gels (Kratz, 1993).

The degree of esterification of high-methoxyl pectins influences the setting behaviour of the used pectin. Pectins with high degree of esterification are more rapidly setting and form gels having a more increased brittleness than pectins with a lower degree of esterification. A minimum setting temperature for high-methoxyl pectins used in the so-called sugar-acid gelling mechanism is seen with pectin having a degree of esterification of abt. 60%. And pectins with a lower degree of esterification show increased setting temperature. Because of the high water binding capacity, the concentration of pectin in the formula influences the firmness of the finished product.

3. Important Product Parameters which Influence Gelation

The type of sugar and the concentration of soluble solids affects the gelling behaviour of pectins. With an increasing amount of sugar the setting temperature of the formula will also increase. Sugar will partly dehydrate pectin molecules, thus supporting their association in cluster zones. The texture of the finished product becomes firmer and more brittle.
A certain proton concentration is necessary for gelation of a formula with high-methoxyl pectins, so the pH-value of the finished product has an influence on the gelling behaviour of pectin. By lowering the pH-value in the formula, the dissociation of the free carboxyl groups is suppressed, thus minimising the electrostatic repulsion between the pectin chains and forming of junction zones stabilised by hydrogen interactions between non-esterified carboxyl groups.
Because of the high soluble solids content and a low pH-value for pectin based confectionery formulae the resulting setting temperature of the formula would make it impossible to deposit the formula without pre-gelling. Therefore a high formula pH-value for the production of gum- and jelly products is needed, which also ensures a long depositing time. To reduce the risk of pre-gelling gel retarding agents, so-called retarders, are used. Retarders are generally buffer salts with monovalent cations, which interfere clustering of pectin molecules during the manufacture process. Depending on the desired effect, the salts of citric acid, tartaric acid or lactic acid can be used. The difference between these salts is their molecular weight, acidity and taste.

Due to dissociation of the buffer salt the pH-value of the formula is raised, thus preventing pre-gelling. By addition of an edible acid solution to the batch subsequently buffer salts and pectin molecules become neutralised leading to a controlled setting of the formula. The setting temperature as well as texture of the finished product may be decisively changed by the type of buffer salt chosen.

4. Sensory Aspects of Gum and Jellies

In the sensory assessment of gum and jelly products between consistency of gels and construction of the gel formation, the gel structure is distinguished. Consistency describes the compactness, firmness of a sample. These properties can be observed sensorily by pressing and spreading. They are often described as firmness. The behaviour during deformation can be rheologically determined, for example with penetration measuring.
The gel structure indicates the composition of a gel. Gels are mainly elastic because of their relatively high rigid gel formation. Easily shifting bonds within the gel structure will always result in a certain amount of viscous phases. The proportion of rigid bonds to shiftings bonds determines the gel structure. The higher the elastic shares in the gel, the more fragile and brittle the gel structure; with increasing viscous portions in the gel, the gel structure becomes smoother.
Texture is the overall image of the sensory feeling of a product, i.e. softness when biting, and the way it disintegrates while chewing, so that the consistency and the gel structure will give the finished product its unique texture.

Rheological analysis will help to get objective information regarding the texture of the manufactured products.

5. Texture Analyser

The texture analysers have been designed to examine the texture of food. The probe of the texture analyser is pressed into the tested sample with preselected speed and distance. The measurement starts when the plunger touches the surface of the sample, then being pressed into the sample. The data obtained is the force needed to press the probe into the sample leading to force-distance curves. After touching the surface the structure gives way with a certain resistance characteristic for its structure. Because of the elastic set the structure of the sample will be destroyed at a certain value, which is characteristic for its breaking strength. After breaking the structure the strain is released leading to a decrease in the force-distance curve. In the following the structure which is not yet destroyed will build up resistance leading to a further breaking with pushing the sample aside. After passing the first breaking point the evaluation of the curves can be difficult because of poor reproducibility (Endress et al, 1996).
The distance between touching the surface of the sample and the breaking point indicates the brittleness of the sample. Brittle gel-structures build up a strain quickly and break earlier than structures having less brittleness (figure 1).

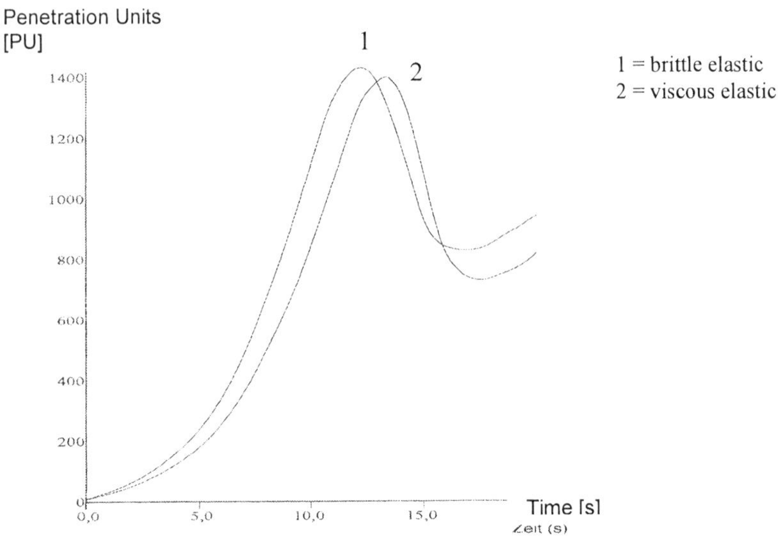

Figure 1: Texture Analysis of jelly gums manufactured with pectin

6. Oscillating Rheometer

Visco-elastic solids can be characterised using an oscillating test. In these tests the substances of interest are subjected to harmonically oscillating shear deformation. This deformation γ is given by a sine function, [$\gamma = \gamma_0 \sin(\omega t)$] by γ_0 as the deformation amplitude and ω the angular velocity. The response of the system is an oscillating shear stress τ with the same angular velocity ω (Endress et al, 1996).
A purely elastic substance is linearly connected with the deformation via Hooke's Law producing a phase displacement δ of 0°. The deformation of the spring is caused by energy. This energy is stored and regained when the applied force is removed. Ideal viscous substances can be described with the shock absorber model of Newton. The energy applied to the viscous body is completely transformed in energy of deformation or dissipated in form of heat. The mathematical description of this problem shows that the shear stress and shear rate curves are in phase and the delaying of these curves to the deformation curve [γ (t)] cause a phase displacement δ of 90 °.
Visco-elastic substances can be described with the spring-shock-absorber model of Kelvin and Voigt, and have phase displacements of 0° to 90°. In analogy to other time dependent processes in physics, the oscillation tests are evaluated with complex arithmetics. Obtained are the complex quantities:

- the complex viscosity (γ^*)
- the complex young modulus (G^*)
- the storage modulus G'
- the loss modulus G"
- the loss factor giving the ratio of the viscous and elastic share (G'/G")
  = dynamic Weissenberg-number

The mathematical dependence applies only in the linear visco-elastic range. In this region the sample can be deformed up to a maximum deformation (γ_0, max.) without destroying the structure of the sample.
The linear visco-elastic range of a sample is obtained using a strain amplitude sweep. In this measurement the deformation amplitude is increased [$\gamma(t) = \gamma_0 \sin(\omega t)$] at a constant angular velocity. The resulting storage modulus (G') is plotted versus the strain giving a horizontal line in a double logarithmic plot within the visco-elastic range of the solid. At a critical strain amplitude the storage modulus (G') starts to fall off with a further increase of the strain amplitude that indicates destroying of the structure.
With a strain frequency test the grade of cross linkages between the polymers can be examined indicating the structure stability related to time-depending effects. A low frequency simulates long time effects, whereas increased frequencies simulate short time effects or fast movements.
The measurement is conducted at a preset strain amplitude, increasing the oscillating frequency in dependence of time. The resulting double-logarithmic plot of the storage modulus (G') versus the angular velocity ω shows characteristic spectra depending on the measured substances.
Within the visco-elastic range an ideal gel structure shows no dependency of the storage modulus (G') on the frequency (f) so a horizontal line will be seen. Polysaccharide gel structures, i.e. pectin, are characterised by a flat dependenc of the

storage modulus (G') over a wide range of frequencies, showing rigid cross linkages between the molecules. The molecules have a small space to oscillate at their resting area (figure 2).

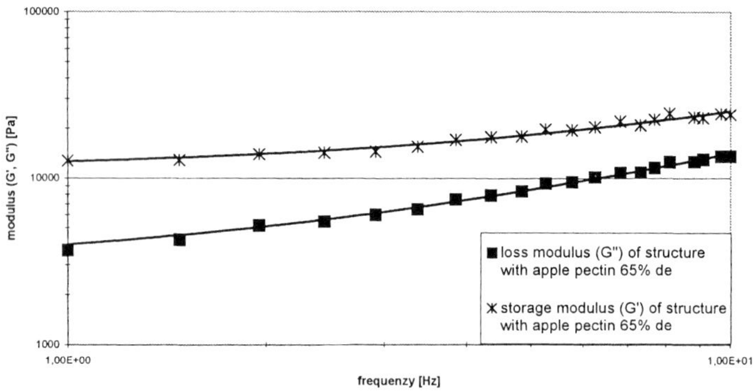

Figure 2: Frequency Sweep Measurement of jelly gums manufactured with pectin

The measurement is influenced by the type of pectin used. Depending on the used pectin characteristic curves are obtained for a characteristic area, but the curves have the same appearance yet shifted depending on the pectin used. Within the used strain and frequency the cross linkages show the same stability. Differences are seen in the responding time of the sample to the strain indicating differences in the texture.
The strain frequency test can be used to obtain the storage modulus (G') and loss modulus (G") value, when carried out in the visco-elastic range of the investigated sample. Though pectin gels show a flat dependency on the frequency a certain frequency has to be used as a basis to compare different structures.
With this oscillating technique the gelling behaviour of the sample can be investigated. In a strain temperature sweep with a preselected amplitude for the applied strain (γ) and a constant frequency (f) the setting temperature of the formula can be determined. The setting temperature is an important factor for the characterization and application of pectins. Confectionery articles must be deposited above the setting temperature of its gelling agent. Pre-gelling occurs, when the network is formed before product depositing, so that further mechanical treatment leads to an irreversible destruction of the three-dimensional network.
If the measurement is carried out at a preselected temperature close to the setting temperature the setting time of the formula for this temperature can be examined. This data can be used in the development for investigating the handling properties of formulae. The setting temperature predicts the minimum temperature needed in the manufacturing of the product. The setting time will indicate the batch size processed under the given conditions. These figures are necessary especially for scale-up trials.

7. Jelly and Gummy Products manufactured with Pectin

To develop new confectionery products it is necessary to know the possibilities to influence the texture of the finished product as well as developing formulae suitable for the given production process. To make life easier for the manufacturer pectins which are already standardised with different buffer salts can be supplied. That takes the responsibility of controlling the setting behaviour of the supplied pectin to the pectin manufacturer to ensure easy handling and standardised gelling behaviour of these pectin types.
The following figures will show possibilities in using already buffered pectin to obtain different confectionery products:

- pectin raw material (apple, citrus),
- degree of esterification (DE 60%, 65 %)
- dosage of buffered pectin (2,00%, 2.90%, 3.85% resp. 1,30%, 1,90% and 2,50% pectin 150°USA-Sag)
  The ratio of 1.90% pectin standardised to 150 °USA Sag to 1% retardor was set to obtain a setting temperature of the formula of 90 °C, when a dosage of 2.90% buffered pectin was used.

Table 1: Basic Confectionery Formula for Trials

x	g	Buffered Pectin
500	g	Sugar
330	g	Glucose Syrup (80 % tss[1)], abt. 42 DE)
220	g	Water
		Colour
		Flavour
y	ml	Citric Acid Anhydrous Solution (50 % m/m) to adjust pH-value

Input:	abt. 1080 g
Output:	1000 g
tss:[1)]	78 %
pH-value:	3.3

[1)] total soluble solids

The formula described in table 1 was used to obtain confectionery jellies with different firmness and texture.

8. Determination of Gel Strength and Breaking Distance

To measure the firmness of the manufactured products and for carrying out texture analysis a Texture Analyser TA-XT 2 i from Stable Microsystems was used. The penetration measurement was carried out using a 1/2" Zylinder (AOAC Standard).

After preparation the jelly mass was poured into the glass jar. The deposited jelly mass was placed in a climate cabinet (25 °C) for 24 hours for gelling. To prepare the samples for the measurement the separated top layer was removed, then placed underneath the probe and the texture analysis was carried out using following parameters:

Probe:	1/2 " Zylinder (AOAC-Standard)
Test Method:	Simple Penetration
	Force for penetration is measured
Pre-velocity:	5 mm/s
Test-velocity:	0.5 mm/s
Post-velocity	5 mm/s
Distance:	10 mm

The firmness of the samples were taken from the value of the maximum force needed to penetrate the probe into the sample, which was a characteristic breaking point. The same point was used to read the distance at which the gel structure broke to obtain the breaking distance.

9. Determination of Storage Modulus and Loss Modulus

Storage modulus and loss modulus of the texture were obtained by running a frequency sweep using an oscillating rheometer Bohlin CS 10. The jelly mass was poured into a metal plate, then placed in a climate cabinet (25 °C) for 24 hours for gelling. To prepare the samples for the measurement the separated top layer was removed and the test was run using the following parameters:

Measurement system:	SP 25 (plate-plate, with profile ∅ 25 mm)
Temperature:	25 °C
Shear stress:	100,000 Pa
Sweep type:	linear ramp
Start frequency:	1 Hz
Stop frequency:	10 Hz
Gap:	3.5 mm

After measurement the storage modulus and loss modulus were read at a frequency of 9.09 Hz.

10. Determination of Setting Temperature

For determination of the formulaes setting temperature the oscillating rheometer Bohlin CS 10 was used. In these tests the pH-value of the formula, described in table 1 was increased to pH 3.5 for better handling. For the measurement the preparation was placed in the test chamber and then cooled under defined conditions.

Measurement system:	SP 25 (plate-plate, with profile ∅ 25 mm)
Frequency:	1 Hz
Shear stress:	0.06 Pa
Sweep type:	linear ramp
Start temperature:	95 °C
Stop temperature:	20 °C
Cooling rate	2 °C / min.
Gap:	1 mm

The setting temperature of the formula was taken as the temperature at which the value of the phase displacement reached 45 °.

11. Gelation of Jelly Products with 1.9% Pectin

Pectin jellies set at a certain temperature, which is a characteristic value for each formula, depending on, i.e. the total soluble solids content, pH-value and type of pectin used. Table 2 shows the setting temperature of jelly products with 1.9% pectin. The formula with apple pectin 60% DE and sodium citrate as buffer salt have a setting temperature of 81 °C, which is increased to 85 °C when citrus pectin is used. It is also increased if apple pectin with an increased degree of esterification is used.

Table 2: Setting temperature of formulae with 1.9% pectin

	Sodium citrate			
Raw material	Apple Pectin		Citrus Pectin	
Degree of esterification	60%	65%	60%	65%
Setting temperature	81 °C	85 °C	85 °C	88 °C
	Sodium tartrate			
Raw material	Apple Pectin		Citrus Pectin	
Degree of esterification	60%	65%	60%	65%
Setting temperature	82 °C	87 °C	90 °C	100 °C

If sodium citrate is replaced by sodium tartrate a similar setting temperature is measured in the formula with apple pectin 60% DE: 82 °C. The setting temperature is increased to 87 °C, when apple pectin with 65% DE is used. The replacement of apple pectin by citrus pectin will further increase the setting temperature to 90 °C for citrus pectin 60% DE and 100 °C for citrus pectin 65% DE. Pectin has a much bigger influence on the setting temperature of formulae in which sodium tartrate as buffer salt is used. The formula with citrus pectin 65% DE and sodium tartrate would be very difficult to deposit, so that further addition of buffer salt is necessary for better handling.

Pectin obtained from apple pomace gives jellified products in comparison to citrus pectin having the same gel strength and degree of esterification but differences in the

gel texture will be obtained. Hence the products in which apple pectins are used will be in sensory assessments described with a smooth gel structure.
In figure 3 the firmness of the jellies with 1.9% pectin is seen. In the first part results of the trials carried out with sodium citrate as a buffer salt are described. The second part shows texture analysis of the products with sodium tartrate in the used formula. Apple pectin with 60% DE gives a visco-elastic set having 826 PU. Products with the same firmness (832 PU) are obtained with citrus pectin, but differences can be seen in the breaking distance. The samples with citrus pectin have an earlier breaking point after 5.9 mm than the samples with apple pectin at 6.0 mm, indicating a more brittle texture. Pectin with an increased degree of esterification give firmer texture seen in an increased breaking strength. Using apple pectin with 65% degree of esterification the firmness of the jellies are measured with 937 PU. If citrus pectin with 65% degree of esterification is used a similar firmness is obtained: 884 PU. The breaking distance of the jellies is decreased: 5.8 mm for the product with apple pectin indicates differences in the texture. Pectin with an increased degree of esterification results in gels with increased brittleness. A similar gelling behaviour is seen in the products in which citrus pectin is used.

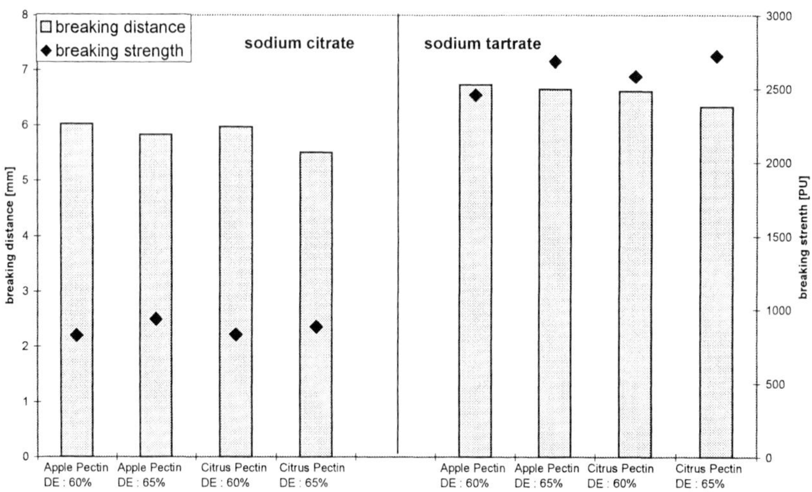

Figure 3: Firmness and texture analysis of jellies manufactured with 1.9% pectin

It is also seen that jelly products with sodium tartrate as buffer salt gel much firmer than products in which sodium citrate is used. The firmness of the products with apple pectin 60% DE is increased from 826 PU to 2457 PU when sodium tartrate is used in the formula. The breaking distance also increases from 6.0 mm to 6.7 mm, because of the increased firmness. Therefore the probe must be penetrated deeper into the gel to break the jelly. But with sodium tartrate the same relationship in firmness and brittleness between apple and citrus pectin and degree of esterification is seen.
Using the oscillating technique to investigate the texture profile of the samples accurate results can be obtained. By measuring the storage and loss modulus of the samples it

can be clearly seen that pectin has a significant influence on the storage modulus (figure 4).

The storage modulus of jellies with sodium citrate and apple pectin 60 % DE was measured having 20800 Pa. The storage modulus was increased, when citrus pectin with 60% DE was used and further increased with apple pectin 65% DE. Citrus pectin with 65% DE resulted in products with a storage modulus of 27500 Pa. In contrary pectin did not show a significant influence on the loss modulus, hence giving the samples a more brittle texture with increased storage modulus. This gelling behaviour can also be described by using the ratio of storage modulus and loss modulus, resp. dynamic Weissenberg Number.

The storage modulus and loss modulus of the samples in which sodium tartrate was used were increased. But using sodium tartrate increased the value of the loss modulus less than the storage modulus, which indicates an increased brittle texture in comparison to the formulae in which sodium citrate was used.

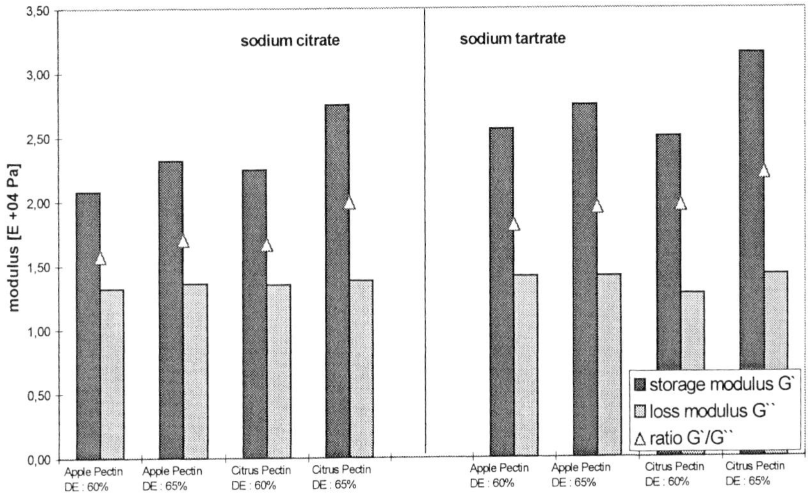

Figure 4: Storage modulus and loss modulus of jellies manufactured with 1.9% pectin

12. Gelation of Jelly Products with 1.3% Pectin

Additional tests were carried out using 1.3% pectin in the basic confectionery formula. The setting temperature of the obtained formulae are seen in table 3. Similar to the results described in the last chapter the used pectin had an influence on the setting temperature. Apple pectin with 60 % DE gave formulae with lowest setting temperature 82 °C for the formula with sodium citrate, 80 °C for the formula with sodium tartrate. The jellies in which citrus pectin with 65 % DE was used had the highest setting temperature 92 °C with sodium citrate, 91 °C with sodium tartrate.

Table 3: Setting temperature of formulae with 1.3% pectin

	Sodium citrate			
Raw material	Apple Pectin		Citrus Pectin	
Degree of esterification	60%	65%	60%	65%
Setting temperature	82 °C	89 °C	91 °C	92 °C
	Sodium tartrate			
Raw material	Apple Pectin		Citrus Pectin	
Degree of esterification	60%	65%	60%	65%
Setting temperature	80 °C	85 °C	83 °C	91 °C

In figure 5 the firmness of the obtained jellies is seen. Similar to the results of the trials with 1.9% pectin samples made with apple pectin, 60% DE had a soft set with 548 PU, when sodium citrate as buffer salt was used but had a firmer texture with sodium tartrate: 1402 PU.

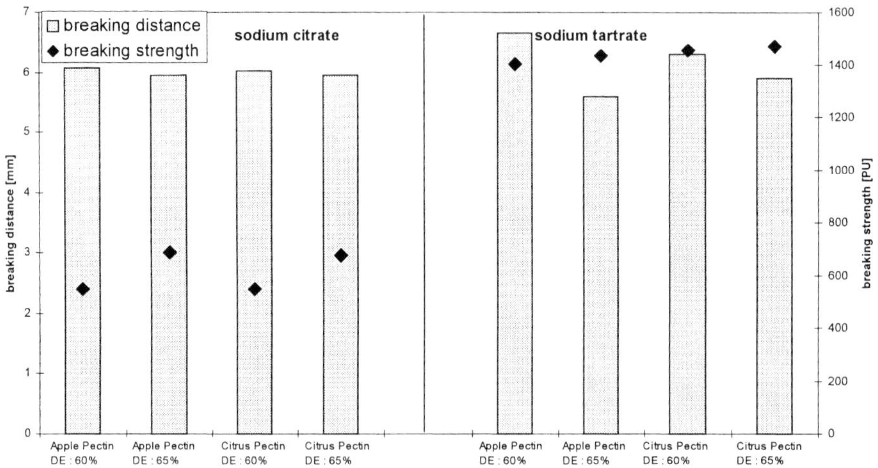

Figure 5: Firmness and texture analysis of jellies manufactured with 1.3% pectin

When citrus pectin with 60% DE was used the firmness of the finished product was not significantly increased. The firmness of the samples with sodium citrate and apple pectin or citrus pectin with 65% DE was significantly increased to 688 PU for apple pectin, and to 675 PU for citrus pectin. A significant change of the firmness of the samples with sodium tartrate as buffer salt could not be measured.

Determination of the texture using the breaking distance gave no significant values for the samples made with sodium citrate. In tendency samples manufactured with pectin either apple or citrus with 65% DE had a shorter breaking distance than the samples made with pectin, 60% DE. Significant values were obtained with samples made with sodium tartrate. The sample manufactured with apple pectin, 60 % showed a breaking distance of 6.7 mm, which was reduced to 6.3 mm, when citrus pectin 60% DE was

used. With apple pectin 65% DE jellies having a breaking distance of 5.60 mm were obtained, with citrus pectin, 65% DE 5.9 mm. Pectin with an increased degree of esterification gave the jellies a more brittle texture.
These results could be confirmed more accurately by measuring the storage modulus and loss modulus of the samples (figure 6). With this technique also small differences of the texture with sodium citrate were obtained. Apple pectin and citrus pectin having 65% DE samples with increased brittleness were obtained compared to the products manufactured with pectin 60% DE. Also significant differences between apple and citrus pectin are obtained. Products manufactured with citrus pectin had an increased storage modulus than apple pectin with the same degree of esterification.

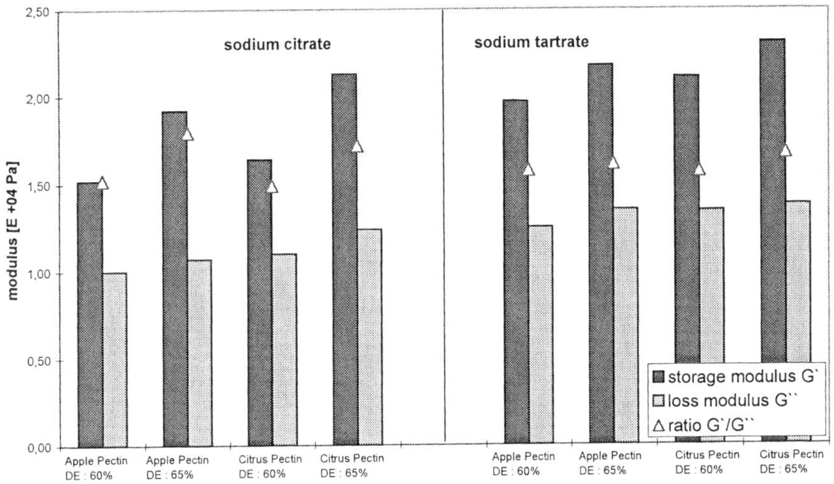

Figure 6: Storage modulus and loss modulus of jellies manufactured with 1.3 % pectin

13. Gelation of jelly products with 2.5% pectin

It was possible to manufacture jellies with a dosage of 2.5% pectin and 1.35% retardator. The setting temperatures of these formulae are seen in table 4 and are decreased in comparison to the formulae with 1.9% pectin. The setting temperature of the formulae with 1.3% pectin and 0.7% retardator were increased in comparison to the values obtained with 2.5% pectin and 1.35% retardator, except for the formulae with sodium tartrate as buffer salt. A constant ratio of the used buffer salt with pectin does not result in formulae with a constant setting temperature when the dosage of the buffered pectin is varied. If the pectin dosage is increased also the setting temperature will increased. The addition of buffer salts reduces the setting temperature of the formulae. By comparing these results it is seen that buffer salts have a bigger influence on the setting temperature than pectin.

For the formulae with 2.5% a high viscosity was seen, but were easy to handle for depositing. So a borderline for pectin dosage is not fixed by the setting temperature but the viscosity of the formulae.

Table 4: Setting temperature of formulae with 2.5% pectin

	Sodium citrate			
Raw material	Apple Pectin		Citrus Pectin	
Degree of esterification	60%	65%	60%	65%
Setting temperature	74 °C	76 °C	80 °C	82 °C
	Sodium tartrate			
Raw material	Apple Pectin		Citrus Pectin	
Degree of esterification	60%	65%	60%	65%
Setting temperature	79 °C	85 °C	89 °C	93 °C

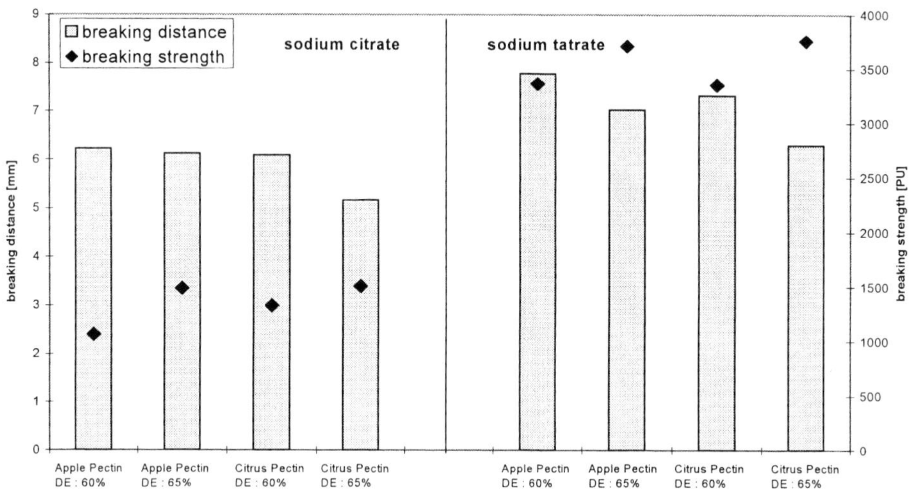

Figure 7: Firmness and texture analysis of jellies manufactured with 2.5% pectin

In figure 7 it can be seen, that with sodium citrate as buffer salt products with a high gel strength (1000 – 1500 PU) were obtained. Very firm products (abt. 3500 PU) were obtained by using sodium tartrate as buffer salt. Also for the formulae with 2.5% pectin the same gelling behaviour of apple pectin 60% DE which forms softer gels and an earlier breaking point than apple pectin, 65% DE is seen. This relationship can also be found when citrus pectin is used and depends not on the used buffer salt.
In the texture analysis using the oscillating rheometer similar texture properties as found in the formulae with 1.9% pectin or 1.3% pectin are seen (figure 8). The storage modulus of the products with sodium citrate were decreased in comparison to the products manufactured with sodium tartrate. The products manufactured with apple

pectin showed a less brittle texture than the products manufactured with citrus pectin having the same degree of esterification.

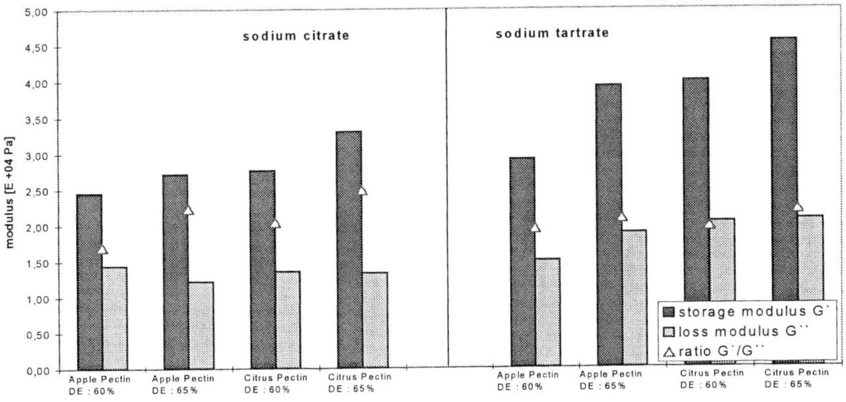

Figure 8: Storage modulus and loss modulus of jellies manufactured with 2.5 % pectin

14. Discussion

Apple and citrus pectin with a degree of esterification of 60% and 65% and a constant ratio of retarding buffer salt (sodium citrate or sodium tartrate) were used for manufacturing jelly products with a standard formula having 78 °Brix, pH 3.3.
The texture of the products was examined using two different techniques. A texture analysis using a penetration method in which the breaking strength as the firmness of the products and breaking distance to determine the brittleness of the gel structure was used.
Additionally the storage modulus and loss modulus of the samples were determined using an oscillating rheometer. The methods could be used to characterise the products objectively. As expected jellies in which apple pectin was used had a less brittle texture than products with citrus pectin (table 5). But due to the soft texture of the products with 1.3% pectin clear differences in the texture measured by texture analysis or oscillating rheometer could not be obtained.
By variation of the dosage of the buffered pectin it was possible to obtain products with different firmness and texture. For all types of pectins used in the formula the firmness of the finished products was increased. With increased firmness the products had an increased brittle texture (table 6)

Table 5: Firmness, storage modulus and loss modulus of jellies with apple pectin 65% DE and citrus pectin 65% DE

Pectin dosage [%]	Retardator	Firmness [PU]	Storage modulus (G') [E + 03 Pa]	Loss modulus (G") [E +03 Pa]	Ratio G':G"
Apple Pectin, 65% DE					
1.3	sodium citrate	688	1.92	1.07	1.79
	sodium tartrate	1432	2.18	1.34	1.62
1.9	sodium citrate	937	2.32	1.36	1.70
	sodium tartrate	2685	2.75	1.48	1.86
2.5	sodium citrate	1487	2.72	1.22	2.22
	sodium tartrate	3708	3.92	1.88	2.09
Citrus Pectin, 65% DE					
Pectin dosage [%]	Retardator	Firmness [PU]	Storage modulus (G') [E + 03 Pa]	Loss modulus (G") [E +03 Pa]	Ratio G':G"
1.3	sodium citrate	675	2.13	1.24	1.72
	sodium tartrate	1468	2.31	1.38	1.67
1.9	sodium citrate	884	2.75	1.38	1.99
	sodium tartrate	2723	3.15	1.42	2.22
2.5	sodium citrate	1507	3.30	1.33	2.48
	sodium tartrate	3790	4.55	2.07	2.20

Table 6: Firmness, storage modulus and loss modulus of jellies with apple pectin 60% DE

Pectin dosage [%]	Retardator	Firmness [PU]	Storage modulus (G') [E + 03 Pa]	Loss modulus (G") [E +03 Pa]	Ratio G':G"
Apple Pectin, 60% DE					
1.3	sodium citrate	548	1.52	1.00	1.52
	sodium tartrate	1402	1.97	1.25	1.58
1.9	sodium citrate	826	2.08	1.32	1.57
	sodium tartrate	2457	2.56	1.41	1.82
2.5	sodium citrate	1060	2.45	1.44	1.70
	sodium tartrate	3362	2.91	1.50	1.94

Big differences in the influence of the used buffer salt was observed. Using sodium citrate a less brittle texture was obtained but also softer gelation (table 5 and table 6). The used buffer salt has a much bigger influence on the firmness of the finished products than the used pectin type, when its degree of esterification is varied between

60% and 65% and raw material. With sodium tartrate products with more brittle texture were obtained than with sodium citrate. Further possibilities to influence the gelling behaviour are variations in the ratio of used buffer salts to pectin.

Apple pectin with 60% degree of esterification is therefore suitable for manufacturing jelly products with a smooth texture and high viscous shares like soft gelled jelly fruits. These properties are rheologically characterised by measuring the storage modulus, which is decreased in comparison to citrus pectin significantly. Also in the texture analysis these characteristics are seen in measuring the firmness which had the same value than products with citrus pectin but with a longer breaking distance. With such a texture flavour transfer is enhanced. Jelly fruits with a brittle texture and brilliant rupture are obtained by using citrus pectin with 65% DE. These products require a high storage modulus, which citrus pectin with 65% DE will provide. If the resulting setting temperature is too high for handling in the production facility a citrus pectin with 60% degree of esterification can be chosen.

For gummy products a very firm texture is needed which can be achieved by using 1.9 to 2.5% pectin. Using apple pectin with 65% DE product with the necessary ratio of storage modulus and loss modulus are obtained not giving a too brittle texture but with the necessary firmness. Also citrus pectin with 65% degree of esterification can be chosen, so that a more brittle texture can be achieved.

To support development work and quality assurance the texture analyser will give objective information on the firmness of the finished products. Information on the texture can be achieved by measuring the breaking distance but the obtained information can only be used in combination with the value of the gel strength measurement. Gels with increased firmness had an increased breaking distance, which could lead to misunderstanding for having a smoother texture. But the texture analysis with the oscillating rheometer showed that these samples had an increased storage modulus and increased ratio of storage modulus to loss modulus. That means these samples have a more brittle texture. For objective texture analysis an oscillating rheometer has to be used to investigate the texture properties of the manufactured products.

15. References

Endress, H.-U.; Döschl-Volle, C.; Dengler, K. (1996): Rheological Methods to Characterize Pectins in Solutions and Gels. Pectins and Pectinases, Elsevier, p. 407 - 423

Kratz, E. (1993): Tailor made pectins for specific influencing of the sensory rheological properties of preserves, jams, jellies, marmalades and other jellified products. Gums and Stabilisers for the Food Industry 7, Oxford University Press, p 403 – 411.

COMPARISON OF THE STABILISATION MECHANISM OF ACID DAIRY DRINKS (ADD) INDUCED BY PECTIN AND SOLUBLE SOYBEAN POLYSACCHARIDE (SSP).

P. BOULENGUER AND M.A. LAURENT
Degussa Texturant Systems SAS
Research Center
50500 Baupte France
patrick.boulenguer@degussa.com

Abstract

The stabilisation mechanism of acid dairy drinks (ADD) induced by pectin and soluble soybean polysaccharides (SSP) has been compared. The attachment of polymers onto the casein particles in both cases is due to electrostatic interactions. The layer of polymer formed around the casein prevents the aggregation of casein particles by steric repulsion. Nevertheless, this phenomenon is not sufficient to obtain an ADD with a long term stability. This stability requires the presence of a network which is only existing in the pectin based ADD and which can be displayed by rheological measurements. The weak gel formed presents thixotropic properties which are useful when the drink is sheared (during pumping or shaking of the drink). In the present work, we have also shown that only a fraction of the pectin is active in the drink. The other fraction is unable to interact with the casein.

1. Introduction

Acid dairy drinks (ADD) cover many types of beverage: fruited milk drinks, yoghurt drinks, buttermilk and whey drinks. The common feature of these different kinds of beverage is that they are diluted, acidified, homogenised milk.
ADD generally comprises between 1-8.5% (more typically between 4 to 8%) of Milk Solid Non Fat (MSNF)

Acidification can be performed:

By direct acidification of the milk with the fruit preparation or with citric acid
-By dilution of a yoghurt.

The final pH is generally between 3.7 and 4.6.

Depending from the desired shelf life (short or long), a thermal treatment can be applied to the ADD.
These drinks can also be enriched in calcium or in whey proteins.
The pH (below the isoelectric point of the casein) and the important dilution stage cause flocculation. This is prevented by homogenisation (generally performed between 100-

F. Voragen et al. (eds.), Advances in Pectin and Pectinase Research, 467-480.
© 2003 *Kluwer Academic Publishers. Printed in the Netherlands.*

200 bars) and by the addition of a polymer. Generally the polymer used is pectin(1-5) but it can be also CMC (6-7) or soluble soybean polysaccharide (SSP) (8,10).
In this article we describe a comparison between pectin and SSP in term of stabilisation mechanisms for ADD with 8% MSNF without homogenisation of the final drink.

2. Materials and methods

2.1. SAMPLES

Pectin A and B are produced by Degussa, Redon (France). The SSP is a Soyafibe-S^{TM} sample produced by Fuji Oil Co. The main characteristics of these samples are presented in Table 1.

TABLE 1. Comparison of chemical characteristics of pectin and SSP

Samples	**pectin A and B**	SSP (10)
Structure	Linear polygalacturonan	Branched structure with a charged main chain and neutral side chain (like a comb)
Galacturonic Acid	84	18
Esterification Degree	71	
Mw (SEC-MALLS)	200 000	>1 MD
Rg (SEC-MALLS)	50 nm	50 nm
Number of meq/g at pH 4.2	0.97 (α=73%)	0.3 (α=41%)
Viscosity at 20°C	500 mPa.s at 1%	3 mPa.s at 2%

SSP (Figure 1) is considered to have a branched structure resembling a comb with a charged back-bone and neutral side chains.
The molecular weight of SSP measured by SEC MALLS is very high but the gyration ratio (Rg) is nearly the same as the Rg of pectin. This observation can be explains by the compact branched structure of the SSP, while pectin is a linear rather stiff molecule. This difference in structure, also explains the difference in observed viscosity.

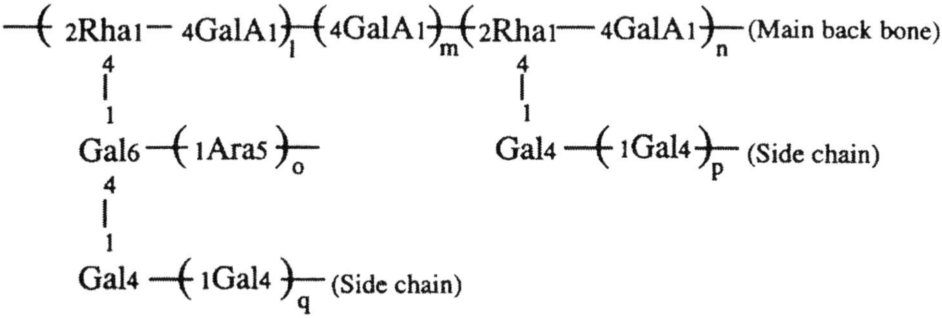

Figure 1: Structure of soluble SSP polysaccharide reproduced from (10)

Table 1 also shows that the number of meq per gram at pH 4.2 (number of meq/g * by the degree of dissociation α at pH 4.2) obtained by titration is different. Pectin contains more charged residues than SSP.

2.2. ADD PREPARATION

A commercial 0% fat yoghurt (Printiligne) was homogenised at 200kg/cm^2. Potassium sorbate (0.075%) was then added to prevent fungal growth and the pH adjusted to 4.2 with lactic acid. 20ml of yoghurt were poured into 10ml of polymer solution, shaken vigorously (there is no homogenisation of the final drink) and left for at least 2 hours in the fridge before use. The final dry material (MSNF) of the ADD was 8%.
-The final pectin A and B concentrations were: 0, 250, 500, 1000,1500, 2000, 2500ppm.
-The final SSP concentrations were: 0, 2000, 3000, 4000, 5000, 6000ppm. The difference between the range of pectin and SSP is due to the dose required in the application.

2.2.1 *Procedure*

Different measurements were performed directly on the ADD with respect to: rheological characteristics, particle size distribution and sedimentation occurring during storage (i.e. long term stability).
In a second step, the ADD was centrifuged at 13000g for 2 hours:

- Viscosity and dissolved polymer measurements were performed on the supernatant.
- Water content and sediment weight were measured on the solid phase.

- Viscosity measurements were performed on a Low shear 30 (Contraves) apparatus fitted with coaxial cylinders (1ml sample) at shear rates close to: 10, 25, 40, 70, 100s^{-1}.
- Laser granulometry was performed using a Malvern Mastersizer with a 300mm focal length lens, after dilution in a 0.066M acetate buffer at pH 4.2, containing 0.973g/l of calcium (corresponding to the ionic strength and calcium content of an ADD) to establish particle size distribution.

- Dynamic measurements of G' and G" were performed on 1.5ml of ADD with a Carrimed CSL^2 100 apparatus fitted with a cone and plate geometry (6cm diameter, 1° angle) at 20°C in the linear viscoelastic domain.
- Thixotropic measurements were performed with the following steps. First a frequency spectrum is realised at 10% of strain between 0.05-5Hz. The gel is then sheared during 1min. at 100s^{-1} and the recovery is measured at 0.3Hz and 10% of strain during 20min. At the end of this time a second frequency spectrum is realised.
- Natural sedimentation during storage was followed using a Turbiscan MA 1000.
- The polymer dissolved in the supernatant of the ADD (100µl) was measured by size exclusion chromatography with two TSK GMPW XL columns at 0.6ml/min, with a sodium acetate 0.1M pH 4.2 solution as mobile phase. The results were expressed as the ratio (in %) between the area of the supernatant and the area of a pure polymer solution, measured by the refractive index apparatus. The results for pectin were confirmed by copper precipitation (9).
- Dry material was obtained by drying over 48 hours (until constant weight) at 60°C (to avoid darkening of the sediment).

3. Results and discussion

3.1. WEIGHT FRACTION ANALYSIS

The weight fraction (% w/w) variation, as a function of the concentration in polymer, is presented in Figure 2a. It was measured after high speed centrifugation (13000g for 2hours) and represents the weight of sediment in 100g of ADD. Therefore, the weight fractions measured were obtained in a compressed state (the result will be different in other conditions). This variation is nearly concentration independent for SSP and this weight is also very close to the weight without polymer. We can conclude that: Firstly the SSP polysaccharide makes a very thin and compressible monolayer around the casein, Secondly there are only casein-SSP interactions. In contrast, in the pectin containing system, the pectin layer is very concentration dependent and not very easily compressed. The thickness variation of the layer is mainly due to its hydration (Figure 2b). Therefore there are pectin-casein interactions and also pectin-pectin interactions. pectin A and B provide different thickness of gel around the casein.

The difference in behaviour of the SSP and pectin containing ADD systems can be explained by the difference in molecular structure of SSP and pectin. Pectin has a linear and charged structure therefore it is possible for it to interact with both casein and pectin by electrostatic interactions (perhaps for pectin-pectin interactions through calcium ions). Conversely the SSP has a comb-like structure. It is bound through the charged main chain to the casein and the neutral side chains prevent other SSP molecules interacting between them. The thickness of the layer is the thickness of the side chain therefore the layer must be far below the Rg.

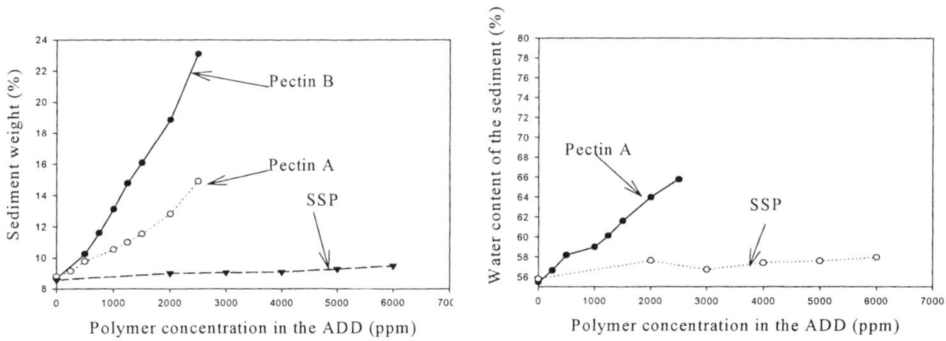

Figure 2: a) Weight fraction of sediment expressed in %
b) Water content of the sediment in (%)

3.2. VISCOSITY MEASUREMENTS

Pectin A and SSP demonstrate Newtonian behaviour in their supernatant solution (Figure 3). The viscosity increases linearly with the concentration showing that a part of the polymer remains in solution.

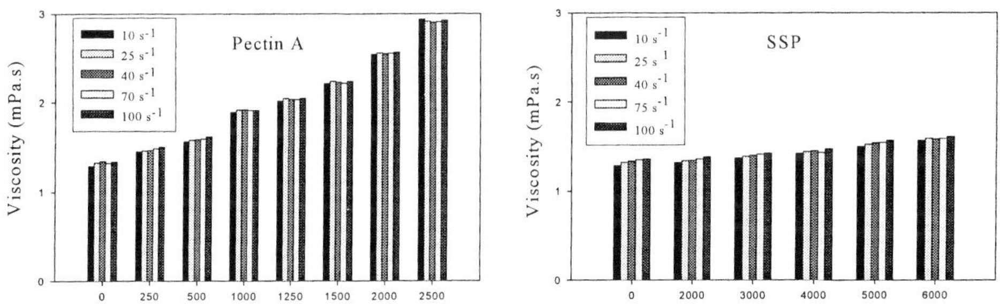

Figure 3: Viscosity of supernatant of pectin and SSP ADD at different shear rates.

If we look at the viscosity of ADD containing pectin A measured at 5 different shear rates (Figure 4), we can distinguish 3 areas:

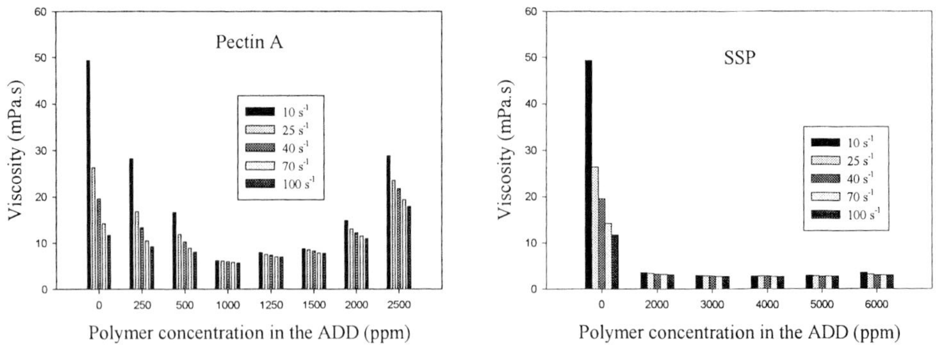

Figure 4: Viscosity of pectin and SSP ADD at different shear rates.

At a concentration below 1000ppm the viscosity decreased and demonstrated a pseudo-plastic (shear thinning) behaviour. There was not enough polymer to stabilise the drink. Some aggregation occurred probably by bridging because the casein was not completely covered by the polymer (11).

At concentrations between 1000-1250ppm, the solution had a very low viscosity and a near Newtonian behaviour. We can conclude that the casein was coated by the pectin (1).

At concentrations over 1250ppm, an increase in viscosity is recorded again as well as a pseudo-plastic behaviour (shear thinning).

According to Kravtchenko et al. (2), the increase in viscosity is related to the increase in volume fraction (which is very close to the weight fraction) under a non-compressed state. Viscosity and volume fraction are bound by the Krieger-Dougherty equation:

$$\eta_{sp} = (1 - \frac{\Psi}{p})^{-[\eta]p}$$

where Ψ is the volume fraction of particles including their adsorbed layer, and $[\eta]$ is the intrinsic viscosity of the ADD and p an adjustable parameter.

If this equation is applied to the viscosity measurements of SSP ADD (Figure 4), it is confirmed that the weight fraction does not increase with concentration (no increase in viscosity with concentration). Newtonian behaviour confirms the absence of interactions between the particles.

If this equation is applied to pectin, it can be deduced that viscosity and sediment weight provided the same types of results and a comparison of these two values demonstrated the compressibility of the polymer layer. But this equation does not explain why the ADD presented a pseudo-plastic behaviour. The shear thinning confirmed the presence of interactions between particles and therefore probably the presence of a gel.

3.3. DYNAMIC MEASUREMENTS

To prove the existence of a gel and to understand what type of interactions occurred in the ADD, dynamic measurements were performed on pectin B.
The strain dependence of G' at 0.3Hz for 1000, 1500, 2000ppm concentration (Figure 5) was first measured.

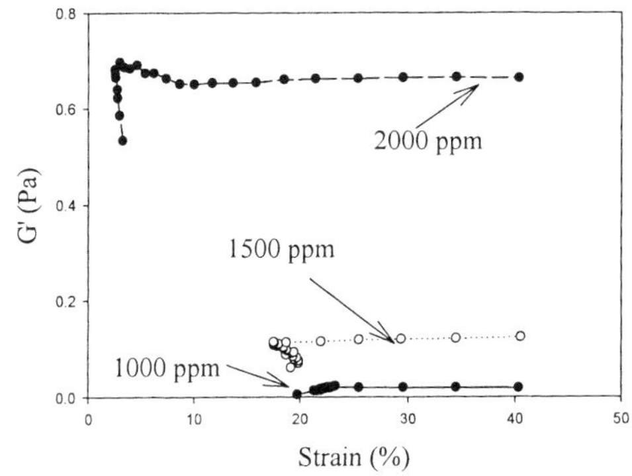

Figure 5: Strain dependence of G' at 0.3Hz on pectin B.

This figure illustrates a linear behaviour between 20-40% of strain. The choice of 30% of strain was made for the frequency spectra, i.e. part of the spectrum where G' and G" are not dependent of the strain.

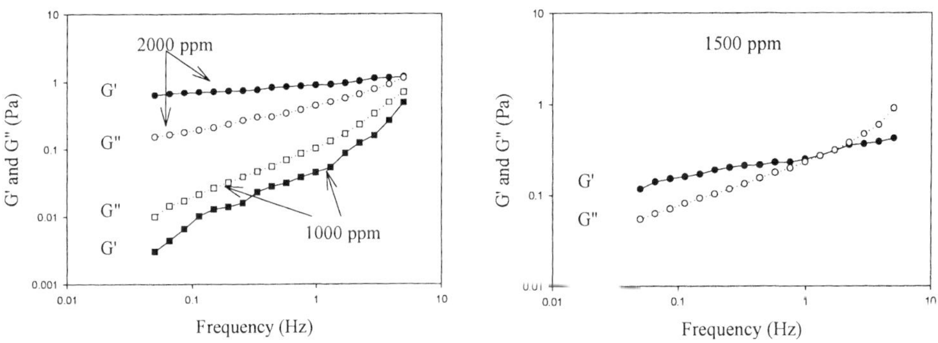

Figure 6: Frequency dependence of G' and G'' at 30% of strain amplitude on pectin B.

Figure 6 illustrates the frequency dependence of G' and G'' for pectin B. We can see at 1000ppm (within the region of Newtonian behaviour) that G'' is greater than G' therefore the ADD is a solution. At 2000ppm (pseudo-plastic domain), G' is greater

than G'' and almost independent of the frequency. Therefore at this concentration there is a weak gel (between 0.1-1 Pa). At 1500ppm, a very weak gel develops since G' and G'' are crossing beyond a frequency of 0.8-0.9 Hz. G' is very close to G'' therefore the gel is not very well organised. At a concentration over 1500ppm, the pectin ADD forms a network which stabilises the isolated casein-pectin particles.
To determine the exact nature of the gel: pectin-pectin gel due to calcium interactions or pectin-casein gel, the following experiment was performed: 2ml of water was added to 20ml of homogenised yoghurt. After high speed centrifugation (same conditions as previously) the supernatant was collected (11ml) and 4ml of a 0.75% pectin B solution was added in order to obtain a "normal" ADD with 2000ppm of pectin but without casein, G' and G" as a function of percentage strain was then measured at 0.3Hz. At all strain amplitude G" is over G' (Figure 7). Therefore no gel occurred without casein, the gel obtained in normal ADD was a mixed casein-pectin gel.

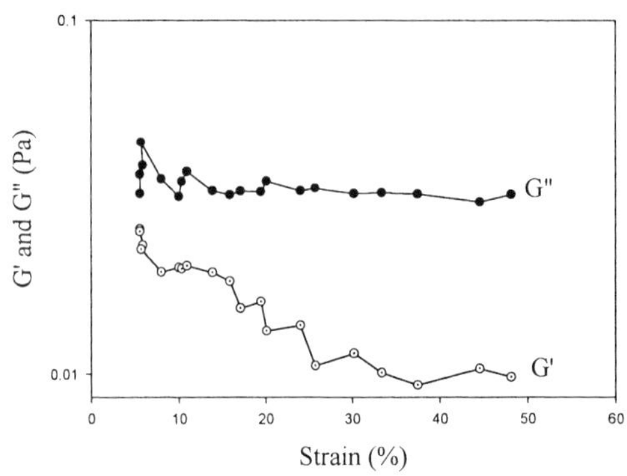

Figure 7: Strain dependence of G', G" at 0.3Hz of the ADD without casein and 2000ppm of pectin.

To determine the physical properties of the gel, thixotropic measurements were performed. The results of the frequency spectra are given in Figure 8a, the recovery in Figure 8b.
From these figures we can determine that the recovery of the gel is very fast (recovery of 20% of the initial G' in less than 40s) and that G' is over G" immediately after stopping the shearing. This property allows a fast immobilisation of the casein particles especially when the ADD are shaken. The recovery is not complete after 20min. (40% of the initial G') but near complete after 1.5h. The same results are obtained for shearing of five min. at $300s^{-1}$ (results not shown).

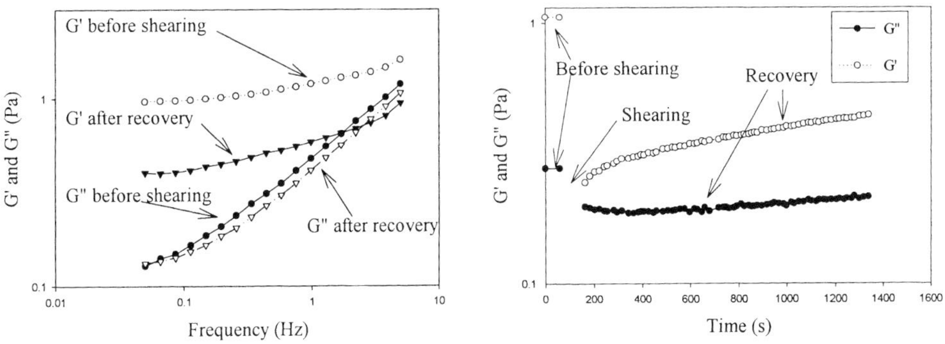

Figure 8: Thixotropic measurement on pectin ADD. a: Frequency spectrum before shearing and after recovery. b: Recovery of the ADD gel after 1min. shearing at $100s^{-1}$ measured at 0.3Hz, 10% of strain.

3.4. POLYMER ADSORBED ON CASEIN

Figure 9 illustrates the pectin adsorbed onto the casein both in weight and in percentage of the polymer added. Again a considerable difference in behaviour appeared between SSP and pectin. The SSP presented a "classical" behaviour. At low concentrations, a significant part of the polymer interacted with the casein. As the concentration increased less and less polymer finds a "place" to stick to the casein, therefore the percentage of the polymer adsorbed decreased and tended towards zero. The pectin presented an atypical behaviour. The more pectin was added the more pectin was bound, which also confirmed the pectin-pectin interactions. Depending on the sample, the percentage of pectin bound can vary from 50% for pectin A to 90% for pectin B. This can be explained by proposing the existence of two types of pectin in each sample: an "active" pectin and an "inactive" one. To test this hypothesis (9), the sample of pectin A containing 2000ppm of pectin was split by high speed centrifugation in an active fraction bound on the sediment and an inactive fraction dissolved in the supernatant (nearly 1000ppm in each fraction). These two fractions were added respectively (see Figure 10) to the supernatant and the sediment of a pectin free ADD and mixed. Only the ADD containing the sediment with Pectin gave a stable drink. The pectin in the supernatant was completely unable to stabilise the casein of pectin free ADD. This result confirmed the hypothesis that two types of molecule are present in pectin sample A. At high concentration the results must be taken cautiously for pectin gel can distort the results.

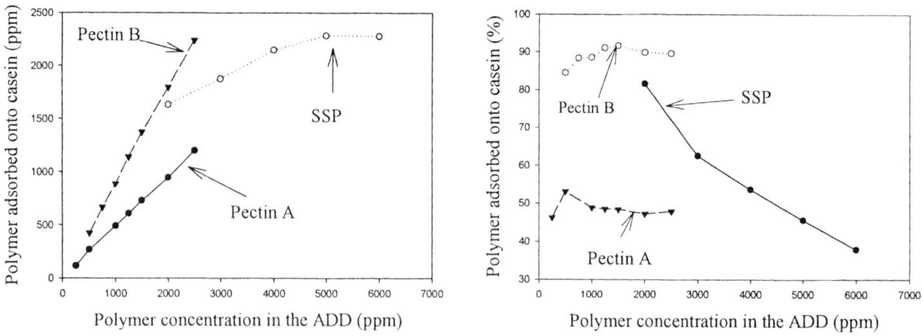

Figure 9: Polymer adsorbed onto casein as a function of polymer concentration in the ADD expressed in concentration and in percentage of the added polymer.

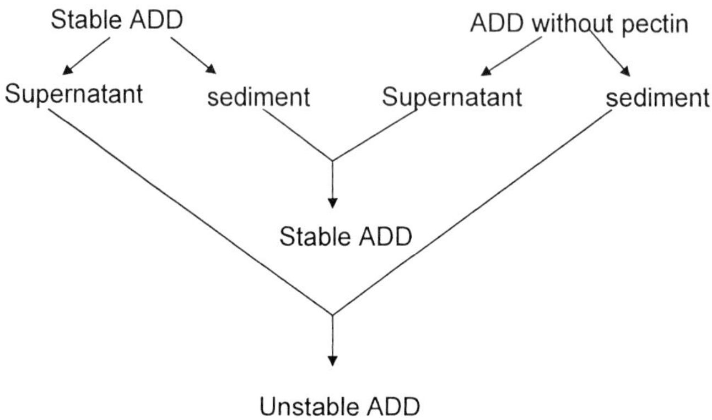

Figure 10: Experiment showing the presence of an active and an inactive pectin in the sample of pectin A

3.5. SIZE PARTICLE ANALYSIS

Figure 11 shows the respective particle size distribution of SSP and pectin A ADD measured by laser granulometry. When increasing polymer concentration of both polymers, particle size first decreased and in a second step (at a concentration below 2000ppm for pectin), the population of particles centred on a diameter size 2-3μm became independent of the concentration. Previously we saw that the weight fraction of the two polymers was different and varied with concentration therefore we can assume that the casein only was visible by this technique (the size of the casein particle is probably defined by the homogenisation pressure). If the concentration was increased to more than 2000ppm in pectin an increasing particle size with a bimodal peak appeared. This is probably as a result of an aggregation of casein micelles due to depletion phenomenon (12) which was confirmed by visual observation of the flocculated appearance of the ADD at high concentration. To prove this hypothesis and observe the aggregation, we tried to melt the gel. Unfortunately G' is independent of the temperature even at 70°C (results not shown). At temperatures over 70°C the casein gel is destroyed, therefore direct observation of the phase separation due to the depletion was not possible. In the same way the concentration of pectin adsorbed on casein must decrease when depletion occurred, but it cannot be observed due to the presence of the gel. No aggregation appeared with SPP.

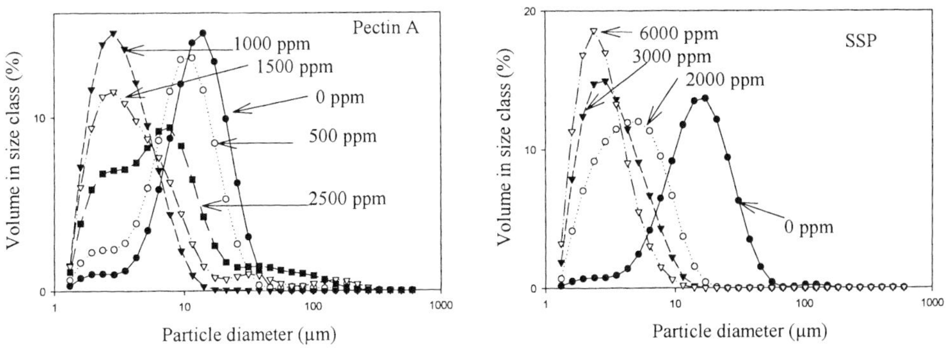

Figure 11: Laser granulometry distribution for pectin and SSP as a function of the polymer present in the ADD.

3.6. SEDIMENTATION DURING STORAGE

The sedimentation was followed with a Turbiscan apparatus which allowed precise measurement of sedimentation occurring in an ADD using back scattering measurements.

Figure 12 shows the typical behaviour of pectin A ADD during storage. Firstly at low concentrations we can see significant sedimentation rates which are more important than in the pectin free ADD. This observation corresponded to the first part of the viscosity curve where pectin was insufficient to coat all the casein, therefore bridging occurred which flocculated the ADD. At 1000ppm the ADD displayed Newtonian characteristics and the particles were no longer aggregated. At this concentration, the

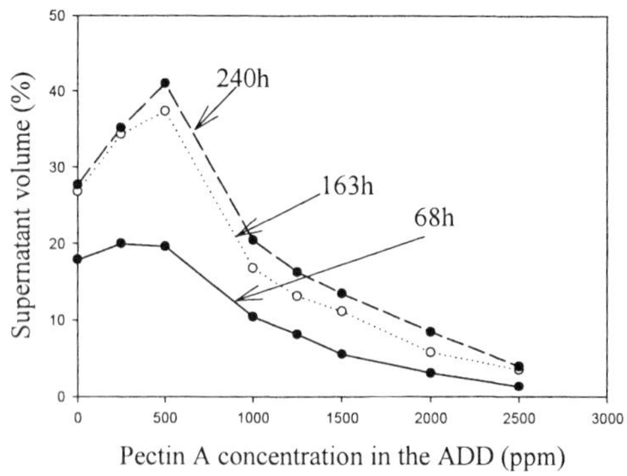

Figure 12: Stability during storage of pectin as a function of the polymer concentration.

sedimentation continued but more slowly than previously due to the smaller size of the particles. When the concentration was increased, the ADD became stable.

Figure 13 compares both sedimentation in SSP and pectin as a function of time. We can see that the ADD with pectin was only stable at the highest concentrations of pectin. At the same time the SSP was completely unable to prevent the sedimentation of the ADD. We can conclude that the disaggregation of the casein micelle is not sufficient to provide a stable ADD. There is a need for a second phenomenon (i.e. the gel formation) to give rise to a stable ADD.

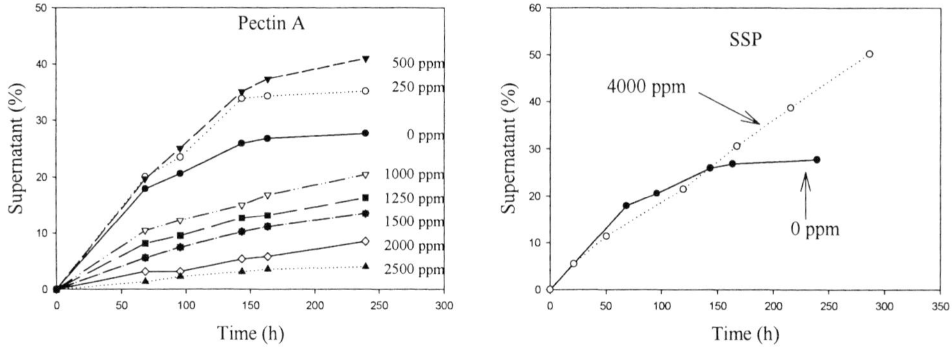

Figure 13: Comparison of stability during storage of SSP and pectin as a function of time.

4. Conclusion

This study demonstrated that the two polymers SSP and pectin were able to prevent the aggregation of casein in acid dairy drinks. Firstly an electrostatic interaction between the casein and the polymer occurred (on all the charged parts of the pectin and only on the charged main chain of SSP), secondly steric stabilisation occurred which prevented

the aggregation of the casein (1-2). The dispersing effect of SSP on casein is caused by a thin layer of neutral side chain of SSP around the casein. No other phenomenon appeared to be responsible for SSP behaviour.
The stabilisation by pectins is caused by a thick layer of pectin around the casein. Pectin presented a more complex situation in that not all the pectin was active. Probably the presence of pectins with too high DE or without galacturonic blocks which were not able to participate in the casein stabilisation. At low concentration of pectin, bridging occurred (probably also for SSP but we have not measured ADD with sufficiently low concentration of SSP to observe this phenomenon). Secondly, the pectin acted as a disegregating agent like SSP by steric repulsion. Thirdly, interactions occurred inducing finally a weak thixotropic gel which prevented the sedimentation of the ADD. The gel is a mixed gel of pectin and casein. At a very high pectin concentration, depletion occurred (perhaps the pectin which was not involved in the stabilisation process accelerated this phenomenon) which induced a flocculation but the sedimentation was prevented by the pectin network. In fact, there must be a kinetic competition between the depletion which tended to give a flocculation and the formation of the network which prevented this phase separation.
Of course these results are only correct for ADD with 8% of MSNF and without homogenisation of the final drink (i.e. with high dry material and great casein particles). The validity of the results for other conditions of MSNF and homogenisation will be discussed in a future publication.

5. Acknowledgements

The authors gratefully acknowledge the staff of the Research Center of Degussa Texturant Systems S.A.S for their help, especially K. Born, A. Bourdais, A. Critchley, V. Langendorff, and P. Poutrel.

6. References

1) Parker, A., Boulenguer, P. and Kravtchenko, T.P. (1994) Effect of the addition of high methoxyl pectin on the rheology and colloidal stability of acid milk drinks, In K. Nishinari and E. Doi (eds.), *Food hydrocolloids : structure, properties and functions*, Plenum Press, New York, pp 307-312.
2) Kravtchenko, T.P., Parker, A. and Trespoey, A. (1995) Colloidal stability and sedimentation of pectin-stabilised acid milk drinks, In E. Dickinson and D. Lorient (eds.), *Food Macromolecules and Colloids,* The Royal Society of Chemistry, Cambridge, pp 349-355.
3) Glahn, P.E. (1982) Hydrocolloid stabilisation of protein suspensions at low pH. In G.O. Phillips, D.J. Wedlock and P.A. Williams (eds.), *Progress in Food and Nutrition Science,* **6** Gums and Stabilisers for the Food Industry, *Interaction of Hydrocolloids*, Pergamon Press, Oxford, pp 171-177.
4) Leskauskaitė, D., Liutkevichius, A. and Valantinaitė, A. (1998) Influence of the level of pectin on the process of protein stabilisation in an acidified milk system, *Milchwissenschaft*, **53** (3), 149-152.
5) Lucey, J.A., Tamenaha, M., Singh, H. and Munro, P.A. (1999) Stability of model acid milk beverage : Effect of pectin concentration, storage temperature and milk heat treatment, *Journal of texture studies,* **30,** 305-318.
6) Qureshi, A.K., Shah, M.A., Mahmood-ul-Hassan and Elahi, M. (1972) Studies on the effect of emulsifiers on milk protein's stability at different pH values, *Science and Industry, Pakistan* **9** (3), 211-215.

7) Shenkenberg, D.R., Chang, J.C. and Edmonson, L.F. (1971) Develops milk orange juice, *Food Eng.*, **25** (4), 97-101.
8) Asai, I., Watari, Y., Iida, H., Masutake, K., Ochi, T., Ohashi, S., Furuta, H.and Maeda, H. (1994) Effect of soluble soybean polysaccharide on dispersion stability of acidified milk protein, In K. Nishinari and E. Doi (eds), *Food hydrocolloids : Structure, properties and functions*, Plenum Press, New York, pp 151-156.
9) Amice-Quemeneur, N. (1995) Etude des interactions caseine-pectine dans les boissons laitières acides, (thesis), Institut National Polytechnique de Lorraine (France).
10) Maeda, H. (2000) Soluble soybean polysaccharide, In: G.O. Phillips and P.A. Williams, Handbook of Hydrocolloids (eds), CRC press, New York, pp 309-320.
11) Hunter, R.J. (1986) *The Foundations of Colloid Science*, **1**, Oxford University Press, Oxford, pp 450-493.
12) Maroziene, A., De Kruif, C.G. (2000) Interaction of pectin and casein micelles, *Food Hydrocolloids*, **14**, 391-394.

RECENT STUDIES ON POSSIBLE FUNCTIONS OF BIOACTIVE PECTINS AND PECTIC POLYSACCHARIDES FROM MEDICINAL HERBS ON HEALTH CARE

H. YAMADA, H. KIYOHARA and T. MATSUMOTO
Kitasato Institute for Life Sciences, Kitasato University and Oriental Medicine Research Center, The Kitasato Institute
5-9-1, Shirokane, Minato-ku, Tokyo 108-8641, Japan
e-mail: yamadaha@kitasato-u.ac.jp

1. Abstract

Many plants have long been used as traditional herbal medicines, and several pharmacological activities have been observed in pectic polysaccharides from medicinal plants. To resolve the biological functions of bioactive pectins and pectic polysaccharides for health care, studies on carbohydrate sequences having pharmacological activities and their modes of actions are required. Present paper describes recent studies on a murine B cell proliferating pectin from *Bupleurum falcatum*, murine intestinal immune system enhancing pectic polysaccharides from *Atractylodes lancea*, and human natural antibody for pectic polysaccharides.

2. Introduction

Over the past 20 years, several pharmacological activities including immunostimulating activity, anti-ulcer activity, anti-metastasis activity, anti-mutagenic activity, hypoglycemic activity and others have been observed in pectic polysaccharides from medicinal plants (Yamada, 1996). Because phytomedicines and functional foods have now become to attract more global attention, scientific elucidation of the role of the polysaccharides as the active ingredient is very important. Recently, pectins have been reported to affect intestinal functions. For example, in animal studies pectins were shown to have immunoregulatory effect in the intestine, and to change the ileal microbial activity and a morphology of small intestinal wall (Lim *et al.*, 1997; Langhout *et al.*, 1999). Recent studies also demonstrated that pectins stimulate intestinal microbes to produce short chain fatty acids, which are involved in several biological actions (Tappenden *et al.*, 1995; Mandal *et al.*, 2001). Our studies indicate that certain pectins and pectic polysaccharides also can be absorbed by the body or affect intestinal immune system directly for expression of pharmacological activities (Sakurai *et al.*, 1996; Yu *et al.*, 2001a, 2001b). To resolve the biological function of bioactive pectins and pectic polysaccharides in human health care, the carbohydrate sequences having pharmacological activities should be identified, and their chemical fine structure and their mode of actions should be elucidated. The present paper deals

F. Voragen et al. (eds.), Advances in Pectin and Pectinase Research, 481-490.
© 2003 *Kluwer Academic Publishers. Printed in the Netherlands.*

with our recent studies of pectins and pectic polysaccharides from medicinal herbs on human health.

3. Mitogenic activity of bioactive pectin from *Bupleurum falcatum* on B lymphocytes

The roots of *Bupleurum falcatum* have been used clinically in Japanese herbal (Kampo) medicines for the treatment of chronic hepatitis, nephrotic syndrome and auto-immune diseases. This medicinal herb contains interesting bioactive pectins (bupleuran 2IIb and 2IIc), which have complement activating activity, anti-ulcer activity, and macrophage Fc receptor up-regulating activity to enhance immune complex clearance (Yamada, 1994; Yamada, 1996; Yamada and Kiyohara, 1999). Bupleuran2IIc, which has a molecular weight of 63,000, is composed of about 86% of α-(1→4)-linked galacturonan region consisting for 70% of galacturonic acid and 30% of carboxymethylated galacturonic acid and branched galacturonic acid at position 2 or 3. Bupleuran 2IIc also contains ramified regions, which consist of a rhamnogalacturonan core with several arabino- and galacto-oligosaccharide side chains attached to positon 4 of 2-linked rhamnosyl residue in the core directly or through 4-linked galacturonic acid (Hirano *et al.*, 1994). Rhamnogalacturonan II (RG-II) like region is also present in bupleuran 2IIc as a minor region. Bupleuran 2IIc also showed potent mitogenic activity against mouse spleen cells and Peyer's patch cells of the small intestine *in vitro*. Citrus pectin, however did not show the activity (Sakurai *et al.*, 1999). Flow cytometric (FCM) analyses indicated that bupleuran 2IIc increased the population of B cells but not T cells. Even when the adherent cells such as macrophages or T cells were removed from the spleen cells, the mitogenic activity of bupleuran 2IIc was similar to the original spleen cells. These results indicate that bupleuran 2IIc expresses its mitogenic activity against B cells independently from adherent cells and T cells. Bupleuran 2IIc proliferated only certain subpopulations of B cells from murine spleen cells. When pectic polysaccharide fraction BR-2 containing bupleuran 2IIc was administrated orally to LPS non-responsive C3H/HeJ mice for 7 consecutive days, proliferative response of spleen cells was enhanced by the presence of bupleuran 2IIc, but another B cell mitogen, LPS did not show this effect *in vivo*.

Endo-α-(1→4)-polygalacturonase digestion of bupleuran 2IIc results galacturonan regions, ramified regions, which contain rhamnogalacturonan core with neutral sugar chains attached, and RG-II like region. Among these fragments only ramified region, PG-1 gave potent mitogenic activity (Fig. 1). Therefore, ramified region may be involved in the activity.

In order to study absorption and tissue distribution of the active polysaccharide after oral administration to mice, anti-polysaccharide antibody (anti-bupleuran 2IIc/PG-1 antibody) against the ramified region containing the active site was prepared (Sakurai *et al.*, 1996). To resolve antigenic epitopes against the anti-polysaccharide antibody, carbohydrate sequence in the ramified region was analysed by using specific carbohydrases such as endo-α-L-arabinofuranosidase, rhamnogalacturonase A, exo-β-(1 →3)-galactanase, endo-β-(1 →6)-galactanase and exo-β-glucuronidase (Sakurai *et al.*, 1998).

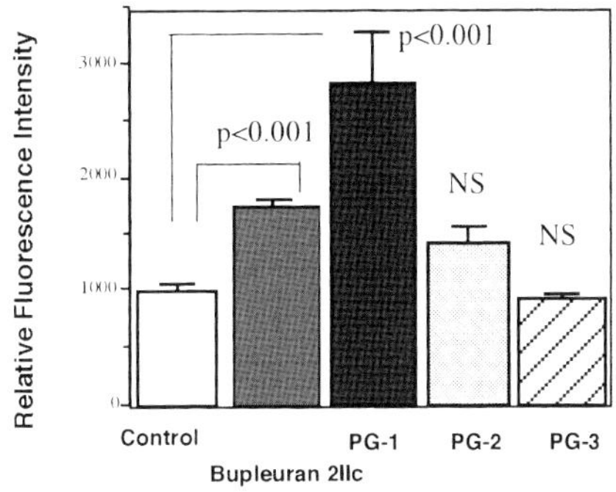

Fig. 1 Mitogenic activity of endo-α-(1→4)-polygalacturonase digested products of bupleuran 2IIc on spleen cells.
PG-1, ramified region (rhamnogalacturonan core attached neutral sugar chains); PG-2, RG-II like region; PG-3, oligogalacturonides

The ramified region consisted of a rhamnogalacturonan core which attached with branched arabinofuranosyl chains and galactosyl chains. In this structure, the antigenic epitope was suggested to be 6-linked β-galactosyl chains containing terminal glucuronic acid and 4-methylglucuronic acid attached to (1 →3)-β-D-galactosyl chains (Sakurai *et al.*, 1998). Because the mitogenic activity of the ramified region, PG-1, was inhibited in the presence of the anti-polysaccharide antibody, the antigenic epitope was also suggested to be involved in the mitogenic activity (Fig. 2) (Sakurai *et al.*, 1999). This result suggests that the receptor molecule, which is able to recognize the ramified region of bupleuran 2IIc, is present on surface of the B cells. Purified B cells were incubated with the ramified region of bupleuran 2IIc at 4 °C for 60 min., then biotynylated anti-bupleuran 2IIc/PG-1 antibody was added to the cells, and the antibody binding cells were detected by the addition of streptavidin-FITC.

Antigenic epitope in bupleuran 2IIc was detected on B cell surface as the capping structure by fluorescene microscopy. This observation suggests that the active carbohydrate segments in bupleuran 2IIc bind to the cell surface receptor on the B cells, and cause capping of the receptor molecule. LPS is known to stimulate Toll-like receptor, TLR4, for further signal transduction, and to activate transcription factor, NF-κB. Therefore we tested whether bupleuran 2IIc affects the LPS receptor or not by using LPS receptor gene transformed cells which were also transformed in the luciferase gene in the down stream of NF-κB binding DNA. As a result, LPS enhanced luciferase activity through LPS receptor, but bupleuran 2IIc did not show significant stimulation suggesting that another receptor molecule is involved in the stimulation of B cells by bupleuran 2IIc. When the membrane fraction from B cell lysate was incubated with ramified region of bupleuran 2IIc, and anti-bupleuran 2IIc/PG-1 antibody was added to the mixture, an immunoprecipitation complex was formed after further addition of Protein G-Sepharose. When this precipitate was analyzed by SDS-PAGE, a protein band which was not observed in untreated cell lysate, was detected in bupleuran 2IIc-treated cell lysate. This result suggests that bupleuran 2IIc/PG-1 binding protein is present on the B cell membrane as candidate for receptor protein.

When pectic polysaccharide fraction BR-2, which contains bupleuran 2IIc, was administrated orally to the mice, the ramified region of the polysaccharide was detected in the T cell area of follicles in Peyer's patches by immunohistochemical staining used anti-polysaccharide antibody (Sakurai *et al.*, 1996). The ramified region of bupleuran 2IIc was also detected in the liver homogenate one week after oral administration by sandwich ELISA using the same antibody. These results suggest that when bupleuran 2IIc was administrated orally, at least a part of bupleuran 2IIc can be absorbed into the body but a part of bupleuran 2IIc affects gastric mucosal immune system through Peyer's patches.

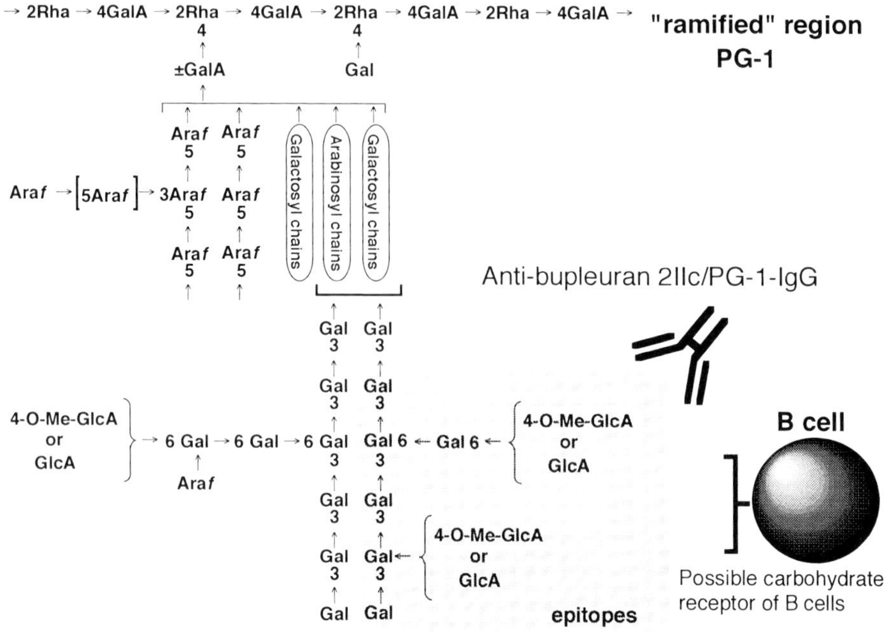

Fig. 2 Proposed structure of the antigenic epitopes in the ramified region of bupleuran 2IIc for anti-bupleuran 2IIc/PG-1-IgG and possible active site for the mitogenic activity.

Mammalian cell cycle begins with the G_0/G_1 phase, and enters to S, G_2 and M phases, and this cycle is repeated during the process of cell division and cell growth. The cell cycle is known to be regulated by several regulatory proteins such as cyclin, cyclin dependent kinase and so on. Flow cytometric (FCM) analysis indicated that bupleuran 2IIc increased the populations of murine B cells entering S and G2/M phases. Bupleuran 2IIc induced increases of cyclin D, cyclin dependent kinase, CDK4 and 6, and phosphorylation of protein Rb in G_0/G_1 phase, followed by induction of cyclin E, CDK2 and finally cyclin A in S phase. Bupleuran 2IIc also induced increase of cyclin B in G_2/M phase. These results indicate that these cell cycle regulatory proteins might be involved in the mitogenic activity of bupleuran 2IIc. This is the first evidence that these regulatory proteins are involved in the action of a mitogenic polysaccharide. Possible signal transduction pathway in bupleuran 2IIc-stimulated B cells were studied by using metabolic inhibitors and direct analysis of each enzyme activity. The results suggest that bupleuran 2IIc enhanced cell cycle regulatory proteins through induction of tyrosin kinase, PI-phospholipase C and protein kinase C for the proliferation of B cells after the recognition of certain receptor molecule for bupleuran 2IIc.

When purified B cells from LPS non-responsive C3H He/J mice or LPS-responsive C3H/HeN mice were cultured with or without bupleuran 2IIc, bupleuran 2IIc increased the IgM secretion of purified B cells from both mice to the same extent (Guo *et al.*, 2000). These results indicate that the enhancement of IgM secretion by bupleuran 2IIc is dependent on B cells but not on T cells and macrophages. The results also suggest that enhanced IgM production is not due to the contamination of LPS. Because IL-6 is one of the major factors for the terminal differentiation of activated B cells into Ig secreting cells and enhances Ig secretion, IL-6 is possibly involved in the Ig secretion-enhancing activity of B cells by bupleuran 2IIc. The presence of anti-IL-6 neutralizing antibody alone did not affect the spontaneous secretion of IgM from B cells, however the enhancement of IgM secretion by bupleuran 2IIc was reduced in the presence of anti-IL-6 antibody (Fig. 3) (Guo *et al.*, 2000).

The results suggest that IL-6 contributes to the enhancement of IgM secretion from B cells by buplenran 2IIc. When purified B cells from murine spleen were cultured in the presence of bupleuran 2IIc, IL-6 content in the conditioned medium was increased in a time-dependent manner compared with control by ELISA and bioassay for IL-6 (Guo *et al.*, 2000). Bupleuran 2IIc also enhanced transcription of IL-6 mRNA in activated B cells, suggesting that IL-6 is newly produced and secreted from B cells by the stimulation of bupleuran 2IIc. This is the first time that it is shown that pectic polysaccharide enhances IL-6 secretion in B cells. The result of flow cytometric (FCM) analysis indicated that the active site in bupleuran 2IIc affects the differentiation from immature B cells to mature and activated B cells (Guo *et al.*, 2000).

Because bupleuran 2IIc induced IL-6 synthesis of B cells, it was expected that the newly produced IL-6 affected mature B cells, followed by further induction of IgG and IgM antibody forming cells by autocline or paracline mechanism.

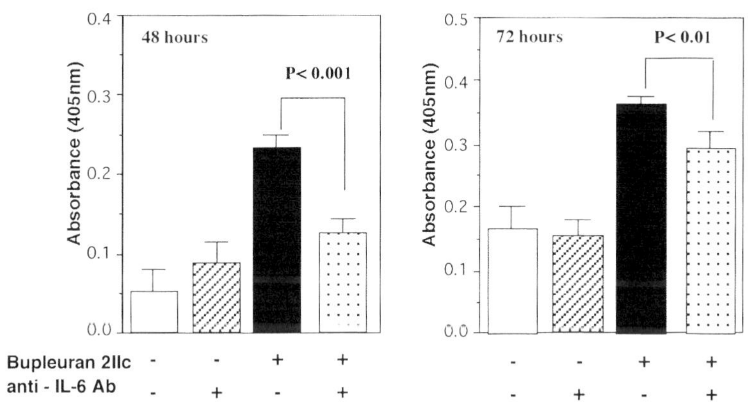

Fig. 3 Effect of anti-IL-6 antibody (Ab) on enhancement of IgM production from bupleuran 2IIc-treated normal B cells.
Purified B cells (2 x 10^6 cells/ml) from C3H/HeJ mice were cultured with bupleuran 2IIc (100 μg/ml), anti-IL-6 Ab (20 μg/ml), anti-IL-6 Ab (20 μg/ml) + bupleuran 2IIc (100 μg/ml) or culture medium at 48 or 72 hr. IgM in the culture supernatants was measured by ELISA. Data were expressed as mean ± s.d. (n=4).

4. Intestinal immune system enhancing activity of bioactive pectic polysaccharides from *Atractylodes lancea*

Most herbal medicines have generally been taken orally, and some active ingredients may be absorbed from intestine to result some pharmacological activities. Mucosal immune system also exists in intestine ("intestinal immune system"), and it plays an important role not only in host defense of intestine but also in regulation of other mucosal immune systems and systmic immune system. Therefore there is a_possibility that active ingredients in herbal medicines may be show pharmacological activities through regulation of intestinal immune system without their direct absorptions from intestine. Hong *et al.* have been reported that one of herbal formulations containing *Atractylodes lancea* affected functions of immune cells from intestinal immune system (1998), and we found three active polysaccharides from rhizomes of *A. lancea* which modulate functions of immune cells in Peyer's patches belonging to intestinal immune system (intestinal immune system modulating activity) (Yu *et al.*, 1998). This intestinal immune system modulating activity was measured by Peyer's patch cells-mediated proliferative response of bone marrow cells (Hong *et al.*, 1998). One of active polysaccharides, arabino-3,6-galactan ALR-5IIa-1-1 was purified from the rhizomes of *A. lancea*, and was analyzed by using specific enzymes (Yu *et al.*, 2001a). ALR5IIa-1-1 has a β-(1→3)-galactan backbone with some inserted α-L-arabinofuranosyl residues, and substutited by several β-(1→6)-galactosyl oligosaccharides which were partly substituted with α-L-arabinofuranosyl residues. Most of the activity appeared in the non-reducing terminal side. *A. lancea* also contained two other pectic polysaccharides having the same intestinal immune system enhancing activity (Yu *et al.*, 2001b). When one of these polysaccharides, ALR-b was digested with endo-polygalacturonase and the digest was fractionated by gel filtration, only an intermediate size-region showed a potent activity among the fragments. Because ALR-b contained several unusual sugars which were observed in rhamnogalacturonan II (RG-II), the active region was suggested to be RG-II. Another active polysaccharide, ALR-a-Bb was also characterized as RG-II-like polysaccharide. These results indicate that RG-II moieties are involved in intestinal immune system enhancing activity. Previously we have found a bioactive RG-II in the leaves of *Panax ginseng*, which has both immune complex clearance enhancing activity and IL-6 production enhancing activity to macrophages (Shin *et al.*, 1997). Ginseng RG-II consisted of typical structural units such as AceA-containing oligosaccharide, uronic acid-rich octasaccharide chains and the disaccharides like Rha→Kdo and Ara*f*→DHA, which were attached to an oligogalacturonide core as similar as sycamore RG-II (O'Neill *et al.*, 1996). But, AceA-containing oligosaccharide chain of RG-II from *P. ginseng* was found to comprise more one rhamnosyl residue as a side chain to make AceA-containing oligosaccharide as nonasaccharide chain (Shin *et al.*, 1997). Several research groups have reported that RG-II molecules are able to form dimers by cross linking with borate ions (O'Neill *et al.*, 1996; Kobayashi *et al.*, 1996; Ishii *et al.*, 1996; Kaneko *et al.*, 1997; Shin *et al.*, 1998). Therefore the intestinal immune system enhancing activity of the dimers was studied. The RG-II like region from ALR-b of *A. lancea* as well as the dimerized RG-II from ginseng leaves showed similar activity, but dimerized RG-II from bupleuran 2IIb of *Bupleurum falcatum* had no activity. These results suggest that RG-II has some microheterogeneity, and certain carbohydrate sequence may be involved in the activity. We are now looking for essential active carbohydrate sequence in RG-II.

5. Pectic polysaccharide-reacting natural antibody in human serum

Complement system is known to exist in blood stream, and consists of over 20 serum proteins including 9 complement components such as C1 to C9 and their regulatory proteins such as factors H and I. These complement proteins can be activated through three cascade pathways named as classical, alternative and lectin pathways (Yamada and Kiyohara, 1999). It has been clarified that several complement activating pectic polysaccharides show their activity through both classical and alternative pathways (Yamada and Kiyohara, 1999). The complement system is essential for the operation of the innate as well as the adaptive immune defence. By the activation of the complement system with pectic polysaccharides, the fragments such as C3a and C5a are produced from C3 and C5, and these fragments are believed to modulate the immune system. The classical pathway is an antibody dependent cascade triggered by antigen-antibody immune complexes. Whereas alternative and lectin pathways are triggered by direct interaction of surface materials of antigens with C3 and serum lectin such as RaRF, respectively (Yamada and Kiyohara, 1999). Analysis of serum components for expression of the activity of complement activating pectic polysaccharide (AAF-IIb-2) isolated from *Artemisia princeps* PAMP led to the postulation that IgG molecules are required for activation of C1, C4 and further complement pathway by pectic polysaccharides (Yamada *et al.*, 1991). This arizes the possibility of the presence of a natural antibody, which can react with complement activating pectic polysaccharides in normal human serum. Five pharmacologically active pectic polysaccharides purified from *Artemisia princeps, Glycyrrhiza uralensis, Angelica acutiloba* and *Bupleurum falcatum* each were coated on ELISA plate (Yamada *et al.*, 1987; Zhao *et al.*, 1991; Yamada, 1994; Yamada, 1996; Yamada and Kiyohara, 1999). When human serum was incubated with the coated polysaccharides, the reacting antibodies were detected by enzyme-labelled anti-immunoglobulins (Igs). Normal human serum contained natural IgG which can react with all active pectic polysaccharides tested, and the degree of the reactivity seemed to be dependent on the carbohydrate structures of the pectic polysaccharides. The serum also contained IgM and IgA classes of antibodies reacting with these pectic polysaccharides. When the normal human serum was tested for reactivity with various kinds of polysaccharides from other origins such as foods, pectins from apple and citrus fruit reacted with only IgA. However, gum arabic and soy bean arabinogalactan mainly reacted with only IgM whereas beet arabinan and yeast mannan mainly with IgG (Fig. 4). Human serum also contained IgA class antibody which reacted with cirus pectin and polygalacturonic acid, suggesting that normal human serum contains several kinds of the polysaccharide-reacting Igs. When the natural IgG class antibody against pectin from *Bupleurum falcatum*, bupleuran 2IIc, was purified from normal human serum by Protein G-Sepharose and bupleuran 2IIc-immobilized-Sepharose, the purified IgG reacted with only bupleuran 2IIc among the polysaccharides listed in Fig. 4. Among the structural units in the pectic polysaccharides, the "ramified" region has been indicated to play a role as active site for expression of complement activating activity (Yamada and

Kiyohara, 1999). Only the "ramified" region was also shown to react with pectic polysaccharide-reacting antibody. The statistically significant positive correlation between complement activating activity and antibody titer was observed by simple linear regression analysis. Therefore it is suggested that the pectic polysaccharide-reacting antibody, especially IgM and IgG classes, contribute to the expression of the complement activating activity of the pectic polysaccharides.

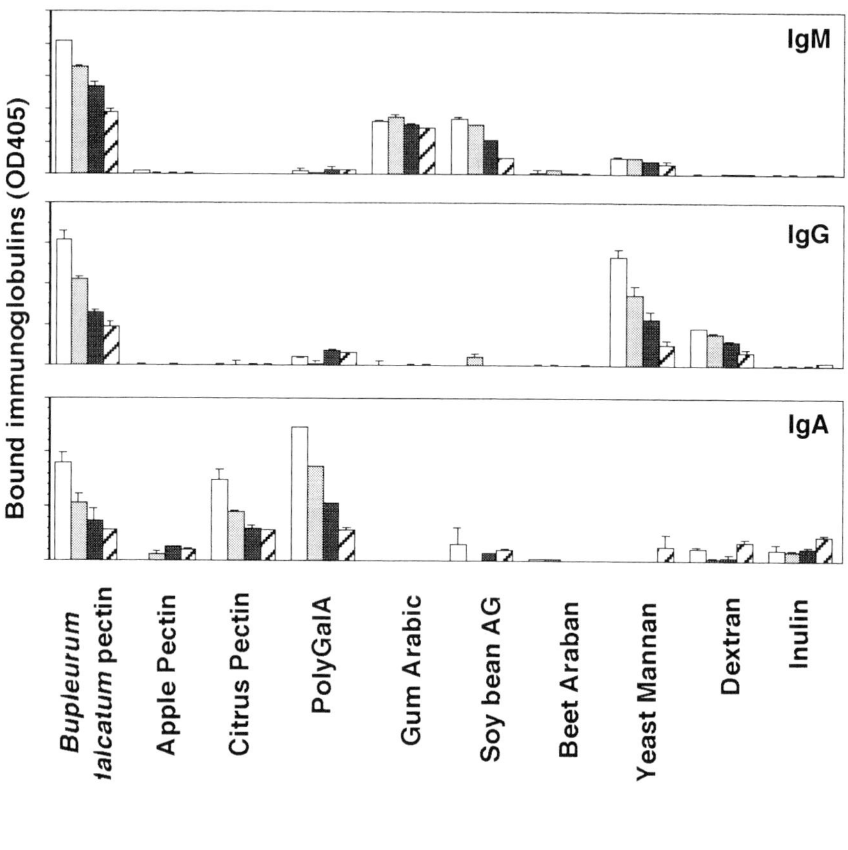

Fig. 4 Presence of natural antibodies against different polysaccharides in human serum. Polysaccharides were coated on microtiter plate at amounts of 1000 (□), 500 (□), 250 (■) and 125 ng polysaccharide/well (▨). Dilutions of normal human serum (1: 200) were reacted with the coated polysaccharides, and the bound antibodies were detected with alkaline phosphatase-labelled anti-immunoglobulins (IgM, IgG and IgA). Enzymic activity were measured by using *p*-nitrophenyphosphate. PolyGalA, polygalacturonic acid; Soy bean AG, soy bean arabinogalactan.

5. Conclusion

The present study indicates that ramified regionof pectins are involved in several pharmacological activities, and that certain RG-IIs are involved in immunostimulating activity. Both regions have microheterogeneity and certain carbohydrate sequences may involve in the activities. The results also strongly suggest that bioactive pectins and pectic polysaccharides are absorbed directly by the body or affect intestinal immune system through some receptor molecule to express pharmacological activity. Natural antibodies, which recognize pectic polysaccharides, are suggested to be involved in the expression of the activity. Therefore elucidation of active carbohydrate sequences and their mode of action is very important in order to understand pharmacological activity of bioactive pectins and pectic polysaccharides.

7. References

Guo, Y-J., Matsumoto, T., Kikuchi, Y., Ikejima, T., Wang, B-X. and Yamada, H. (2000) Effects of a pectic polysaccharide from a medicinal herb, the roots of *Bupleurum falcatum* L. on interleukin 6 production of murine B cells and B cell line, Immunopharmacol., **49**, 307-316.

Hirano, M., Kiyohara, H., Matsumoto, T. and Yamada, H. (1994) Structural studies of endo-polygalacturonase-resistant fragments in anti-ulcer pectin from the roots of *Bupleurum falcatum* L., Carbohydr. Res., **251**, 145-162.

Hong, T., Matsumoto, T., Kiyohara, H. and Yamada, H. (1998) Enhanced production of hematopoietic growth factors through T cell activation in Peyer's patches by oral administration of Kampo (Japanese herbas) medicines, 'Juzen-taiho-to', Phytomedicine, **5**, 353-360.

Ishii, T. and Matsunaga, T (1996) Isolation and characterization of a boron-rhamnogalacturonan-II complex from cell walls of sugar beet pulp, Carbohydr. Res., **284**, 1-9.

Kaneko, S., Ishi, T. and Matsunaga, T. (1997) A boron-rhamnogalacturonan-II complex from bamboo shoot cell walls, Phytochem., **44**, 243-248.

Kobayashi, M., Matoh, T. and Azuma, J. (1996) Two chains of rhamnogalacturonan II are cross-linked by borate-diol ester bonds in higher plant cell walls. Plant Physiol., **110**, 1017-1020.

Langhout, D. J., Schutte, J. B., Van Leeuwen, P., Wiebenga, J. and Tamminga S. (1999) Effect of dietary high- and low-methylated citrus pectin on the activity of the ileal microflora and morphology of the small intestinal wall of broiler chicks, British Poultry Sci., **40**, 340-347.

Lim, B. O., Yamada, K., Nonaka, M., Kuramoto, Y., Hung, P. and Sugano, M. (1997) Dietary fibers modulate indices of intestinal immune function in rats, J. Nutr., **127**, 663-667.

Mandal, M., Olson, D. J., Sharma, T., Vadlamudi, R. K. and Kumar, R. (2001) Butyric acid induces apoptosis by upregulating Bax expression via stimulation of the c-Jun N-terminal kinase/activation protein-1 pathway in human colon cancer cells, Gastroenterology, **120**, 71-78.

O'Neill, M. A., Warrenfeltz, D., Kates, K., Pellerin, P., Doco, T., Darvill, A. G. and Albersheim, P. (1996) Rhamnogalacturonan-II, a pectic polysaccharide in the walls of growing plant cell, forms a dimer that is covalently cross-linked by a borate ester. *In vitro* conditions for the formation and hydrolysis of the dimer., J. Biol. Chem., **271**, 22923-22930.

Sakurai, M. H., Matsumoto, T., Kiyohara, H. and Yamada, H. (1996) Detection and tissue distribution of anti-ulcer polysaccharides from *Bupleurum falcatum* L. by polyclonal antibody, Planta Med., **62**, 341-346.

Sakurai, M. H., Kiyohara, H., Matsumoto, T., Tsumuraya, Y., Hashimoto, Y. and Yamada, H. (1998) Characterization of antigenic epitopes in anti-ulcer pectic polysaccharides from *Bupleurum falcatum* L. using several carbohydrases, Carbohydr. Res., **311**, 218-229.

Sakurai, M. H., Matsumoto, T., Kiyohara, H. and Yamada, H. (1999) B cell proliferation activity of pectic polysaccharide from a medicinal herb, the roots of *Bupleurum falcatum* L. and its structural requirement, Immunology, **97**, 540-547.

Shin, K-S., Kiyohara, H., Matsumoto, T. and Yamada, H. (1997) Rhamnogalacturonan II from the leaves of *Panax ginseng* C. A. Meyer as a macrophage Fc receptor expression-enhancing polysaccharide, Carbohydr. Res., **300**, 239-249.

Shin, K-S., Kiyohara, H., Matsumoto, T. and Yamada, H. (1998) Rhamnogalacturonan II dimers cross-linked by borate diesters from the leaves of *Panax ginseng* C. A. Meyer are responsible for expression of their IL-6 production enhancing activities, Carbohydr. Res., **307**, 97-106.

Tappenden, K. A., Pratt, V. C., Goruk, S. D., Field, C. J. and McBurney, M. C. (1995) Short chain fatty acid (SCFA) supplementation of total parenteral nutrition (TPN) improves whole body response to intestinal resection, FASEB J. 9, 862A.

Yamada, H., Kiyohara H. and Otsuka, Y. (1987) Further structural studies of an anti-complementary acidic heteroglycan from the leaves of *Artemisia princeps*, Carbohydr. Res., **170**, 181-191.

Yamada, H., Nagai, T., Cyong J-C. and Otsuka, Y. (1991) Mode of complement activation by acidic heteroglycans from the leaves of *Artemisia princeps* PAMP, Chem. Pharm. Bull., **39**, 2077-2081.

Yamada, H., (1994) Pectic polysaccharides from Chinese herbs: Structure and biological activity, Carbohydr. Polymers, **25**, 269-276.

Yamada, H., (1996) Contribution of pectins on health care, in J. Visser and A.G.J. Voragen (eds) Pectin and Pectinase, Progress in Biotechnology, Elsevier, Amsterdam., Vol. 4, pp. 173-190.

Yamada, H. and Kiyohara, H. (1999) Complement activating polysaccharides from medicinal herbs, in H. Wagner (eds), Immunomodulatory Agents from Plants, Birkhäuser Publishing Ltd., Basel., pp. 161-202.

Yu, K-W., Kiyohara, H., Matsumoto, T., Yang, H-C. and Yamada, H. (1998) Intestinal immune system modulating polysaccharides from rhizomes of *Atractylodes lancea*, Planta Med., **64**, 714-719.

Yu, K.-W., Kiyohara, H., Matsumoto, T., Yang, H.-C. and Yamada, H. (2001a) Structural characterization of intestinal immune system modulating new arabino-3, 6-galactan from rhizomes of *Atractylodes lancea* DC., Carbohydr. Polymer, **46**, 147-156.

Yu, K.-W., Kiyohara, H., Matsumoto, T., Yang, H.-C. and Yamada, H. (2001b) Characterization of pectic polysaccharides having intestinal immune system modulating activity from rhizomes of *Atractylodes lancea* DC., Carbohydr. Polymer, **46**, 125-134.

Zhao, J-F., Kiyohara, H., Yamada, H., Takemoto N. and Kawamura, H. (1991) Heterogeneity and characterization of mitogenic and anti-complementary pectic polysaccharides from the roots of *Glycyrrhiza uralensis* Fisch. *et* DC,Carbohydr. Res., **219**, 149-172.

EFFECT OF PECTOLYTIC AND CELLULOLYTIC ENZYME TREATMENTS ON FUNCTIONAL AND NUTRITIONAL PROPERTIES OF CELL WALL MATERIALS FROM APPLES

G. DONGOWSKI[1], S. FÖRSTER[2], H. KUNZEK[2]
[1]*German Institute of Human Nutrition, Department of Food Science and Preventive Nutrition, Arthur-Scheunert-Allee 114-116, 14558 Potsdam-Rehbrücke, Germany, Dongo@www.dife.de*
[2]*Technical University Berlin, Institute of Food Technology, Food Functionality, Königin-Luise-Strasse 22, 14195 Berlin, Germany*

Abstract

Cell wall materials prepared from apple parenchyma tissue by treatment with commercial pectolytic and cellulolytic enzyme preparations were characterised in model experiments. During maceration and mash fermentation, apple tissue was decomposed mainly into single cells whereas enzyme action under conditions of liquefaction led to cell fragments. The macrostructure of the plant material was preserved after maceration and mash fermentation due to water–ethanol exchange as pre-treatment for drying. The porosity decreased and the solid density increased with advancing enzymatic degradation of the cell walls. This increasing degradation decreased the galacturonan content and the degree of methylation of the pectin component. It was possible to remove pectin from the cell wall matrix completely under optimal conditions of the enzymatic liquefaction process. Cell wall material prepared by maceration showed high water binding and excellent rheological properties, which were however reduced with increasing enzymatic degradation of the materials. The thermal stability of the pectin matrix as well as the cellulose network were nearly independent on a previous enzyme treatment. The advancing enzymatic degradation, however, results in an decreasing relative weight loss of the pectin matrix (first degradation step) and an increasing weight loss of the cellulose network (second degradation step) during thermogravimetry. During *in-vitro* fermentation of the cell wall materials with human faeces flora a spectrum of short-chain fatty acids (SCFA) was formed. With decreasing amounts of pectin in the cell wall materials, the total concentration of SCFA decreased whereas the portion of butyrate increased. Cell wall materials prepared from parenchyma tissue by specific enzymatic treatments have beneficial functional properties and are dietary fibre materials with interesting physiological effects.

1. Introduction

Cell wall materials (CWM) can be regarded as a model substrate for the dispersed phase of crushed fruit or vegetable products [1]. During processing, CWM undergo

F. Voragen et al. (eds.), Advances in Pectin and Pectinase Research, 491-504.
© 2003 *Kluwer Academic Publishers. Printed in the Netherlands.*

changes of their composition and structure, as well as structure-dependent changes in their physico-chemical and physiological properties [1, 2]. Besides the determination of bulk density solid density, porosity and particle size distribution, the physico-chemical properties of CWM can be evaluated by characterisation of their hydration properties, the rheological behaviour of their water suspensions and by their thermal properties [1]. Especially the thermal analysis of CWM preparations can be helpful to describe the process-dependent state transitions of the pectin matrix as well as the cellulose network of these materials [3]. A treatment of apple parenchyma with commercial pectolytic enzyme preparations results in a solubilisation of pectins and hemicelluloses, respectively, a degradation of the cell wall structure and a decrease of the porosity and of the water binding capacity of the degraded CWM [1, 4]. However, there is not much known about the rheological properties of the suspensions formed from enzymatically degraded cell wall preparations and about the state transitions occurring during processing.

Besides their functional properties, CWM have physiological importance as dietary fibre (DF) rich materials. Most of DF are cell wall polysaccharides which are resistant to digestion by the alimentary enzymes of humans. They may interact with steroids and drugs, act on glycaemic response or on absorption of nutrients in the upper parts of intestinal tract [5-8]. In large intestine, DF regulates water binding and transit time and they are the main substrate for the intestinal microflora. Most important end-products of the fermentation are the short-chain fatty acids (SCFA). Butyrate plays an physiologically outstanding role due to its importance as main energy source for the colonic epithelium, regulation effects in cell cycle and protection against colon cancer [9, 10].

In this study, the aim was to modify apple CWM by treatment with commercial pectolytic and cellulolytic enzymes. Further objective was to evaluate how the increasing enzymatic degradation influences the composition, structure, state of matter and the properties of the modified CWM. Moreover, tests should be carried out to see if there exists essential dependencies between the enzymatic treatments and the changes of structure and state transitions, correspondingly, as well as between physico-chemical and physiological properties.

2. Material and Methods

2.1. MATERIALS

All investigated materials were prepared from apples cultivar "Granny Smith" (South Africa). The preparation (Fig. 1) was similar to the procedures for CWM previously described [4, 11, 12]. Basic material (BM) was obtained according to Müller *et al.* [11]. For the preparation of macerated material (MA) [4, 12] 300 mg/l Rohament PL (Röhm Enzyme GmbH, Darmstadt, Germany), for mash fermented material (MF) [12] 62.5 mg/l Rohapect D5L special (Röhm) and for total liquefied material (TL) [4, 12] a mixture of 100 mg/l Rohapect B1L (Röhm), 100 mg/l Rohalase 7069 (Röhm) and 200 mg/l Rohapect AP1 (Röhm) were used. The application of the enzyme preparation was carried out as recommended by the manufacturer.

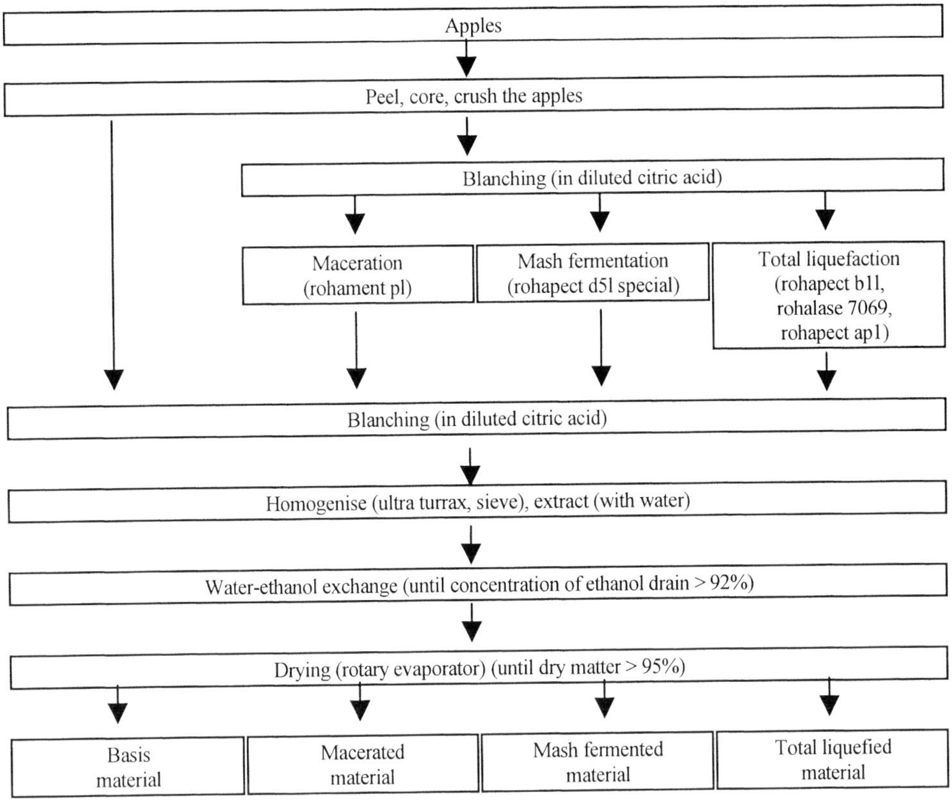

Figure 1. Preparation scheme of apple cell wall materials

2.2. METHODS

The galacturonan content was determined photometrically with *m*-hydroxybiphenyl [13]. The degree of methylation was analysed by titration of carboxyl groups before and after de-esterification [14]. Protein was determined by a Kjeldahl method (N × 6.25). The glucose content was estimated after hydrolysis of 250 mg CWM first in 72 % H_2SO_4 at 20 °C under stirring for 2 h, followed by refluxing in 1 M H_2SO_4 for 2 h, neutralisation and using the enzyme kit from Boehringer (Mannheim, Germany).

The bulk density was measured as previously described [15]. A multivolume pycnometer 1305 (Micrometrics Instruments Corp., U.S.A.) was used to determine the solid density [11]. The porosity was calculated according to Kabbert *et al.* [15]. The particle size distribution was determined with a Galai cis-1 particle size analyser (L.O.T.-Oriel GmbH, Germany) [12, 16].

The swelling of the CWM in water was measured as previously described [4, 17] (0.1 g sample in 25 ml water at 20 °C for 96 h; result expressed in ml hydrated sample/g dry matter). The water retention capacity [18] was examined by a filtration method according to Kunzek *et al.* [19]. The water retention capacity was determined after keeping the sample in a water excess at 20 °C for 2 h and removing the surplus water by the influence of gravity (result in g water/g sample dry matter). The water uptake of the CWM was determined using a modified Baumann apparatus [20, 21] (10 mg sample; glass filter G 2).

The CWM-water suspensions were examined rheologically using an oscillation method (Universal Dynamic Spectrometer UDS 200, Paar Physica, Stuttgart, Germany) as described in [22] (33.3 g sample dry matter/l distilled water).

Simultaneous thermal analysis (thermogravimetry and differential scanning calorimetry) was carried out using a STA 409 C-device (Netzsch, Selb, Germany) and 85 µl crucible without lid, without DSC reference material. The sample weight ranged from 10 to 17 mg. The temperature program consisted of a linear heating rate of 5 K/min from 20 to 450 °C under dynamic nitrogen atmosphere (flow rate 100 ml/min). Temperature calibration of the thermal analyses system was carried out following the recommendations of the Gesellschaft für Thermische Analyse (Society of Thermal Analysis, GEFTA) [23, 24].

For *in-vitro* fermentation, 15 mg of the CWM were incubated anaerobically with 1 ml of suspension from fresh human faeces flora (15 g/60 ml 0.1 M phosphate buffer; pH 6.5) at 37 °C. At different times up to 24 h samples were taken under sterile conditions. SCFA were determined in the samples using a modified GLC method [25, 26] (Hewlett-Packard 5890A Series II with a capillary column HP-FFAP (30 m × 0.53 mm; 1µm) HP 7673 GC/SFC Injector; HP GC Auto Sampler Controller; Flame Ionisation Detector; Software - HP Chemstation; He as carrier gas).

All preparations and measurements were made at least in duplicate. The mean values, standard deviations (SD) and sample size (n) are reported. Statistically significant differences were given with a 95 % level of significance ($P \leq 0.05$) using the Student's *t*-test.

3. Results and Discussion

3.1. STRUCTURE, SIZE AND COMPOSITION

The chemical parameters of the CWM depend on the degree of cell wall degradation caused by processing steps, especially the enzymatic treatments (Table 1). The increased cell wall degradation in the order BM ~ MA, MF and TL was shown by the decrease of the galacturonan content and the increase of the glucose content as well as increasing solid densities and reduced porosities (Table 1). BM and MA mainly consisted of single cells (scanning electron microscopy, not shown). The single cells were formed by a maceration step caused by heating and mechanical stress on the one hand (BM) [11] and by enzymatic treatment on the other hand (MA) [4, 12]. Because of the application of mash fermentation enzymes, the pectin content in the MF was reduced and the portion of single cells was slightly decreased in comparison to BM and MA (Table 1). Furthermore, in some cases the single cells of MF were partly destroyed at the edges (scanning electron microscopy, not shown).

Table 1. Composition and structural parameters of the cell wall materials

Parameter	Bm	Ma	Mf	Tl
Galacturonan content (%)	16.57±0.32	13.57±0.41	5.26±0.49	0
Degree of methylation (%)	87.72±0.26	85.67±0.45	73.04±0.54	0
Glucose content (%)	42.43±0.06	44.53±0.46	51.90±2.11	66.77±2.76
Protein content (%)	2.62±0.12	3.71±0.14	3.44±0.02	2.54±0.11
Bulk density (g/cm^3)	0.115±0.005	0.080±0.000	0.095±0.005	0.449±0.008
Solid density (g/cm^3)	1.531±0.022	1.524±0.023	1.728±0.049	1.822±0.045
Porosity (%)	92.5±0.5	94.8	94.5±0.6	75.4±0.6
Portion of single cells (wet) (%)	61.63±1.26	63.63±0.01	55.65±0.98	16.79±0.39

Bm, basis material; ma, macerated material; mf, mash fermented material; tl, total liquefied material; Values are mean ± sd, n = 2-9.

The application of total liquefaction enzymes resulted in a drastic structural degradation. So, a pectin-free TL consisting mainly of aggregates of cell wall fragments could be prepared. On the contrary, the application of various combination of liquefaction enzymes for apple mash or apple cell wall material often lead to the formation of preparations containing residues of pectin components [4, 12, 27, 28]. Obviously in the case of the preparation of TL, the high efficiency of the enzymes was caused by a suitable combination of the enzyme preparations as well as probably due to the relatively high degree of maturity of the used raw material (cell wall softening). The enzymatic treatment mainly resulted in the formation of pectin-free cell wall fragments (< 50 μm; portion determined by laser particle analysis 81 %). These fragments, however, aggregated during the drying process forming large particles (> 500 μm; scanning electron microscopy, not shown). Further, it was observed that the aggregates resulting from drying can be re-dissolved in an excess of water by supplying mechanical energy (for instance stirring at room temperature for 15 h) [4].

3.2. HYDRATION PROPERTIES

For the determination of hydration properties of the CWM, different methods were chosen according to the recommendation of EU concerted action group Profibre [18]: the swelling using the bed volume technique [17], the water retention capacity measured by a filtration method [19] and the water uptake determined using the capillary suction method [20, 21]. The hydration properties of the CWM are generally higher than those of apple fibre materials from other research groups [2, 18]. This may be resulted mainly from the water-ethanol exchange carried out during the preparation of the materials (Fig. 1). The water-ethanol exchange can be supposed to reduce the extent of shrinkage and structural collapse during the following drying process [1, 15, 29].

Reduced hydration properties were found in the order of increasing cell wall degradation (Fig. 2). MA showed the highest water binding, independent of the method applied. In comparison to MA, the hydration properties of BM were reduced. It can be assumed, that the maceration process causes a cell wall softening and therefore improved swelling properties [11, 27]. Furthermore, the high porosity and the intact single cell structure can favour the hydration properties, especially for MA but also for BM. A significant loss of galacturonan content and porosity, however, resulted in reduced hydration properties of MF and TL. Especially growing shrinkage tendency, cell collapse and cell aggregation caused strongly reduced water binding properties of TL [4, 27] (Fig. 2). However, it should be considered that a suitable step of dis-aggregation can lead to an important improvement of the hydration properties [6]. Such a step of dis-aggregation of the dried preparations during the rehydration process can be carried out via state transitions caused by the plasticisation effect (water as outstanding plasticiser), the temperature effect and mechanical stress (stirring) [30].

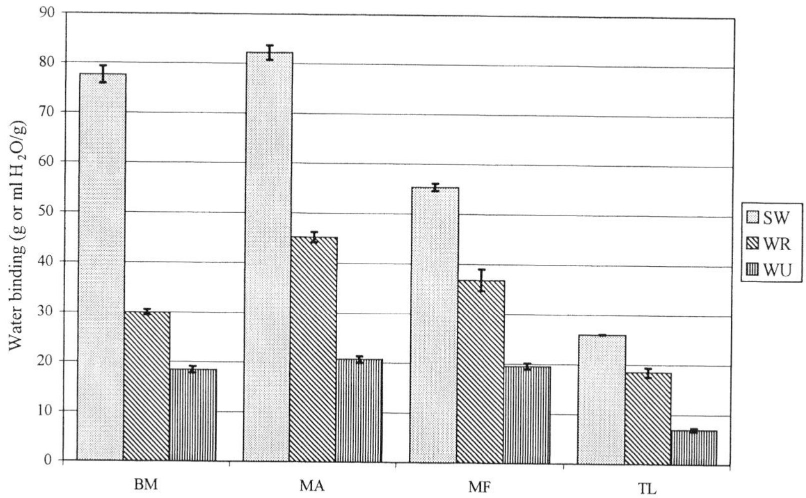

Figure 2. Hydration properties of the cell wall materials
(SW, swelling at 20 °C for 96 h; WR, water retention capacity, filtration method at 20 °C; WU, water uptake, capillary suction method; BM, basis material; MA, macerated material; MF, mash fermented material; TL, total liquefied material).

3.3. RHEOLOGICAL BEHAVIOUR

The rheological investigations of the CWM-water suspensions were carried out under standardised conditions (33.3 g dry matter/l H_2O) using oscillation measurements [22]. The examinations took place within the linear visco-elastic region. So it is possible to obtain useful results without causing disruption to the structure of the suspension. Within the frequency range applied, the storage modulus G' was much greater than the loss modulus G" for all suspensions studied, indicating dominant elastic properties.

It can be assumed that all CWM-water suspensions, prepared before drying process, are concentrated dispersed systems containing stable and strong interactions of the swollen particles (cell cluster, single cells or cell fragments). The G' values as well as the G" values of these suspensions decreased in the order TL > BM ≥ MA ≥ MF (Fig. 3). It is remarkable that the cell fragments of TL can also form such stable suspensions. However, the loss angle tan δ = G"/G' of the TL suspension was higher than those of all other suspensions. Therefore, the TL suspension showed a higher portion of dissipated energy due to viscous flowing (diagram not shown) [4].

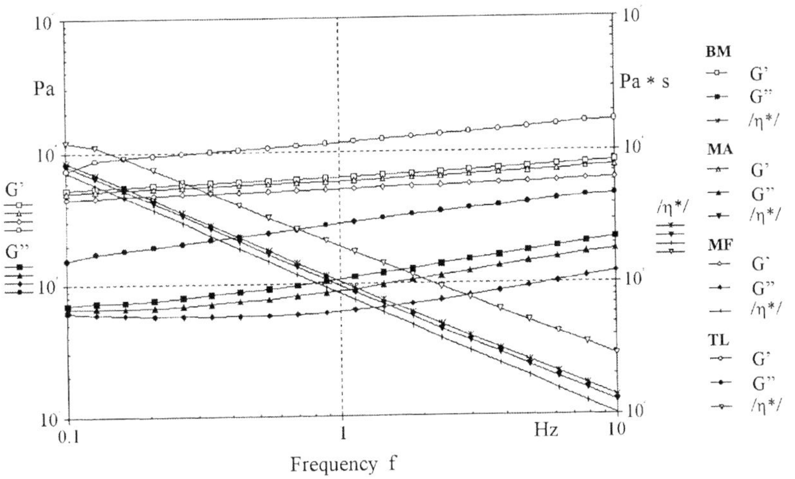

Figure 3. Frequency sweep of CWM-water suspensions before drying
(G', storage modulus; G'', loss modulus, [η*], complex viscosity. Conditions: concentration 33.3 g solid dry matter/l; cylinder measuring system: Z3-DIN, profiled bob; temperature 20 °C. Procedure: shearing 30 s 25 s^{-1}; 2 min rest; frequency sweep f = 10-0.1 Hz; deformation γ 10^{-3}. Further abbreviations see Figure 2).

The rheological investigations of the suspensions, formed by rehydration of dried CWM preparations, showed drastically changed results. The G' and the G'' values of the suspensions decreased in the order MA ≥ MF > BM > TL (Fig. 4). While MA, MF and BM formed stable suspensions, TL suspension doesn't represent a stable dispersion at the applied concentration of 33.3 g dry matter/l H_2O [4]. It can be regarded as a diluted solution of rehydrated aggregates. This thin dispersion is *per se* not measurable using oscillation technique. Nevertheless, the graph was plotted in Figure 4 to illustrate its bad rheological properties compared to those of the other suspensions. In comparison to the TL suspension before drying, the TL suspension of the rehydrated sample has strongly reduced visco-elastic properties. Also in this case, the TL suspension exhibited the highest loss angle tan δ compared with those of the other suspension formed after rehydration. Obviously, the small water binding capacity of the aggregated TL preparation inhibit the interaction between the less swollen particles and so the formation of elastic properties in the suspension. However, if the aggregates are re-dissolved by mechanical stress during rehydration, stable dispersions can be formed [4].

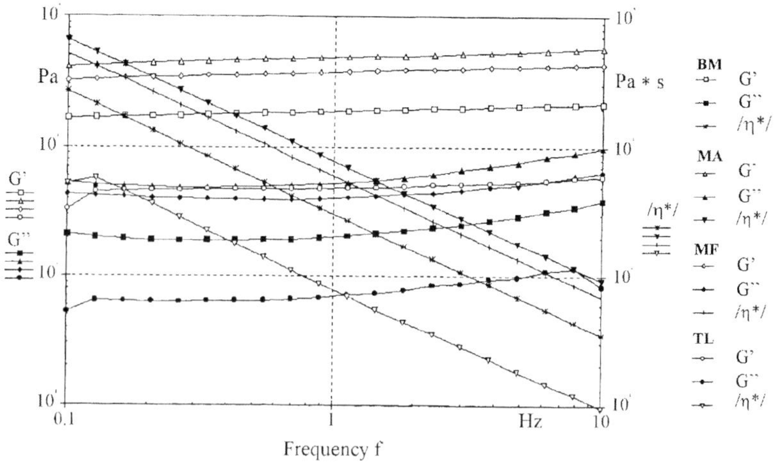

Figure 4. Frequency sweep of CWM-water suspensions prepared by rehydration of dried CWM samples (Conditions see Figure 3. Further abbreviations see Figure 2).

3.4. THERMOANALYTICAL BEHAVIOUR

Simultaneous thermal analysis is a powerful method to evaluate structural changes and state transitions occurring during processing. The interpretation of the thermoanalytical measurements includes the evaluation of the thermogravimetry curve (TG curve), the differential thermogravimetry curve (DTG curve) and the differential scanning calorimetry curve (DSC curve). The TG curve describes the weight loss in dependence on heating and the DTG curve the velocity of the degradation. The temperature T_{exon} characterises the beginning of the thermal degradation and T_{max} gives the point of the maximal degradation velocity. The DSC curve allows to observe changes of the reaction enthalpy during the thermal analysis.

A typical thermogram of a CWM is that of MF presented in Figure 5. The DTG curve clearly shows a two step degradation of the cell wall. In a first step the pectin matrix is degraded. Therefore, the extrapolated onset-temperature of the pectin matrix, T_{exon} (DTG), can be regarded as an indicator of the thermal stability of the CWM. In a second step, the cellulose network is degraded. The weight loss of the pectin degradation Δm_1 and of the cellulose degradation Δm_2 can be defined as the loss between the extrapolated temperature at the beginning of the reaction T_{exon} and the extrapolated temperature at the end of the reaction T_{exoff} in the relevant DTG signal. Furthermore, the difference between T_{exoff} and T_{exon} (DTG signal) is used for the determination of the signal width. The signal width characterises the duration of the thermal degradation of the corresponding matrix.

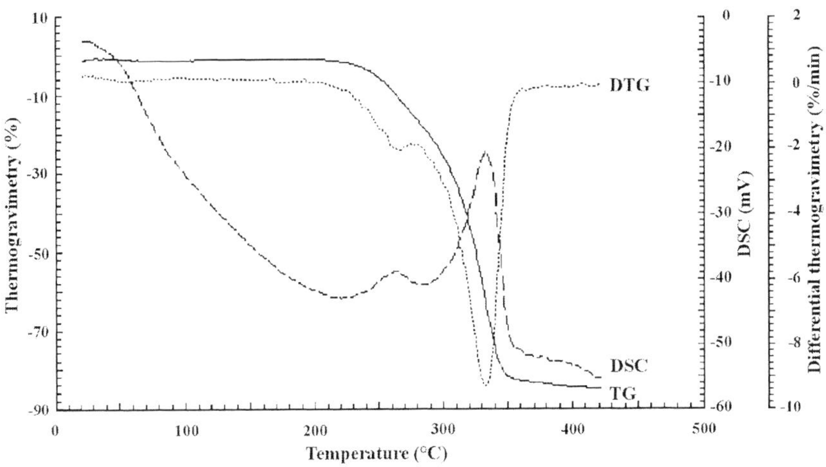

Figure 5. Thermogravimetry (TG), differential TG (DTG) and differential scanning calorimetry (DSC) curves of mash fermented cell wall material, MF (without temperature correction).

The thermal stability is slightly increased in the order BM, MA, MF (Fig. 6a and b; increasing values for T_{exon} pectin matrix and for T_{max} pectin matrix, respectively). The weight loss and the signal width of the pectin matrix, however, are decreased in the same order (Fig. 6c and d). Probably, due to preferable solubilisation of energy-rich components during the enzymatic treatment, a more stable uniform pectin matrix was formed. It can be assumed that such change of the pectin matrix leads to an increasing thermal stability and a faster thermal degradation (reduced signal width of the pectin matrix, Fig. 6d) [3].

In accordance with analytical data (Table 1), the thermogram of TL showed only the signal for the cellulose degradation (thermogram not shown). Therefore the thermal stability of TL was much higher than those of the other CWM.

As expected, the weight loss of the pectin matrix was reduced and that of the cellulose network was growing in the order of increasing enzymatic degradation: BM ~ MA, MF, TL (Figure 6c). The signal widths of the pectin degradation (BM, MA, MF) were markedly higher than those of the cellulose degradation. Also in this case can be assumed that the more uniform matrix was degraded faster.

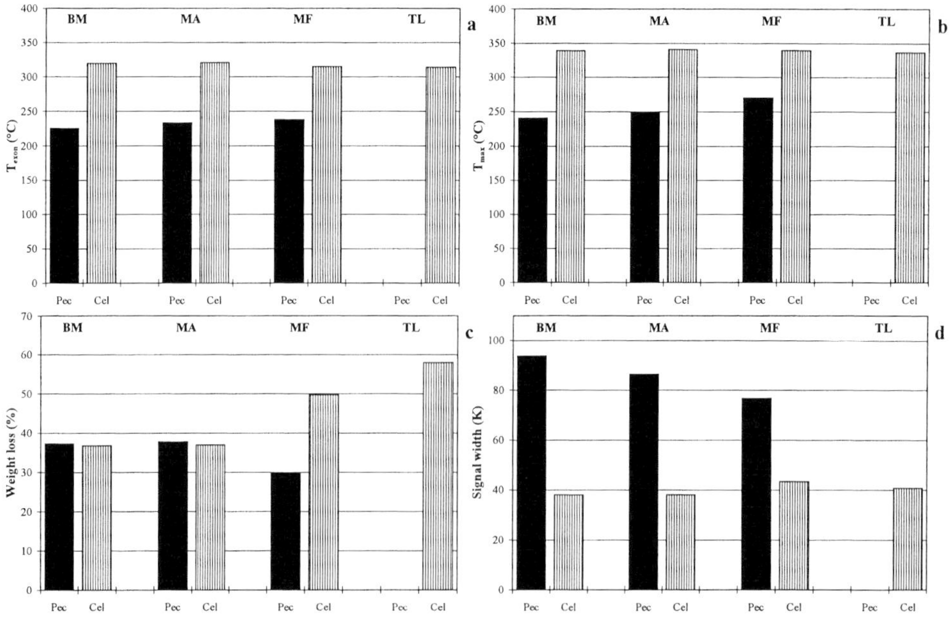

Figure 6. Parameters of the thermal analysis of the cell wall materials (with temperature correction). 6a, T_{exon} (DTG); 6b, T_{max} (DTG); 6c, weight loss; 6d, signal width (Pec, pectin matrix; Cel, cellulose network. Further abbreviations see Figure 2).

3.5. *IN-VITRO* FERMENTATION

Dietary fibres are fermented by the intestinal microflora partially or completely. Plant cell wall polysaccharides are first depolymerised to their monosaccharide units which are then metabolised via the Embden-Meyerhof-Parnas pathway in multi-step procedures to SCFA besides gases, EtOH, etc. [31]. Amount and composition of the SCFA formed in colon are dependent on both the structural features of the polysaccharides and the bacteria involved preferentially in the fermentation. It is well known that faeces flora isolated from different individuals produce a different composition of SCFA. Therefore, *in-vitro* fermentation experiments were made with fresh faeces flora from 3 persons with a distinct different fermentation behaviour.

Formation of SCFA was followed up to 24 h. Under the used conditions, SCFA were produced most intensively within the starting period (up to 4 h); later the reaction velocity decreased as shown in Table 2 for person 1. Because of the higher contents of pectin and hemicelluloses, highest concentrations of SCFA were produced from BM and MA (approximately 1800 μmol/g dry matter after 24 h) during the *in-vitro* fermentation. The formation of SCFA was lower if MF or TL were used as substrate. In these preparations, low amounts of pectin or no pectin were present. In contrast to most of isolated cellulose preparations [32], cellulose was obviously not in the lowest energy state in TL – a result of the intense enzymatic attacks during its preparation. On the other hand, results from indirect calorimetry in rats points to a partial fermentation of cellulose [33, 34].

TABLE 2. Formation of short-chain fatty acids during fermentation of cell wall materials (CWM) with faeces flora (person 1)

CWM	Incubation time (h)	Acetate	Propionate	Butyrate	Sum
		Concentration (μmol/g dry matter)			
BM	4	725.9 ± 76.2	180.8 ± 25.9	209.6 ± 19.5	1116.3
	8	828.1 ± 95.9	260.7 ± 12.4	308.6 ± 36.0	1397.4
	24	1009.8 ± 85.2	347.0 ± 7.0	441.0 ± 7.9	1797.8
MA	4	742.9 ± 145.1	214.7 ± 34.4	231.9 ± 32.2	1189.5
	8	758.6 ± 88.5	253.2 ± 20.9	307.3 ± 29.2	1319.1
	24	1044.4 ± 103.1	338.4 ± 8.2	424.1 ± 20.9	1806.9
MF	4	622.8 ± 81.4	216.3 ± 10.4	194.8 ± 8.4	1033.9
	8	739.0 ± 164.6	272.5 ± 14.7	309.5 ± 7.2	1321.0
	24	890.7 ± 95.4	379.2 ± 14.3	446.1 ± 3.4	1716.0
TL	4	561.7 ± 214.4	185.2 ± 62.9	268.8 ± 81.3	1015.7
	8	511.5 ± 43.9	182.8 ± 9.4	290.4 ± 9.9	984.7
	24	774.6 ± 56.0	293.1 ± 15.1	568.2 ± 26.0	1635.9

BM, basis material; MA, macerated material; MF, mash fermented material; TL, total liquefied material; values are mean ± SD, n = 4.

Acetate was the main SCFA produced during the fermentation experiments with the CWM. In each case, more butyrate was found then propionate. Valeric and *iso*-valeric acid appeared only in low concentrations (not shown). It is interesting that with increasing incubation time both the absolute amount and the percentage of butyrate increased whereas acetate decreased (Fig. 7). No significant changes were found in percentage of propionate. After 24 h, highest percentage of butyrate (approximately 30 mol%) appeared in fermentation experiments with TL. It is known that during fermentation of glucans like resistant starch high amounts of butyrate were produced [35].

Fermentation of apple fibre with human faecal bacteria resulted in an increase of butyrate levels from 9.4% (3 h) up to 12.6% (24 h) [36]. Galacturonic acid and arabinose containing polymers were the most extensively fermented saccharides while xylose and glucose polymers were less degraded. Similar results were reported by Stevens *et al.* [37]. They found that pectins, arabinans and galactans were readily degraded by human microflora while highly branched rhamnogalacturonans, xyloglucans and especially cellulose were fermented later.

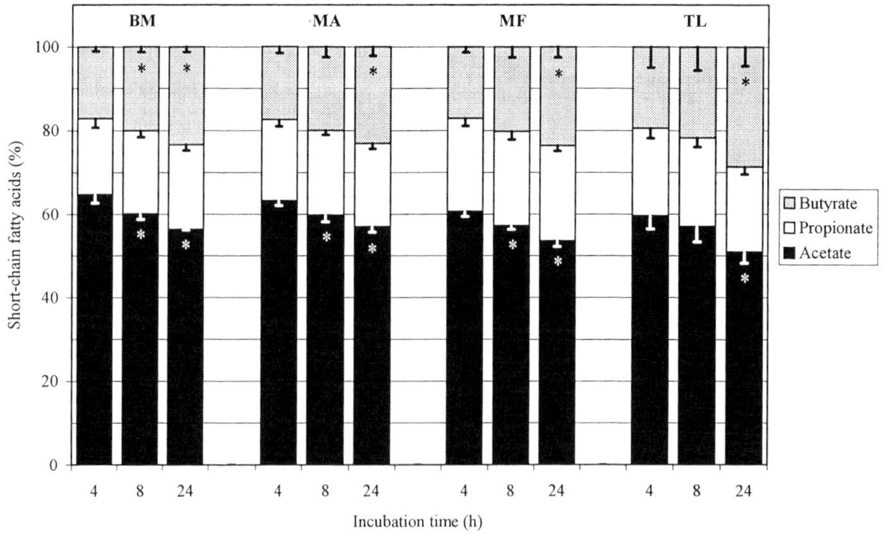

Figure 7. Molar percentage of short-chain fatty acids formed during fermentation of basis material (BM), macerated material (MA), mash fermented material (MF) and total liquefied material (TL) with human faeces flora from 3 persons. Values are mean ± SD; n = 12; * $P \leq 0.05$ (related to 4 h).

Both effects found in *in-vitro* fermentation experiments are physiologically important: the higher concentration of total SCFA formed from pectin-rich CWM and the higher portion of butyrate formed from the cellulose-rich CWM.

4. Conclusions

The results show dependencies between process-dependent changes of composition, structure and state of matter of the cwm and their physico-chemical as well as physiological properties. It can be assumed that the functional properties, such as hydration properties and rheological properties of the cwm-water suspensions as well as the physiological properties, are influenced by the composition (pectin content, cellulose content, protein content, ...), The structure (particle size and distribution, portion of cell clusters, single cells, cell fragments; porosity, ...) And the state of matter (shrinkage, cell wall collapse, ...) On the one hand, and by the application conditions of the materials on the other hand. So an increasing enzymatic degradation of the cwm in the order ma, mf, tl caused a growing loss of pectin, a decreasing portion of single cells, decreasing porosity and advancing aggregation of cell wall fragments as well as reduced water binding properties (tested by standard methods), reduced visco-elastic properties of the suspensions and reduced fermentation rates, but a slight shift of the acetate : propionate : butyrate ratio to a higher butyrate portion. These changes of the properties can be partly explained by state transitions occurring during drying of the cwm preparation. For instance, in the case of the tl preparation, the poor water binding capacity (measured by standard methods) and the limited visco-elastic properties of the suspensions can result from shrinkage, structure collapse and aggregation. However, these state transitions can be reversible if the conditions during the rehydration process are favourable for dis-aggregation.

The results show that the properties of the cwm do not simply correlate with single parameters such as pectin content or porosity. Obviously, the functional properties are influenced multivariately by many material parameters on the one hand and by the application conditions on the other hand. Further investigations are necessary for a better understanding of the physico-chemical and physiological properties of cwm and their interrelations.

5. References

1. Kunzek, H., Kabbert, R., and Gloyna, D. (1999) Aspects of material science in food processing: changes in plant cell walls of fruits and vegetables, *Z. Lebensm. Unters. Forsch. A* 208, 233-250.
2. Guillon, F. and Champ, M. (2000) Structural and physical properties of dietary fibres, and consequences of processing on human physiology, *Food Res. Int.* 33, 233-245.
3. Godeck, R., Kunzek, H., and Kabbert, R. (2001) Thermal analysis of plant cell wall materials depending on the chemical structure and pre-treatment prior to drying, *Eur. Food Res. Technol.* (in press).
4. Vetter, S., Kunzek, H., and Senge, B. (2001) The influence of the pre-treatment of apple cell wall samples on their functional properties, *Eur. Food Res. Technol.* 212, 630-635.
5. McCleary, B.V. and Prosky, L. (eds.) (2001) *Advanced Dietary Fibre Technology*, Blackwell Science, Oxford.
6. Kritchevsky, D. and Bonfield, C. (eds.) (1995) *Dietary Fiber in Health and Disease*, Eagan Press, St. Paul.
7. Dongowski, G., Schnorrenberger, B., Plätzer, M., Schwarz, M., and Neubert, R. (1997) Interactions between food components and drugs. Part 5: Effect of acetylation and amidation of pectin on the interaction with drugs, *Int. J. Pharm.* 158, 99-107.
8. Dongowski, G. (1995) Influence of pectin structure on the interaction with bile acids under *in vitro* conditions, *Z. Lebensm. Unters. Forsch.* 201, 390-398.
9. Jacobasch, G. and Dongowski, G. (2000) Ballaststoffe mit spezifischen Wirkungen, in H.F. Erbersbobler and A.H. Meyer (eds.) *Praxishandbuch Functional Foods*, Behr's Verlag, Hamburg, pp. 1-34.
10. Smith, J.G., Yokoyama, W.H., and German, J.B. (1998) Butyric acid from the diet: Actions at the level of gene expression, *Crit. Rev. Food Sci.* 38, 259-297.
11. Müller, S. and Kunzek, H. (1998) Material properties of processed fruit and vegetables I. Effect of extraction and thermal treatment on apple parenchyma, *Z. Lebensm. Unters. Forsch. A* 206, 264-272.
12. Kabbert, R., Goworek, S., and Kunzek, H. (1997) Preparation of single cell material from enzymatically disintegrated apple tissue - changes in structure and functional properties, *Z. Lebensm. Unters. Forsch. A* 205, 380-387.
13. Blumenkrantz, N. and Asboe-Hansen, G. (1973) New method for quantitative determination of uronic acids. *Anal. Biochem.* 54, 484-489.
14. (Anon.) Pectin (1981) in Committee on Codex Specifications (ed.) *Food Chemical Codex*, National Academy Press, Washington, D.C., 3rd ed., pp.215-217.
15. Kabbert, R., Kotte, K., Cech, B., and Kunzek, H. (1996) Einfluß von Ethanol auf den Trocknungsverlauf und die Makrostruktur eines Einzel-Zellmaterials aus Apfelparenchym (Sorte Granny Smith), *Z. Lebensm. Unters. Forsch.* 203, 512-516.
16. Kabbert, R. and Kunzek, H. (1995) Charakterisierung der Größenverteilung von Einzel-Zellmaterial mit der Laser-Partikel-Analyse, *Z. Lebensm. Unters. Forsch.* 201, 387-389.
17. Kuniak, L. and Marchessault, R. (1972) Study of the crosslinking reaction between epichlorhydrin and starch, *Stärke* 24, 110-116.
18. Robertson, J.A., De Monredon, F.D., Dysseler, P., Guillon, F., Amadó, R., and Thibault, J.-F. (2000) Hydration properties of dietary fibre and resistant starch: a European collaborative study, *Lebensm. Wiss. Technol.* 33, 72-79.
19. Kunzek, H., Loewe, B., and Renger, S. (1993) Herstellung und Charakterisierung von zellstrukturiertem Apfelmaterial mit unterschiedlichem Pectingehalt, *Z. Lebensm. Unters. Forsch.* 197, 353-357.

20. Baumann, H. (1967) Apparatur nach Baumann zur Bestimmung der Flüssigkeitsaufnahme von pulvrigen Substanzen, *GIT Fachz. Lab.* 11, 540-542.
21. Heinevetter, L. and Kroll, J. (1982) Zur Bestimmung der Wasserbindung pulverförmiger quellfähiger Substanzen mittels einer Kapillarsaugmethode, *Nahrung* 26, K17-K18.
22. Kunzek, H., Opel, H., and Senge, B. (1997) Rheological examination of material with cellular structure. II. Creep and oscillation measurements of apple material with cellular structure, *Z. Lebensm. Unters. Forsch. A* 205, 193-203.
23. Höhne, G.W.H., Cammenga, H.K., Eysel, W., Gmelin ,E., and Hemminger, W. (1990) Die Temperaturkalibrierung dynamischer Kalorimeter, *PTB-Mitteilungen* 100, 25-31.
24. Cammenga, H.K., Eysel, W., Gmelin, E., Hemminger, W., Höhne, G.W.H., and Sarge, S.M. (1992) Temperaturkalibrierung dynamischer Kalorimeter II. Kalibriersubstanzen, *PTB-Mitteilungen* 102, 13-18.
25. Brighenti, F. (1997) Simple method for quantitative analysis of short chain fatty acids in serum by gas-liquid chromatography, in F. Guillon, G. Abraham, R. Amadò, H. Andersson, N.G. Asp, K.E. Bach Knudsen, M. Champ, and J. Robertson (eds.) *Plant Polysaccharides in Human Nutrition: Structure, Function, Digestive Fate and Metabolic Effects*, INRA, Nantes, pp. 114-119.
26. Jacobasch, G., Schmiedl, D., Kruschewski, M., and Schmehl, K. (1999) Dietary resistant starch and chronic inflammatory bowel disease. *Int. J. Colorectal Dis.* 14, 201-211.
27. Kunzek, H. and Vetter, S. (2001) Funktionelle Eigenschaften von Lebensmittelbestandteilen und die Entwicklung innovativer Produkte, *Dtsch. Lebensm. Rundsch.* 97, 12-22.
28. Renard, C.M.G.C., van Leenwen, M.J.F.S., Voragen, A.G.J., Thibault, J.-F., and Pilnik, W. (1991) Studies on apple protopectin. II Apple cell wall degradation by pure polysaccharidases and their combinations, *Carbohydr. Polym.* 14, 295-314.
29. Kabbert, r., Herrmuth, k., And kunzek, h. (1993) wasserbindungskapazität und makrostruktur von apfelgewebepartikeln, *z. Lebensm. Unters. Forsch.* 196, 219-223.
30. Tolstoguzov, V.B. (2000) The importance of glassy biopolymer components in food, *Nahrung* 44, 76-84.
31. Macfarlane, G.T. and Gibson, G.R. (1996) Carbohydrate fermentation, energy transduction and gas metabolism in the human large intestine, in R.I.M. Mackie and B.A. White (eds.) *Gastrointestinal Microbiology. Vol. 1. Gastrointestinal Ecosystems and Fermentations*, Chapman and Hall, New York, pp. 269-318.
32. Barry, J.L., Hoebler, C., Macfarlane, G.T., Macfarlane, S., Mathers, J.C., Reed, K.A., Mortensen, P.B., Nordgaard, I., Rowland, I.R., and Rumney, C.J. (1995) Estimation of the fermentability of dietary fibre *in vitro*: a European interlaboratory study, *Br. J. Nutr.* 74, 303-322.
33. Hsu, J.-C. and Penner, M.H. (1989) Influence of cellulose structure on its digestibility in the rat, *J. Nutr.* 119, 457-468.
34. Aust, L., Dongowski, G., Frenz, U., Täufel, A., and Noack, R. (2001) Estimation of available energy of dietary fibres by indirect calormetry in rats, *Eur. J. Nutr.* 40, 23-29.
35. Asp, N.-G., van Amelsvoort, J.M.M., and Hautvast, J.G.A.J. (1996) Nutritional implications of resistant starch, *Nutr. Rev. Intern.* 9, 1-31.
36. Guillon, F., Renard, C.M.G.C., Hospers, J., Thibault, J.-F., Barry, J.-L. (1995) Characterisation of residual fibres from fermentation of pea and apple fibres by human faecal bacteria, *J. Sci. Food Agric.* 68, 521-529.
37. Stevens, B.J.H., Selvendran, R.R., Bayliss, C.E., and Turner, T. (1988) Degradation of cell wall material of apple and wheat bran by human faecal bacteria *in vitro*, *J. Sci. Food Agric.* 44, 155-166.